CONSERVATION EQUATIONS AND MODELING OF CHEMICAL AND BIOCHEMICAL PROCESSES

CHEMICAL INDUSTRIES

A Series of Reference Books and Textbooks

Founding Editor

HEINZ HEINEMANN

ADDITIONAL VOLUMES IN PREPARATION

CONSERVATION EQUATIONS AND MODELING OF CHEMICAL AND BIOCHEMICAL PROCESSES

Said S. E. H. Elnashaie
Parag Garhyan

Auburn University
Auburn, Alabama, U.S.A.

Library of Congress Cataloging-in-Publication Data
A catalog record for this book is available from the Library of Congress.

The publisher offers discounts on this book when ordered in bulk quantities. For more information, write to Special Sales/Professional Marketing at the headquarters address above.

Preface

We would like readers—instructors and students—to read this preface carefully before using this book. This preface is classified into three parts:

1. **Background and Basic Ideas** explains the fundamentals of using a system approach as a more advanced approach to teaching chemical engineering. It also discusses very briefly how this approach allows compacting the contents of many chemical engineering subjects and relates them with one another in a systematic and easy-to-learn manner. More details on this aspect of the book are given in Chapter 1.
2. **Review of Chapters and Appendices** briefly describes the contents of each chapter and the educational philosophy behind choosing these materials.
3. **Relation of the Book Contents to Existing Chemical Engineering Courses** shows how this book can be used to cover a number of courses in an integrated manner that unfortunately is missing in many curricula today. The relation of the contents of the book to existing courses is discussed. Although our frame of reference is the curricula of the Chemical Engineering Department at Auburn University, the discussion can be applied to many curricula worldwide.

1. BACKGROUND AND BASIC IDEAS

We have adopted a novel approach in the preparation of this rather revolutionary undergraduate-level chemical engineering textbook. It is based on the use of *system theory* in developing mathematical models (rigorous design equations) for different chemical and biochemical systems. After a brief introduction to system theory and its applications, the book uses the generalized modular conservation equations (material and energy balances) as the starting point.

This book takes as its basis the vision of chemical engineering transformed, as expressed in the Amundson report of 1989, in which areas new to the traditional subject matter of the discipline are explored. These new areas include biotechnology and biomedicine, electronic materials and polymers, the environment, and computer-aided process engineering, and encompass what has been labeled the BIN- Bio, Info, Nano—revolution. The book addresses these issues in a novel and imaginative way and at a level that makes it suitable for undergraduate courses in chemical engineering.

This book addresses one of the most important subjects in chemical engineering—modeling and conservation equations. These constitute the basis of any successful understanding, analysis, design, operation, and optimization of chemical and biochemical processes. The novel system approach used incorporates a unified and systematic way of addressing the subject, thus streamlining this difficult subject into easy-to-follow enjoyable reading.

By adopting a system approach, the book deals with a wide range of subjects normally covered in a number of separate courses—mass and energy balances, transport phenomena, chemical reaction engineering, mathematical modeling, and process control. Students are thus enabled to address problems concerning physical systems, chemical reactors, and biochemical processes (in which microbial growth and enzymes play key roles).

We strongly believe that this volume strikes the right balance between fundamentals and applications and fills a gap in the literature in a unique way. It efficiently transmits the information to the reader in a systematic and compact manner. The modular mass/energy balance equations are formulated, used, and then transformed into the design equations for a variety of systems in a simple and systematic manner.

In a readily understandable way, this book relates a wide spectrum of subjects starting with material and energy balances and ending with process dynamics and control, with all the stages between. The unique system approach shows that moving from generalized material and energy balance equations to generalized design equations is quite simple for both lumped and distributed systems. The same has been applied to homogeneous and heterogeneous systems and to reacting and nonreacting systems as well as to steady- and unsteady-state systems. This leads the reader gracefully and with great ease from lumped to distributed systems, from homogeneous to heterogeneous systems, from reacting to nonreacting systems, and from steady-state to unsteady-state systems.

Although steady-state systems are treated, we have provided enough coverage of transient phenomena and unsteady-state modeling for students to appreciate the importance of dynamic systems. While the early part of the book is restricted to homogeneous systems, a later chapter introduces a novel systems approach and presents, in an easy-to-understand manner, the modeling of heterogeneous systems for both steady-state and unsteady-state conditions, together with a number of practical examples.

Chemical and biochemical units with multiple-input multiple-output (MIMO) and with multiple reactions (MRs) for all of the above-mentioned systems are also covered. Nonreacting systems and single-input single-output (SISO) systems are treated as special cases of the more general MIMO, MR cases. The systems approach helps to establish a solid platform on which to formulate and use these generalized models and their special cases.

As the book covers both steady- and unsteady-state situations, it logically includes a chapter on process dynamics and control that is an excellent introduction to a more advanced treatment of this topic, with special emphasis on the industrially more relevant digital control systems design.

Given that all chemical/biochemical engineering processes and systems are highly nonlinear by nature, the book discusses this nonlinear behavior in some detail. All the necessary analytical and numerical tools required are included. Matrix techniques are also covered for large-dimensional systems that are common in chemical/biochemical engineering. The book also covers, in a manner that is clear and easy to understand for undergraduate chemical engineers, advanced topics such as multiplicity, bifurcation, and chaos to further broaden the student's perspective. It is increasingly important for undergraduate students to think outside the conventional realm of chemical engineering, and we have shown that these phenomena are relevant to many important chemical/biochemical industrial systems. It is also shown that these phenomena cannot be neglected while designing these systems or their control loops. In the past these subjects—multiplicity, bifurcation, and chaos—have tended to be relegated to advanced research treatises. We treat them here in a manner that undergraduate students can understand and appreciate.

In our fast-changing world the chemical/biochemical industry is also rapidly changing. Today's chemical/biochemical engineering graduates should be exposed to training in creativity as applied to these systems. Therefore a chapter on novel configurations and modes of operations for two important processes is presented in the form of detailed exercises. This important chapter requires a special effort from the instructor to make it the exercise on creativity that it is meant to be.

2. REVIEW OF CHAPTERS AND APPENDICES

This book presents a unified approach to the analysis of a wide range of chemical and biochemical systems. It begins with a summary of the fundamental principles governing thermodynamics and material and energy balances and proceeds to consider the mathematical modeling of a range of systems from homogeneous steady state to heterogeneous unsteady state. A novel feature is the inclusion of the concepts surrounding chaotic systems at undergraduate level—an area of growing importance but one sadly neglected in most texts of this kind. The last chapter deals with two industrial processes—reforming and fermentation—in which the foregoing principles are applied and illustrated for novel configurations and modes of operation. The useful appendices deal with many of the mathematical techniques such as matrix algebra, numerical methods, and the Laplace transform that are utilized in the book.

Chapter 1:   System Theory and Chemical/Biochemical Engineering Systems

This chapter, one of the most important, introduces the main components of the philosophy governing the entire book. It covers in a simple manner the main ideas regarding system theory and its application to chemical and biochemical systems. These systems are classified according to the principles of system theory, and this more novel classification is related to the more classical classifications. This chapter also covers the main differences between material and energy balances (inventory) and design equations, the concepts of rate processes together with their relation to state variables, and the general modeling of processes. The thermodynamic limitation of rate processes in relation to modeling and simulation is examined. A brief discussion of the new approach adopted in this book in connection with recent advances in the profession based on the Amundson report is also presented.

Chapter 2:   Material and Energy Balances

This chapter addresses materials and energy balances for reacting (single as well as multiple reactions) and nonreacting systems in a compact way. It also covers SISO as well as MIMO systems. A generalized material and energy balance equation for a MIMO system with MRs is rigorously developed. All other cases can be easily considered as special cases of this general case. A large number of solved illustrative examples are provided, and unsolved problems are given as exercises at the end of the chapter. Chapter 2 is sufficient for a solid course on material and energy balances. The modular system approach used in this chapter ultimately requires the reader to know only two generalized equations (material and energy balances), with all other cases being special cases. This approach makes the subject easy to comprehend and utilize in a short time, and will also prove extremely useful in preparing the reader to modify these equations into design equations (mathematical models).

Chapter 3: Mathematical Modeling (I): Homogeneous Lumped Systems

This chapter covers in an easy and straightforward manner the transformation of the material and energy balance equations to design equations (mathematical models). It explores closed, isolated, and open lumped homogeneous systems. Steady-state as well as unsteady-state models are developed and solved for both isothermal and nonisothermal systems. Both chemical and biochemical systems are addressed. Again, generalized design equations are developed with all other cases treated as special cases of the general one. This approach helps to achieve a high degree of efficiency regarding rational transformation of knowledge in a concise and clear manner. We concentrate our efforts on reacting systems for two reasons: the first is that for homogeneous systems the nonreacting systems are rather trivial, and the other is that the nonreacting system can be considered a special case of reacting systems when the rates of reactions are set equal to zero. A good number of solved and unsolved problems are given in this chapter.

Chapter 4: Mathematical Modeling (II): Homogeneous Distributed Systems and Unsteady-State Behavior

This chapter covers the transformation of the material and energy balance equations to design equations (mathematical models) for distributed systems. Steady-state as well an unsteady-state models are developed and solved for both isothermal and nonisothermal systems. Again, generalized design equations are developed with all other cases treated as special cases of the general one, and this approach facilitates efficient transformation of knowledge. We concentrate on reacting systems for the same reasons previously discussed. Chapter 4 gives detailed coverage of the mathematical modeling and analytical as well as numerical solution of the axial dispersion model for tubular reactors as an illustrative example for diffusion/reaction homogeneous systems. The same example is extended to provide the solution of the two-point boundary value differential equations and its associated numerical instability problems for nonlinear systems. Several unsolved problems are provided at the end of this chapter.

Together, Chapters 3 and 4 provide systematic, easy-to-understand coverage of all types of homogeneous models, both lumped/distributed and isothermal/nonisothermal systems. Both chapters can also be used as the necessary materials for a thorough course on chemical reaction engineering based on a well-organized approach utilizing system theory.

Chapter 5: Process Dynamics and Control

In the last 20 years, digital control has completely replaced analog control in industry and even in experimental setups. It is our strong belief that the classic complete course on analog control is no longer necessary. Control courses should be directed mainly toward digital control systems, which are beyond the scope of this book. It is useful, however, for readers to have a basic background in analog control (only one well-chosen chapter, not necessarily an entire course) to prepare them for a next course on digital control. Chapter 5 aims to do this by introducing the basic principles of process dynamics and classical control, including the various forms of process dynamic models formulation, basic process control concepts, the use of Laplace transformation and its utilization, the transfer function concepts, ideal forcing functions, block diagram algebra, components of the control loop, and a limited number of simple techniques for choosing the control constants for PID controllers. All these important concepts are supplemented with useful solved examples and unsolved problems.

Chapter 6: Heterogeneous Systems

Most chemical and biochemical systems are heterogeneous (formed of more than one phase). The modular system approach we adopt in this book makes the development of material and energy balances, as well as design equations for heterogeneous systems quite straightforward. Heterogeneous systems are treated as just a number of homogeneous systems (each representing one phase), and these systems are connected to each other through material and energy exchange. This approach proves to be not only rigorous and general but also easy to comprehend and apply to any heterogeneous system utilizing all the knowledge and experience gained by the reader through the previous chapters on homogeneous systems. Chapter 6 introduces these concepts and develops generalized material and energy balance equations as well as design equations for all types of systems—isothermal/nonisothermal, lumped/distributed, and steady-/unsteady-state. A number of chemical and biochemical examples of varying degrees of complexity and unsolved problems are presented for better understanding of the concepts.

Chapter 7: Practical Relevance of Bifurcation, Instability, and Chaos in Chemical and Biochemical Systems

This chapter covers the basic principles of multiplicity, bifurcation, and chaotic behavior. The industrial and practical relevance of these phenomena is also explained, with reference to a number of important industrial processes. Chapter 7 covers the main sources of these phenomena for both isothermal and nonisothermal systems in a rather pragmatic manner and with a minimum of mathematics. One of the authors has published a more detailed book on the subject (S. S. E. H. Elnashaie and S. S. Elshishini, *Dynamic Modelling, Bifurcation and Chaotic Behavior of Gas-Solid Catalytic Reactors*, Gordon & Breach, London, 1996); interested readers should consult this reference and the other references given at the end of Chapter 7 to further broaden their understanding of these phenomena.

Chapter 8: Novel Designs for Industrial Chemical/ Biochemical Systems

As discussed in the introduction of this preface, it is now important to develop creative talents in chemical engineers. Chapter 8 aims to do this by offering two examples of novel processes: one for the efficient production of the ultraclean fuel hydrogen and the other for the production of the clean fuel ethanol through the biochemical path of utilizing lingo-cellulosic wastes. Readers can expect to use the tools provided earlier in this book in order to develop these novel processes and modes of operation without the need of the expensive pilot plant stage.

Appendices

Although it is difficult to make a book completely comprehensive, we tried to make this one as self-contained as possible. The six appendices cover a number of the critical mathematical tools used in the book. Also included is a short survey of essential available software packages and programming environments. These appendices include analytical as well as numerical tools for the handling and solution of the different types of design equations, including linear and nonlinear algebraic and ordinary differential and partial differential equations.

3. RELATION OF THE BOOK CONTENTS TO EXISTING CHEMICAL ENGINEERING COURSES

Chapters 1 and 7 should always be included in any usage of this book.

Chapter 2 can be used for a course on material and energy balance (CHEN 2100, Principles of Chemical Engineering, which covers the application of multicomponent material and energy balances to chemical processes involving phase changes and chemical reactions).

Chapter 3 can be used as the basis for CHEN 3650, Chemical Engineering Analysis (which covers mathematical modeling and analytical, numerical, and statistical analysis of chemical processes). Statistical process control (SPC) is not, of course, covered in this book and the course reading should be supplemented by another book on SPC (e.g., Amitava Mitra, *Fundamentals of Quality Control and Improvement*, Prentice Hall, New York, 1998).

Chapters 4, 5, 6, and 8 are suitable for a senior class on modeling of distributed systems and process dynamics and control (CHEN 4160, Process Dynamics and Control, which covers steady-state and dynamic modeling of homogeneous and heterogeneous distributed chemical processes, feedback systems, and analog controller tuning and design) prior to the course on digital control (CHEN 6170, Digital Process Control).

Chapters 3 and 4 and the first part of Chapter 8 can be used for an undergraduate course on chemical reaction engineering (CHEN 3700, Chemical Reaction Engineering, which covers design of chemical reactors for isothermal and nonisothermal homogeneous reaction systems).

Acknowledgments

I would like to express my appreciation and thanks to many colleagues and friends who contributed directly and indirectly to the successful completion of this book, namely: Professor Robert Chambers, the head of the Chemical Engineering Department at Auburn University, and Professors Mahmoud El-Halwagi and Chris Roberts of the same department. I also appreciate the support I received from Professor Nabil Esmail (Dean of Engineering, Concordia University, Montreal, Canada), Professor John Yates (Chairman of the Chemical and Biochemical Engineering Department, University College, London), Professor John Grace (University of British Columbia, Canada), and Professor Gilbert Froment (Texas A & M

University). I also thank Professor A. A. Adesina (University of New South Wales, Australia) and Professor N. Elkadah (University of Alabama, Tuscaloosa).

Last but not least, I express my love and appreciation for the extensive support and love I receive from my wife, Professor Shadia Elshishini (Cairo University, Egypt), my daughter Gihan, and my son Hisham.

Said Elnashaie

I would like to express my sincere thanks to Dr. Said Elnashaie for giving me the opportunity to work with him as his graduate student and later offering me the chance to be the coauthor of this book. I express my gratitude to my grandfather, Shri H. P. Gaddhyan, an entrepreneur from Chirkunda (a small township in India) for always being an inspiration to me. Without the motivation, encouragement, and support of my parents Smt. Savita and Shri Om Prakash Gaddhyan, I would have not been able to complete this book. A special note of thanks goes to my brother Anurag and his wife Jaishree. Finally, I express my love and thanks to my wife Sangeeta for her delicious food, endurance, and help; her smile always cheered me up and provided the impetus to continue when I was busy working on the manuscript.

Parag Garhyan

Contents

1

System Theory and Chemical/ Biochemical Engineering Systems

1.1 SYSTEM THEORY

1.1.1 What Is a System?

The word *system* derives from the Greek word "systema" and means an assemblage of objects united by some form of regular interaction or interdependence. A simpler, more pragmatic description regarding systems includes the following:

- The system is a whole composed of parts (elements).
- The concept of a system, subsystem, and element is relative and depends on the degree of analysis; for example, we can take the entire human body as a system, and the heart, the arms, the liver, and so forth as the elements. Alternatively, we can consider these elements as subsystems and analyze them with respect to smaller elements (or subsystems) and so on.
- The parts of the system can be parts in the physical sense of the word or they can be processes. In the physical sense, the parts of the body or of a chair form a system. On the other hand, for chemical equipment performing a certain function, we consider the various processes taking place inside the system as the elements which are almost always interacting with each other to give the function of the system. A simple chemical engineering example is a

1

chemical reactor in which processes like mixing, chemical reaction, heat evolution, heat transfer, and so forth take place to give the function of this reactor, which is the changing of some reactants to some products.

- The properties of the system are not the sum of the properties of its components (elements), although it is, of course, affected by the properties of its components. The properties of the system are the result of the nonlinear interaction among its components (elements). For example, humans have consciousness which is not a property of any of its components (elements) alone. Also, mass transfer with chemical reaction has certain properties which are not properties of the chemical reaction or the mass transfer alone (e.g., multiplicity of steady states, as will be shown later in this book).

This is a very elementary presentation of system theory. We will revisit the subject in more detail later.

1.1.2 Boundaries of a System

The system has boundaries distinguishing it from the surrounding environment. Here, we will develop the concept of environment. The relation between the system and its environment gives one of the most important classifications of a system:

1. An Isolated System does not exchange matter or energy with the surroundings. Thermodynamically it tends to the state of thermodynamic equilibrium (maximum entropy). An example is a batch adiabatic reactor.
2. A Closed System does not exchange matter with the surroundings but exchanges energy. Thermodynamically it tends to the state of thermodynamic equilibrium (maximum entropy). An example is a batch nonadiabatic reactor.
3. An Open System does exchange matter and energy with the surroundings. Thermodynamically, it does not tend to the thermodynamic equilibrium, but to the steady state or what should be called the "stationary non equilibrium state," characterized by minimum entropy generation. An example is a continuous stirred tank reactor.

This clearly shows that the phrase we commonly use in chemical engineering, "steady state," is not really very accurate, or at least it is not distinctive enough. A better and more accurate phrase should be "stationary nonequilibrium state," which is a characteristic of open systems and distinguishes it from the "stationary equilibrium state," associated with an isolated and closed systems.

1.2 STEADY STATE, UNSTEADY STATE, AND THERMODYNAMIC EQUILIBRIUM

As briefly stated above, the steady state and unsteady state are concepts related to open systems (almost all continuous chemical engineering processes are open systems). Steady state is when the state of the system does not change with time, but the system is not at thermodynamic equilibrium (i.e., the process inside the system did not stop and the stationary behavior with time is due to the balance between the input, output, and processes taking place in the system). The thermodynamic equilibrium is stationary with time for isolated and closed systems because all processes have stopped, and the nonequilibrium system is changing with time but tending to the thermodynamic equilibrium state. We will come back to these concepts with more details later in this book.

1.2.1 The State of the System

We have used the term "state of the system" many times; what is the state of a system? The state of a system is rigorously defined through the state variables of the system. The state variables of any system are chosen according to the nature of the system. The state of a boiler can be described by temperature and pressure, a heat exchanger by temperature, a nonisothermal reactor by the concentration of the different components and temperature, an isothermal absorption tower by the concentration of different components on different plates, a human body by blood pressure and temperature, flow through a pipe by the velocity as a variable varying radially and axially, and so on.

Thus, state variables are variables that describe the state of the system, and the state of the system is described by the state variables.

1.2.2 Input Variables

Input variables are not state variables; they are external to the system, but they affect the system or, in other words, "work on the system." For example, the feed temperature and composition of the feed stream to a distillation tower or a chemical reactor or the feed temperature to a heat exchanger are the input variables. They affect the state of the system, but are not affected by the state of the system (except when there is a feedback control, and in this case, we distinguish between control variables and disturbances or input variables).

Later, we will discuss the distinction between different types of variables in more detail; for example, design and operating variables and the difference between variables and parameters.

1.2.3 Initial Conditions

These are only associated with unsteady-state systems. An unsteady-state system is a system in which its state variables are changing with time. Unsteady-state open systems will change with time, tending toward the "stationary nonequilibrium state", usually called the "steady state" in the chemical engineering literature. On the other hand, for closed and isolated systems, the unsteady-state behavior tends toward thermodynamic equilibrium.

For these unsteady-state systems, whether open or closed, the system behavior cannot be defined without knowing the initial conditions, or the values of the state variables at the start (i.e., time $= 0$). When the initial conditions are defined, the behavior of the system is uniquely defined. The multiplicity (nonunique) phenomena that appear in some chemical engineering systems are related to the steady state of open systems and not to the unsteady-state behavior with known initial conditions. The trajectory describing the change of the state variables with time starting at a specific initial condition is unique.

1.3 MODELING OF SYSTEMS

The simplest definition of modeling is the following: putting physical reality into an acceptable mathematical form.

A model of a system is some form of mathematical representation that gives, in the final analysis, a relation between the inputs and outputs of the system. The simplest and least reliable are empirical models, which are based more or less on the black-box concept. As shown in Figure 1.1, an *empirical* model may have the following form:

$$O = f(I)$$

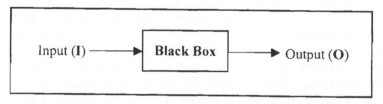

Figure 1.1   The black-box concept.

where O is the output and I is the input. On the other hand, a *physical* model is a model based on the understanding of what is happening inside the system and the exchange between the system and the surrounding. There is no model which is completely empirical and there is no model which is completely physical, but we name the model according to its dominant feature.

1.3.1 Elementary Procedure for Model Building

An elementary procedure for model building is as follows:

1. Define the boundaries of the system (Fig. 1.2).
2. Define the type of system: open, closed, or isolated.
3. Define the state variables.
4. Define the input variables (sometimes called input parameters).
5. Define the design variables (or parameters).

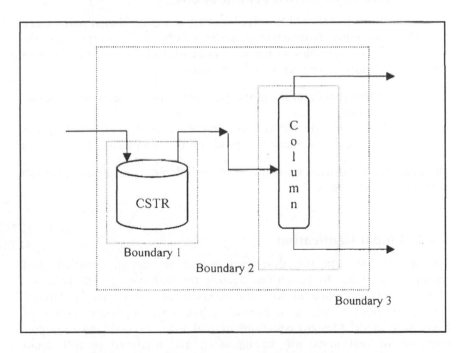

Figure 1.2 Boundaries of a system and subsystems.

6. Define the nature of the interaction between the system and the surroundings.
7. Define the processes taking place within the boundaries of the system.
8. Define the rate of the different processes in terms of the state variables and rate parameters, and introduce the necessary equations of state and equilibrium relations between the different phases. *Note*: Usually, equilibrium relations between certain variables are used instead of rates as an approximation when the rate is quite high and the process reaches equilibrium quickly.
9. Write mass, heat (energy), and momentum balance equations to obtain the necessary equations (or model equations) relating the input and output through the state variables and parameters. These equations give the variation of state variables with time and/or space.

1.3.2 Solution of the Model Equations

The model developed should be solved for certain inputs, design parameters, and physicochemical parameters in order to obtain the output and the variation of state variables within the boundaries of the system. To solve the model equations, we need two main items:

1. Determination of the model parameters (to be determined experimentally)
2. A solution algorithm, the complexity of it, and whether it is analytical or numerical depends on the complexity of the system

Of course, we will discuss the solution of the different types of model in full detail later in this book.

1.3.3 Model Verification

To be able to use the developed model in design, control, and optimization, it has to be verified against the real system. The behavior of the system is compared with the behavior of a real system (laboratory scale, pilot plant, or industrial units). This important aspect of developing model (design) equations of real value by verifying its representation of real units will be discussed and analyzed in full detail later.

1.4 FUNDAMENTAL LAWS GOVERNING THE PROCESSES IN TERMS OF THE STATE VARIABLES

Here, we give a very simple presentation of the necessary components for developing model (design) equations for chemical/biochemical processes. Full details and generalization will be given in Chapter 3.

1.4.1 Continuity Equations for Open Systems

Total continuity equations (mass balances):

{(Mass flow into system) − (Mass flow out of system)}

= {Time rate of change of mass inside the system}

Component continuity equations (component mass balances):

(Flow of moles of jth component into the system)

− (Flow of moles of jth component out the system)

+ (Rate of formation of moles of jth component by chemical reactions)

= (Time rate of change of moles of jth component inside the system)

1.4.2 Diffusion of Mass (Transport Law)

Fick's law gives that the mass transfer (diffusion) is proportional to the concentration gradient (Fig. 1.3):

$$N_A = -D_A \frac{dC_A}{dZ}$$

where, N_A is the molar flux in moles/(unit area)(unit time), D_A is the diffusion coefficient, dC_A is the concentration driving force, and dZ is the distance in the direction of diffusion. The same equation over a certain thickness (film, interface, etc.) can be approximated as

$$N_A = K_L \Delta C_A$$

where K_L is the overall mass transfer coefficient and N_A is the flux.

Hence, we can write

$$K_L = \frac{D_A}{\delta}$$

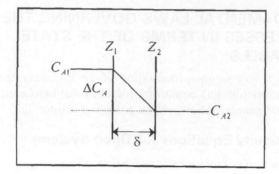

Figure 1.3 Concentration gradient for diffusive mass transfer.

1.4.3 Energy Equation (Conservation of Energy, First Law of Thermodynamics for an Open System)

$$
\begin{pmatrix}
\text{Flow of internal,} \\
\text{kinetic and potential} \\
\text{energies into the} \\
\text{system by convection} \\
\text{or diffusion}
\end{pmatrix}
-
\begin{pmatrix}
\text{Flow of internal,} \\
\text{kinetic and potential} \\
\text{energies out of the} \\
\text{system by convection} \\
\text{or diffusion}
\end{pmatrix}
$$

$$
+
\begin{pmatrix}
\text{Heat added} \\
\text{to the system} \\
\text{by conduction,} \\
\text{radiation and} \\
\text{reaction}
\end{pmatrix}
-
\begin{pmatrix}
\text{Work done} \\
\text{by the system} \\
\text{on surroundings,} \\
\text{i.e. shaft work} \\
\text{and } PV \text{ work}
\end{pmatrix}
$$

$$
= \{\text{Time rate of change of internal, kinetic and potential} \\
\text{energies inside the system}\}
$$

In most chemical engineering systems that we will study, the above general form reduces to essentially an enthalpy balance, as will be shown later.

Heat Transfer

Fourier's law describes the flow of heat in terms of temperature gradient as follows:

$$
q = -\lambda \frac{dT}{dZ}
$$

The above relation can be approximated in terms of the temperature difference between two points as follows:

$$q = h\Delta T$$

where q is the heat flux in units of J/cm^2s, λ is the thermal conductivity, dT is the temperature driving force, dZ is the distance in the direction of heat transfer, and $h = \lambda/\delta$ (analogous to the mass transfer case shown above and in Fig. 1.3).

1.4.4 Equations of Motion

The force is given by

$$F = \frac{1}{g_C}(Ma)$$

where F = force (lb_f), g_C is the conversion constant needed to keep units consistent [32.2 (lb_m ft/lb_f s^2)], M is the mass (lb_m), and a is the acceleration (ft/s^2). Alternatively, we can write that

$$\text{Force} = \text{Mass} \times \text{Acceleration}$$

In this case, the mass is considered to be constant.

When the mass varies with time, the equation will have the following general form:

$$\frac{1}{g_C}\frac{d(Mv_i)}{dt} = \sum_{j=1}^{N} F_{ji}$$

where v_i is the velocity in the i direction (ft/s or m/s) and F_{ji} is the jth force acting in the i direction (lb_f or N).

Momentum Transfer (Fig. 1.4)

$$\tau_{xy} = \text{Momentum Flux} = \frac{\dot{P}_y}{A_x}$$

where $\dot{P}_y$ is the time rate of change of momentum caused by the force in the y direction F_y (the momentum transfer is in the x direction) and A_x is the area perpendicular to x, where x is the direction of momentum transfer.

Shear Stress

$$\sigma_{xy} = \frac{F_y}{A_x}$$

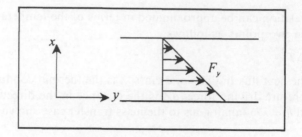

Figure 1.4 Momentum transfer.

where F_y is the scalar y-component of the force vector and A_x is the area perpendicular to the x axis. We can write

$$F = k\dot{P}$$

For one particular direction,

$$F_y = k\dot{P}_y$$

where

$$k = \left(\frac{1}{g_C}\right) \quad \text{for the British system}$$

$$= 1 \quad \text{for the SI system}$$

For shear stress, we can write

$$\sigma_{xy} = \frac{k\dot{P}_y}{A_x} = k\tau_{xy}$$

and

$$\tau_{xy} = \frac{1}{k}\sigma_{xy}$$

where τ_{xy} is the momentum flux.

Newton's Law of Viscosity (or Momentum Transfer)

This is given by

$$\tau_{xy} = \frac{1}{k}\sigma_{xy} = -\mu\frac{\partial v_y}{\partial x}$$

1.4.5　Equations of State

To write mathematical models, in addition to material and energy balances and rate of different processes taking place within the boundaries of the system, we need equations that tell us how the physical properties, primarily density and enthalpy, change with temperature.

$$\text{Liquid density } = \rho_L = f_1(P, T, x_i)$$
$$\text{Vapor density } = \rho_V = f_2(P, T, y_i)$$
$$\text{Liquid enthalpy } = h = f_3(P, T, x_i)$$
$$\text{Vapor enthalpy } = H = f_4(P, T, y_i)$$

For simplicity, h is related to $C_p T$ and H is related to $C_p T + \lambda_V$.

If C_p is taken as function of temperature and we consider that the reference condition is T_o (at which $h = 0$), then

$$h = \int_{T_0}^{T} C_p(T)\,dT = \text{sensible enthalpy change}$$

If

$$C_p(T) = A_1 + A_2 T$$

then

$$h = A_3 + A_4 T + A_5 T^2$$

where A_3, A_4, and A_5 are defined in terms of T_0, A_1 and A_2.

For Mixture of Components (and Negligible Heat of Mixing)

The enthalpy of the liquid mixture can be expressed as

$$h = \frac{\sum_{j=1}^{J} x_j h_j M_j}{\sum_{j=1}^{J} x_j M_j}$$

where x_j is the mole fraction of the jth component, M_j is the molecular weight of the jth component (g/gmol), and h_j is the enthalpy of the pure component j (J/gmol). The denominator depicts clearly the average molecular weight of the mixture.

Density

Liquid densities are usually assumed constant (unless large changes in composition and temperature occur).

Vapor densities can be obtained from the relation

$$PV = nRT \tag{1.1}$$

which gives

$$\rho = \frac{MP}{RT}$$

where P is the absolute pressure, V is the volume, n is the number of moles, R is the universal gas constant, T is the absolute temperature, and M is the molecular weight.

Notes

1. The reader must be cautious about the use of consistent units for R and other variables.
2. For a high-pressure and/or high-temperature system, the compressibility factor (z factor) should be introduced, which is obtained from the knowledge of the critical temperature and critical pressure of the system (the reader is advised to refer to a thermodynamics book; for example, Ref. 1).
3. For an open (flow) system, Eq. (1.1) becomes

$$Pq = \bar{n}RT$$

where q is the volumetric flow rate (L/min) and $\bar{n}$ is the molar flow rate (mol/min).

1.4.6 Rate of Reaction

Homogeneous reaction rates are, in general, functions of concentrations, temperature, and pressure:

$$r = f(T, C, P)$$

where r is the rate of reaction (gmol/L s), T is the temperature, usually the dependence of the rate of reaction constant on the temperature has the form $k = k_0 e^{(-E/RT)}$, P is the pressure (usually used for gas-phase reactions), and C represents the concentrations.

In many cases, we can write the rate of reaction as the product of three functions, each a function in one of the variables (temperature, concentrations, or pressure.)

$$r = f_1(T) f_2(C) f_3(P)$$

Gas–solid catalytic reactions will have the same form, but will refer to the weight of catalyst rather than the volume:

$$r = f(T, C, P)$$

where r is the rate of reaction (gmol/g catalyst s). Note the difference in the units of the rate of reaction for this gas–solid catalytic reaction compared with the homogeneous reaction.

1.4.7   Thermodynamic Equilibrium

Chemical Equilibrium (for Reversible Reactions)

Chemical equilibrium occurs in a reacting system when

$$\sum_{j=1}^{J} \sigma_j \mu_j = 0$$

where σ_j is the stoichiometric coefficient of jth component with the sign convention that reactants have negative sign and products have positive signs (this will be discussed in full detail later) and μ_j is the chemical potential of the jth component.

Also, we have the following important relation:

$$\mu_j = \mu_j^0 + RT \ln P_j$$

where μ_j^0 is the standard chemical potential (or Gibbs free energy per mole) of the jth component, P_j is the partial pressure of the jth component, R is the universal gas constant, and T is the absolute temperature.

For the reaction

$$\sigma_a A \overset{k_1}{\underset{k_2}{\Longleftrightarrow}} \sigma_B B$$

with the forward rate of reaction constant being k_1 and the backward being k_2. Then, at equilibrium,

$$\sigma_B \mu_B - \sigma_A \mu_A = 0$$

which can be written as

$$\sigma_B (\mu_B^0 + RT \ln P_B) - \sigma_A (\mu_A^0 + RT \ln P_A) = 0$$

which gives

$$\ln P_B - \ln P_A = \frac{\sigma_A \mu_A^0 - \sigma_B \mu_B^0}{RT}$$

which leads to

$$\ln \underbrace{\left[\frac{P_B}{P_A}\right]}_{K_p} = \underbrace{\frac{\sigma_A \mu_A^0 - \sigma_B \mu_B^0}{RT}}_{\text{Function of temperature only}}$$

where, K_p is the equilibrium constant; we can write

$$\ln K_p = f(T).$$

Finally, we can write

$$K_p = e^{f(T)} = K_{p0}e^{-(\Delta H/RT)}$$

This equilibrium constant of a reversible reaction is very important in order to determine the limit of conversion of reactants under any given design and operating conditions.

Phase Equilibrium

Equilibrium between two phases occurs when the chemical potential of each component is the same in the two phases:

$$\mu_j^I = \mu_j^{II}$$

where, μ_j^I is the chemical potential of the jth component in phase I and μ_j^{II} is the chemical potential of the jth component in phase II.

For Vapor–Liquid Systems

This phase equilibrium leads to the satisfaction of our need for a relationship which permits us to calculate the vapor composition if we know the liquid composition or vice versa when the two phases are at equilibrium. We will give an example regarding the bubble-point calculation.

Bubble-Point Calculation (Fig 1.5)

Given the pressure P of the system and the liquid composition x_j, we calculate the temperature of the system T and the vapor composition y_j. This calculation usually involves trial and error (for some other cases, the situation will be that we know x_j and T and want to find P and y_j, or we know P and y_j and want to find x_j and T).

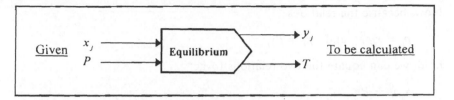

Figure 1.5 Calculation of bubble point.

For Ideal Vapor-Phase Behavior (Needs Correction at High Pressures)

Dalton's Law

Dalton's law applies to the ideal vapor-phase behavior and states that the partial pressure of component j in the vapor phase is the mole fraction of the component multiplied by the total pressure:

$$p_j = Py_j$$

where p_j is the partial pressure of component j in the vapor phase, P is the total pressure, and y_j is the mole fraction of component j in the vapor phase.

Rault's Law

Rault's law states that the total pressure is the summation of the vapor pressure of each component (p_j^0) multiplied by the mole fraction of component j in the liquid phase:

$$P = \sum_{j=1}^{J} x_j p_j^0$$

We can write the partial pressure in the vapor phase of each component as related to the vapor pressure as follows:

$$p_j = x_j p_j^0$$

Hence,

$$P = \sum_{j=1}^{J} p_j$$

where P is the total pressure, x_j is the mole fraction of component j in liquid, and p_j^0 is the vapor pressure of the pure component j.

Now, because the relations

$$p_j = x_j p_j^0 \quad \text{and} \quad p_j = P y_j$$

hold, we can equate the two relations to get

$$x_j p_j^0 = P y_j$$

Therefore, we get

$$y_j = \frac{p_j^0}{P} x_j$$

The vapor pressure p_j^0 is a function of temperature only, as shown by the relation

$$\ln p_j^0 = \frac{A_j}{T} + B_j$$

Therefore, the vapor–liquid equilibrium computation can be performed according to the above relation.

The Utilization of Relative Volatility

The relative volatility of component i to component j is defined as

$$\alpha_{ij} = \frac{y_i/x_i}{y_j/x_j} = \frac{\text{Volatility of } i}{\text{Volatility of } j}$$

For binary systems, we have

$$\alpha = \frac{y/x}{(1-y)/(1-x)}$$

which gives

$$y = \frac{\alpha x}{1 + (\alpha - 1)x}$$

The K-Values

Equilibrium vaporization ratios or K values are widely used, particularly in the petroleum industry:

$$K_j = \frac{y_j}{x_j}$$

K_j is a function of temperature and, to a lesser extent, of composition and pressure.

Activity Coefficient

For nonideal liquids, Rault's law may be easily modified to the form:

$$p_j = x_j p_j^0 \gamma_j$$

Also,

$$P = \sum_{j=1}^{J} x_j p_j^0 \gamma_j$$

where γ_j is the activity coefficient (a function of composition, pressure, and temperature).

The Z-Factor

For nonideal gases, the z factor (compressibility factor) is introduced. It is a function of the critical temperature and pressure.

This very brief review of phase equilibrium gives the reader the principles for dealing with two-phase systems. More background on the thermodynamic equilibrium of multiphase systems will sometimes be needed; it is not covered in this book. The reader is advised to consult a multiphase thermodynamics book for that purpose.

1.5 DIFFERENT CLASSIFICATIONS OF PHYSICAL MODELS

Before we move to a more intellectual discussion of the history of chemical engineering and the role and position of system theory and mathematical modeling, we present the different basis for classification of mathematical models.

I. **Classification according to variation or constancy of the state variables with time**
   - Steady-state models: described by algebraic equations or ODEs (ordinary differential equations) or PDEs (partial differential equations).
   - Unsteady-state models: described by ODEs or PDEs.
II. **Classification according to the spatial variation of the state variables**
   - Lumped models (usually called lumped parameter models, which is wrong terminology because it is the state variables that are lumped together not the parameters): described by algebraic equations for the steady state and ODEs for the unsteady state.

- Distributed models (usually called distributed parameter models, which is wrong terminology because it is the state variables that are distributed, not the parameters): described by ODEs or PDEs for the steady state and PDEs for the unsteady state.

III. **According to the functional dependence of the rate governing laws on the state variables**
  - Linear models: described by linear equations (can be algebraic equations, ODEs, or PDEs). These models can be solved analytically.
  - Nonlinear models: described by nonlinear equations (can be algebraic equations ODEs, or PDEs). These models need to be solved numerically.

IV. **According to the type of processes taking place within the boundaries of the system**
  - Mass transfer (example: isothermal absorption).
  - Heat transfer (example: heat exchangers).
  - Momentum transfer (example: pumps or compressors).
  - Chemical reaction (example: homogeneous reactors).
  - Combination of any two or more of the above processes (example: heterogeneous reactors).

V. **According to the number of phases in the system**
  - Homogeneous models (formed of one phase).
  - Heterogeneous models (more than one phase).

VI. **According to the number of stages in the system**
  - Single stage.
  - Multistage.

VII. **According to the mode of operation of the system**
  - Batch (closed or isolated system).
  - Continuous (open system).

VIII. **According to the system's thermal relation with the surroundings**
  - Adiabatic (neither heating nor cooling).
  - Nonadiabatic (example: cocurrent or countercurrent cooling or heating).

IX. **According to the thermal characteristics of the system**
  - Isothermal.
  - Nonisothermal.

All of these classifications are, of course, interactive, and the best approach is to choose a main classification and then have the other classification as subdivisions of this main classification.

1.6 THE STORY OF CHEMICAL ENGINEERING IN RELATION TO SYSTEM THEORY AND MATHEMATICAL MODELING

Chemical engineering as we witness today at the beginning of the 21st century has evolved into a very demanding discipline extending to a wide spectrum of industries and requiring a wide spectrum of knowledge in many diversified fields. The easy days of chemical engineers (or chemical technologists), being simply mechanical engineers with some background in industrial chemistry, have long gone. Norton, at MIT, started, in 1888, what may be called the first chemical engineering curriculum. However, it was more of a chemical technology than a chemical engineering curriculum as we have known it for the last three to four decades. It consisted mostly of descriptive courses in industrial chemistry and chemical technology. In about 1923 came the preliminary steps in the direction of classifying chemical engineering equipment and processes, on a higher level of generality. This was represented by the concept of unit operations, where, for example, distillation and extraction are taught as unified courses not necessarily related to a specific industry.

This conceptual approach spread worldwide and started to give birth to more of these generalized subjects. Systematic, though simple, methods of design were developed (e.g., McCabe–Thiele diagrams) and the concept of equilibrium stages became well established. The emphasis at this stage, which extended to the 1960s, was on the overall behavior of the chemical equipment without real involvement into the details of the microscale processes.

The second milestone in chemical engineering came in 1960 with the publication of *Transport Phenomena*, by Bird et al. [2]. Their new approach emphasized the microscale processes and the analogy among mass, heat, and momentum transfer in different processes.

This stage witnessed an explosion of changes in chemical engineering both in research and education, giving birth to new well-defined fields. A prime example is chemical engineering kinetics, which gradually evolved into the rich discipline under the title "Chemical Reaction Engineering," which emphasizes the design, analysis, and optimization of different types of chemical reactors.

The chemical engineering discipline started to "bifurcate" faster and faster year after year. Process dynamics and control (PDC) became an important branch of chemical engineering. Plant design, economics, and other specialized disciplines within chemical engineering started to grow at a rapid rate.

These developments were accompanied by a steady growth in productivity, sophistication, and a high level of competition in the chemical industry. In order to meet these demands, a typical chemical engineering curriculum had to be crowded with many subjects. In addition to basic science courses (mathematics, physics, chemistry, biology, thermodynamics, etc.), the student had to be taught the new chemical engineering disciplines: mass and heat balance, mass and heat transfer, chemical reaction engineering, multistage operations, process dynamics and control, plant design and economics, and so forth, in addition to supplementary subjects from other engineering disciplines such as mechanical, civil, and electronic engineering. The training of chemical engineers left little room for what James Wei calls the "Third Paradigm," which should emphasize a "global outlook to the relation between Engineering and Society" [3].

Despite this crowding of the chemical engineering curriculum, a practicing engineer or an experienced professor can look at any curriculum and find a lot to be desired. Students are learning about chemical reaction engineering principles, which is a must, but they do not learn much about actual industrial reactors; they learn classical control theory, which is a must, but they do not know enough about digital control, in a time when most plants are discarding analog control and installing digital control systems. The list is endless and offers a strong temptation to add more and more courses, but because the students are *incompressible* (as Levenspiel once put it [4]), it is very difficult to add more courses in response to the legitimate desires of experienced industrialists and professors.

The situation becomes even more acute with the expansion of chemical engineering into new fields, especially biotechnology, the electronic industries, new materials, and composite membranes. In addition to that, chemical engineers are qualified and obliged to play a leading role in the environmental challenge that is facing the human society in a dangerous and complicated manner.

In a nutshell, we do have a problem that cannot be solved by quantitative measures; it is so acute that it needs, actually, a change of concept. This change of concept requires looking at chemical engineering from a system theory point of view.

1.7 THE PRESENT STATUS OF CHEMICAL INDUSTRY AND UNDERGRADUATE CHEMICAL ENGINEERING EDUCATION

This book is designed for undergraduate chemical engineering students. The system approach is utilized in order to structure undergraduate chemical engineering education in a new fashion suitable for the present state of chemical, biochemical, and related industries.

The chemical industry has revolutionized human life to such an extent that it invaded every domain of modern life in our society. The industrial development and the innovative work of many pioneering chemical engineering researchers, coupled with the advancement in computer technology, makes the training of chemical engineers for the future quite a challenging task. A chemical engineer graduating today is expected in his/her career to deal with a wide range of problems that need a sound fundamental basis as well as an arsenal of practical knowledge. Of course, much of the practical knowledge is acquired in the industry after graduation; however, undergraduate training is the critical factor that determines the degree of success of the trajectory of the chemical engineers after graduation. The socioeconomic, safety, and environmental challenges, together with the fast expansion in the use of digital computers, make the task even more difficult. Is it plausible today to produce a chemical engineer who is not fluent in the application of computer power to chemical engineering problems? Is it possible to be satisfied with a graduate who knows enough about chemical engineering, but who is illiterate with regard to some of the basic computer software and hardware necessary for computer-controlled experimentation or operation of equipment? Considering the socioeconomic implications of chemical, biochemical, and related industries, is it wise to produce chemical engineers who are not sufficiently socially aware of the problems of their society and the connections between these problems and their profession? Of course, we should be aware that the 4 or 5 years of chemical engineering education is not and should not be a substitute for the whole life experience that he or she will acquire before university education or after graduation. However, these 4–5 years are crucial for the production of our hoped-for "Chemical Engineer Citizen," a chemical engineer who can grasp the growing fascinating opportunities of our modern society as well as coping with its difficulties and problems.

1.8   SYSTEM THEORY AND THE MATHEMATICAL MODELING APPROACH USED IN THIS BOOK

The challenges facing the chemical engineering profession in its process of healthy and fruitful intercourse with society, as well as with first and second natures, cannot be advanced to higher levels with one single idea. Certainly, man has learned that life is much more complicated than the deterministic views of the 18th and 19th centuries. A more complex view of nature, society, and man-made processes is emerging, from simple monotony to complex bifurcations which were thought to represent the highest degree of complexity before the revolutionary discovery of chaos, strange attractors, and fractals structures. Scientific development in the last few years should teach us that, most probably, nobody will say the last word in anything and that the most that anybody can hope for, which is very honorable, is to achieve one's part in a successful iteration of the continuous human rise to higher levels of civilization and intellect with all its dynamical beauty and its enjoyable new challenges and difficulties. Our argument regarding the role of systems theory and mathematical modeling in undergraduate chemical engineering education will, hopefully, represent one small part of a complex set of changes that are needed to take the chemical engineering science forward, through a tortuous route of developments, to match the revolutionary changes the human society is witnessing scientifically, technologically, socially, and politically. We argue, in this section of the book, that the extraordinary expansion in the domain of chemical engineering requires a new step in the direction of generalized classification that will open new and faster expanding horizons for this important discipline. This can be achieved through a radical change of the undergraduate chemical engineering syllabus, in order to make it based on system theory and the mathematical modeling approach, which is a very effective step forward in organizing knowledge and achieving a much higher level of economy of information. Before getting deeper into the subject, it may be worthwhile to present a description of system theory and mathematical modeling in chemical engineering, which will add further to the simpler understanding we presented in the previous section.

1.8.1   Systems and Mathematical Models

In this subsection, we discuss very briefly the basic principles of systems and mathematical modeling theories, with special emphasis on chemical biochemical engineering problems, in a manner that enriches the very simple explanation we presented earlier. System theory is the more general, more abstract part, whereas mathematical modeling is more applied and less abstract.

As explained earlier, a system is a whole consisting of elements or subsystems. The concept of systems–subsystems and elements is relative and depends on the level of analysis. The system has a boundary that distinguishes it from the environment. The system may exchange matter and/or energy with the environment depending on the type of system from a thermodynamical point of view. A system (or subsystem) is described by its elements (or subsystems), the interaction between the elements, and its relation with the environment. The elements of the system can be material elements distributed topologically within the boundaries of the system and giving the configuration of the system, or they can be processes taking place within the boundaries of the system and defining its function. They can also be both, together with their complex interactions. An important property of the system wholeness is related to the principle of the irreducibility of the complex to the simple, or of the whole to its elements; the whole system will possess properties and qualities not found in its constituent elements. This does not mean that certain information about the behavior of the system cannot be deduced from the properties of its elements, but, rather, it adds something to them.

The systems can be classified on different bases as briefly explained earlier. The most fundamental classification is that based on thermodynamic principles, and on this basis, systems can be classified into the following classes:

Isolated Systems

Isolated systems are systems that exchange neither energy nor matter with the environment. The simplest chemical engineering example is the adiabatic batch reactor. These systems tend toward their thermodynamic equilibrium with time, which is characterized by maximum entropy (highest degree of disorder).

Closed Systems

Closed systems are systems that exchange energy with the environment through their boundaries, but do not exchange matter. The simplest example is a nonadiabatic batch reactor. These systems also tend toward a thermodynamic equilibrium with time, characterized by maximum entropy (highest degree of disorder).

Open Systems

Open Systems are systems that exchange both energy and matter with the environment through their boundaries. The most common chemical engineering example is the continuous-stirred tank reactor (CSTR). These systems do not tend toward their thermodynamic equilibrium with time, but rather toward a "stationary nonequilibrium state," which is characterized by minimum entropy production.

It is clear from the above classification that batch processes are usually of the isolated or closed type, whereas the continuous processes are usually of the open type. As can be seen from the above definitions, the system theory is very abstract and, in general, it treats any system regardless of whether a mathematical model for this system can be built or not; mathematical modeling, on the other hand, is less abstract and more applied than the system concept.

For continuous processes, a classification of systems from a mathe-
matical point of view is very useful for both model formulation and
algorithms for model solution. According to this basis, systems can be
classified as shown briefly earlier.

Lumped Systems

Lumped systems are systems in which the state variables describing the
system are lumped in space (invariant in all space dimensions). The simplest
example is the perfectly mixed continuous-stirred tank reactor. These systems are described at steady state by algebraic equations, whereas the
unsteady state is described by initial-value ordinary differential equations
for which time is the independent variable.

Distributed Systems

Distributed systems are systems in which the state variables are varying in
one direction or more of the space coordinates. The simplest example is the
plug flow reactor. These systems are described at steady state, either by
ordinary differential equations [where the variation of the state variables
is only in one direction of the space coordinates (i.e., one-dimensional systems), and the independent variable is this space direction] or partial differential equations [where the variation of the state variables is in more than
one direction of space coordinates (i.e., two- or three-dimensional systems)
and the independent variables are these space directions]. The ordinary
differential equations describing the steady state of the one-dimensional
distributed systems can be either initial-value differential equations (e.g.,
plug flow systems) or two-point boundary-value differential equations
(e.g., systems with superimposed axial dispersion). The equations describing
the dynamic behavior of distributed systems are invariably partial differential equations.

Another classification of systems, which is important for deciding the
algorithm for model solution, is that of linear and nonlinear systems.
The equations of linear systems can usually be solved analytically,
whereas the equations of nonlinear systems are almost always solved
numerically. In this respect, it is important to recognize the significant
fact that physical systems are almost always nonlinear and that linear
systems are either an approximation that should be justified or are
intentionally linearized in the neighborhood of a certain state of the
system and are strictly valid only in this neighborhood. This local
linearization is usually applied to dynamical system to investigate
local stability characteristics of the system.

A third classification, which is relevant and important in chemical engineering, is the classification based on the number of phases involved within the boundaries of the system. According to this classification, the systems are divided as follows:

Homogeneous Systems

Homogeneous systems are systems in which only one phase is involved in the processes taking place within the boundaries of the system. In reaction systems, the behavior of these systems is basically governed by the kinetics of the reactions taking place, without the interference of any diffusion processes.

Heterogeneous Systems

Heterogeneous systems are systems in which more than one phase is involved in the processes taking place. In reaction systems, the behavior of these systems is not only governed by the kinetics of the reactions taking place, but also by the complex interaction between the kinetics and diffusion processes. When the system does not involve a chemical reaction, then the system behavior is not governed by processes taking place in one phase, but, rather, by the totality of the processes taking place in the different phases and the interaction between them. The modeling and analysis of these systems is obviously much more complicated than for homogeneous systems. Heterogeneous systems are globally divided into two- and three-phase systems. In more detail, they are divided as follows:

1. Liquid–liquid systems
2. Gas–liquid systems
3. Gas–solid systems: catalytic and noncatalytic
4. Liquid–solid systems: catalytic and noncatalytic
5. Gas–solid–liquid systems: catalytic and noncatalytic

1.8.2 Mathematical Model Building: General Concepts

The process of classification and building of mathematical models (design equations) has been simply discussed earlier. Here, we give more details about this process which add to the buildup of knowledge for the reader in this direction.

Building a mathematical model for a chemical engineering system depends to a large extent on the knowledge of the physical and chemical laws governing the processes taking place within the boundaries of the system. This includes the different rates of mass, heat, and momentum transfer, rates of reactions and rates of adsorption–desorption, and so forth. It also includes the thermodynamic limitations that decide the feasibility of the process to start with, as well as heat production and absorption.

Mass and heat transfer rates are both dependent on the proper description of the fluid flow phenomena in the system. The ideal case is when all of these processes are determined separately and then combined into the system's model in a rigorous manner. However, very often this is quite difficult in experimental measurement; therefore, special experiments need to be devised, coupled with the necessary mathematical modeling, in order to decouple the different processes implicit in the measurements.

In the last two decades, there has been a considerable advancement in the development of mathematical models of different degrees of sophistication for chemical engineering processes. These models are taking their part in directing design procedures as well as in directing scientific research. It is important in this respect to recognize the fact that most mathematical models are not completely based on rigorous mathematical formulation of the physical and chemical processes taking place within the boundaries of the system. Every mathematical model, contains a certain degree of empiricism. The degree of empiricism limits the generality of the model, and as our knowledge of the fundamentals of the processes taking place increases, the degree of empiricism decreases and the generality of the model increases. The existence of models at any stage, with their appropriate level of empiricism, helps greatly in the advancement of the knowledge of the fundamentals, and, therefore, helps to decrease the degree of empiricism and increase the level of rigor in the mathematical models. Models will always contain certain simplifying assumptions believed by the model-builder not to affect the predictive nature of the model in any manner that sabotages the purpose of it. Mathematical models (design equations) are essential for design, optimization, and control of different equipments for chemical, biochemical, and related industries.

Different models with different degrees of sophistication can be built. Models which are too simplified will not be reliable and will not serve the purpose, whereas models which are too sophisticated will present an unnecessary and sometimes expensive overburden. In undergraduate chemical engineering education, the concept of model sophistication and its relation to the purpose of the model building should be emphasized. This book will explain and discuss this extremely important issue at every stage of development, starting from material and energy balance to heterogeneous distributed models.

1.8.3 Outline of the Procedure for Model Building

This issue has been discussed earlier in a simplified manner. It is useful for the gradual buildup of the reader's understanding of modeling and simulation to visit the problem again on a higher level of rigor and to distinguish clearly the difference between modeling and simulation.

First, we start with procedure for model building, followed by the distinction between modeling and simulation, followed by the Amundson report and its emphasis on modern chemical engineering education, and finally ending up by the relation among systems theory, mathematical modeling, and modern chemical engineering education.

The procedure for model building can be summarized in the following steps:

1. The identification of the system configuration, its environment, and the modes of interaction between the system and its environment.

2. The introduction of the necessary justifiable simplifying assumptions.

3. The identification of the relevant state variables that describe the system.

4. The identification of the processes taking place within the boundaries of the system.

5. The determination of the quantitative laws governing the rates of the processes in terms of the state variables. These quantitative laws can be obtained from information given in the literature and/or through an experimental research program coupled with the mathematical modeling program.

6. The identification of the input variables acting on the system.

7. The formulation of the model equations based on the principles of mass, energy, and momentum balances appropriate to the type of system.

8. The development of the necessary algorithms for the solution of the model equations.

9. The checking of the model against experimental results (laboratory, pilot plant, or commercial units) to ensure its reliability and carrying out a re-evaluation of the simplifying assumptions. This re-evaluation process may result in imposing new simplifying assumptions or relaxing some of them.

It is clear that these steps are interactive in nature and the results of each step should lead to a reconsideration of the results of all previous ones. In many instances, steps 2 and 3 are interchanged in the sequence, depending on the nature of the system and the degree of knowledge regarding the processes taking place.

Of course the modeling of an existing process will differ from the modeling to design a new plant based on known technology. And both will be different from using modeling to aid in the development of a novel technology.

1.9 MODELING AND SIMULATION IN CHEMICAL ENGINEERING

Aris [5], in his 1990 Dankwarts Memorial Lecture entitled "Manners Makyth Modellers," distinguishes between modeling and simulation in a special manner which follows the reasoning of Smith [6] in ecological models. This reasoning is very useful for chemical engineering students to comprehend early in their career. The reasoning goes as follows (having the beautiful and very intellectual characteristics of the writings of Professor Aris, the readers are advised to read the classical papers referred to in this reasoning):

> It is an essential quality in a model that it should be capable of having a life of its own. It may not, in practice, need to be sundered from its physical matrix. It may be a poor thing, an ill-favored thing when it is by itself. But it must be capable of having this independence. Thus Liljenroth (1918) in his seminal paper on multiplicity of the steady states can hardly be said to have a mathematical model, unless a graphical representation of the case is a model. He works out the slope of the heat removal line from the ratio of numerical values of a heat of reaction and a heat capacity. Certainly he is dealing with a typical case, and his conclusions are meant to have application beyond this particularity, but the mechanism for doing this is not there. To say this is not to detract from Liljenroth's paper, which is a landmark of the chemical engineering literature, it is just to notice a matter of style and the point at which a mathematical model is born. For in the next papers on the question of multiple steady states, those of Wagner (1945), Denbigh (1944, 1947), Denbigh et al. (1948) and Van Heerden (1953), we do not find more general structures. How powerful the life that is instinct in a true mathematical model can be seen from the Fourier's theory of heat conduction where the mathematical equations are fecund of all manner of purely mathematical developments. **At the other end of the scale a model can cease to be a model by becoming too large and too detailed a simulation of a situation whose natural line of development is to the particular rather than the general. It ceases to have a life of its own by becoming dependent for its vitality on its physical realization** [the emphasis is ours]. Maynard Smith (1974) was, I believe, the first to draw the distinction in ecological models between those that aimed at predicting the population level with greater and greater accuracy (simulation) and those that seek to disentangle the factors that affect population growth in a more

general way (model). The distinction is not a hard and fast one, but it is useful to discern these alternatives [5].

The basis of the classification given by Aris is very interesting, true, and useful. It is typical of the Minnesota group founded by Amundson some 50 years ago. This important research group in the history of chemical engineering has almost never verified the models (or the variety of interesting new phenomena resulting from them) against experiments or industrial units. Experimental verifications of the new and interesting steady-state and dynamic phenomena discovered by the Minnesota group were carried out at other universities, mostly by graduates from Minnesota. The most interesting outcome is the fact that not a single phenomenon, which was discovered theoretically using mathematical models by the Minnesota group, was not experimentally confirmed later. This demonstrates the great power of the mathematical modeling discipline as expressed by Aris [5], where the model is stripped of many of its details in order to investigate the most fundamental characteristics of the system. It also demonstrates the deep insight into physical systems that can be achieved using mathematics, as Ian Stewart puts it: "Perhaps mathematics is effective in organizing physical existence because it is inspired by physical existence .... The pragmatic reality is that mathematics is the most effective method that we know for understanding what we see around us" [7]. In this undergraduate book, a different, more pragmatic definition for mathematical modeling and simulation than that of Aris [5] will be adopted. The definition we adopt is that mathematical modeling will involve the process of building up the model itself, whereas simulation involves simulating the experimental (or pilot plant or industrial) units using the developed model. Thus, simulation in this sense is closely linked to the verification of the model against experimental, pilot plant, and industrial units. However, it will also include the use of the verified models to simulate a certain practical situation specific to a unit, a part of a production line, or an entire production line.

However, because universities are responsible also for the preparation of the students for research careers, this book will also include the elements discussed by Aris. In this case, the definition by Aris [5] must come strongly into play. Because much of the chemical engineering research work today is interdisciplinary in nature, especially in the relatively new fields such as biotechnology and microelectronics, it should be clear that mathematical modeling and system theory are the most suitable and efficient means of communication between the different disciplines involved. Therefore, chemical engineering education based on system theory and mathematical modeling as portrayed in this book seems to be the best approach to prepare the chemical engineers for this interdisciplinary research. The fact that

Henry Poincaré, the undisputed first discoverer of Chaos, was trained first in engineering, following the Napoleonic traditions, helps to emphasize the need for a broad education of the engineering researchers of tomorrow, which is hard to achieve through the present structure of engineering syllabi.

Another very general definition of models is given by Stephen Hawking [8], who relates models to theories of the universe:

> A theory is just a model of the universe, or a restricted part of it, and a set of rules that relate quantities in the model to observations that we make. It exists only in our minds and does not have any other reality (whatever that might mean). A theory is a good theory if it satisfies two requirements: it must accurately describe a large class of observations on the basis of the model that contains only a few arbitrary elements, and it must make definite predictions about the results of future observations.

One of the major findings of the Minnesota group, using mathematical modeling, is the discovery of a wide variety of static and dynamic bifurcation phenomena in chemical reactors. Although theoretical studies on the bifurcation behavior have advanced considerably during the last three decades, the industrial appreciation of these phenomena remains very limited. It is of great importance that practically oriented chemical engineers dealing with the mathematical modeling of industrial units become aware of them. These phenomena are not only of theoretical and academic importance, but they have very important practical implications (e.g., for industrial Fluid Catalytic Cracking Units [9–12]). Most chemical engineers complete their undergraduate education without knowing almost anything about these phenomena. A large percentage of chemical engineers finish their postgraduate education without knowing anything about these phenomena. This book will introduce undergraduate students to these phenomena in a very simple manner early in their education.

1.10 AMUNDSON REPORT AND THE NEED FOR MODERN CHEMICAL ENGINEERING EDUCATION

The report of the committee on "Chemical Engineering Frontiers: Research Needs and Opportunities" [13], known as the Amundson report, outlines beautifully the picture of chemical engineering in the next decades. The report is quite optimistic about the future of the profession, and the authors of this book strongly share this optimism. Chemical engineering played an important part in human development in the last few decades and it is expected to play an even larger role in the future.

Although the "Amundson Report" emphasizes the challenges facing the American chemical and related industries, the report should not be looked upon from this narrow point of view. In fact, the report is far more reaching than that and is to a great extent relevant to the worldwide chemical engineering profession. The report, without ignoring classical chemical engineering problems, stresses a number of relatively new chemical engineering fields:

1. Biotechnology and Biomedicine
2. Electronic, Photonic, and Recording Materials and Devices
3. Polymers, Ceramics, and Composites
4. Processing of Energy and Natural Resources
5. Environmental Protection, Process Safety, and Hazardous Waste Management
6. Computer-Assisted Process and Control Engineering
7. Surfaces, Interfaces, and Microstructures

Emphasizing these fields without being able to ignore classical chemical engineering problems helps only to emphasize the view that chemical engineering is entering a new era. In addition, the chemical engineering community cannot ignore its fundamental scientific responsibilities toward the revolution in scientific knowledge created by the discovery of chaos, strange attractors, and fractal structures, especially that many manifestations of these phenomena are evident in typical chemical engineering systems.

The report is now more than 13 years old and the issues addressed in this report must reflect themselves on chemical engineering undergraduate education. This can only be achieved through the adoption of system theory and mathematical modeling in undergraduate chemical engineering education.

These considerations lead to the inevitable conclusion that a higher level of organization of thinking and economy of knowledge is needed in undergraduate chemical engineering education, as briefly discussed earlier and will be detailed in the rest of the book.

1.11 SYSTEM THEORY AND MATHEMATICAL MODELING AS TOOLS FOR MORE EFFICIENT UNDERGRADUATE CHEMICAL ENGINEERING EDUCATION

In order to achieve a higher level of organization and economy of knowledge, it will be extremely useful for the students to be introduced early in their undergraduate education to the basic principles of system theory. The best and most general classification of systems is based on thermodynamic principles. Therefore, it is possible that the students, after passing their basic science courses, be exposed to a course in thermodynamics that emphasizes the basic concepts of thermodynamics. The basic principles of system theory can either be integrated into this course or may be taught in a separate course. This course should emphasize not only the basic principles of system theory but also its relevance to chemical engineering systems. In fact, such a course can be used as an elegant and efficient tool for introducing the students to chemical engineering systems. The students should learn how to classify chemical engineering systems into their main categories: isolated, closed, and open systems. They should learn how to divide the system into its subsystems or elements depending on the level of analysis. They should learn not only that distillation is a unit operation regardless of the specific kind of distillation, as the unit operation paradigm teaches us, but to extend their mind further and learn that continuous, heterogeneous, multistage processes are systems that are open, formed of more than one phase and more than one stage, that there are certain processes taking place within their boundaries that need to be expressed in terms of state variables, and that input variables should be specified and parameters should be identified (regardless of whether the process is distillation, extraction, drying, or multistage catalytic reactors).

Of course, at this early stage, students will not be able yet to develop specific models for specific processes because they would not yet have studied the laws governing the rates of these processes and their form of dependence on the state variables. They even may not be able, at this stage, to identify completely the state variables of the system. However, this early training in system theory will orient the students mind and their further education in the framework of the system approach.

The students can then be ready to view their other chemical engineering courses in a new light; those courses should also be changed to emphasize the system approach, where all of the rate processes are treated in a unified fashion, emphasizing the laws governing the rates of these processes and its dependence on state variables, rather than dividing it into mass transfer, heat transfer, momentum transfer, and rates of reactions. The transport phenomena paradigm can be easily extended in this direction. This will also allow higher emphasis on the interaction between these processes and the effects resulting from these interactions (e.g., the interaction between mass transfer and chemical reaction and its implication for the behavior of the system).

It is natural that, at this stage, students be introduced to the formulation of some simple mathematical models (design equations) for certain systems. When the students are familiar with the laws governing the rates of different processes in terms of the state variables, they will be ready for an intensive applied course on mathematical modeling of chemical engineering systems. This course should cover the basic principles of mathematical modeling theory, the main classifications of mathematical models, the procedures for building mathematical models, application to the development of mathematical models for a large number of chemical engineering processes in the petrochemical, petroleum refining, biochemical, electronic industries, as well as some mathematical models for biological systems.

It is important to note that material and energy balances represent the basis of any rational and useful chemical engineering education. In this book, we use an approach that generalizes the material and energy balance equations in modular forms for the most general cases with multiple inputs, multiple outputs, and multiple reactions. These generalized material and energy balance equations will represent the basis for all material and energy balance problems, with less complicated cases treated as special cases of these generalized mathematical formulation. These material and energy balance equations will themselves be used to develop both lumped and distributed mathematical models (design equations). This approach will represent a serious achievement with regard to economy of knowledge, organization of thinking, and systematic optimization of information for chemical engineering students. The extension of design equations (steady-state models) to dynamic equations (unsteady-state models) will also be shown.

So far, we have emphasized the practical usefulness of the approach. In this last sentence, we should also emphasize that the approach is beautiful and elegant: "He gets full marks who mixes the useful with the beautiful" [14].

In this preliminary section, we have explained to the reader that chemical engineering is expanding very quickly and that a new approach based on system theory and mathematical modeling is needed for a higher level of organization of thinking and economy of information. The concept has been briefly introduced and discussed and the influence of this approach on the structure of chemical engineering teaching has been discussed in a limited fashion, only to show how it can be integrated into a modern and progressive chemical engineering syllabus and how it does affect the approach to teaching material and energy balance and rate processes in an integrated manner. This approach is adopted in this book, and the approach will be clearer as more details and applications are given throughout the book.

1.12 SUMMARY OF THE MAIN TOPICS IN THIS CHAPTER

It is useful for the reader to summarize all of the ideas and classifications. Although each of the following subjects will be discussed in more detail, it is very useful for the reader to have an "eagle's-eye view" to know where it all fits.

1.12.1 Different Types of Systems and Their Main Characteristics

What Is a System?

It is a whole, formed of interactive parts.

Isolated Systems

Do not exchange matter or energy with the surrounding. Example is adiabatic batch reactors.

Most Basic Characteristics. Tend toward thermodynamic equilibrium characterized by maximum entropy [it is stationary with time, stationary in the sense of "death" (i.e., nothing happens)].

Closed Systems

Do not exchange matter with the environment or surrounding, but do exchange energy. Example is a non-adiabatic batch reactor.

Most Basic Characteristics. Tend toward thermodynamic equilibrium characterized by maximum entropy [it is stationary with time, stationary in the sense of "death" (i.e., nothing happens)].

Open Systems

Exchange matter and energy with the environment or surrounding (actually stating that it exchanges matter is sufficient because this implies an exchange of the energy in the matter transferred). Example is a CSTR.

Most Basic Characteristics. Do not tend toward thermodynamic equilibrium, but rather toward what may be called a "stationary nonequilibrium state"; it is stationary with time but not in the sense of being "dead." Things do happen; the stationary nature comes from the balance between what is happening within the boundaries of the system and the exchange of matter with the surrounding.

Important note A stationary nonequilibrium state can also be called a "point attractor." We will see later that other types of attractor are also possible.

Relative Nature of Systems, Subsystems, and Elements

- Subsystems are parts of the system, that together with their interactions define the system.

- The element is the smallest "subsystem" of the system, according to level of analysis.
- Subsystems and elements are relative and depend on the level of analysis.
- Subsystems, elements, and "whole" system can be defined on different basis.

Linear and Nonlinear Systems

Linear and nonlinear systems are usually defined on the basis of the form of the appearance of the state variables in the process. However, what are "state variables"?

State Variables and Parameters

State variables are the variables that describe the state of the system (i.e., temperature of the fluid in a tank, or concentration of a component in a reactor or on a tray of a distillation column).

Parameters are constants (or variables) that are imposed (or chosen) for the system and determine the state of the system through their effect on the processes. We can classify them as follows:

1. Input parameters: feed flow rate, feed concentration, and so forth.
2. Design parameters: diameter, height, and so forth of a vessel.
3. Process parameters: mass transfer coefficients, kinetic rate constants, and so forth.
4. Physical parameters: density, viscosity, and so forth.

The classical definition of linear/nonlinear systems is related to when the system is described by a black box with empirical relation(s) between input(s) (I) and output(s) (O) and the relation between I and O is linear/nonlinear (refer to Fig. 1.1). However, with more models developed on a physicochemical basis, the usual definition is as follows: The model in linear if all state variables appear to the power 1 and without two state variables appearing as multiplied with each other. The model is nonlinear otherwise.

For example,

$$y = ax$$

is a linear model, where y and x are state variables and a is a parameter. However,

$$y = ax^2$$

is a nonlinear model. Also,

$$\frac{dx}{dt} = ax$$

is a linear model and

$$\frac{dx}{dt} = ax^2$$

is a nonlinear model.

1.12.2   What Are Models and What Is the Difference Between Models and Design Equations?

Models in the sense used in this book are equations describing the process; in other words, a model is some kind of mathematical equation(s) (of different degrees of rigor and complexity) that is (are) able to predict what we do not know using the information that we know. In most cases, it is as follows:

- The output is unknown, and the input is known.
- The input is known, the desired output is fixed and it is the design parameters which are to be computed to obtain the desired output.
- It can be that the system design parameters are known, the desired output is fixed, and it is the input parameters which are to be calculated.

As a matter of fact,

Mathematical models = Design equations

When the equations are based on fundamentals of the different processes and are rather rigorous and complex, they are called mathematical models. However, when they are simple and highly empirical, they are called design equations. In the present age of powerful computers, the computer is able to solve complicated nonlinear equations in a short time using suitable solution techniques. Thus, the boundaries between mathematical models and design equations should disappear.

Types of Model

There are different bases on which models are classified; none of them is sufficient in itself. The integral of all definitions and classifications gives the complete picture.

Examples are as follows:

Homogeneous and heterogeneous models
Lumped and distributed models
Rigorous (relatively rigorous) and empirical models
Linear and nonlinear models

Steady-state and unsteady-state models
Deterministic and stochastic models.

1.12.3 Summary of Numerical and Analytical Solution Techniques for Different Types of Model

The various solution techniques are presented in Figures 1.6–1.10.

Most of the efficient solution techniques (whether analytical or numerical) will be presented in this book.

Programming

For most chemical engineering problems, the solution algorithm is quite complex and needs to be solved using a computer program/software. Students can use whatever programming language or computational environment they want (a short list of some programming environments and softwares is given in Appendix F). Examples are as follows:

- Fortran with IMSL subroutines library
- C and C++ with necessary libraries
- Matlab
- Polymath
- Mathcad
- Mathematica.

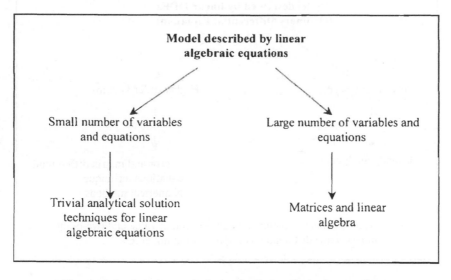

Figure 1.6 Solution techniques for linear algebraic equations.

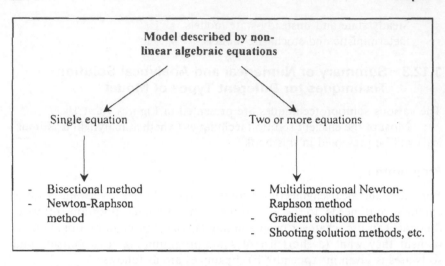

Figure 1.7   Solution techniques for nonlinear algebraic equations.

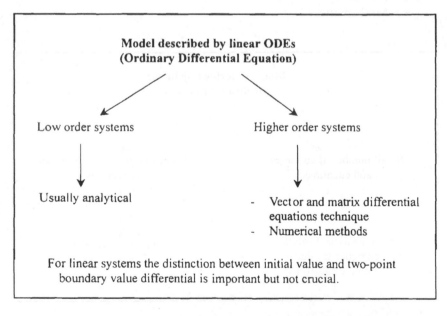

Figure 1.8   Solution techniques for linear ordinary differential equations.

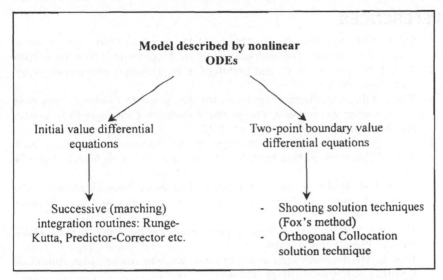

Figure 1.9 Solution techniques for nonlinear ordinary differential equations.

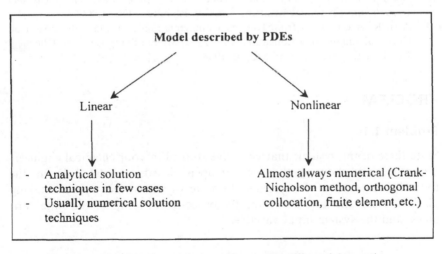

Figure 1.10 Solution techniques for partial differential equations.

REFERENCES

1. Smith, J.M., Van Hess, H.C., Abbott, M.M., and Van Hess. *Introduction to Chemical Engineering Thermodynamics*, 6th ed. McGraw-Hill, New York, 2000.
2. Bird, R.B., Stewart, W.E., and Lightfoot, E.N. *Transport Phenomena*. Wiley, New York, 1960.
3. Wei, J. Educating Chemical Engineers for the Future. In *Chemical Engineering in a Changing Environment, Engineering Foundation Conference* (S.L. Sandler and B.A. Finlayson, eds.), 1988, pp. 1–12.
4. Levenspiel, O. Private discussion during AIChE meeting, San Francisco, 1989.
5. Aris, R. Manners makyth modellers. *Chem. Eng. Res. Des.* 69(A2), 165–174, 1991.
6. Smith, J.M. *Models in Ecology*. Cambridge University Press, Cambridge, 1974.
7. Stewart, I. *Does God Play Dice: The New Mathematics of Chaos*, Penguin, London, 1989.
8. Hawking, S.W. *A Brief History of Time: From the Big Bang to Black Holes*. Bantam Press, London, 1989.
9. Iscol, L. The dynamics and stability of a fluid catalytic cracker. Joint Automatic Control Conference, 1970, pp. 602–607.
10. Elnashaie, S.S.E.H. and El-Hennawi, I.M. Multiplicity of the steady states in fluidized bed reactors, IV. Fluid catalytic cracking. *Chem. Eng. Sci.* 34, 1113–1121, 1979.
11. Edwards, W.M. and Kim, H.N. Multiple steady states in FCC unit operations. Chem. Eng. Sci., 43, 1825–1830, 1998.
12. Elshishini, S.S. and Elnashaie, S.S.E.H. Digital simulation of industrial fluid catalytic cracking units, bifurcation and its implications. Chem. Eng. Sci., 45, 553–559, 1990.
13. Amundson, N.R. *Frontiers in Chemical Engineering, Research Needs and Opportunities*, National Academy Press, Washington, DC, 1988.
14. Aris, R. A quotation from Horace. Comments made during a discussion on Chemical Engineering Education, 10th International Symposium on Chemical Reaction Engineering, (ISCRE 10), 1988.

PROBLEM

Problem 1.1

State three of the systems that you have studied in your chemical engineering courses and show whether they are open, closed, or isolated. Define the following for each of the systems: the state variables, the processes taking place in them, the laws governing the processes in terms of the state variables, and the system input variables.

2

Material and Energy Balances

2.1 MATERIAL AND ENERGY BALANCES

2.1.1 A Simple, Systematic, and Generalized Approach

Material and energy (M&E) balances are the basis of almost all chemical engineering calculations. M&E balances, in themselves, are used for the inventory of overall materials, different species and energy for single units, and total process flowsheets. It is essential, as a first step, for the design of single units as well as overall processes formed of a number of units to carry out M&E balance calculations, in order to determine the basic characteristics of unit/processes and is a prelude for the detailed design of these units/ processes. The tightly controlled inventory for units/process using M&E balances, which is as rigorous as possible, is not only important for design, operation and control but also for tight control over pollution. By definition, the polluting components can be escaping reactants or impurities with the feed or escaping products or side products. The tight control over the material balances and the comparison with continuous measurements after the plant is built is one of the best and most reliable means to predict pollution emissions from the plant (specially fugitive emissions).

The rigorous M&E balances are also the basis for the rigorous design and control equations as will be shown in Chapter 5. M&E balances seem so simple so that it can look intuitive; for example, we gave you six oranges,

41

you have eaten two of them, then the remaining are certainly four oranges. We gave you $200, you spent $50 and gave a friend $30, so the remaining dollars are certainly $120. This intuitive approach was used in M&E balances for a long time.

However, it is important in our present age of large complicated processes and large effective computers to systematize the process of M&E balances into generalized equations, with all simpler cases coming out of these equations as special cases. The first textbook using this approach is the book by Reklaitis [1]; however he did not explain this approach fully. This approach not only allows the systematic computerization of M&E balance calculations for small as well as large complicated processes but also presents a simple, clear, and very useful link between M&E balance equations on the one hand and design/control equations on the other, as shown in Chapters 3–6.

In order to make the subject exceedingly easy and smooth for the reader, we will develop this generalized and systematic approach and these generalized equations starting from the intuitive approach (which is very dear to a number of generations of chemical engineers). Our approach here will make the subject very easy for the reader and will make the final generalized equation almost obvious.

2.1.2 Development of Material Balance Relations

Semi-Intuitive Approach for Nonreacting Systems

Consider the system in Figure 2.1; it is obviously an open system. So after start-up, it reaches the state rigorously called the "stationary nonequilibrium state" (see Chapter 1) and commonly known in chemical engineering as the "steady state." It is a separation unit with no reactions taking place.

Note that

X_i (Weight fraction of component i)

$$= \frac{\text{Mass of component } i \text{ in the stream}}{\text{Total mass of the stream}}$$

This is a completely defined mass balance diagram for this single unit; in other words, there are no unknowns.

Let us now start training the reader on mass balance by hiding some of the known variables. How many variables can we hide and refind? (We will answer this question in stages, coming at the end in a very logical systematic way to the rules of degrees of freedom analysis for single unit and complete flow sheets).

So let us have the following problem (Fig. 2.2) by hiding some of the known variables in Figure 2.1. Counting the number of unknowns (denoted

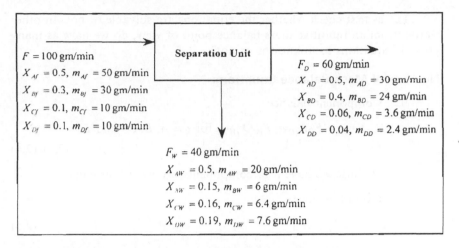

Figure 2.1   Complete material balance over a separation unit.

by ?) shows that it has seven unknowns. Can we find out those seven unknowns through the laws of mass balance? Shall we start solving directly or we can check if it is solvable or not?

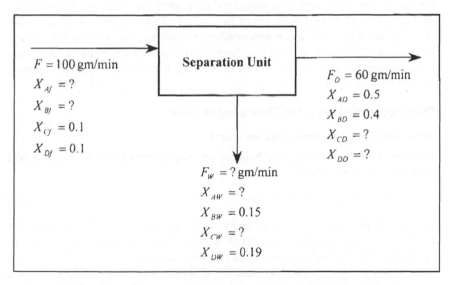

Figure 2.2   Material balance diagram with some missing values.

Let us first check whether the equations are solvable or not. In other words, from an inuititive mass balance point of view, do we have as many possible equations as unknowns or not?

Number of Mass Balance Equations

1. Component balance

$$FX_{if} = F_D X_{iD} + F_W X_{iW} \quad \text{for } i = A, B, C, \text{ and } D$$

$$(2.1)\text{-}(2.4)$$

Because we have four components, we have four equations

2. Overall balance

$$F = F_D + F_W \tag{2.5}$$

3. Summation of weight fractions for each stream is equal to 1:

$$\sum_i X_{if} = 1.0 \tag{2.6}$$

$$\sum_i X_{iD} = 1.0 \tag{2.7}$$

$$\sum_i X_{iW} = 1.0 \tag{2.8}$$

Are all of the above eight equations independent? No, because the summing of Eqs. (2.1)–(2.4) and then using Eqs. (2.6)–(2.8) results in Eq. (2.5). So, actually, we have only seven independent equations, but we also have seven unknowns. This leads to that the degrees of freedom are equal to zero and the problem is solvable. This is one of the simplest examples for the degrees-of-freedom analysis.

Strategy of Solution for This Simple Case

With which mass balance shall we start?

If we start with a component balance on, say, component A (this is a bad choice); then we will have

$$FX_{Af} = F_D X_{AD} + F_W X_{AW} \tag{2.9}$$

In this equation, we have three unknowns (X_{Af}, F_W, and X_{AW}). Also, the product $F_W X_{AW}$ can be considered a type of nonlinearity, which is not desirable when we solve a sequence of linear equations.

Certainly, it is very clear that Eq. (2.9) is a very bad choice because it has to be solved simultaneously with at least two other equations.

So, What Is a Good Choice for Starting the Solution?

It is clear that the overall balance is the only starting balance that gives an equation with one unknown. It is clear that this is the best starting step. Thus, we start by using the overall balance Eq. (2.5),

$$F = F_D + F_W$$

which becomes

$$100 = 60 + F_W$$

thus giving

$$F_W = 40 \text{ min/g}$$

What Is the Next Best Step?

After F_W has been calculated, we have other components that will give an equation in one variable. Try component B:

$$FX_{Bf} = F_D X_{BD} + F_W X_{BW}$$

On substituting the corresponding values, we get

$$100 \, X_{Bf} = (60)(0.4) + (40)(0.15)$$

Therefore,

$$X_{Bf} = 0.3$$

Similarly, try component D:

$$FX_{Df} = F_D X_{DD} + F_W X_{DW}$$

On substituting the corresponding values, we get

$$(100)(0.1) = (60)X_{DD} + (40)(0.19)$$

Therefore,

$$X_{DD} = 0.04$$

Until now we have used the overall mass balance equation and two of the component mass balance equations (we are left with only one) and we still have three $\sum_i X_i = 1.0$ relations.

What Is the Next Best Move?

If we choose to use a component mass balance on component A, we will have an equation with two unknowns. If we choose to use a component mass balance on component C, we will have an equation with two unknowns. For example, the best choice is to start to use the $\sum_i X_i = 1.0$ relations:

$$X_{Af} + X_{Bf} + X_{Cf} + X_{Df} = 1.0$$

Thus,

$$X_{Af} + 0.3 + 0.1 + 0.1 = 1.0$$

and we get

$$X_{Af} = 0.5$$

Now, if we use a component mass balance on component A, we will have an equation with only one unknown:

$$FX_{Af} = F_D X_{AD} + F_W X_{AW}$$

Substitution of values gives

$$(100)(0.5) = (60)(0.5) + 40 X_{AW}$$

Calculation gives

$$X_{AW} = 0.5$$

We have thus exhausted the component balances; we can then directly use the remaining two $\sum_i X_i = 1.0$ relations. Using

$$\sum_i X_{iD} = 1.0$$

we get

$$X_{AD} + X_{BD} + X_{CD} + X_{DD} = 1.0$$

Substituting the values gives

$$0.5 + 0.4 + X_{CD} + 0.04 = 1.0$$

Thus,

$$X_{CD} = 0.06$$

Also, for the other stream,

$$\sum_i X_{iW} = 1.0$$

we get

$$X_{AW} + X_{BW} + X_{CW} + X_{DW} = 1.0$$

Substituting the values gives

$$0.5 + 0.15 + X_{CW} + 0.19 = 1.0$$

Thus,

$$X_{CW} = 0.16$$

Note: By the proper sequence of solution, we went through the problem without solving any simultaneous equations. Each equation we have written in the proper sequence had only one unknown.

Suppose now that we have the following arrangement of the known and unknown variables (keeping the total number of unknowns the same) as shown in Figure 2.3. Counting the number of unknowns, there are obviously the same number of unknowns as earlier (seven unknowns). The unaware chemical engineer who does not use the degrees-of-freedom concept fluently will immediately go into trying to solve this problem, will spend a long time and lots of frustration, and will end up not being able to solve it, whereas the chemical engineer with good awareness regarding the use of the degrees-of-freedom analysis will realize in a few seconds that the problem is not solvable. Why? Because, actually, with the given structure of unknowns, we have only six relations not seven. Why? This is because for the feed stream the relation

$$\sum_i X_{if} = 1.0$$

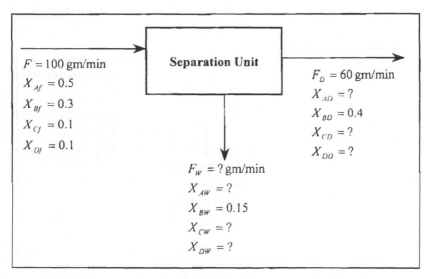

Figure 2.3 Material flow diagram with unknown values.

is redundant, because all of the weight fractions of the four components are given in this stream.

   Now let us pose the same mass balance problem in a third fashion (Fig. 2.4). Here, we have eight unknowns and seven relations. Thus, the degree of freedom is 1, so strictly speaking, the problem is not solvable. However, actually, it can be solved for an assumed value of F or F_D or F_W, which will be called the *basis* and we will notice a simple linear relation between the solution and the chosen basis. This linear relation makes the solution quite useful.

1. **Assume $F=100$ min/g (basis)**

   Here, we have seven unknowns and seven relations.

   Step 1
   $$100 = F_D + F_W$$
   which gives
   $$F_W = 100 - F_D$$
   Step 2

   Use
   $$\sum_i X_{if} = 1.0$$

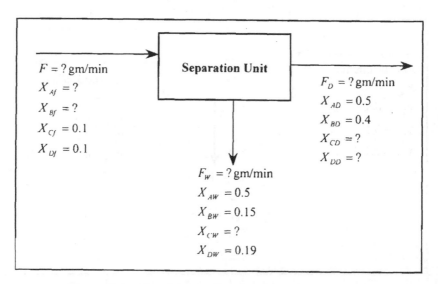

$F = ?$ gm/min
$X_{Af} = ?$
$X_{Bf} = ?$
$X_{Cf} = 0.1$
$X_{Df} = 0.1$

Separation Unit

$F_D = ?$ gm/min
$X_{AD} = 0.5$
$X_{BD} = 0.4$
$X_{CD} = ?$
$X_{DD} = ?$

$F_W = ?$ gm/min
$X_{AW} = 0.5$
$X_{BW} = 0.15$
$X_{CW} = ?$
$X_{DW} = 0.19$

Figure 2.4   Material flow diagram with unknown values.

To get

$$X_{Af} + X_{Bf} + X_{Cf} + X_{Df} = 1.0$$

which gives

$$X_{Af} + X_{Bf} = 0.8$$

Step 3

The mass balance on component A is

$$FX_{Af} = F_D X_{AD} + F_W X_{AW}$$

which gives

$$100 X_{Af} = F_D(0.5) + (100 - F_D)(0.5)$$

Thus,

$$X_{Af} = 0.5$$

Step 4

From the relation in step 2, we get

$$X_{Bf} = 0.8 - 0.5 = 0.3$$

Step 5

The mass balance on component B is

$$FX_{Bf} = F_D X_{BD} + F_W X_{BW}$$

Substituting the values gives

$$(100)(0.3) = F_D(0.4) + (100 - F_D)(0.15)$$

Thus,

$$F_D = 60 \text{ min/g}$$

Then, using the relation in step 1, we get,

$$F_W = 40 \text{ min/g}$$

obtaining the rest of the variables is really straightforward. The reader is advised to compute the remaining unknown values.

2. **Assume a basis of $F = 50$ g/min**

   Step 1

   $$50 = F_D + F_W$$

   which gives

   $$F_W = 50 - F_D$$

   Step 2

   Use

   $$\sum_i X_{if} = 1.0$$

   to get

   $$X_{Af} + X_{Bf} + X_{Cf} + X_{Df} = 1.0$$

   which gives

   $$X_{Af} + X_{Bf} = 0.8$$

   Step 3

   The mass balance on component A is

   $$FX_{Af} = F_D X_{AD} + F_W X_{AW}$$

   which gives

   $$50 X_{Af} = F_D(0.15) + (50 - F_D)(0.5)$$

   Thus,

   $$X_{Af} = 0.5$$

   Step 4

   From the relation in step 2, we get

   $$X_{Bf} = 0.8 - 0.5 = 0.3$$

   Step 5

   The mass balance on component B is

   $$FX_{Bf} = F_D X_{BD} + F_W X_{BW}$$

Substituting the values gives

$$(50)(0.3) = F_D(0.4) + (50 - F_D)(0.15)$$

Thus,

$$F_D = 30 \text{ min/g}$$

Then, using the relation in step 1, we get

$$F_W = 20 \text{ min/g}$$

Thus, the difference in the values by taking different basis can be summarized as in Table 2.1. Because

$$\frac{F(\text{first basis})}{F(\text{second basis})} = 2$$

we have

$$\frac{F_D(\text{first basis})}{F_D(\text{second basis})} = 2 \quad \text{and} \quad \frac{F_W(\text{first basis})}{F_W(\text{second basis})} = 2$$

It should be noted that in both cases the weight fractions of all components in all streams remain the same.

This is the essence for the wide and very useful use of *basis* with regard to mass or molar flow rates.

Cases Where Taking Basis Does Not Solve the Problem of the Number of Unknowns Being Larger Than the Number of Relations

What will be the situation when we have eight unknowns (all of them weight fractions, as shown in Figure 2.5)? In this case, let us proceed a few steps in the calculations. We can obtain X_{Af} by

$$FX_{Af} = F_D X_{AD} + F_W X_{AW}$$

Table 2.1 Comparison of Values for Different Bases

Basis: $F_1 = 100$ g/min	Basis: $F_2 = 50$ g/min
$F_{D1} = 60$ g/min	$F_{D2} = 30$ g/min
$F_{W1} = 40$ g/min	$F_{W2} = 20$ g/min

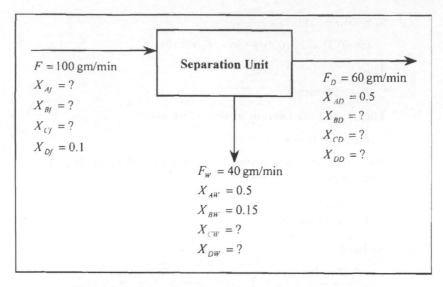

Figure 2.5   Mass flow diagram where basis does not work.

On substituting the values, we get

$$100X_{Af} = (60)(0.5) + (40)(0.5)$$

Thus, we get

$$X_{Af} = 0.5$$

After that, we cannot go any further with the computations; the remaining equations are six in number and the unknowns are seven in number and no basis can be taken because F, F_D, and F_W have specified values.

2.2   SINGLE AND MULTIPLE REACTIONS: CONVERSION, YIELD, AND SELECTIVITY

Now, we move one important step forward toward the mass balance of units with chemical reactions (chemical reactors). In order to reach this stage in a systematic way, we start by analyzing the different types of reaction and introduce the necessary additional information needed for the mass balance of units with chemical reactions.

2.2.1 Single Reactions

A single reaction is a reaction formed of one stoichiometric relation regardless of the number of components (species) involved as reactants or products.

Example of Single Reactions

$A \rightarrow B$ (two components, irreversible)
$A \Leftrightarrow B$ (two components, reversible)
$A + B \rightarrow C + D$ (four components, irreversible)
$A + B \Leftrightarrow C + D$ (four components, reversible)
$5A + 8B + 9C \rightarrow 3D + 6F$ (five components, irreversible)
$3A + 2B \Leftrightarrow 5C + 7D$ (four components, reversible)

A reversible reaction such as

$$A \Leftrightarrow B$$

is strictly depicting two reactions. However, it has many relations to single reactions because of the special nature of the two reactions involved. However, in this book, for straightforward consistency we will always treat it as two reactions.

The number in front of each component is called the stoichiometric number and it is an indication of the proportion (in moles) of the component entering the reaction in comparison with other components (reactants or products).

The following stoichiometric relations are strictly equivalent:

$$5A + 10B \rightarrow C + 20D$$

$$A + 2B \rightarrow \tfrac{1}{5}C + 4D$$

$$50A + 100B \rightarrow 10C + 200D$$

Conversion and Yield for a Single Reaction

A reactor with a single reaction can be completely defined (and solved) in terms of one variable (conversion of any one of the reactants, or the yield of any one of the products, or the concentration of one of the components, or the rate of reaction of one of the components) because all other quantities can be computed in terms of this single variable as long as the stoichiometric equation is fully defined. This will be clearly known after we define conversion and yield.

The conversion of a reactant component is defined as the number of moles reacted of the specific component divided by the original number of moles of the same specific component. For continuous operation, this statement will read: Conversion of a reactant component is defined as the number of moles reacted per unit time of the specific component divided by the feed numbers of moles per unit time of the same specific component (i.e., molar flow rate).

For a reactor as the one shown in Fig. 2.6, the reaction is

$$A + B \rightarrow C + D \quad \text{(four components, irreversible)}$$

The conversion of component A is defined as

$$x_A = \frac{n_{Af} - n_A}{n_{Af}}$$

The conversion of component B is defined as

$$x_B = \frac{n_{Bf} - n_B}{n_{Bf}}$$

The relation between x_A and x_B depends on the stoichiometric numbers of each component. This will be shown and discussed later. Even if the stoichiometric numbers are different,

$$2A + 5B \rightarrow 7C + 8D$$

the definitions still remain strictly the same:

$$x_A = \frac{n_{Af} - n_A}{n_{Af}}$$

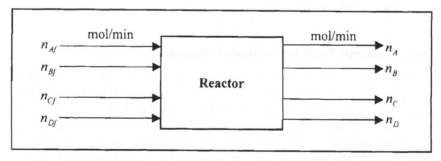

Figure 2.6   Molar flow diagram across a reactor.

and

$$x_B = \frac{n_{Bf} - n_B}{n_{Bf}}$$

Only the relation between x_A and x_B will be different because of the stoichiometric numbers, as will be explained later.

Note: It is possible to use the same basis for the two conversions, such as

$$x_A = \frac{n_{Af} - n_A}{n_{Af}}$$

and

$$x_B = \frac{n_{Bf} - n_B}{n_{Af}}$$

Of course, the relation between x_A and x_B will be different because of the different ways we define the basis for each component.

Yield of a Product (e.g., *C*)

The yield of a product C is the number of moles of C formed (per unit time for continuous processes) divided by the original (or feed for continuous processes) number of moles of reactant A (per unit time for continuous processes); that is,

$$Y = \frac{n_C - n_{Cf}}{n_{Af}}$$

or we can reference the yield to reactant B; that is,

$$\overline{Y} = \frac{n_C - n_{Cf}}{n_{Bf}}$$

Also, the yield can be referred to the number of moles of A (or B) that reacted:

$$Y' = \frac{n_C - n_{Cf}}{n_{Af} - n_A} \quad \text{or} \quad Y'' = \frac{n_C - n_{Cf}}{n_{Bf} - n_B}$$

Mass Balance Calculations for a Reactor with a Single Reaction

As we mentioned earlier, the mass balance of the reactor can be defined completely in terms of only one variable (x_A, x_B, Y, Y', Y'', etc.) when there is a single reaction.

2.2.2 Degrees-of-Freedom Analysis

Solved Example 2.1

For the reactor shown in Figure 2.7, with known feed conditions (n_{Af}, n_{Bf}, n_{Cf}, n_{Df}), determine the relations giving the output variables (n_A, n_B, n_C, n_D) in terms of one variable (x_A or x_B or Y) for the following reaction:

$$2A + 3B \rightarrow 3C + 5D$$

If the number of unknowns is four (n_A, n_B, n_C, n_D) plus one extra variable (the rate of reaction, conversion, or yield; the relation among them will be given later), then the total number of unknowns is five.

The number of mass balance relations for four components is four. We require an additional relation in terms of conversion (or yield). Suppose that the conversion is given, then this is an extra given "known variable." Thus, the degrees of freedom is zero and the problem is solvable.

Solution for This Continuous Process

$$x_A = \frac{n_{Af} - n_A}{n_{Af}}$$

where $n_{Af} - n_A$ is the number of moles of component A reacted per minute ($\overline{R}_A = n_{Af} x_A$), where $\overline{R}_A$ is the rate of consumption of component A (moles of A consumed per unit time).

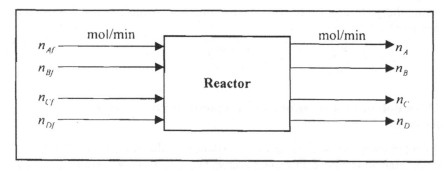

Figure 2.7 Molar flow diagram for solved example 2.1.

From the stoichiometric equation, we have the number of moles of component B reacted per unit time ($\overline{R}_B = \frac{3}{2}$ number of moles of A reacted per unit time):

$$\overline{R}_B = \tfrac{3}{2}(n_{Af} - n_A) = \tfrac{3}{2} n_{Af} x_A$$

where $\overline{R}_B$ is the rate of consumption of component B.

From Stoichiometric Equations

The number of moles of C produced per unit time ($R_C = \frac{3}{2}$ number of moles of A reacted per unit time) is

$$R_C = \tfrac{3}{2}(n_{Af} - n_A) = \tfrac{3}{2} n_{Af} x_A$$

The number of moles of D produced per unit time ($R_D = \frac{5}{2}$ number of moles of A reacted per unit time) is

$$R_D = \tfrac{5}{2}(n_{Af} - n_A) = \tfrac{5}{2} n_{Af} x_A$$

Now, the simplest rational thinking will tell us that

$$n_A = n_{Af} - R_A$$
$$n_B = n_{Bf} - \overline{R}_B$$
$$n_C = n_{Cf} + R_C$$
$$n_D = n_{Df} + R_D$$

Thus,

$$n_A = n_{Af} - n_{Af} x_A = n_{Af}(1 - x_A)$$
$$n_B = n_{Bf} - \tfrac{3}{2} n_{Af} x_A$$
$$n_C = n_{Cf} + \tfrac{3}{2} n_{Af} x_A$$
$$n_D = n_{Df} + \tfrac{5}{2} n_{Af} x_A$$

Thus, all the exit variables from the reactor are given in terms of the known input variables n_{Af}, n_{Bf}, n_{Cf}, n_{Df}, and one conversion (i.e., x_A).

Numerical Values for Example 2.1

$$n_{Af} = 10 \text{ mol/min}$$
$$n_{Bf} = 30 \text{ mol/min}$$
$$n_{Cf} = 2 \text{ mol/min}$$
$$n_{Df} = 1.5 \text{ mol/min}$$

Then, for any given conversion, we can obtain n_A, n_B, n_C, and n_D. Say, for example, the conversion is 80%; then,

$$x_A = 0.8$$

From the above equations,

$$n_A = 10(1 - 0.8) = 10(0.2) = 2 \text{ mol/min}$$

$$n_B = 30 - \tfrac{3}{2}(10)(0.8) = 30 - 12 = 18 \text{ mol/min}$$

$$n_C = 2 + \tfrac{3}{2}(10)(0.8) = 2 + 12 = 14 \text{ mol/min}$$

$$n_D = 1.5 + \tfrac{5}{2}(10)(0.8) = 1.5 + 20 = 21.5 \text{ mol/min}$$

Exactly the same can be done in terms of x_B or in terms of Y.

The reader should practice obtaining similar relations once in terms of x_B and once in terms of Y.

Rate of Reaction

The rate of reaction in the context of material and energy balance is defined as the number of moles produced of a component per unit time. Of course, produced or consumed in the definition is a matter of convention. We will choose the convention that R is a rate of production. However, this is arbitrary, and the reader should practice the derivation with $\overline{R}$, which is the rate of consumption.

According to the above sign convention, the rate of reaction (production) of component A is obviously negative, because actually A is a reactant (it is being consumed):

$$R_A = (n_A - n_{Af})$$

If we write the rate of consumption of A and call it $\overline{R}_A$, then it is positive and defined as

$$\overline{R}_A = (n_{Af} - n_A)$$

This simple fact should be very clear in the mind of the reader. It is really very simple; it is like saying that John lost $100, which is exactly the same as saying that John gained $-$$100.

However, of course, we have to follow a certain convention. Do we work with the rates of production (and in this case, rates of consumption are only negative rates of production) or do we work with the rates of consumption (and in this case, rates of production are only negative rates of consumption)? The choice is arbitrary; however, the chemical engineer should be very clear about the above meanings because in the complex process of design in the chemical industry, different people (and different books and manuals) may choose the convention differently, which may cause confusion if the chemical engineer is not completely aware of the above simple and fundamental facts.

We will choose to work with the rates of production. Thus, for the reaction

$$2A + 3B \rightarrow 3C + 5D$$

we will have the following:

Rate of production of $A = R_A = n_A - n_{Af}$ (negative)

Rate of production of $B = R_B = n_B - n_{Bf}$ (negative)

Rate of production of $C = R_C = n_C - n_{Cf}$ (positive)

Rate of production of $D = R_D = n_d - n_{Df}$ (positive)

Therefore, we have four rates of reaction: R_A, R_B, R_C and R_D. Does this mean that the single reaction is defined by four different rates of reactions? Of course not! The four rates are related by the stoichiometric numbers. Intuitively, we can easily note that

$$R_B = \tfrac{3}{2}R_A$$

$$R_C = -\tfrac{3}{2}R_A$$

$$R_D = -\tfrac{5}{2}R_A$$

Thus, one rate of reaction (e.g., R_A; it can of course be R_B or R_C or R_D, but only one rate of reaction) defines automatically, together with the stoichiometric numbers, the other rates of reactions as long as it is a single reaction system.

We can obviously define the rates of reaction in terms of R_B or R_C or R_D (but only one rate) as follows:

1. The above relations are in terms of component A.

2. In terms of R_B,

$$R_A = \tfrac{2}{3}R_B$$

$$R_C = -R_B$$

$$R_D = -\tfrac{5}{3}R_B$$

3. In terms of R_C,

$$R_A = -\tfrac{2}{3}R_C$$

$$R_B = -R_C$$

$$R_D = \tfrac{5}{3}R_C$$

4. In terms of R_D,

$$R_A = -\tfrac{2}{5}R_D$$

$$R_B = -\tfrac{3}{5}R_D$$

$$R_C = \tfrac{3}{5}R_D$$

Thus, the rates of reaction of a single reaction can be all expressed in terms of one rate of reaction as long as the stoichiometric numbers for this single reaction are all known. We will define the generalized rate of reaction (r) in a later section.

2.2.3 Relations Among Rate of Reaction, Conversion, and Yield

We have made it very clear earlier that the single reaction system is fully defined (and solvable) in terms of any conversion for one of the reactants or any yield of one of the products. We have also made it very clear that all rates of reaction of the components of a single reaction are fully defined in terms of one rate of reaction.

Does this mean that we need one conversion (or yield) plus one rate of reaction to define (and solve) the system? Of course not! This will contradict the strict statement that the system is completely defined (and solvable) in terms of one variable. In fact, we need only either one conversion (or yield) or one rate of reaction. This is because rates of reaction are directly related to conversions (or yields), as will be shown in the next few lines.

It also has to be very clear to the reader that if he/she has a degree of freedom in a problem that is not zero, he/she cannot use a conversion (or yield) and rate of reaction as two given relations. If conversion (or yield) is used, any rate of reaction information is redundant from the degree-of-freedom point of view and vice versa. This simple fact will be made even clearer after we show the relations between conversion (or yield) and rate of reaction.

Rate of Reaction, Conversions, and Yields

Consider the reaction

$$2A + 3B \rightarrow 3C + 5D$$

taking place in the reactor in Figure 2.8. It is straightforward to define R_A as follows:

Rate of production of component A,

$$R_A = n_A - n_{Af}$$

which is negative because A is consumed not produced.

The fractional conversion x_A is defined as

$$x_A = \frac{n_{Af} - n_A}{n_{Af}} = \frac{\text{Number of moles of } A \text{ reacted}}{\text{Number of moles of } A \text{ fed}}$$

Therefore, from the above two relations, it is very clear that

$$x_A = -\frac{R_A}{n_{Af}}$$

or, equivalently,

$$R_A = -n_{Af} x_A$$

Thus, there is a direct relation between R_A and x_A.

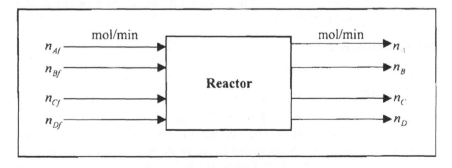

Figure 2.8 Molar flow diagram.

Similar relations can be found between R_A and Y (yield) as follows. Because

$$R_A = n_A - n_{Af}$$

Y_C (yield of component C) can be defined as

$$Y_c = \frac{n_C - n_{Cf}}{n_{Af}} = \frac{\text{Number of moles of } C \text{ produced}}{\text{Number of moles of } A \text{ fed}}$$

However, we know from the stoichiometry of the reaction that the number of moles (per unit time for continuous process) produced of C is $\frac{3}{2}$ times the number of moles of A reacted; that is,

$$n_C - n_{Cf} = \tfrac{3}{2}(n_{Af} - n_A)$$

Therefore,

$$Y_c = \frac{\tfrac{3}{2}(-R_A)}{n_{Af}}$$

which gives

$$R_A = -\tfrac{2}{3} Y_C n_{Af}$$

Similar relations can be developed between any rate of reaction (R_A, R_B, R_C, or R_D) and any reactant conversion (x_A or x_B) or any yield (Y_C or Y_D). The reader should practice deriving these relations.

The following should be clear:

1. For a single reaction, the definition of one conversion (or yield) makes any other definition of any other conversion (or yield) or any other rate of reaction redundant.
2. For a single reaction, the definition of one rate of reaction makes any other definition of any other rate of reaction or conversion (or yield) redundant.

Solved Example 2.2

Again consider the reaction

$$2A + 3B \rightarrow 3C + 5D$$

taking place in the reactor in Figure 2.9 with the shown feed molar flow rates of various components. The number of unknowns for this reactor is four (n_A, n_B, n_C, n_D), but because there is a single reaction, there is an extra unknown (a rate of reaction or a conversion for the above equation). This will add up to five unknowns.

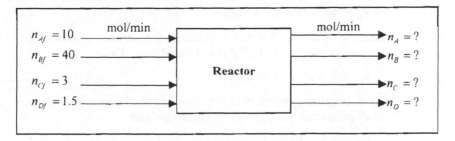

Figure 2.9 Molar flow diagram for solved example 2.2.

Relations

We have four independent mass balance relations. Thus, the problem as shown is not solvable. Why? The reason is that if we define the degrees of freedom as the difference between number of unknowns and the number of relations, we have

$$\text{Degrees of freedom} = 5 \text{ unknowns} - 4 \text{ mass balance relations} = 1$$

(The reader can obviously define the degree of freedom in the opposite sense; i.e. as the difference between number of relations and the number of unknowns. This gives the degrees of freedom to be $4 - 5 = -1$; this should not cause any problem to the reader). Now, if this problem is to be solved, something else should be defined, let us consider the following cases:

1. **Conversion of A defined as 80%**

    In this case, the problem is solvable because the degree of freedom now is zero. For

    $$x_A = \frac{n_{AF} - n_A}{n_{Af}}$$

    we get

    $$0.8 = \frac{10 - n_A}{10}$$

    giving

    $$n_A = 2 \text{ min/mol}$$

    Now, because 8 mol of A was consumed, the number of moles of B consumed is equal to $(3/2)(8) = 12$ mol/min. Thus,

$$n_B = 40 - 12 = 28 \text{ mol/min}$$

Because, 8 moles of A was consumed, the number of moles of C produced is equal to $(3/2)(8) = 12$ mol/min. Thus,

$$n_C = 3 + 12 = 15 \text{ mol/min}$$

Also, because 8 moles of A was consumed, the number of moles of D produced is $(3/2)(8) = 20$ mol/min and

$$n_D = 1.5 + 20 = 21.5 \text{ mol/min}$$

Making the Above Calculations More Systematic

As we agreed earlier, we define R as the rate of production; therefore,

$$n_A = n_{Af} + R_A$$
$$n_B = n_{Bf} + R_B$$
$$n_C = n_{Cf} + R_C$$
$$n_D = n_{Df} + R_D$$

However, we also agreed earlier that all the rates of reaction can be expressed in terms of one rate of reaction and the stoichiometric numbers; thus,

$$R_B = \tfrac{3}{2}R_A$$

$$R_C = -\tfrac{3}{2}R_A$$

$$R_D = -\tfrac{5}{2}R_A$$

Therefore, the four mass balance equations can be written in terms of R_A:

$$n_A = n_{Af} + R_A$$

$$n_B = n_{Bf} + \tfrac{3}{2}R_A$$

$$n_C = n_{Cf} - \tfrac{3}{2}R_A$$

$$n_D = n_{Df} - \tfrac{5}{2}R_A$$

However, we also agreed that the rates of reaction can be expressed in terms of conversion (or yield) and vice versa. Thus, we have

$$R_A = -n_{Af}x_A$$

This will give $R_A = -8$ mol/min, which can be substituted in the previous equation (or we write the equations in terms of x_A).

Therefore, the mass balance equations can be written as

$$n_A = n_{Af} - n_{Af}x_A$$

$$n_B = n_{Bf} - \tfrac{3}{2}n_{Af}x_A$$

$$n_C = n_{Cf} + \tfrac{3}{2}n_{Af}X_A$$

$$n_D = n_{Df} + \tfrac{5}{2}n_{Af}x_A$$

Substituting the numbers of the specific example in the above relations give the same results as we obtained earlier:

$$n_A = 10 - (10)(0.8) = 10 - 8 = 2 \text{ min/mol}$$

$$n_B = 40 - \tfrac{3}{2}(10)(0.8) = 40 - 12 = 28 \text{ min/mol}$$

$$n_C = 3 + \tfrac{3}{2}(10)(0.8) = 3 + 12 = 15 \text{ min/mol}$$

$$n_D = 1.5 + \tfrac{5}{2}(10)(0.8) = 1.5 + 20 = 21.5 \text{ min/mol}$$

2. **Rate of reaction of A defined as $R_A = -8$ mol/min**

In this case, we use the relations

$$n_A = n_{Af} + R_A$$

$$n_B = n_{Bf} + \tfrac{3}{2}R_A$$

$$n_C = n_{Cf} - \tfrac{3}{2}R_A$$

$$n_D = n_{Df} - \tfrac{5}{2}R_A$$

Therefore, we again obtain the same results we obtained earlier:

$$n_A = 10 - 8 = 2 \text{ min/mol}$$

$$n_B = 40 - \tfrac{3}{2}(8) = 28 \text{ min/mol}$$

$$n_C = 3 - \tfrac{3}{2}(-8) = 15 \text{ min/mol}$$

$$n_D = 1.5 - \tfrac{5}{2}(-8) = 21.5 \text{ min/mol}$$

The reader is advised to solve for cases where other conversion (x_B) is defined or one of the yields (Y_C or Y_D) or one other rate of reaction (R_B, R_C, or R_D) is defined.

An Important Question

Can other relations (other than one conversion or one yield or one rate of reaction) be used to solve the problem?

To answer this question, let us consider the same problem (Fig. 2.10). The reaction is again the same,

$$2A + 3B \rightarrow 3C + 5D$$

and we have the same reactor with the same feeds and feed flow rates. Instead of giving one conversion (or one yield) or one rate of reaction we give one relation,

$$n_B = 14n_A$$

Can we solve the problem? Yes, of course, because the degrees of freedom is zero.

We have the following mass balance (with a single chemical reaction) relations:

$$n_A = n_{Af} + R_A \tag{2.10}$$

$$n_B = n_{Bf} + \tfrac{3}{2}R_A \tag{2.11}$$

$$n_C = n_{Cf} - \tfrac{3}{2}R_A \tag{2.12}$$

$$n_D = n_{Df} - \tfrac{5}{2}R_A \tag{2.13}$$

we also have

$$n_B = 14n_A \tag{2.14}$$

On substituting the feed conditions, Eq. (2.10) becomes

$$n_A = 10 + R_A \tag{2.15}$$

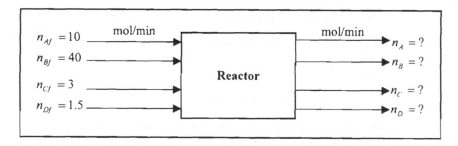

Figure 2.10   Molar flow diagram.

On substituting the feed conditions and using Eq. (2.14), Eq. (2.11) gives

$$14n_A = 40 + \tfrac{3}{2}R_A \tag{2.16}$$

Substituting n_A from Eq. (2.15) into Eq. (2.16) gives

$$14(10 + R_A) = 40 + \tfrac{3}{2}R_A$$

which becomes

$$100 = -12.5R_A$$

Therefore, the rate of reaction of component A is obtained as

$$R_A = -8 \text{ min/mol}$$

Then, substituting R_A in Eqs. (2.10)–(2.13) gives

$$n_A = 10 - 8 = 2 \text{ min/mol}$$

$$n_B = 40 - \tfrac{3}{2}(8) = 28 \text{ min/mol}$$

$$n_C = 3 + \tfrac{3}{2}(8) = 15 \text{ min/mol}$$

$$n_D = 1.5 + \tfrac{5}{2}(8) = 21.5 \text{ min/mol}$$

Thus, from the given relation ($n_B = 14n_A$), we are able to calculate R_A and then the solution proceeds as before. However, the reader cannot consider this relation together with a given rate of reaction or conversion as two independent relations. To make this point even clearer, let us consider the same problem but with five flow rates unknown (meaning it is a six unknowns problem because of the single reaction) (Fig. 2.11).

The reaction is the same as earlier:

$$2A + 3B \rightarrow 3C + 5D$$

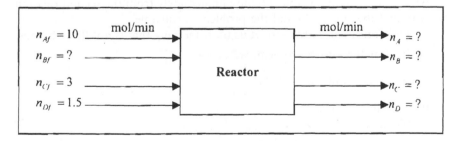

Figure 2.11 Molar flow diagram.

Total number of unknowns = 6 unknowns (5 molar flow rates
+1 rate of reaction)

Mass balance reactions = 4

Degrees of freedom = 2

Thus, this problem is not solvable.
    If we define any one relation,

$$x_A = 0.8 \quad \text{or} \quad R_A = -8 \quad \text{or} \quad n_B = 14n_A$$

we will have degree of freedom = $6 - 5 = 1$. Still the problem is unsolvable!
    What if we define any two of the above relations, say,

$$R_A = -8 \quad \text{and} \quad n_B = 14n_A$$

Now, the absent-minded chemical engineer or the one who is not strictly clear
about the degrees-of-freedom concept (the one who does not know at all
about the degree-of-freedom-concept will be in even worse condition) will
count as follows: I have six unknowns, I have four mass balance relations,
and two extra relations ($R_A = -8$ and $n_B = 14n_A$), then the degree of freedom
is equal to zero and the problem is solvable! This is completely wrong!!! $R_A =
-8$ and $n_B = 14n_A$ are not two independent relations and the problem is still
not solvable. The reader should try to solve it to see why it is unsolvable.

Another Important Question

Can two other independent relations be defined that will make the above
problem solvable? The answer is a conditional yes, depending on the true
independence of these relations.

Case 1:   For the above problem, we will define

$$R_A = -8 \quad \text{and} \quad \frac{n_{Bf}}{n_{Af}} = 4$$

Now, the problem is solvable, because immediately from the second relation
we can find that $n_{Bf} = 40$ and the problem is solvable.
    This appears too obvious, let us try something that is not that obvious.

Case 2:   For the above problem, we define

$$R_A = -8 \quad \text{and} \quad n_C = 7.5n_A$$

Is this solvable? Let us try.
    We have

$$n_A = n_{Af} + R_A \tag{2.10}$$

$$n_B = n_{Bf} + \tfrac{3}{2}R_A \tag{2.11}$$

$$n_C = n_{Cf} - \tfrac{3}{2}R_A \tag{2.12}$$

$$n_D = n_{Df} - \tfrac{5}{2}R_A \tag{2.13}$$

Using $R_A = -8$ and $n_C = 7.5n_A$, we get

$$n_A = 10 - 8 = 2 \text{ min/mol}$$

$$n_B = n_{Bf} - \tfrac{3}{2}(8) = (n_{Bf} - 12) \text{ min/mol}$$

$$n_C = 3 + \tfrac{3}{2}(8) = 15 \text{ min/mol}$$

$$n_D = 1.5 + \tfrac{5}{2}(8) = 21.5 \text{ min/mol}$$

The problem is not solvable, as we still have n_B and n_{Bf} as unknowns.
We notice from the above that automatically by applying

$$R_A = -8$$

we get

$$\frac{n_C}{n_A} = 7.5$$

The relation we gave as $n_C = 7.5n_A$ is actually redundant and cannot be used to obtain the unknowns n_B and n_{Bf}.

Conclusion

Because for given feed conditions, the output conditions (for the case of a single reaction) are completely defined in terms of one and only one variable (conversion of any reactant, or yield of any product, or rate of reaction of any component), then only one relation related to these variables or one relation relating the output variables together can be used in the solution of the problem (and in the determination of the degrees of freedom).

2.3 GENERALIZED MATERIAL BALANCE

In the previous section, we have seen that, for a single reaction, the rates of reaction for different components can all be expressed in terms of the rate of reaction of one component (together with the stoichiometric numbers), or the conversion of one of the reactants, or the yield of one of the products (of course, together with the stoichiometric numbers). These information and relations for the single reaction are adequate for the solution of any mass balance problem with a single reaction and can be easily extended to multiple reaction systems, as will be shown later. However, in this section, we will try to make the calculations even more systematic. This will require, first, that we introduce the sign convention for the stoichiometric numbers, as we introduced earlier the sign convention for the rates of reaction (positive for production and negative for consumption).

2.3.1 Sign Convention for the Stoichiometric Numbers

The sign convention is arbitrary, and the reader should be very clear about this and should not memorize certain relations under a certain sign convention. The equations you use should be always correct under the chosen sign convention. However, it is best to adhere to the sign convention used by most chemical engineers and textbooks.

Most books use the following sign convention:

R_i = rate of production of component i
Rate of production = positive
Rate of consumption = negative
Stoichiometric number of products = positive
Stoichiometric number of reactants = negative
Stoichiometric numbers for inerts = zero

We will use different sign conventions and show that as long as it is consistent it is correct.

I. First Sign Convention (Not Commonly Used)

R_i = rate of production of component i = negative
Stoichiometric number of reactants (σ_R) = positive
Stoichiometric number of products (σ_P) = negative

We use the reaction given earlier,

$$2A + 3B \rightarrow 3C + 5D$$

Using this uncommon sign convention, we get

R_A = rate of production of component A = negative
$\sigma_A = 2$
$\sigma_B = 3$
$\sigma_C = -3$
$\sigma_D = -5$

The balance equations developed earlier are

$$n_A = n_{Af} + R_A \qquad (2.10)$$

$$n_B = n_{Bf} + \tfrac{3}{2}R_A \tag{2.11}$$

$$n_C = n_{Cf} - \tfrac{3}{2}R_A \tag{2.12}$$

$$n_D = n_{Df} - \tfrac{5}{2}R_A \tag{2.13}$$

and the relations between R's are

$$R_B = \tfrac{3}{2}R_A \tag{2.17}$$

$$R_C = -\tfrac{3}{2}R_A \tag{2.18}$$

$$R_D - \tfrac{5}{2}R_A \tag{2.19}$$

Of course, we could have used R_B, R_C, or R_D as the basis rather than R_A. To avoid this multiplicity of choices, we can define a rate of reaction that does not depend on the component. It is only related to the reaction; let us call this rate "the generalized rate of reaction, r":

$$r = \frac{R_i}{\sigma_i}$$

We can easily show that r does not depend on the choice of i. If $i = A$, then

$$r_A = \frac{R_A}{\sigma_a} = \frac{R_A}{2} \tag{2.20}$$

If $i = B$, then

$$r_B = \frac{R_B}{\sigma_B} = \frac{R_B}{3} = \frac{1}{3}\left(\frac{3}{2}R_A\right) = \frac{R_A}{2} \tag{2.21}$$

If $i = C$, then

$$r_C = \frac{R_C}{\sigma_C} = \frac{R_C}{-3} = \frac{1}{-3}\left(\frac{-3}{2}R_A\right) = \frac{R_A}{2} \tag{2.22}$$

If $i = D$, then

$$r_D = \frac{R_D}{\sigma_D} = \frac{R_D}{-5} = \frac{1}{-5}\left(\frac{-5}{2}R_A\right) = \frac{R_A}{2} \tag{2.23}$$

Therefore, it is clear that

$$r_A = r_B = r_C = r_D = r = \text{generalized rate of reaction.}$$

Thus, for the given uncommon sign convention (which is usually not used),

$$\boxed{r = \frac{R_i}{\sigma_i}} \quad \Rightarrow \quad \text{always negative} \tag{2.24}$$

Therefore, the mass balance equations can be written as

$$n_A = n_{Af} + \frac{\sigma_A}{\sigma_i} R_i = n_{Af} + \sigma_A r \tag{2.25}$$

$$n_B = n_{Bf} + \frac{\sigma_B}{\sigma_i} R_i = n_{Bf} + \sigma_B r \tag{2.26}$$

$$n_C = n_{Cf} + \frac{\sigma_C}{\sigma_i} R_i = n_{Cf} + \sigma_C r \tag{2.27}$$

$$n_D = n_{Df} + \frac{\sigma_D}{\sigma_i} R_i = n_{Df} + \sigma_D r \tag{2.28}$$

where R_i is the rate of production of the component i used as a basis for r and σ_i is its stoichiometric number according to the sign convention chosen. Equations (2.25)–(2.28) can be rewritten in general as

$$
\boxed{
\begin{aligned}
&n_i = n_{if} + \sigma_i r \\
&\sigma_i > 0 \text{ for reactants} \\
&\sigma_i < 0 \text{ for products} \\
&\text{and} \\
&r = \frac{R_i}{\sigma_i} \quad \Rightarrow \quad \text{always negative}
\end{aligned}
}
\tag{2.29}
$$

II. Second Sign Convention (Commonly Used)

R_i = rate of production of component i = positive
Stoichiometric number of reactants (σ_R) = negative
Stoichiometric number of products (σ_P) = positive

Relation (2.29) remains the same, but r is now always positive (even when R_i is negative),

$$
\boxed{
\begin{aligned}
&n_i = n_{if} + \sigma_i r \\
&\sigma_i < 0 \text{ for reactants} \\
&\sigma_i > 0 \text{ for products} \\
&\text{and} \\
&r = \frac{R_i}{\sigma_i} \Rightarrow \text{always positive}
\end{aligned}
}
\tag{2.30}
$$

Relations Among R_i, r, x_i, and Limiting Components

$$R_A = \text{rate of production} = n_A - n_{Af}$$

and

$$x_A = \frac{n_{Af} - n_A}{n_{Af}} \tag{2.31}$$

Thus,

$$x_A = -\frac{R_A}{n_{Af}} \tag{2.32}$$

or

$$R_A = -n_{Af}x_A = -\frac{\sigma_A}{\sigma_A}n_{Af}x_A$$

$$R_B = \frac{\sigma_B}{\sigma_A}R_A = -\frac{\sigma_B}{\sigma_A}n_{Af}x_A$$

$$R_C = \frac{\sigma_C}{\sigma_A}R_A = -\frac{\sigma_C}{\sigma_A}n_{Af}x_A$$

$$R_D = \frac{\sigma_D}{\sigma_A}R_A = -\frac{\sigma_D}{\sigma_A}n_{Af}x_A$$

In general, we can write

$$\boxed{R_i = -\frac{\sigma_i}{\sigma_A}n_{Af}x_A} \tag{2.33}$$

Therefore, the relation between the rate of reaction of any component and the conversion for a specific equation does not depend on the stoichiometric number sign convention. Of course, we can choose any other component other than A as a base for conversion (if it is another reactant, then it will be conversion, but if it is a product, then it will be called yield, not conversion). Thus, the general form of Eq. (2.33) is

$$\boxed{R_i = -\frac{\sigma_i}{\sigma_k}n_{kf}x_k} \tag{2.34}$$

where n_{kf} is the feed number of moles on which basis x (conversion) is defined.

Now, the generalized rate of reaction is

$$r = \frac{R_i}{\sigma_i} = -\frac{n_{kf}x_k}{\sigma_k} \tag{2.35}$$

When we substitute into the mass balance relation, we have

$$n_i = n_{if} + \sigma_i r \tag{2.36}$$

which is actually

$$n_i = n_{if} - \frac{\sigma_i}{\sigma_A} n_{Af} x_A \tag{2.37}$$

2.3.2 The Limiting Component

This is a very important concept regarding the rational definition of the conversion for a certain feed component; thus, this definition should not violate the natural laws of the reaction itself. When the stoichiometric numbers for all reactants are equal, the problem is rather trivial; it is even more trivial when the stoichiometric numbers of the reactants are equal and the feed is equimolar. However, when the stoichiometric numbers of the reactants are not equal and/or the feed molar flows of the different components are not equal, the problem is not trivial, although it is very simple.

Consider, first, the following reaction with equal stoichiometric numbers for the different components:

$$A + B \rightarrow C + D$$

taking place in the reactor in Figure 2.12 with the shown feed molar flow rates for the four components.

If we specify the conversion of A as $x_A = 90\%$, then we find that

$$n_A = n_{Af} - n_{Af} x_A = 10 - 9 = 1 \text{ min/mol}$$

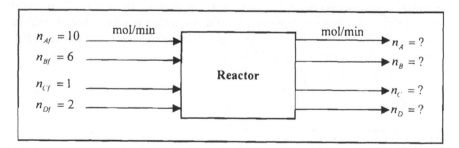

Figure 2.12 Molar flow diagram.

Now,

$$n_B = n_{Bf} - n_{Bf}x_A = 6 - 9 = -3 \text{ min/mol!!!!}$$

Obviously, this is impossible and it means that the molar flow rate of component B does not allow 90% conversion of component A; however, if we specify the conversion of B as $x_B = 90\%$. Then, we have

$$x_B = \frac{n_{Bf} - n_B}{n_{Bf}} \quad \text{and} \quad n_B = n_{Bf} - n_{Bf}x_B = 6 - 6(0.9) = 0.6 \text{ min/mol}$$

Therefore,

$$n_A = n_{Af} - n_{Bf}x_B = 10 - 6(0.9) = 4.6 \text{ min/mol}$$

Thus, B is the limiting component, which means that it is the component which gets exhausted or reacted completely before the other reactant is exhausted. In other words, its conversion can be specified as high as 100% without violating the natural laws of the reaction.

Therefore, for reactions in which all of the reactants have the same stoichiometric numbers, the limiting component is the reactant with the lowest number of the moles in feed. For this case of equal stoichiometric numbers for the reactants, if we additionally have equimolar feed, than any reactant can be the limiting reactant.

2.3.3 Reactions with Different Stoichiometric Numbers for the Reactants

Consider the following chemical reaction with different stoichiometric numbers for different components:

$$2A + 10B \rightarrow 6C + 3D \tag{2.38}$$

Now, if we take for this reaction that the limiting component is the component with the lowest number of moles in the feed, we will fall into a serious mistake. How?

Considering the case shown in Figure 2.13, if we wrongly use the previous statement (the limiting component is the reactant with the lowest number of moles in the feed), we will strongly say that A is the limiting component. Let us check, say, the conversion of component A, $x_A = 90\%$, then

$$n_A = n_{Af} - n_{Af}x_A = 10 - 9 = 1 \text{ min/mol}$$

Now,

$$n_B = n_{Bf} - \frac{10}{2}n_{Bf}x_A = 20 - (5)(10)(0.9) = -25 \text{ min/mol!!!!}$$

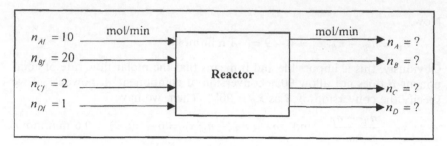

Figure 2.13   Molar flow diagram.

Obviously this is wrong. A is not the limiting component (or reactant); in fact, B is the correct limiting component (despite the fact that $n_{Bf} > n_{Af}$; this is because $|\sigma_B| >> |\sigma_A|$). The reader could find that out directly from looking at the ratio of the stoichiometric numbers. However, for complex systems, it is not that obvious.

A Deceptive Wrong Procedure for the Determination of the Limiting Component

Let us refer to the generalized form of the mass balance equations for reactants:

$$n_i = n_{if} + \sigma_i r$$

which can be written as

$$n_i = n_{if} + \frac{\sigma_i}{\sigma_j} R_j$$

where R_j = rate of production (negative for reactants).
Let us make

$$n_i = n_{if} + \frac{\sigma_i}{\sigma_j} R_{j_{max}}$$

the limit of this. The reactant i with the largest absolute value of $(\sigma_i/\sigma_j)R_{j_{max}}$ is the limiting component.

However, now, what is $R_{j_{max}}$? It is R_j at 100% conversion (obvious, isn't it?).

Therefore may be it is better to write the above relation in terms of conversion x_j as follows. Because

$$R_j = n_j - n_{jf}$$

and

$$x_j = \frac{n_{jf} - n_j}{n_{jf}}$$

therefore

$$x_j = -\frac{R_j}{n_{jf}}$$

From the above relation, we get

$$R_j = -x_j n_{jf}$$

and, therefore,

$$R_{j_{max}} = -(1.0)n_{jf} \quad \text{(as maximum conversion is 100\% or } x_j = 1.0)$$

Therefore, the mass balance relation will be reduced to

$$n_i = n_{if} - \frac{\sigma_i}{\sigma_j} n_{jf} \quad \text{(at maximum conversion } x_j = 1.0)$$

Thus, the quantity

$$\boxed{\tilde{x}_i = \frac{\sigma_i}{\sigma_j} n_{jf}}$$

is the measure for the controlling component i. The component i with the largest value of $\tilde{x}_i$ is the rate-limiting component. Here, reactant i with the largest value of $\tilde{x}_i$ is the limiting component. (Be careful, this is a wrong conclusion!! Although, it will be correct in the next deceiving example).

Application to the Previous Problem

Now, for the case shown in Figure 2.13 and for reaction (2.38), we want to find the reactant i (A or B) which represents the limiting component. The base component j is arbitrary, we can actually choose whatever we like. Choose component j to be A. Then,

$$\tilde{x}_i = \frac{\sigma_A}{\sigma_A} n_{Af}$$

which gives

$$\tilde{x}_A = \left(\frac{\sigma_A}{\sigma_A}\right)10 = 10$$

and

$$\tilde{x}_B = \left(\frac{\sigma_B}{\sigma_A}\right)10 = \left(\frac{10}{2}\right)10 = 50$$

Thus, the limiting components is B (as $\tilde{x}_B$ is higher than $\tilde{x}_A$). Therefore, B, again, is the limiting component regardless of the basis j taken (A or B). Is the procedure correct, or can it cause errors? The reader should think about it and check this question before proceeding to the next part, where a counterexample is given.

Counterexample

The previous procedure for the determination of the limiting reactant, although it "sounds" logical and when applied to the previous example gave correct results, it is actually wrong!! It is important for the reader to realize that he/she cannot develop a procedure and check for one case and from the result of this single case conclude that it is correct. Things should be thought about much deeper.

We will now give a counterexample (Fig. 2.14) showing that the previous procedure is not correct. We will only change the feed conditions; the reaction remains the same as in the Eq. (2.38) case.

Let us apply the previous procedure:

1. Choose component A as j.

$$\tilde{x}_i = \frac{\sigma_A}{\sigma_A} \, n_{Af}$$

which gives

$$\tilde{x}_A = \left(\frac{\sigma_A}{\sigma_A}\right) 10 = 10$$

and

$$\tilde{x}_B = \left(\frac{\sigma_B}{\sigma_A}\right) 10 = \left(\frac{10}{2}\right) 10 = 50$$

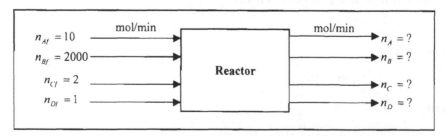

$$n_{Af} = 10 \quad \text{mol/min}$$
$$n_{Bf} = 2000$$
$$n_{Cf} = 2$$
$$n_{Df} = 1$$

Reactor

$$n_A = ? \quad \text{mol/min}$$
$$n_B = ?$$
$$n_C = ?$$
$$n_D = ?$$

Figure 2.14 Molar flow diagram.

Thus, B is the limiting component according to this (wrong) procedure.

2. Choose component B as j.

$$\tilde{x}_i = \frac{\sigma_B}{\sigma_B} n_{Bf}$$

which gives

$$\tilde{x}_B = \left(\frac{\sigma_B}{\sigma_B}\right) n_{Bf} = \left(\frac{10}{10}\right) 2000 = 2000$$

and

$$\tilde{x}_A = \left(\frac{\sigma_A}{\sigma_B}\right) n_{Bf} = \left(\frac{2}{10}\right) 2000 = 400$$

Thus, B is the limiting component according to this procedure (wrong again!!).

In fact, B is certainly not the limiting component. For if B is the limiting component, then we should be able to define $x_B = 100\%$ without violating the natural laws of the reaction.

Let us do that and see. Just take $x_B = 90\% = 0.9$. Now, calculate

$$n_A = n_{Af} - \frac{\sigma_A}{\sigma_B} x_B n_{Bf}$$

Thus,

$$n_A = 10 - \left(\frac{2}{10}(0.9)(2000)\right) = -350!!!$$

Obviously wrong!! So although the procedure gave the correct conclusion when $n_{Bf} = 20$, it gave the wrong conclusion when $n_{Bf} = 2000$; therefore, it is not a correct procedure.

The Correct Procedure for Determining the Limiting Component

From the mass balance,

$$n_i = n_{if} + \sigma_i r$$

The limiting situation for any component i is when it is exhausted (i.e., 100% conversion and therefore $n_i = 0$). Thus, for this situation, we get

$$0 = n_{if} + \sigma_i r$$

which gives

$$r_{Li} = -\frac{n_{if}}{\sigma_i}$$

If we adhere to the more common sign convention for stoichiometric numbers, which is $\sigma_i < 0$ for reactants, then r_{Li} will always be positive. It is obvious now that the reactant which will deplete first will be the reactant i with the smallest r_{Li}.

Solved Example 2.3

Let us apply this simple (and hopefully correct) criterion to the previous cases:

$$2A + 10B \rightarrow 6C + 3D$$

Case 1 (Fig. 2.15)

$$r_{LA} = \frac{10}{2} = 5$$

$$r_{LB} = \frac{20}{10} = 2$$

The limiting component is B, which is a correct conclusion.

Case 2 (Fig. 2.16)

$$r_{LA} = \frac{10}{2} = 5$$

$$r_{LB} = \frac{2000}{10} = 200$$

The limiting component is A, which is a correct conclusion.

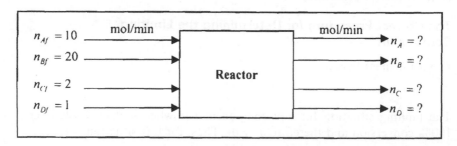

Figure 2.15   Molar flow diagram.

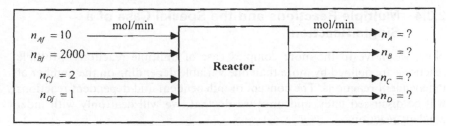

Figure 2.16 Molar flow diagram.

In the previous examples, we used cases which are simple and almost obvious in order to check and develop a procedure. Now, as we have developed a simple and nice procedure, let us use it to find the limiting reactant for a slightly complex case:

$$3A + 8B + 13C + 2D \rightarrow \text{Products}$$

The feeds are as shown in Figure 2.17.

What is the limiting reactant (component)?

$$r_{LA} = \frac{15}{3} = 5$$

$$r_{LB} = \frac{22}{8} = 2.75$$

$$r_{LC} = \frac{43}{13} = 3.308$$

$$r_{LD} = \frac{8}{2} = 4$$

Thus, the limiting reactant is B (correct), as it has the lowest r_{Li} value.

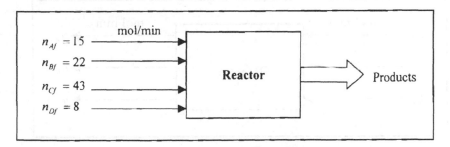

Figure 2.17 Molar flow diagram.

2.3.4   Multiple Reactions and the Special Case of a Single Reaction

Now, we move to the more complex case of multiple reactions. Multiple reactions are defined by more than one variable depending on the number of independent reactions. The concept of independent and dependent reactions will be discussed later, and until we do that, we will deal only with independent reactions.

Consider the two independent consecutive reactions:

$$A \rightarrow B \rightarrow C$$

It can also be written as two separate reactions:

$$A \rightarrow B \quad \text{and} \quad B \rightarrow C$$

If we define the conversion of A in Figure 2.18 as

$$x_A = \frac{n_{Af} - n_A}{n_{Af}}$$

then we can find n_A in terms of conversion x_A (if it is given) as follows:

$$n_A = n_{Af} - n_{Af} x_A$$

However, we cannot find n_B or n_C. Why?

The reason is that we cannot know how much of the converted A ($n_{Af} x_A$) has changed to B and how much to C. Therefore, we need to define one more variable. For example, we can define the yield of one of the other components (B or C):

$$\text{Yield of } B = Y_B = \frac{n_B - n_{Bf}}{n_{Af}}$$

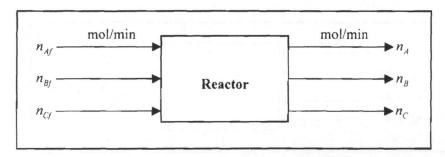

Figure 2.18   Molar flow diagram for a multiple-reaction system.

or

$$\text{Yield of } C = Y_C = \frac{n_C - n_{Cf}}{n_{Af}}$$

To completely specify the problem, we need to specify any two of the following three variables: x_A, Y_B, or Y_C. A specification of the third variable is redundant because it can be deducted from the other two variables.

Solved Example 2.4

In this example, we prove that if x_A and Y_B are defined, then Y_C is automatically defined because

$$n_A = n_{Af} - n_{Af} x_A$$

$$n_B = n_{Bf} + n_{Af} Y_B$$

and

$$n_C = n_{Cf} + n_{Af} x_A - n_{Af} Y_B$$

which means that if x_A and Y_B are defined, then we can compute n_C and then Y_C. Also, if x_A and Y_C are defined, then Y_B is automatically defined, and if Y_B and Y_C are defined, then x_A is also automatically defined.

Different Definitions of Yield

1. Yield of component i, Y_i:

$$Y_i = \frac{n_i - n_{if}}{n_{Af}}$$

   This is the amount of i produced as related to initial amount (or rate) of reactant fed.

2. Yield of component i, $\overline{Y}_i$:

$$\overline{Y}_i = \frac{n_i - n_{if}}{n_{Af} x_A}$$

   This is the amount (or rate) of i produced as related to the amount (or rate) of the reactant that has reacted.

Selectivity

Selectivity has many definitions; one of the most popular one is

$$S_i = \frac{\text{Number of moles of } i \text{ produced}}{\text{Number of moles of all products produced}}$$

In the above case, it is defined as

$$S_B = \frac{n_B}{n_B + n_C}$$

Multiple Reactions in Terms of Rates of Reaction

The simplest case is the above case, where the reactions are independent and also all of the stoichiometric numbers are equal to unity.

$$A \to B \qquad R_1$$
$$B \to C \qquad R_2$$

R_1 is defined as the rate of production of A by reaction 1 (negative, of course). Then the rate of production of B by reaction 1 is R_1; this is to say,

$$R_{A1} = R_1 \qquad \text{(negative)}$$
$$R_{B1} = -R_1 \qquad \text{(positive)}$$

The rate of production of B by reaction 2 is $R_{B2} = R_2$ (negative) and the rate of production of C by reaction 2 is $R_{C2} = -R_2$ (positive). Therefore, for the reactor shown in Figure 2.19, we can write

$$n_A = n_{Af} + R_{A1} = n_{Af} + R_1 \quad \text{where } R_1 = R_{A1} \text{ is negative} \qquad (2.39)$$

$$n_B = n_{Bf} + R_{B1} + R_{B2}$$

$$\qquad = n_{Bf} - R_1 + R_2 \qquad\qquad \text{where } R_1 \text{ and } R_2 \text{ are negative} \qquad (2.40)$$

$$n_C = n_{Cf} + R_{C2} = n_{Cf} - R_2 \quad \text{where } R_2 \text{ is negative} \qquad (2.41)$$

Note that for this reaction,

$$\sum_i n_{if} = \sum_i n_i, \text{ where i} \equiv A, B, \text{ and } C$$

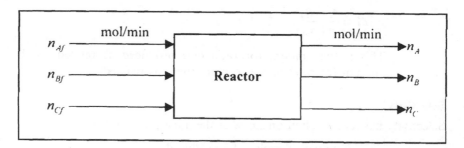

Figure 2.19 Molar flow diagram.

that is, the reaction is not accompanied by any change in the number of moles.

Relations Among Conversions, Yields, and Rates of Reaction

1. As

$$x_A = \frac{n_{Af} - n_A}{n_{Af}} \quad \text{and} \quad R_1 = n_A - n_{Af}$$

Thus,

$$x_A = \frac{-R_1}{n_{Af}}$$

finally giving

$$\boxed{R_1 = -n_{Af} x_A}$$

2. As $Y_B = \dfrac{n_B - n_{Bf}}{n_{Af}}$

From relation (2.40),

$$n_B - n_{Bf} = R_2 - R_1$$

Thus,

$$Y_B = \frac{R_2 - R_1}{n_{Af}} = \frac{R_2 + n_{Af} x_A}{n_{Af}}$$

Finally giving

$$\boxed{R_2 = n_{Af}(Y_B - x_A)}$$

3. As

$$Y_C = \frac{n_C - n_{Cf}}{n_{Af}}$$

We get

$$Y_C = -\frac{R_2}{n_{Af}}$$

finally giving

$$\boxed{R_2 = -n_{Af} Y_C}$$

Therefore, when R_1 and R_2 are defined, then x_A, Y_B, and Y_C can be deduced easily from them. When x_A and Y_B, or x_A and Y_C, or Y_B and Y_C are defined, then R_1 and R_2 can be easily deduced from them. Thus, the system is fully defined in terms of x_A and Y_B, or x_A and Y_C, or Y_B and Y_C, or R_1 and R_2.

For N independent reactions, the system is completely defined in terms of N conversions and yields or N rates of reaction.

A Case of Multiple Reactions with Slightly More Complex Stoichiometry

In this case, let us consider the following two reactions:

$$3A + 5B \rightarrow 6C + 8D$$
$$4C + 8K \rightarrow 4F + 7L$$

These are two independent reactions and we are given two rates of reaction:

R_{A1} (rate of production of A in reaction 1, negative)

R_{C2} (rate of production of C in reaction 2, negative)

Consider the reactor in Figure 2.20 with the shown feed molar flow rates of the seven components involved in the two reactions.

For the two values for R_{A1} and R_{C2},

$$R_{A1} = -6 \text{ mol/min}$$
$$R_{C2} = -2 \text{ mol/min}$$

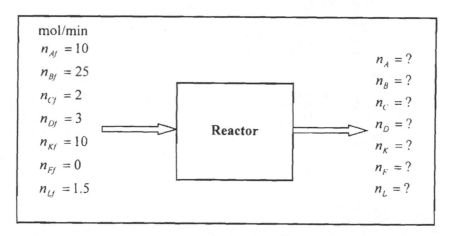

Figure 2.20 Molar flow diagram.

the mass balance relations are

$$n_A = n_{Af} + R_{A1} = n_{4f} + \tfrac{3}{3}R_{A1}$$

$$n_B = n_{Bf} + \tfrac{5}{3}R_{A1}$$

$$n_C = n_{Cf} + \tfrac{-6}{3}R_{A1} + \tfrac{4}{4}R_{C2}$$

$$n_D = n_{Df} + \tfrac{-8}{3}R_{A1}$$

$$n_K = n_{Kf} + \tfrac{8}{4}R_{C2}$$

$$n_F = n_{Ff} + \tfrac{-4}{4}R_{C2}$$

$$n_L = n_{Lf} + \tfrac{-7}{4}R_{C2}$$

Therefore, on using the given values of the rates of reactions ($R_{A1} = -6$ mol/min and $R_{C2} = -2$ mol/min), we get

$$n_A = 4 \text{ mol/min}$$
$$n_B = 15 \text{ mol/min}$$
$$n_C = 12 \text{ mol/min}$$
$$n_D = 19 \text{ mol/min}$$
$$n_K = 6 \text{ mol/min}$$
$$n_F = 2 \text{ mol/min}$$
$$n_L = 5 \text{ mol/min}$$

Note that this reaction is accompanied by a change in the number of moles:

$$\sum n_{if} = 10 + 25 + 2 + 3 + 10 + 0 + 1.5 = 51.5 \text{ min/mol}$$
$$\sum n_i = 4 + 15 + 12 + 19 + 6 + 2 + 5 = 63 \text{ min/mol}$$

Let us try to make the relations more systematic by using stoichiometric numbers (with its sign conventions, of course).
The following common sign convention is used:

$$\sigma_i = positive \text{ for products}$$
$$\sigma_i = negative \text{ for reactants}$$
$$\sigma_i = zero \text{ for inerts in the specific reaction}$$

σ_{ij} is the stoichiometric number of component i in reaction j, when i is a reactant $\sigma_{ij} < 0$ and when i is a product, then $\sigma_{ij} > 0$, and, finally, $\sigma_{ij} = 0$ when i is not involved in the reaction.

The mass balance relations become

$$n_A = n_{Af} + \frac{\sigma_{A1}}{\sigma_{A1}} R_{A1} + \frac{\sigma_{A2}}{\sigma_{C2}} R_{C2}$$

$$n_B = n_{Bf} + \frac{\sigma_{B1}}{\sigma_{A1}} R_{A1} + \frac{\sigma_{B2}}{\sigma_{C2}} R_{C2}$$

$$n_C = n_{Cf} + \frac{\sigma_{C1}}{\sigma_{A1}} R_{A1} + \frac{\sigma_{C2}}{\sigma_{C2}} R_{C2}$$

$$n_D = n_{Df} + \frac{\sigma_{D1}}{\sigma_{A1}} R_{A1} + \frac{\sigma_{D2}}{\sigma_{C2}} R_{C2}$$

$$n_K = n_{Kf} + \frac{\sigma_{K1}}{\sigma_{A1}} R_{A1} + \frac{\sigma_{K2}}{\sigma_{C2}} R_{C2}$$

$$n_F = n_{Ff} + \frac{\sigma_{F1}}{\sigma_{A1}} R_{A1} + \frac{\sigma_{F2}}{\sigma_{C2}} R_{C2}$$

$$n_L = n_{Lf} + \frac{\sigma_{L1}}{\sigma_{A1}} R_{A1} + \frac{\sigma_{L2}}{\sigma_{C2}} R_{C2}$$

Now, according to the sign convention we use for this problem, $\sigma_{A1} = -3$, $\sigma_{A2} = 0$, $\sigma_{B1} = -5$, $\sigma_{B2} = 0$, $\sigma_{C1} = 6$, $\sigma_{C2} = -4$, $\sigma_{D1} = 8$, $\sigma_{D2} = 0$, $\sigma_{K1} = 0$, $\sigma_{K2} = -8$, $\sigma_{F1} = 0$, $\sigma_{F2} = 4$, $\sigma_{L1} = 0$, and $\sigma_{L2} = 7$. Substituting the values in the above mass balance relations gives

$$n_A = 4 \text{ mol/min}$$
$$n_B = 15 \text{ mol/min}$$
$$n_C = 12 \text{ mol/min}$$
$$n_D = 19 \text{ mol/min}$$
$$n_K = 6 \text{ mol/min}$$
$$n_F = 2 \text{ mol/min}$$
$$n_L = 5 \text{ mol/min}$$

This is the same result as we obtained earlier.

From the above relation, we note that we can define generalized rates for each reaction:

$$r_1 = \frac{R_{A1}}{\sigma_{A1}} \quad \text{and} \quad r_2 = \frac{R_{C2}}{\sigma_{C2}}$$

where r_1 is the generalized rate of the first reaction and r_2 is the generalized rate of the second reaction.

Thus, we can write for the above two reactions that

$$n_i = n_{if} + \sum_{j=1}^{2} (\sigma_{ij} r_j), \text{ where } i \equiv A, B, C, D, K, F, \text{ and } L.$$

Generalizing and using summation relation, we can write for N reactions:

$$n_i = n_{if} + \sum_{j=1}^{N} (\sigma_{ij} r_j)$$

(for the balance of any component i in a system of N reactions), where

$$r_j = \frac{R_{ij}}{\sigma_{ij}}$$

where i is the component for which the rate of production in the reaction j is given, R_{ij} is the rate of production of component i in reaction j, and σ_{ij} is the stoichiometric number of component i in reaction j.

Thus, the generalized mass balance equation for any system with single-input, single-output and N reactions (as shown in Fig. 2.21) is given by

$$n_i = n_{if} + \sum_{j=1}^{N} \sigma_{ij} r_j \qquad (2.42)$$

for any number of components i and any number of reactions N.

If we have a single reaction, then the above expression becomes

$$n_i = n_{if} + \sigma_i r$$

For no reactions (nonreacting systems), it becomes

$$n_i = n_{if}$$

For a system with multiple-inputs, multiple-outputs and multiple reactions (the most general case), as shown in Figure 2.22, the mass balance equation is

$$\sum_{k=1}^{K} n_{ik} = \sum_{l=1}^{L} n_{if_l} + \sum_{j=1}^{N} \sigma_{ij} r_j, \quad i = 1, 2, \ldots, M \qquad (2.43)$$

This equation is the most general mass balance equation. It applies to all possible mass balance cases. If there is only one input and one output, then

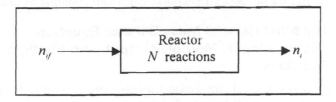

Figure 2.21 Molar flow diagram.

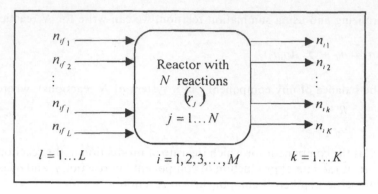

Figure 2.22 Molar flow diagram for multiple-inputs, multiple-outputs, and multiple reactions.

it reduces to Eq. (2.42). If there are no reactions, then it reduces to the following form:

$$\sum_{k=1}^{K} n_{ik} = \sum_{l=1}^{L} n_{if_1}$$

and so on.

2.3.5 The Algebra of Multiple Reactions (Linear Dependence and Linear Independence of Multiple Reactions)

The number of variables (whether conversions and yields or rates of reaction) to define a system of reactions is equal to the number of independent reactions. Therefore, for a system of M reactions, we should determine the number of linearly independent reactions N.

Because this process requires a little knowledge of linear algebra (determinants and matrices), a limited review is given in Appendix A (matrix algebra). Here, we introduce the concept through a simple example.

2.3.6 The Most General Mass Balance Equation (Multiple-Input, Multiple-Output, and Multiple Reactions)

This is the most general one-phase (homogeneous) chemical engineering lumped system. It has F input streams (and we use L as the counter for the input streams ($n_{if_1}, n_{if_2}, \ldots, n_{if_l}, \ldots, n_{if_L}$)) and K output streams (we use k

as the counter for output streams $(n_{i_1}, n_{i_2, i_k}, \ldots, n_{i_K}))$ with M $(i = 1, 2, \ldots, M)$ components and N reactions $(r_1, r_2, \ldots, r_j, \ldots, r_N)$. Therefore the generalized material balance equation is

$$\sum_{k=1}^{K} n_{ik} = \sum_{l=1}^{L} n_{if_l} + \sum_{j=1}^{N} \sigma_{ij} r_j, \quad i = 1, 2, \ldots, M$$

There are M equations having N reactions $(r_j, j = 1, 2, \ldots, N)$ and σ_{ij} is the stoichiometric number of component i in reaction j. The generalized rate of reaction for reaction j is given by r_j. Note that r_j is the overall rate of reaction for the whole unit (it is not per unit volume or per unit mass of catalyst and so on; that is why our equations are mass balance equations, not design equations as shown later).

Introductory Example

Consider the following system of reversible isomerization reactions:

$$1\text{-}butene \quad \Leftrightarrow \quad cis\text{-}2\text{-}butene$$
$$cis\text{-}2\text{-}butene \quad \Leftrightarrow \quad trans\text{-}2\text{-}butene$$
$$trans\text{-}2\text{-}butene \quad \Leftrightarrow \quad 1\text{-}butene$$

Let us call 1-butene component A, cis-2-butene component B, and trans-2-butene component C. Then the reaction network can be rewritten as

$$A \Leftrightarrow B$$
$$\text{Rate} = r_1 = \frac{R_1}{\sigma_{A1}}$$

$$B \Leftrightarrow C$$
$$\text{Rate} = r_2 = \frac{R_2}{\sigma_{B2}}$$

$$C \Leftrightarrow A$$
$$\text{Rate} = r_3 = \frac{R_3}{\sigma_{C3}}$$

There are apparently three reactions and three components. Let us write the mass balance equations in terms of the rates of reaction $(r_1, r_2,$ and $r_3)$ for a feed of $n_{Af}, n_{Bf},$ and n_{Cf}:

$$n_A = n_{Af} + \sigma_{A1} r_1 + \sigma_{A3} r_3$$
$$n_B = n_{Bf} + \sigma_{B1} r_1 + \sigma_{B2} r_2$$
$$n_C = n_{Cf} + \sigma_{C2} r_2 + \sigma_{C3} r_3$$

When we introduce the values (with correct signs) for the stoichiometric numbers, we get

$$n_A = n_{Af} - r_1 + r_3$$
$$n_B = n_{Bf} + r_1 - r_2$$
$$n_C = n_{Cf} + r_2 - r_3$$

If n_{Af}, n_{Bf}, and n_{Cf} are known, then in the above three equations we have six unknowns: n_A, n_B, n_C, r_1, r_2, and r_3. Without thorough examination we would say that we have to define r_1, r_2, and r_3 (three rates of reaction) in order to solve for n_A, n_B, and n_C. This is actually not true and the problem is solvable with only two specifications.

Why Is That?

Because the given three reactions are not linearly independent. The system actually contains only two independent reactions. The mass balance equations could have been written in terms of the two modified rate variables

$$r_1' = r_1 - r_3 \quad \text{and} \quad r_2' = r_1 - r_2$$

We can rewrite the mass balance relations as

$$n_A = n_{Af} - r_1'$$
$$n_B = n_{Bf} + r_2'$$
$$n_C = n_{Cf} + r_1' - r_2'$$

Therefore, there are only two unknown variables (r_1' and r_2') other than n_A, n_B, and n_C. How can we tell beforehand that the reactions are linearly independent or not? Let us illustrate that first for the previous example.

$$A \Leftrightarrow B \tag{2.44}$$

$$B \Leftrightarrow C \tag{2.45}$$

$$C \Leftrightarrow A \tag{2.46}$$

We will write them in the following algebraic form:

$$A - B + 0 \times C = 0 \tag{2.47}$$

$$0 \times A + B - C = 0 \tag{2.48}$$

$$-1 \times A + 0 \times B + C = 0 \tag{2.49}$$

The matrix of the coefficients is

$$A = \begin{pmatrix} 1 & -1 & 0 \\ 0 & 1 & -1 \\ -1 & 0 & 1 \end{pmatrix}$$

It is a 3×3 matrix; if its determinant is zero (singular matrix), then the rank of this matrix is less than 3, which means there is linear dependence:

$$\text{Det}[A] = 1(1) + 1(-1) + 0 = 1 - 1 = 0$$

This means that the matrix is singular and the above three equations are linearly dependent. In fact addition of Eqs (2.47) and (2.48) gives Eq. (2.49).

Consider another reaction set; it is the steam reforming of methane. In this set of reactions, there are five components and three reactions:

$$CH_4 + H_2O \Leftrightarrow CO + 3H_2$$

$$CO + H_2O \Leftrightarrow CO_2 + H_2$$

$$CH_4 + 2H_2O \Leftrightarrow CO_2 + 4H_2$$

We will write these three reactions in algebraic form. Let us call CH_4 component A, H_2O component B, CO component C, H_2 component D, and CO_2 component E. We get

$$1A + 1B - 1C - 3D - 0E = 0 \tag{2.50}$$

$$0A + 1B + 1C - 1D - 1E = 0 \tag{2.51}$$

$$1A + 2B + 0C - 4D - 1E = 0 \tag{2.52}$$

The matrix of the coefficients is

$$\begin{pmatrix} 1 & 1 & -1 & -3 & 0 \\ 0 & 1 & 1 & -1 & -1 \\ 1 & 2 & 0 & -4 & -1 \end{pmatrix}$$

For these three reactions to be linearly independent, we must have at least one 3×3 submatrix which is nonsingular. This is not achieved in the above matrix because all 3×3 submatrices are singular and therefore the rank is less than 3 and the system is linearly dependent. Therefore, Eqs. (2.50)–(2.52) are linearly dependent.

2.4 SOLVED PROBLEMS FOR MASS BALANCE

Readers are advised to solve the problem on their own before inspecting the given solution.

Solved Example 2.5

In the Deacon Process for the manufacture of chlorine, HCl and O_2 react to form Cl_2 and H_2O. Sufficient air (21% O_2, 79% N_2) is fed to the reactor to supply 30% excess oxygen, and the fractional conversion of HCl is 70%. Calculate the molar composition of the product stream.

Solution

$$4HCl + O_2 \rightarrow 2H_2O + 2Cl_2$$

For simplification, we can write

$$4A + B \rightarrow 2C + 2D$$

where $HCl \equiv A$, $O_2 \equiv B$, $H_2O \equiv C$, $Cl_2 \equiv D$, and $N_2 \equiv E$

We can write the generalized mass balance (see Fig. 2.23) as

$$\sum_{k=1}^{K} n_{i_k} = \sum_{l=1}^{L} n_{if_l} + \sum_{j=1}^{N} \sigma_{ij} r_j$$

where $L = 1$, $K = 1$, and $N = 1$, giving

$$n_i = n_{if} + \sigma_i r$$

The fractional conversion of component A is

$$x_A = \frac{n_{A_f} - n_A}{n_{A_f}} = 0.7$$

thus giving

$$n_A = 0.3 n_{A_f}$$

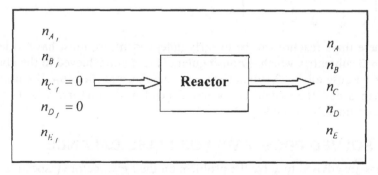

Figure 2.23 Molar flow diagram for solved example 2.5.

We have

$$n_{B_f}(\text{required}) = \tfrac{1}{4}n_{A_f}$$

As 30% excess is fed, we get

$$n_{B_f} = (1.3)\tfrac{1}{4}n_{A_f}$$

thus giving

$$n_{B_f} = 0.325\,n_{A_f}$$

Also, we have

$$n_{E_f} = \frac{79}{21}n_{B_f}$$

On substituting the values, we get

$$n_{E_f} = 1.223\,n_{A_f}$$

Let R_A be the rate of production of component A; then, we get

$$R_A = -(n_{A_f} - n_A)$$

which is equal to

$$R_A = -0.7\,n_{A_f}$$

Now, we have

$$r = \frac{-0.7\,n_{A_f}}{\sigma_A}$$

As $\sigma_A = -4$, we get

$$r = 0.175\,n_{A_f}$$

From the mass balance relations, we get, for component A,

$$n_A = n_{A_f} + \sigma_A r$$

On substituting the values, we get

$$n_A = 0.3\,n_{A_f} \tag{2.53}$$

For component B,

$$n_B = n_{B_f} + \sigma_B r$$

As $\sigma_B = -1$ and $n_{B_f} = 0.325 n_{A_f}$, we get

$$n_B = 0.15 n_{A_f} \tag{2.54}$$

For component C,

$$n_C = n_{C_f} + \sigma_C r$$

As $\sigma_C = 2$ and $n_{C_f} = 0$, we get

$$n_C = 0.35 n_{A_f} \tag{2.55}$$

For component D,

$$n_D = n_{D_f} + \sigma_D r$$

As $\sigma_D = 2$ and $n_{D_f} = 0$, we get

$$n_D = 0.35 n_{A_f} \tag{2.56}$$

For component E,

$$n_E = n_{E_f} + \sigma_E r$$

As $\sigma_E = 0$ and $n_{E_f} = 1.223 n_{A_f}$, we get

$$n_E = 1.223 n_{A_f} \tag{2.57}$$

Thus, the output mixture is as follows:

$$n_A = 0.3 n_{A_f}, \qquad n_B = 0.15 n_{A_f}, \qquad n_C = 0.35 n_{A_f}, \qquad n_D = 0.35 n_{A_f},$$
$$n_E = 1.223 n_{A_f}$$

The total output mixture is

$$n_{\text{Total}_{\text{out}}} = (0.3 + 0.15 + 0.35 + 0.35 + 1.223) n_{A_f}$$

which is equal to

$$n_{\text{Total}_{\text{out}}} = 2.373 n_{A_f}$$

Now, to calculate the composition of exit stream (mole fraction),

$$\text{Mole fraction of } A = X_A = \frac{n_A}{n_{\text{Total}_{\text{out}}}} = \frac{0.3\,n_{A_f}}{2.373\,n_{A_f}} = \frac{0.3}{2.373} = 0.1264$$

$$\text{Mole fraction of } B = X_B = \frac{n_B}{n_{\text{Total}_{\text{out}}}} = \frac{0.15\,n_{A_f}}{2.373\,n_{A_f}} = \frac{0.15}{2.373} = 0.06321$$

$$\text{Mole fraction of } C = X_C = \frac{n_C}{n_{\text{Total}_{\text{out}}}} = \frac{0.35\,n_{A_f}}{2.373\,n_{A_f}} = \frac{0.35}{2.373} = 0.14749$$

$$\text{Mole fraction of } D = X_D = \frac{n_D}{n_{\text{Total}_{\text{out}}}} = \frac{0.35\,n_{A_f}}{2.373\,n_{A_f}} = \frac{0.35}{2.373} = 0.14749$$

$$\text{Mole fraction of } E = X_E = \frac{n_E}{n_{\text{Total}_{\text{out}}}} = \frac{1.223\,n_{A_f}}{2.373\,n_{A_f}} = \frac{1.223}{2.373} = 0.51538$$

Just as a check, the sum of the mole fractions of exit stream should be equal to 1. Thus,

$$\sum_{p=1}^{5} X_p = 0.1264 + 0.06321 + 0.14749 + 0.14749 + 0.51538$$
$$= 0.99997 \cong 1.0$$

Hence, our calculations are correct.

Solved Example 2.6

Chlorine oxide gas is used in the paper industry to bleach pulp and it is produced by reacting sodium chlorate, sulfuric acid, and methanol in lead-lined reactors:

$$6NaClO_3 + 6H_2SO_4 + CH_3OH \rightarrow 6ClO_2 + 6NaHSO_4 + CO_2 + 5H_2O$$

(a) Suppose 14 moles of an equimolar mixture of $NaClO_3$ and H_2SO_4 are added per 1 mole of CH_3OH per hour; determine the limiting reactant.

(b) For the above given molar ratios, calculate the reactant flows required to produce 10 metric tons per hour of ClO_2, assuming 90% conversion (of the limiting component) is achieved.

Molecular weights of the components are $Cl_2 = 70.906$, $Na = 23.0$, $S = 32.0$, $C = 12.0$, $H = 1.0$, and $O = 16.0$.

Solution

The reaction (Fig. 2.24) can be written as

$$6A + 6B + C \rightarrow 6D + 6F + M + 5K$$

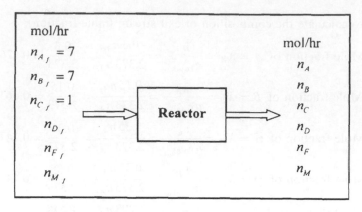

Figure 2.24 Molar flow diagram for example 2.6.

Part a: The limiting rates are given by

$$r_{L_A} = \frac{-n_{A_f}}{\sigma_A} = \frac{-7}{-6} = 1.16666$$

$$r_{L_B} = \frac{-n_{B_f}}{\sigma_B} = \frac{-7}{-6} = 1.16666$$

$$r_{L_C} = \frac{-n_{C_f}}{\sigma_C} = \frac{-1}{-1} = 1$$

As r_{L_C} has the smallest numeral value, component C is the limiting agent. It is required to produce 10,000 kg/h of component D. The molecular weight of $D(M_D)$ is 67 kg/kmol. Therefore,

$$n_D = \frac{W_D}{M_D} = \frac{10,000}{67} = 149.253 \text{ kmol/h} \tag{2.58}$$

Also, it is given in the problem that the conversion of component C is equal to 0.9:

$$x_C = 0.9 \tag{2.59}$$

From the stoichiometric relations in the feed stream,

$$n_{B_f} = n_{A_f} \tag{2.60}$$

$$n_{C_f} = \tfrac{1}{7} n_{A_f} \tag{2.61}$$

Moreover, we have

$$n_{D_f} = 0 \qquad (2.62)$$

$$n_{F_f} = 0 \qquad (2.63)$$

$$n_{M_f} = 0 \qquad (2.64)$$

$$n_{k_f} = 0 \qquad (2.65)$$

Degrees-of-freedom analysis (i.e., the problem is solvable or not)

Total number of unknowns:

n_{if} for $i = A, B, C, D, F, M$ and K:	7 unknowns
n_i for $i = A, B, C, D, F, M$ and K:	7 unknowns
Rate of reaction (single reaction):	1 unknown
Total :	15 Unknowns

Known and specified relations/values:

Number of mass balance relations:	7 for 7 components
Number of given relations (equations 2.59–2.65):	8
Total:	15 known relations
Degrees of freedom $= 15 - 15 = 0$	

Hence, the problem is correctly specified and solvable.

Let us solve it.

Part b: Because we know n_D, from the balance for component D we can get r (because we know that $n_{D_f} = 0$) as follows:

$$n_D = n_{D_f} + \sigma_D r$$

On substituting the values, we get

$$n_D = 0 + 6r$$

Finally, we get

$$149.253 = 6r$$

thus giving

$$r = \frac{149.253}{6} = 24.8756 \text{ kmol/h} \qquad (2.66)$$

Now, we know the conversion of component C, $x_C = 0.9$. We can write

$$x_C = \frac{n_{C_f} - n_C}{n_{C_f}} = \frac{n_{A_f}/7 - n_C}{n_{A_f}/7} = 0.9 \qquad (2.67)$$

Also, we have

$$n_C = n_{C_f} + \sigma_C r \qquad (2.68)$$

On substituting the values ($\sigma_C = -1$), we get

$$n_C = \frac{n_{A_f}}{7} - r$$

Finally, we get

$$n_C = \frac{n_{A_f}}{7} - 24.8756 \qquad (2.69)$$

Using Eqs. (2.69) and (2.67), we get

$$\frac{n_{A_f}/7 - (n_{A_f}/7 - 24.8756)}{n_{A_f}/7} = 0.9$$

On solving, we get

$$n_{A_f} = 193.477 \text{ kmol/h} \qquad (2.70)$$

From Eq. (2.60), we get

$$n_{B_f} = n_{A_f}$$

Hence,

$$n_{B_f} = 193.477 \text{ kmol/h} \qquad (2.71)$$

From Eq. (2.61), we get

$$n_{C_f} = \frac{n_{A_f}}{7}$$

Thus,

$$n_{C_f} = 27.6396 \text{ kmol/h} \qquad (2.72)$$

Another (easier) route of solution

Instead of trying to obtain n_{A_f} and then obtain n_{B_f} and n_{C_f}, we can actually compute n_{C_f}, and then from it, n_{A_f} and n_{B_f} as follows:
We compute r as we did earlier:

$$n_D = n_{D_f} + \sigma_D r \qquad (2.65)$$

thus giving

$$r = 24.8756 \text{ kmol/h} \qquad (2.66)$$

We have

$$x_C = \frac{n_{C_f} - n_C}{n_{C_f}} = \frac{n_{A_f}/7 - n_C}{n_{A_f}/7} = 0.9 \qquad (2.67)$$

From the balance equation for component C, we have

$$n_C = n_{C_f} + \sigma_C r$$

Thus, we get

$$n_{C_f} - n_C = r \qquad (2.68)$$

From Eqs. (2.67) and (2.68), we get

$$x_C = 0.9 = \frac{r}{n_{C_f}} = \frac{24.8756}{n_{C_f}} \qquad (2.69)$$

Thus,

$$n_{C_f} = \frac{24.8756}{0.9} = 27.63958 \text{ kmol/h}$$

Therefore,

$$n_{A_f} = 7 n_{C_f} = 7(27.63958) = 193.477 \text{ kmol/h}$$

Finally,

$$n_{B_f} = n_{A_f} = 193.477 \text{ kmol/h}$$

Solved Example 2.7

Acetaldehyde, CH_3CHO, can be produced by catalytic dehydrogenation of ethanol, C_2H_5OH, via the reaction

$$C_2H_5OH \rightarrow CH_3CHO + H_2$$

There is, however, a parallel reaction producing ethyl acetate, $CH_3COOC_2H_5$:

$$2C_2H_5OH \rightarrow CH_3COOC_2H_5 + 2H_2$$

Suppose that, in a given reactor, the conditions are adjusted so that a conversion of 95% ethanol is obtained with a 80% yield of acetaldehyde based on the consumption of C_2H_5OH. Calculate the mole fraction of H_2 in the product stream.

Solution

$$C_2H_5OH \rightarrow CH_3CHO + H_2$$

and

$$2C_2H_5OH \rightarrow CH_3COOC_2H_5 + 2H_2$$

For simplification, we can write

$A \rightarrow B + C$, and the generalized rate of reaction is r_1

$2A \rightarrow D + 2C$, and the generalized rate of reaction is r_2

where $C_2H_5OH \equiv A$, $CH_3CHO \equiv B$, $H_2 \equiv C$, and $CH_3COOC_2H_5 \equiv D$.
We can write the generalized mass balance equation (Fig. 2.25) as

$$\sum_{k=1}^{K} n_{ik} = \sum_{l=1}^{L} n_{if_l} + \sum_{j=1}^{N} \sigma_{ij} r_j$$

where $N = 2$, $K = 1$, and $L = 1$.
Thus, in a more simplified form, we can write

$$n_i = n_{if} + \sum_{j=1}^{2} \sigma_{ij} r_j$$

Fractional conversion of component A $= x_A = \dfrac{n_{A_f} - n_A}{n_{A_f}} = 0.95$

thus giving

$$n_A - n_{A_f} = -0.95 \, n_{A_f} \tag{2.70}$$

thus, $n_A = 0.05 \, n_{A_f}$ \hfill (2.71)

Yield of B$= Y_B = \dfrac{n_B - n_{B_f}}{x_A n_{A_f}} = 0.8$

thus giving

$$n_B - n_{B_f} = (0.8)(0.95) n_{A_f}$$

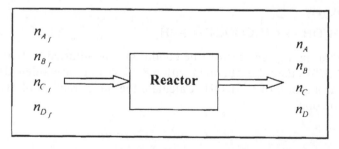

Figure 2.25 Molar flow diagram for solved example 2.7.

Further calculation gives

$$n_B - n_{B_f} = 0.76 n_{A_f} \qquad (2.72)$$

As $n_{B_f} = 0$, we get

$$n_B = 0.76 n_{A_f} \qquad (2.73)$$

From the mass balance equation for component B,

$$n_B = n_{B_f} + \sigma_{B_1} r_1 + \sigma_{B_2} r_2$$

We get

$$(n_B - n_{B_f}) = r_1 + 0$$

thus giving

$$n_B = r_1 = 0.76 n_{A_f} \qquad (2.74)$$

From the mass balance equation for component A,

$$n_A = n_{A_f} + \sigma_{A_1} r_1 + \sigma_{A_2} r_2$$

We get

$$(n_A - n_{A_f}) = -r_1 - 2r_2$$

Using Eqs. (2.70) and (2.74) in the above relation gives

$$-0.95 n_{A_f} = -0.76 n_{A_f} - 2r_2$$

Finally, we get

$$r_2 = 0.095 n_{A_f} \qquad (2.75)$$

From the simplified generalized mass balance relation, we have the following.

For component C,

$$n_C = n_{C_f} + \sigma_{C_1} r_1 + \sigma_{C_2} r_2$$

As $\sigma_{C_1} = 1$, $\sigma_{C_2} = 2$, and $n_{C_f} = 0$, we get

$$n_C = 0.76 n_{A_f} + 0.19 n_{A_f}$$

On solving, we get

$$n_C = 0.95 n_{A_f} \qquad (2.76)$$

For component D,

$$n_D = n_{D_f} + \sigma_{D_1} r_1 + \sigma_{D_2} r_2$$

As $\sigma_{D_1} = 0$, $\sigma_{D_2} = 1$, and $n_{D_f} = 0$ we get

$$n_D = 0 + 0 + r_2$$

On substituting the value, we get

$$n_D = 0.095 \, n_{A_f} \tag{2.77}$$

Thus, the output mixture is

$$n_A = 0.05 \, n_{A_f}, \qquad n_B = 0.76 \, n_{A_f}, \qquad n_C = 0.95 \, n_{A_f},$$
$$n_D = 0.095 \, n_{A_f}$$

The total output mixture is

$$n_{\text{Total}_{\text{out}}} = (0.05 + 0.76 + 0.95 + 0.095) \, n_{A_f}$$

which is equal to

$$n_{\text{Total}_{\text{out}}} = 1.855 \, n_{A_f}$$

Now, to calculate the composition of exit stream (mole fraction),

$$X_A = \frac{n_A}{n_{\text{Total}_{\text{out}}}} = \frac{0.05 \, n_{A_f}}{1.855 \, n_{A_f}} = 0.027$$

$$X_B = \frac{n_B}{n_{\text{Total}_{\text{out}}}} = \frac{0.76 \, n_{A_f}}{1.855 \, n_{A_f}} = 0.410$$

$$X_C = \frac{n_C}{n_{\text{Total}_{\text{out}}}} = \frac{0.95 \, n_{A_f}}{1.855 \, n_{A_f}} = 0.512$$

$$X_D = \frac{n_D}{n_{\text{Total}_{\text{out}}}} = \frac{0.095 \, n_{A_f}}{1.855 \, n_{A_f}} = 0.051$$

Just as a check, the sum of the mole fractions of exit stream should be equal to 1. Thus,

$$\sum_{p=1}^{4} X_p = 0.027 + 0.410 + 0.512 + 0.051 \cong 1.0$$

Hence, our calculations are correct.

2.5 HEAT EFFECTS

The only difference between simple linear heat balance problems and the heat balance problems, which are rather complex and nonlinear, is the existence of chemical reactions [i.e., the conversion of one species or more to other species causing release of heat (exothermic reactions) or absorption of heat (endothermic reactions)]. If the heat of reaction is negligible, then the complex nonlinear heat balance equation reduces to the same simple linear heat balance equations.

2.5.1 Heats of Reactions

It is very important for the reader to understand the concept and the methods of calculating the heat of reaction before he/she starts heat balance problems for systems with chemical reactions. This is due to the fact that the heat of reaction is the heart of the problem here.

The heat of reaction is usually defined as

Heat of reaction = Enthalpy of products − Enthalpy of reactants

Heats of reaction as defined above are:

Positive for endothermic reactions

Negative for exothermic reactions

The enthalpies of products and reactants here are meant to be in the same molar proportions given in the stoichiometric equation. For example, for the hypothetical reaction,

$$2A + 3B \rightarrow 7C + 8D \tag{2.78}$$

the heat of reaction for this reaction (as written) should be calculated as

$$\Delta H_R = 7H_C + 8H_D - 2H_A - 3H_B \tag{2.79}$$

This heat of reaction is not per mole of A or B or C or D; actually, it is per the reaction as stoichiometrically written. Let us suppose that H_A is in J/mol of A, H_B is in J/mol of B, H_C is in J/mol of C, and H_D is in J/mol of D; then, it is clear that ΔH_R in Eq. (2.79) is not calculated per mole of any of the components. If we insist at this stage to refer to the heat of reaction per mole of A, B, C, or D, then we can just for the sake of clarity write it as follows:

$$\Delta H_R = \text{joules/2 moles of } A \text{ reacted}$$

$$= \text{joules/3 moles of } B \text{ reacted}$$

$$= \text{joules/7 moles of } C \text{ produced}$$

$$= \text{joules/8 moles of } D \text{ produced}$$

Therefore, if we want the heat of reaction per mole of any of the components, then we will have to do the following simple division by the absolute value of the stoichiometric number of the corresponding component:

$$\Delta H_A = \frac{\Delta H_R}{|\sigma_A|} \text{ J/mol of } A \text{ reacted}$$

$$\Delta H_B = \frac{\Delta H_R}{|\sigma_B|} \text{ J/mol of } B \text{ reacted}$$

$$\Delta H_C = \frac{\Delta H_R}{|\sigma_C|} \text{ J/mol of } C \text{ produced}$$

$$\Delta H_D = \frac{\Delta H_R}{|\sigma_D|} \text{ J/mol of } D \text{ produced}$$

Actually, in any problem we always need the rate of heat production or absorption (J/time); therefore, we should always combine the heat of the reaction with the rate of reaction.

For reaction (2.78), which is a single reaction, we can express this amount of heat as follows:

$$Q = \Delta H_A |R_A| \text{ J/time}$$

where R_A is the rate of reaction of A. It can also be written as

$$Q = \Delta H_B |R_B| \text{ J/time}$$

or

$$Q = \Delta H_C |R_C| \text{ J/time}$$

or

$$Q = \Delta H_D |R_D| \text{ J/time}$$

Q = Positive for endothermic reactions (heat is absorbed by the reaction)

Q = Negative for exothermic reactions (heat is produced by the reaction)

All of these relations give the same answer, of course. This can easily be demonstrated by inspecting the relation between R_A and R_B as well as ΔH_A and ΔH_B; for example,

$$|R_B| = |R_A| \frac{|\sigma_B|}{|\sigma_A|}$$

and

$$\Delta H_A = \frac{\Delta H_R}{|\sigma_A|}, \qquad \Delta H_B = \frac{\Delta H_R}{|\sigma_B|}$$

Thus,

$$\Delta H_B = \Delta H_A \frac{|\sigma_A|}{|\sigma_b|}$$

Therefore,

$$\Delta H_B |R_B| = \Delta H_A \frac{|\sigma_A|}{|\sigma_B|} |R_A| \frac{|\sigma_B|}{|\sigma_A|} = \Delta H_A |R_A|$$

We can also express Q in terms of ΔH_R and the generalized rate of reaction (r) as

$$Q = \Delta H_A |R_A| = \frac{\Delta H_R}{|\sigma_A|} |R_A| = \Delta H_R r$$

Similarly,

$$Q = \Delta H_B |R_B| = \frac{\Delta H_R}{|\sigma_B|} |R_B| = \Delta H_R r$$

and so on.

Important Notes

1. Note that for this irreversible single reaction, we always take the absolute values for rates of reaction and stoichiometric numbers, because the sign of Q is determined by the sign of the heat of reaction (whether the reaction is exothermic or endothermic) and not by whether we are referring to a product or a reactant. However, when we use the generalized rate of reaction (r), this issue is automatically addressed because r is always positive, as we have shown in the previous section.

2. Heats of reactions obtained from tables or from the literature will have the units and the basis given to them in the data source; for example, if you find ΔH of a reaction given as J/mol of A reacted, then this ΔH actually is ΔH_A and is defined as

$$\Delta H_A = \frac{\Delta H_R}{|\sigma_A|}$$

where ΔH_R is for the reaction as stoichiometrically written.

3. The sign convention of ΔH_R is defined by

$$\Delta H_R = \sum H_{i\text{products}} - \sum H_{i\text{reactants}} \qquad (2.80)$$

This results in

$\Delta H_R = $ Positive for endothermic reaction

$\Delta H_R = $ Negative for exothermic reaction

Of course, scientifically there is nothing wrong in reversing the sign convention to be

$$\Delta H_R = \sum H_{i\text{reactants}} - \sum H_{i\text{products}} \qquad (2.81)$$

In this case, we will have

$\Delta H_R = $ Positive for exothermic reaction

$\Delta H_R = $ Negative for endothermic reaction

However, the first sign convention (2.80) is the one adopted worldwide. In some cases, $-\Delta H_R$ is used instead of ΔH_R. In this case, it is obvious that

$-\Delta H_R = $ Positive for exothermic reaction

$-\Delta H_R = $ Negative for endothermic reaction

2.5.2 Effects of Temperature, Pressure, and Phases on Heat of Reaction

We can now write the heat of reaction as

$$\Delta H_R = \sum_i [|\sigma_i|_{\text{products}} H_i(T, P, \pi_i)] - \sum_i [|\sigma_i|_{\text{reactants}} H_i(T, P, \pi_i)]$$

where π_i defines the phase of component i.

With the sign convention discussed earlier, $\sigma_i > 0$ for products and $\sigma_i < 0$ for reactants. We can write

$$\Delta H_R = \sum_i \sigma_i H_i(T, P, \pi_i)$$

where i is the number of all components (reactants and products) involved in the reaction.

Because the enthalpies are function of T, P, and π_i then ΔH_R is obviously also a function of T, P, and π_i.

Important note: T and P are numbers, and π_i represents definition of phases (gas or liquid or solid). The change of ΔH_R with T or π_i (effect of P is usually negligible except for processes at very high pressure) can be obtained from the change of enthalpies with T or π_i. Suppose that the heat of reaction of a certain reaction is given at $T°$, $P°$, and $\pi_i°$ and is represented by $\Delta H_R(T°, P°, \pi_i°)$. We want to compute the heat of reaction at T, $P°$, and π_i; this can be achieved through the fact that

$$\Delta H_R(T, P°, \pi_i) - \Delta H_R(T°, P°, \pi_i°)$$
$$= \sum_i \{\sigma_i H_i(T, P°, \pi_i) - \sigma_i H_i(T°, P^c, \pi_i°)\}$$

The right-hand side is a difference between enthalpies and thus can be easily calculated for both changes from $T° \rightarrow T$ (sensible heat) and from $\pi°_i \rightarrow \pi_i$ (change of phase, latent heat).

For the change of temperature only

$$\Delta H_R(T, P°, \pi_i°) - \Delta H_R(T°, P°, \pi_i°)$$
$$= \sum_i \{\sigma_i H_i(T, P°, \pi_i°) - \sigma_i H_i(T°, P°, \pi_i°)\}$$
$$= \sum_i \sigma_i \int_{T_0}^{T} C_{Pi} \, dT$$

For the change of phase, if $\pi_i°$ and π_i represent different phases, then the heats of phase transition and the temperatures of phase transitions must be involved as clearly illustrated in the following example.

An Illustrative Example

Given the heat of the reaction

$$4NH_3(g) + 5O_2(g) \rightarrow 4NO(g) + 6H_2O(l)$$

at 1 atm and 298 K to be -279.33 kcal/mol. Calculate the heat of reaction at 920°C and 1 atm with water in the vapor phase.

Solution

Given

$$\Delta H_R(T°, P°, \pi_i^c) = -279.33 \text{ kcal}$$

where

$$T° = 298 \text{ K}, P° = 1 \text{ atm}$$

and

$$\pi_i° = \begin{cases} NH_3 & \text{gas} \\ O_2 & \text{gas} \\ NO & \text{gas} \\ H_2O & \text{liquid} \end{cases}$$

What is required to compute is

$$\Delta H_R(T, P°, \pi_i) = ?$$

where

$$T = 920 + 273 = 1193 \text{ K}, \qquad P° = 1 \text{ atm}$$

and

$$\pi_i = \begin{cases} NH_3 & \text{gas} \\ O_2 & \text{gas} \\ NO & \text{gas} \\ H_2O & \text{vapor} \end{cases}$$

Now,

$$\Delta H_R(T°, p°, \pi_i°) = \sum_i \sigma_i H_i(T°, P°, \pi_i°)$$

and

$$\Delta H_R(T, P°, \pi_i) = \sum_i \sigma_i H_i(T, P°, \pi_i)$$

Then,

$$\Delta H_R(T, P°, \pi_i) - \Delta H_R(T°, P°, \pi°_i) = \sum_i \{\sigma_i H_i(T, P°, \pi_i) - \sigma_i H_i(T°, P°, \pi°_i)\}$$

$$= -4\{H_{NH_3}(1193, 1, \text{gas}) - H_{NH_3}(298, 1, \text{gas})\} - 5\{H_{O_2}(1193, 1, \text{gas}) - H_{O_2}(298, 1, \text{gas})\} + 4\{H_{NO}(1193, 1, \text{gas}) - H_{NO}(298, 1, \text{gas})\} + 6\{H_{H_2O}(1193, 1, \text{vapor}) - H_{H_2O}(298, 1, \text{liquid})\}$$

Thus, for the right-hand side, we get

$$- 4 \int_{298}^{1193} C_{P_{NH_3}} dT - 5 \int_{298}^{1193} C_{P_{O_2}} dT + 4 \int_{298}^{1193} C_{P_{NO}} dT$$

$$+ 6 \left[\int_{298}^{373} C_{P_{H_2O}}(liq) dT + \Delta H_{H_2O}(vap) + \int_{373}^{1193} C_{P_{H_2O}}(vap) dT \right]$$

Finally, we get

$$\Delta H_R(T, P^\circ, \pi_i) = \Delta H_R(T^\circ, P^\circ, \pi_i^\circ) - 4 \int_{298}^{1193} C_{P_{NH_3}} dT - 5$$

$$\int_{298}^{1193} C_{P_{O_2}} dT + 4 \int_{298}^{1193} C_{P_{NO}} dT$$

$$+ 6 \left[\int_{298}^{373} C_{P_{H_2O}}(liq) dT + \Delta H_{H_2O}(vap) \right.$$

$$\left. + \int_{373}^{1193} C_{P_{H_2O}}(vap) dT \right]$$

where $\Delta H_{H_2O}(vap)$ is the latent heat of vaporization.

2.5.3 Heats of Formation and Heats of Reaction

What is the heat of formation? The heat of formation of a compound is the heat of reaction evolved (or absorbed) during the formation of this compound from its elementary constituents; for example, for the reaction

$$C + 2H_2 \rightarrow CH_4$$

the heat of this reaction is actually the heat of formation of CH_4 (methane).

The heats of formation of large number of compounds are tabulated at standard conditions (heat of formation tables available in Perry's *Chemical Engineer's Handbook*, 1997). The standard conditions are usually defined at 25°C and 1 atm and the phase that is normal for compounds under these conditions (carbon is solid, H_2O is liquid, CO_2 is gas, etc.).

The heats of reaction under standard conditions can be computed from the heats of formation of the different components in this reaction (reactants and products) under standard conditions through the following formula, which can be easily proven:

$$\Delta H_R^\circ = \sum_i \sigma_i \Delta H_{Fi}^\circ$$

Here, i is the numbering for the species (reactants and products) involved in the reaction: $\sigma_i > 0$ for products and $\sigma_i < 0$ for reactants. ΔH_R° is negative for exothermic reactions and positive for endothermic reactions.

2.5.4 Heats of Combustion and Heats of Reaction

The heat of combustion is the heat evolved from oxidizing the component completely with a stoichiometric amount of oxygen; for example,

$$CH_4 + 2O_2 \rightarrow CO_2 + 2H_2O$$

The heat evolved from the above reaction is the heat of combustion of methane.

$$S + O_2 \rightarrow SO_2$$

The heat evolved from this reaction is the heat of combustion of sulfur.

The heat of reaction can be computed from the heat of combustion using the formula

$$\Delta H_R^\circ = -\sum \sigma_i \Delta H_{Ci}^\circ$$

Any heat of reaction is computed at standard conditions. $\Delta H_R^\circ(T^\circ, P^\circ, \pi_i^\circ)$ can be used to compute the heats of reaction at any other condition ΔH_R (T, P°, π_i) using the previously discussed simple procedure.

2.6 OVERALL HEAT BALANCE WITH SINGLE AND MULTIPLE CHEMICAL REACTIONS

The heat balance for a nonadiabatic reactor is given in Figure 2.26. i represents the numbering of all components involved in the reaction (reactants and products). If any product does not exist in the feed, then we put its $n_{if} = 0$; if any reactant does not exist in the output, then we put its $n_i = 0$. Enthalpy balance is

$$\sum_i n_{if} H_i(T_f, P_f, \pi_{if}) + Q = \sum_i n_i H_i(T, P, \pi_i) \tag{2.82}$$

where Q is the heat added to the system (the reactor)

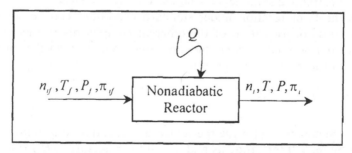

Figure 2.26 Heat balance for a nonadiabatic reactor.

Note: We did not include any heats of reactions; it is automatically included, as will be clear in the next few lines.

Define a reference state, and add and subtract the following terms from Eq. (2.82):

$$\sum_i n_{if} H_{ir}(T_r, P_r, \pi_{ir}) \quad \text{and} \quad \sum_i n_i H_{ir}(T_r, P_r, \pi_{ir})$$

This addition of two terms and subtracting of the same two terms together with rearrangement makes the heat balance equation have the following form:

$$\sum_i n_{if}\{H_{if}(T_f, P_f, \pi_{if}) - H_{ir}(T_r, P_r, \pi_{ir})\} + Q$$

$$= \sum_i n_i\{H_i(T, P, \pi_i) - H_{ir}(T_r, P_r, \pi_{ir})\} + \sum_i (n_i - n_{if})H_{ir}(T_r, P_r, \pi_{ir})$$

From the mass balance of a single reaction, we have

$$n_i = n_{if} + \sigma_i r \tag{2.83}$$

where

$$r = \frac{R_i}{\sigma_i}$$

Equation (2.83) can be rearranged as

$$n_i - n_{if} = \sigma_i r$$

Therefore, the heat balance equation becomes

$$\sum_i n_{if}\{H_{if}(T_f, P_f, \pi_{if}) - H_{ir}(T_r, P_r, \pi_{ir})\} + Q$$

$$= \sum_i n_i\{H_i(T, P, \pi_i) - H_{ir}(T_r, P_r, \pi_{ir})\} + \sum_i \sigma_i r H_{ir}(T_r, P_r, \pi_{ir})$$

This equation can be rearranged to give

$$\sum_i n_{if}\{H_{if}(T_f, P_f, \pi_{if}) - H_{ir}(T_r, P_r, \pi_{ir})\} + Q$$

$$= \sum_i n_i\{H_i(T, P, \pi_i) - H_{ir}(T_r, P_r, \pi_{ir})\} + r \sum_i \sigma_i H_{ir}(T_r, P_r, \pi_{ir})$$

It is clear that, by definition,

$$\sum_i \sigma_i H_{ir}(T_r, P_r, \pi_{ir}) = (\Delta H_R)_r$$

where $(\Delta H_R)_r$ is the heat of reaction at the reference conditions. Thus, the heat balance equation becomes

$$\sum_i n_{if}\{H_{if}(T_f, P_f, \pi_{if}) - H_{ir}(T_r, P_r, \pi_{ir})\} + Q$$

$$= \sum_i n_i\{H_i(T, P, \pi_i) - H_{ir}(T_r, P_r, \pi_{ir})\} + r(\Delta H_R)_r$$

In other words,

$$\{\text{Enthalpy in (above reference conditions)} + \text{Heat added}\}$$
$$= \left\{ \begin{array}{l} \text{Enthalpy out (above reference conditions)} \\ + \text{Heat absorbed by reaction (at reference conditions)} \end{array} \right\}$$

The terms of heat absorbed by reaction (for a single reaction) is $r(\Delta H_R)_r$

Note: If $r = 0$ (i.e., a nonreacting system), then we obtain a simple heat balance relation for the nonreacting system. The heat of reaction $(\Delta H_R)_r$ is evaluated at the reference state (T_r, P_r, π_r).

The term $r(\Delta H_R)_r$ needs some more clarification. For the sake of brevity, let us use:

Enthalpy in (at above reference conditions) $= I_{in}$
Enthalpy out (at above reference conditions) $= I_{out}$

Then,

$$I_{in} + Q = I_{out} + r(\Delta H_R)_r$$

The sign convention ensures that r is always positive and this is more convenient in the heat balance; $(\Delta H_R)_r$ is the heat of reaction for the reaction as stoichiometrically written as discussed earlier and re-explained in the following notes:

Note 1: If the reaction is endothermic, then $(\Delta H_R)_r$ is positive and it is the heat absorbed or consumed; therefore, it is clear that it must be added to the right-hand side with I_{out} as shown above.

If the reaction is exothermic, then $(\Delta H_R)_r$ is negative and the heat is evolved; then, it will be effectively added as positive quantity to the left-hand side.

Note 2: Suppose that the reaction we are dealing with is in the form

$$2A + 4B \rightarrow 5C + 6D$$

Then, $(\Delta H_R)_r$ is evaluated from the enthalpies as follows:

$$(\Delta H_R)_r = \sum_i \sigma_i H_{ir}$$

This $(\Delta H_R)_r$ is for the reaction stoichiometry as written; if R_A is given, then we must divide $(\Delta H_R)_r$ by σ_A or R_A by σ_A, giving

$$(\Delta H_R)_r \frac{R_A}{\sigma_A} = (\Delta H_R)_r r$$

Suppose that the heat of reaction is given from experiment or tabulated in the units of r, ΔH_{RA} ($= $ J/mol of A); then, r to be used should be based on the above reaction rewritten as

$$A + 2B \rightarrow \tfrac{5}{2}C + 3D$$

However, never use ΔH_{RA} ($= $ J/mol of A); use $r = R_A/\sigma_A$ with $|\sigma_A| = 2$ from the first stoichiometric equation! This will mean that you have actually divided all of the stoichiometric numbers by the stoichiometric coefficient of A twice: once when you used ΔH_{RA} which is equal to $(\Delta H_R)_r/\sigma_A$ and once when you computed r to be equal to R_A/σ_A. This mistake is very common and many books and chemical engineers make this serious mistake. The reader should understand this point very clearly.

Note 3: The heat generation (or absorption) term will always have the sign of $(\Delta H_R)_r$ or ΔH_{RA}, because it is always multiplied by the generalized rate of reaction r, which is always positive.

2.6.1 Heat Balance for Multiple Reactions and the Special Case of a Single Reaction

Consider that we have a number of reactions taking place simultaneously. The heat of reaction in the heat balance equation should account for all the heats produced/absorbed by all the reactions.

If the number of components is M, the number of reactions is N; we use i as the component counter and j as the reaction counter. Therefore, the heat balance equation becomes

$$\sum_{i=1}^{M} n_{if}\{H_{if}(T_f, P_f, \pi_{if}) - H_{ir}(T_r, P_r, \pi_{ir})\} + Q$$

$$= \sum_{i=1}^{M} n_i\{H_i(T, P, \pi_i) - H_{ir}(T_r, P_r, \pi_{ir})\} + \sum_{j=1}^{N} r_j(\Delta H_R)_{rj} \qquad (2.84)$$

It is convenient to write these heat balance equations in a shorter form by realizing that the H_{if} is the enthalpy at T_f, P_f, π_{if} without writing them explicitly between brackets, and so on for other terms. Therefore, Eq. (2.84) can be rewritten as

$$\sum_{i=1}^{M} n_{if}(H_{if} - H_{ir}) + Q = \sum_{i=1}^{M} n_i(H_i - H_{ir}) + \sum_{j=1}^{N} r_j(\Delta H_R)_{rj}$$

An Illustrative Example

Steam reforming of methane includes the following three reactions:

$$CH_4 + H_2O \Leftrightarrow CO + 3H_2, \qquad \text{generalized rate of reaction is } r_1$$
$$CO + H_2O \Leftrightarrow CO_2 + H_2, \qquad \text{generalized rate of reaction is } r_2$$
$$CH_4 + 2H_2O \Leftrightarrow CO_2 + 4H_2, \quad \text{generalized rate of reaction is } r_3$$

We can label these components as follows: $CH_4 \equiv A$, $H_2O \equiv B$, $CO \equiv C$, $H_2 \equiv D$, $CO_2 \equiv E$, and Inerts $\equiv I$. Consider single-input, single-output. The mass balance equations are all obtained from the general equations discussed earlier:

$$n_i = n_{if} + \sum_{j=1}^{3} \sigma_{ij} r_j, \quad i \equiv A, B, C, D, E, I$$

Thus, the mass balance equations are simply

$$n_A = n_{Af} + \sigma_{A1} r_1 + \sigma_{A2} r_2 + \sigma_{A3} r_3 = n_{Af} - r_1 - r_3$$
$$n_B = n_{Bf} + \sigma_{B1} r_1 + \sigma_{B2} r_2 + \sigma_{B3} r_3 = n_{Bf} - r_1 - r_2 - 2r_3$$
$$n_C = n_{Cf} + \sigma_{C1} r_1 + \sigma_{C2} r_2 + \sigma_{C3} r_3 = n_{Cf} + r_1 - r_2$$
$$n_D = n_{Df} + \sigma_{D1} r_1 + \sigma_{D2} r_2 + \sigma_{D3} r_3 = n_{Df} + 3r_1 + r_2 + 4r_3$$
$$n_E = n_{Ef} + \sigma_{E1} r_1 + \sigma_{E2} r_2 + \sigma_{E3} r_3 = n_{Ef} + r_2 + r_3$$
$$n_I = n_{If} + \sigma_{I1} r_1 + \sigma_{I2} r_2 + \sigma_{I3} r_3 = n_{If}$$

The heat balance general equation is

$$\sum_{i=1}^{6} n_{if}(H_{if} - H_{ir}) + Q = \sum_{i=1}^{6} n_i(H_i - H_{ir}) + \sum_{j=1}^{3} r_j(\Delta H_R)_j$$

Thus, the heat balance equation (long and tedious, but very straightforward) is

$$n_{Af}(H_{Af} - H_{Ar}) + n_{Bf}(H_{Bf} - H_{Br}) + n_{Cf}(H_{Cf} - H_{Cr})$$
$$+ n_{Df}(H_{Df} - H_{Dr}) + n_{Ef}(H_{Ef} - H_{Er}) + n_{If}(H_{If} - H_{Ir})$$
$$+ Q = n_A(H_A - H_{Ar}) + n_B(H_B - H_{Br}) + n_C(H_C - H_{Cr})$$
$$+ n_D(H_D - H_{Dr}) + n_E(H_E - H_{Er}) + n_I(H_I - H_{Ir})$$
$$+ [r_1 \Delta H_{R1} + r_2 \Delta H_{R2} + r_3 \Delta H_{R3}]$$

2.6.2 The Most General Heat Balance Equation (for Multiple Reactions and Multiple-Input and Multiple-Output System Reactor with Multiple Reactions)

In this case, we just sum up the enthalpies of the input streams and the output streams. If we have F input streams (and we use l as the counter for the input streams) and K output streams (we use k as the counter for output streams) with L components and N reactions, then we have the following most general heat balance equation:

$$\left\{ \sum_{l=1}^{L} \left(\sum_{i=1}^{M} n_{if_l}(H_{if_l} - H_{ir}) \right) \right\} + Q = \left\{ \sum_{k=1}^{K} \left(\sum_{i=1}^{M} n_{i_k}(H_{i_k} - H_{ir}) \right) \right\}$$

$$+ \sum_{j=1}^{N} r_j (\Delta H_R)_{rj}$$

This is the most general heat balance equation for a multiple-input, multiple-output (MIMO), multiple reactions (and, of course, multicomponents) system.

2.7 SOLVED PROBLEMS FOR ENERGY BALANCE

The reader is advised first to solve each problem before inspecting the given solution.

Solved Example 2.8

Two liquid streams of carbon tetrachloride are to be evaporated to produce a vapor at 200°C and 1 atm. The first feed is at 1 atm and 30°C with a flow rate of 1000 kg/h and the second feed stream is at 70°C and 1 atm and has a flow rate of 500 kg/h. Calculate the heat that must be supplied to the evaporator.

The data given are as follows:

Carbon tetrachloride $\equiv CCl_4 \equiv$ component A
Molecular weight of $A = M_A = 153.84$
Normal boiling of $A = T_{BP_A} = 349.7$ K $= 76.7°C$
Latent heat of vaporization of A at boiling point $-\lambda_{.1} = 36,882.1$ J/mol
Specific heat in the vapor phase

$$C_{P_{AV}} = a' + b'T + c'T^2 + d'T^3 + e'T^4 \text{ J/mol K}$$

where $a' = 8.976$, $b' = 0.42$, $c' = -7.516 \times 10^{-4}$, $d' = 6.237 \times 10^{-7}$, and $e' = -1.998 \times 10^{-10}$

Specific heat in the liquid phase

$$C_{P_{AL}} = a + bT + cT^2 + dT^3 \, \text{J/mol K}$$

where $a = 12.284$, $b = 1.09475$, $c = -3.183 \times 10^{-3}$, and
$d = 3.425 \times 10^{-6}$.

Solution (see Fig. 2.27)

The heat balance on the evaporator gives

$$n_{Af_1} H_{Af_1} + n_{Af_2} H_{Af_2} + Q = n_A H_A \tag{2.85}$$

From the mass balance, we have

$$n_A = n_{Af_1} + n_{Af_2} \tag{2.86}$$

Substitution from Eq. (2.86) into Eq. (2.85) gives

$$n_{Af_1} H_{Af_1} + n_{Af_2} H_{Af_3} + Q = (n_{Af_1} + n_{Af_2}) H_A$$

Rearrangement gives

$$Q = n_{Af_1}(H_A - H_{Af_1}) + n_{Af_2}(H_A - H_{Af_2}) \tag{2.87}$$

where H_A is the enthalpy of carbon tetrachloride in the vapor phase at
200°C (473 K) and 1 atm, H_{Af_1} is the enthalpy of carbon tetrachloride in
the liquid phase at 30°C (303 K) and 1 atm; and H_{Af_2} is the enthalpy of
carbon tetrachloride in the liquid phase at 70°C (343 K) and 1 atm.
Therefore,

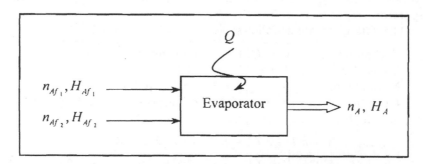

Figure 2.27   Heat flow diagram for solved example 2.8.

$$(H_A - H_{Af_1}) = \int_{303}^{349.7} C_{P_{AL}} \, dT + \lambda_A + \int_{349.7}^{473} C_{P_{AV}} \, dT$$

$$= \int_{303}^{349.7} (a + bT + cT^2 + dT^3) \, dT + \lambda_A$$

$$+ \int_{349.7}^{473} (a' + b'T + c'T^2 + d'T^3 + e'T^4) \, dT$$

$$= \left[aT + b\frac{T^2}{2} + c\frac{T^3}{3} + d\frac{T^4}{4} \right]_{303}^{349.7} + \lambda_A$$

$$+ \left[a'T + b'\frac{T^2}{2} + c'\frac{T^3}{3} + d'\frac{T^4}{4} + e'\frac{T^5}{5} \right]_{349.7}^{473}$$

Thus, we get

$$(H_A - H_{Af_1}) = a(349.7 - 303) + \frac{b}{2}[(349.7)^2 - (303)^2] + \frac{c}{3}[(349.7)^3 - (303)^3]$$

$$+ \frac{d}{4}[(349.7)^4 - (303)^4] + \underbrace{36882.1}_{\lambda_A} + a'(473 - 349.7)$$

$$+ \frac{b'}{2}[(473)^2 - (349.7)^2] + \frac{c'}{3}[(473)^3 - (349.7)^3]$$

$$+ \frac{d'}{4}[(473)^4 - (349.7)^4] + \frac{e'}{5}[(473)^5 - (349.7)^5]$$

By substituting the values of a, b, c, d a', b', c', d', and e' and performing the straightforward computation, compute the value of $H_A - H_{Af_1}$.

The reader is requested to perform this computation and obtain the value of $H_A - H_{Af_1}$; we will label it $\bar{A}$ here, that is,

$$H_A - H_{Af_1} \equiv \bar{A} \text{ J/mol}$$

Similarly, we can compute $H_A - H_{Af_2}$ (the reader is requested to perform this computation), we will label the value as $\bar{B}$,

$$H_A - H_{Af_2} \equiv \bar{B} \text{ J/mol}$$

Now, after computing the values of $\bar{A}$ and $\bar{B}$, the heat balance equation (2.87) can be rewritten as

$$Q = n_{Af_1}\bar{A} + n_{Af_2}\bar{B}$$

$$n_{Af_1} = \frac{1000}{M_A} = \frac{1000}{153.84} = 6.5 \text{ kmol/h} \Rightarrow n_{Af_1} = 6500 \text{ mol/h}$$

and

$$n_{Af_2} = \frac{500}{M_A} = \frac{500}{153.84} = 3.25 \text{ kmol/h} \Rightarrow n_{Af_2} = 3250 \text{ mol/h}$$

Thus,

$$Q = 6500 \, \bar{A} + 3250 \, \bar{B} \text{ J/h}$$

or

$$Q = 6.5 \, \bar{A} + 3.25 \, \bar{B} \text{ kJ/h}$$

For More Practice

The reader is encouraged to carry out the following tasks to get a better understanding of the concepts used in this section:

1. Solve the above problems by numbers and obtain the numerical values of $\bar{A}$, $\bar{B}$, and Q.
2. Solve the above problem if the second feed stream is fed at 85°C instead of 70°C. Compare the Q obtained with that of first case.

Solved Example 2.9

Compute the heat of reactions for the following reactions:

(a) Oxidation of sulfur at 600 K

$$S(s) + O_2(g) \rightarrow SO_2(g)$$

(b) Decomposition of propane at 500 K

$$C_3H_8 \rightarrow C_2H_2 + CH_4 + H_2$$

Solution

Part a: The heat of reaction at 600 K can be calculated as follows:

The reaction is:

$$S(s) + O_2(g) \rightarrow SO_2(g)$$

For simplicity write the reaction as, $A + B \rightarrow C$.

The heat of reaction can be written in terms of the heat of formation as follows:

$$(\Delta H_r)_{298} = \sum_{i=1}^{3} \sigma_i (\Delta H_{f_i})_{298_i}, \quad i = A, B, \text{ and } C$$

Thus, we get

$$(\Delta H_r)_{298} = \sigma_A (\Delta H_{f_A})_{298} + \sigma_B (\Delta H_{f_B})_{298} + \sigma_C (\Delta H_{f_C})_{298}$$

(We will no longer write 298 because it is well known and it is the temperature at which the standard tables are tabulated.) The superscript degree sign means at 298 K and 1 atm here and in Part b. Thus, we can write

$$\Delta H_r = - \underbrace{\Delta H_{f_A}^\circ}_{\text{zero}} - \underbrace{\Delta H_{f_B}^\circ}_{\text{zero}} + \Delta H_{f_C}^\circ$$

Thus, we get

$$\Delta H_r^\circ = \Delta H_{f_C}^\circ \quad \text{at } 25^\circ C \text{ (298 K)}$$

If we also want it evaluated at 600 K, we derive the relations

$$(\Delta H_r)_{298} = \sum_{i=1}^{3} \sigma_i (\Delta H_{f_i})_{298}$$

and

$$(\Delta H_r)_{600} = \sum_{i=1}^{3} \sigma_i (\Delta H_{f_i})_{600}$$

On subtracting the above two relations, we get

$$(\Delta H_r)_{600} - (\Delta H_r)_{298} = \sum_{i=1}^{3} \sigma_i \left\{ (\Delta H_{f_i})_{600} - (\Delta H_{f_i})_{298} \right\}$$

We can write

$$(\Delta H_r)_{600} = \Delta H_r^\circ + \sigma_A [(\Delta H_{f_A})_{600} - (\Delta H_{f_A})_{298}] + \sigma_B [(\Delta H_{f_B})_{600} \\ - (\Delta H_{f_B})_{298}] + \sigma_C [(\Delta H_{f_C})_{600} - (\Delta H_{f_C})_{298}]$$

which can be rewritten as

$$(\Delta H_r)_{600} = \Delta H_r^\circ - \left[\underbrace{\int_{298}^{T_{mp}} C_{PS_A} \, dT}_{\text{solid}} + \underbrace{\lambda_A}_{\text{transition}} + \underbrace{\int_{T_{mp}}^{600} C_{Pl_A} \, dT}_{\text{liquid}} \right] \\ - \int_{298}^{600} C_{P_B} \, dT + \int_{298}^{600} C_{P_C} \, dT$$

Part b: The heat of reaction at 500 K can be calculated as:

The reaction is:

$$C_3H_8 \rightarrow C_2H_2 + CH_4 + H_2$$

For simplicity write the reaction as, $A \rightarrow B + C + D$.

The heat of reaction at standard conditions can be written in terms of heats of formation as follows:

$$(\Delta H_r)_{298} = \sum_{i=1}^{4} \sigma_i (\Delta H_{f_i})_{298}, \quad i = A, B, C, \text{ and } D$$

Thus we get

$$\Delta H_r^\circ = \sigma_A \Delta H_{f_A}^\circ + \sigma_B \Delta H_{f_B}^\circ + \sigma_C \Delta H_{f_C}^\circ + \sigma_D \Delta H_{f_D}^\circ$$

On substituting the values, we get

$$\Delta H_r^\circ = -\Delta H_{f_A}^\circ + \Delta H_{f_B}^\circ + \Delta H_{f_C}^\circ + \Delta H_{f_D}^\circ$$

If we want it at 500 K, then we get

$$(\Delta H_r)_{500} - (\Delta H_r)_{298} = \sum_{i=1}^{4} \sigma_i\{(\Delta H_{f_i})_{500} - (\Delta H_{f_i})_{298}\}$$

We can write

$$(\Delta H_r)_{500} = \Delta H^\circ_r + \sigma_A[(\Delta H_{f_A})_{500} - (\Delta H_{f_A})_{298}) + \sigma_B[(\Delta H_{f_B})_{500}$$
$$- (\Delta H_{f_B})_{298}] + \sigma_C[(\Delta H_{f_C})_{500} - (\Delta H_{f_C})_{298})] + \sigma_D[(\Delta H_{f_D})_{500}$$
$$- (\Delta H_{f_D})_{298}]$$

which can be rewritten as

$$(\Delta H_r)_{600} = \Delta H^\circ_r - \int_{298}^{500} C_{P_A}\, dT + \int_{298}^{500} C_{P_B}\, dT + \int_{298}^{500} C_{P_C}\, dT$$
$$+ \int_{298}^{500} C_{P_D}\, dT$$

Solved Example 2.10

Ethylene is made commercially by dehydrogenating ethane:

$$C_2H_6(g) \Leftrightarrow C_2H_4(g) + H_2(g)$$
$$\Delta H_R(200°C) = 134.7 \text{ kJ/mol}$$

Ethane is fed to a continuous adiabatic reactor at $T_F = 2000°C$. Calculate the exit temperature that would correspond to 100% conversion ($x_A = 1.0$, where $A \equiv C_2H_6$). Use the following data for the heat capacities in your calculations:

$$C_{P_A} = 49.37 + 0.1392T$$
$$C_{P_B} = 40.75 + 0.1147T$$
$$C_{P_C} = 28.04 + 4.167 \times 10^{-3}T$$

where $A \equiv C_2H_6(g)$, $B \equiv C_2H_4(g)$, and $C \equiv H_2(g)$; C_{P_j} is in J/mol °C, and T is the temperature (in °C).

Solution (see Fig. 2.28)

For simplicity, let us represent the reaction as

$$A \Leftrightarrow B + C$$

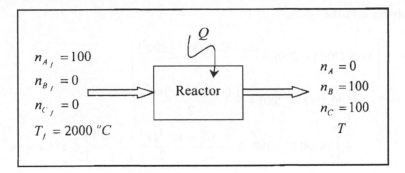

Figure 2.28 Energy flow diagram for solved example 2.10.

We take the basis as 100 mol/min. The generalized heat balance equation is

$$\left\{\sum n_{i_f}(H_{i_f} - H_{ir})\right\} + Q = \left\{\sum n_i(H_i - H_{ir})\right\} + (\Delta H)_R r$$

Based on the basis and the rate of reaction definition in mass balance, $r = 100$ mol/min. Take the reference temperature as 200°C; thus, we get

$$100 \int_{200}^{2000} C_{P_A} \, dT + \underbrace{Q}_{\text{zero}} = 100 \int_{200}^{T} C_{P_B} \, dT + 100 \int_{200}^{T} C_{P_C} \, dT$$

$$+ (134.7 \times 10^3)(100)$$

On substituting the values, we get

$$100 \int_{200}^{2000} (49.37 + 0.1392T) \, dT = 100 \int_{200}^{T} (40.75 + 0.1147T) \, dT$$

$$+ 100 \int_{200}^{T} (28.04 + 4.167 \times 10^{-3} T) \, dT$$

$$+ (134.7 \times 10^3)(100)$$

Solving the integrals gives

$$\left[49.37T + \frac{0.1392T^2}{2}\right]_{200}^{2000} = \left[40.75T + \frac{0.1147T^2}{2}\right]_{200}^{T}$$

$$+ \left[28.04T + \frac{4.167 \times 10^{-3} T^2}{2}\right]_{200}^{T}$$

$$+ 134.7 \times 10^3$$

Solving it further gives

$$
\left[49.37(2000 - 200) + \frac{0.1392(2000^2 - 200^2)}{2} \right]
$$
$$
= \left[40.75(T - 200) + \frac{0.1147(T^2 - 200^2)}{2} \right]
$$
$$
+ \left[28.04(T - 200) + \frac{4.167 \times 10^{-3}(T^2 - 200^2)}{2} \right] + 134.7 \times 10^3
$$

Mathematical calculations give

$$
(5.93835 \times 10^{-5})T^2 + 0.06879T - 232.15 = 0
$$

Solution of this quadratic equation gives

$$
T = \frac{-0.06879 \pm \sqrt{(0.06879)^2 + 4(5.93835 \times 10^{-5})(232.15)}}{2(5.93835 \times 10^{-5})}
$$

Neglecting the negative solution, we get

$$
T = \frac{-0.068709 + 0.2446734}{1.1876} \times 10^4
$$

Thus, $T = 1480.998°C$ (the answer)

Solved Example 2.11

Acetic acid is cracked in a furnace to produce the intermediate ketene via the reaction

$$
CH_3COOH(l) \rightarrow CH_2CO(g) + H_2O(g)
$$

The reaction

$$
CH_3COOH(l) \rightarrow CH_4(g) + CO_2(g)
$$

also occurs to an appreciable extent. It is desired to carry out cracking at 700°C with a conversion of 80% and a fractional yield of methane (CH_4) of 0.0722. Calculate the required furnace heating rate for a furnace feed of 100 kmol/h acetic acid. The feed is at 300°C.

The data given are as follows:
Heats of formation at standard conditions (25°C, 1 atm):

$$\Delta H^{\circ}_{f_{CH_2CO}} = -14.6 \frac{kcal}{g\ mol}$$

$$\Delta H^{\circ}_{f_{H_2O}} = -57.8 \frac{kcal}{g\ mol}$$

$$\Delta H^{\circ}_{f_{CH_3COOH}} = -103.93 \frac{kcal}{g\ mol}$$

$$\Delta H^{\circ}_{f_{CH_4}} = -17.89 \frac{kcal}{g\ mol}$$

$$\Delta H^{\circ}_{f_{CO_2}} = -94.05 \frac{kcal}{g\ mol}$$

The specific heats are given by the following relations (J/gmol K):

$$C_{P_{H_2O(l)}}(\text{liquid water}) = 18.2964 + 0.04721T - 0.001338T^2$$
$$- 1.3 \times 10^{-6}\ T^3$$

$$C_{P_{H_2O(g)}} = 34.047 + 9.65 \times 10^{-3}T - 3.2988 \times 10^{-5}T^2$$
$$- 2.044 \times 10^{-8}T^3 + 4.3 \times 10^{-12}T^4$$

$$C_{P_{CO_2(g)}} = 19.02 + 7.96 \times 10^{-2}T - 7.37 \times 10^{-5}T^2$$
$$+ 3.745 \times 10^{-8}T^3 - 8.133 \times 10^{-12}T^4$$

$$C_{P_{CH_4(g)}} = 3.83 - 7.366 \times 10^{-2}T + 2.909 \times 10^{-4}T^2 - 2.638$$
$$\times 10^{-7}T^3 + 8 \times 10^{-11}T^4$$

$$C_{P_{CH_3COOH(l)}} = -36.08 + 0.06T - 3.9 \times 10^{-5}T^2 - 5.6 \times 10^{-7}T^3$$

$$C_{P_{CH_3COOH(g)}} = 14.6 + 2.3 \times 10^{-2}T - 11.02 \times 10^{-5}T^2 + 2.6$$
$$\times 10^{-9}T^3 - 2.8 \times 10^{-13}T^4$$

$$C_{P_{CH_2CO(g,l)}} = 4.11 + 2.966 \times 10^{-2}T - 1.793 \times 10^{-5}T^2 + 4.22$$
$$\times 10^{-9}T^3$$

Boiling point of acetic acid = 118.1°C
Boiling point of ketene = T_K
Latent heat of vaporization of acetic acid = 96.76 cal/g
Latent heat of vaporization of ketene = λ_k cal/g

Solution (see Fig. 2.29)

In order to carry out the necessary heat balance, we must first compute the input and output streams and their composition. This means that we should formulate and solve the mass balance problem first.

Mass Balance   The two reactions can be written as

$$A \to B + C \text{ with the generalized rate, } r_1$$

and

$$A \to D + E \text{ with the generalized rate, } r_2$$

The feed is pure acetic acid with a feed rate of $n_{Af} = 100$ kmol/h.

Degrees of Freedom   To make sure that the problem is solvable we first check the degree of freedom.

The number of unknowns $= 5(n_A, n_B, n_C, n_D, n_E) + 2(r_1, r_2) = 7$
Number of mass balance equations $= 5$
Number of specifications $(x_A, \overline{Y}_D) = 2$
Total number of equations and specifications available $= 5 + 2 = 7$
Degree of freedom $= 7 - 7 = 0$

Thus, the mass balance problem is solvable. Note that n_A can be easily computed from n_{Af} and the given conversion:

$$x_A = \frac{n_{Af} - n_A}{n_{Af}}$$

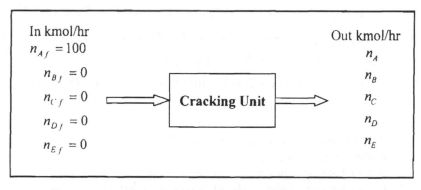

Figure 2.29   Molar flow diagram for solved example 2.11.

Thus, we get

$$0.8 = \frac{100 - n_A}{100}$$

and the value of n_A is equal to

$$n_A = 20 \frac{kmol}{h}$$

n_D can be easily computed from the fractional yield of methane, $\overline{Y}_D$ (kmol of D produced per kmol of A reacted):

$$\overline{Y}_D = 0.0722 = \frac{n_D - n_{Df}}{n_{Af} x_A} = \frac{n_D}{80}$$

which gives

$$n_D = 80 \times 0.0722 = 5.776 \text{ kmol/h}$$

The rest of the components are calculated from their mass balance equations

$$n_B = n_{Bf} + \sigma_{B1} r_1 + \sigma_{B2} r_2$$

which gives

$$n_B = r_1$$

Also,

$$n_C = r_1$$

and, similarly,

$$n_E = r_2$$

r_1 and r_2 can be easily obtained from the mass balance equations for components A and D. The mass balance for A gives

$$n_A = n_{Af} + \sigma_{A1} r_1 + \sigma_{A2} r_2$$

On substituting the values, we get

$$20 = 100 - r_1 - r_2$$

Therefore,

$$r_1 + r_2 = 80 \text{ kmol/h} \tag{2.88}$$

The mass balance for D gives

$$n_D = n_{Df} + \sigma_{D1} r_1 + \sigma_{D2} r_2$$

On substituting the values, we get

$$5.776 = 0 + 0 + r_2$$

or we can write

$$r_2 = 5.776 \text{ kmol/h}$$

Using this value of r_2 in Eq. (2.88), we get

$$r_1 = 80 - 5.776 = 74.224 \text{ kmol/h}$$

Now, all input streams and output streams are known and we can carry out the heat balance to find Q.

Heat Balance (see Fig. 2.30)   The heat balance equation is given by

$$\left\{ \sum n_{if} \left[H_{if}(T_f, P_f, \pi_f) - H_{ir}(T_r, P_r, \pi_r) \right] \right\} + Q$$

$$= \left\{ \sum n_i [H_i(T, P, \pi) - H_{ir}(T_r, P_r, P_r, \pi_r)] \right\} + \sum_{j=1}^{2} r_j (\Delta H_R)_{rj}$$

Take the reference condition to be 25°C and 1 atm (standard conditions). The first term on the left-hand side of the above equation can thus be written as

$$\sum n_{if}[H_{if} - H_{ir}] = n_{Af}(H_{Af} - H_{Ar}) + 0 + 0 + 0 + 0$$

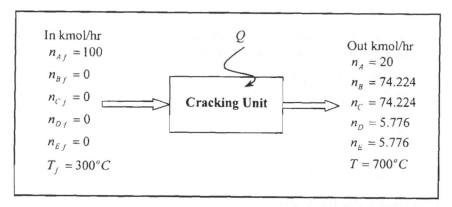

Figure 2.30   Molar flow diagram (with temperature shown) for heat balance for solved example 2.11.

Which can be rewritten as

$$(100 \times 1000)\left[\left[\int_{25+273}^{118.1+273} C_{P_{CH_3COOH(l)}}\, dT + \lambda_{CH_3COOH}\right.\right.$$

$$\left.+ \int_{118.1+273}^{300+273} C_{P_{CH_3COOH(g)}}\, dT\right] \equiv I_f$$

Now, also, the first term on the right-hand side of the heat balance equation can be written as:

$$\sum n_i[H_i - H_{ir}] = \left\{ \begin{array}{l} \underbrace{n_A}_{\substack{20 \quad 1000}} \left[\int_{25+273}^{118.1+273} C_{P_{CH_3COOH(l)}}\, dT + \lambda_{CH_3COOH} + \int_{118.1+273}^{700+273} C_{P_{CH_3COOH(g)}}\, dT\right] \\[4pt] + n_B \left[\int_{25+273}^{118.1+273} C_{P_{CH_2CO(l)}}\, dT + \lambda_{CH_2CO} + \int_{T_K}^{700+273} C_{P_{CH_2CO(g)}}\, dT\right] \\[4pt] + n_C \left[\int_{25+273}^{100+273} C_{P_{H_2O(l)}}\, dT + \lambda_{H_2O} + \int_{100+273}^{700+273} C_{P_{H_2O(g)}}\, dT\right] \\[4pt] + n_D \left[\int_{25+273}^{700+273} C_{P_{CH_4(g)}}\, dT\right] + n_E \left[\int_{25+273}^{700+273} C_{P_{CO_2(g)}}\, dT\right] \end{array} \right\} \equiv I$$

Let the right-hand side of the above equation be equal to some value I. After computing I_f and I as shown above (note that reader has to do the necessary units conversion to keep all the units consistent), the heat balance equation is

$$I_f + Q = I + r_1(\Delta H_{R1}) + r_2(\Delta H_{R2})$$

ΔH_{R1} and ΔH_{R2} are to be computed from the heats of formation:

$$\Delta H_{R1} = \Delta H^\circ_{f_B} + \Delta H^\circ_{f_C} - \Delta H^\circ_{f_A}$$

and

$$\Delta H_{R2} = \Delta H^\circ_{f_D} + \Delta H^\circ_{f_E} - \Delta H^\circ_{f_A}$$

Thus, Q can be easily computed:

$$Q = I - I_f + r_1(\Delta H_{R1}) + r_2(\Delta H_{R2})$$

The reader should perform the above computation to get the numerical value of the result.

More Practice for the Reader

Solve the above problem for 90% acetic acid conversion and a fractional yield of methane of 0.08 and a reactor temperature of 800°C.

Solved Example 2.12

Consider the two-stage ammonia synthesis loop shown in Figure 2.31. In the flowsheet, a cold stoichiometric feed of N_2 and H_2 is introduced between the

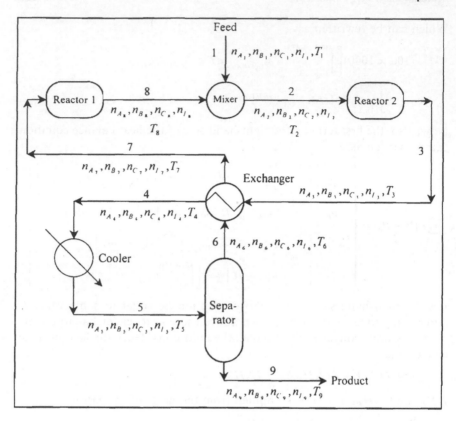

Figure 2.31 Flowsheet of solved example 2.12.

reactor stages. In each reactor, the following synthesis reaction takes place with specific conversion.

$$N_2 + 3H_2 \Leftrightarrow 2NH_3$$

or we can represent it by $A + 3B \Leftrightarrow C$ for simplification.

The effluent of the second reactor stage is cooled by exchanging heat with the input stream to the first reactor stage. After further cooling, stream 5 is separated to recover a product stream containing all of the NH_3 (C) and some $N_2(A)$ and $H_2(B)$. The following specification are given:

1. The feed rate (in stream 1) is 97.8 mol/h with 15% N_2 (A) and 75% H_2 (B) and the feed temperature is 50°C.
2. The input stream to the first reactor stage (stream 7) is at 425°C.

3. Input stream to separator (stream 5) is at 50°C and the temperatures of the outlet streams (streams 6 and 9) are equal to each other.
4. The recycle stream (stream 6) is to contain no NH_3 (C) and 0.9945 each of the N_2 (A) and H_2(B) fed to the separator. This stream contains no inerts.
5. The conversion of N_2 (A) is 10% in stage 1 and 12.33% in stage 2.
6. All of the units are adiabatic except the cooler, which is non-adiabatic.

Assume all streams are gas phase and that pressure effects are neglected.

   a. Show that the problem is correctly specified.
   b. Outline a calculation procedure.
   c. Solve the problem to obtain the temperatures and molar flow rates of all components in all streams.
   d. Construct a flowchart for a computer program for your solution.

Solution

We will start the solution of this problem with the degrees-of-freedom analysis.

First, let us identify all of the given values and specified relations for the problem:

1. $n_{Total_1} = n_{T_1} = 97.8$ mol/h
2. $Y_{A_1} = 0.15$ (molar fraction of component A in stream 1)
3. $Y_{B_1} = 0.75$ (molar fraction of component B in stream 1)
4. $T_1 = 50 + 273 = 323$ K
5. $n_{C_1} = 0$
6. $T_7 = 425 + 273 = 698$ K
7. $T_5 = 323$ K
8. $T_6 = T_9$
9. $n_{C_6} = 0$
10. $n_{A_6} = 0.9945\, n_{A_5}$
11. $n_{B_6} = 0.9945\, n_{B_5}$
12. $x_{A_1} = \dfrac{n_{A_7} - n_{A_8}}{n_{A_7}} = 0.1$
13. $x_{A_2} = \dfrac{n_{A_2} - n_{A_3}}{n_{A_2}} = 0.1233$
14. $n_{I_6} = 0$, where $I \equiv$ inerts

Thus, we have 14 specified relations.

Process Degrees of Freedom   Total number of unknowns,

Reactor 1:   11
Mixer: 10
Reactor 2:   6
Exchanger: 10
Cooler: 6
Separator: 5
Total: 48

Mass and Heat Balance Relations

Reactor 1:   5
Mixer: 5
Reactor 2:   5
Separator: 5
Cooler: 5
Exchanger: 9
Total: 34

Note that the mass balance relations for the cooler (four in number) and exchanger (eight in number) are described by the following relations:
For one side of the exchanger,

$$n_{i_3} = n_{i_4}, \quad i = A, B, C, \text{and } I$$

Thus, giving four different relations.
For the other side of the exchanger,

$$n_{i_6} = n_{i_7}, \quad i = A, B, C, \text{and } I$$

Thus, giving four different relations.
For the cooler,

$$n_{i_4} = n_{i_5}$$
$$i = A, B, C, \text{ and } I$$

Thus, giving four relations.
Given specified relations = 14(9 + 5 as shown above)

So we have

Total known/specified relations = 34 + 14 = 48
Total unknowns = 48

Thus,

$$\text{Degrees of freedom} = 48 - 48 = 0$$

As the degrees of freedom is equal to zero, this problem is correctly specified and hence is solvable.

Degrees of Freedom (DF) for Each Unit Separately

Reactor 1

> Number of unknowns $= 10 + 1 = 11$
> Number of specified relations

$$T_7 = 698 \text{ K} \quad \text{and} \quad x_{A_1} = \frac{n_{A_7} - n_{A_8}}{n_{A_7}} = 0.1$$

> so specified relations $= 2$
> Mass and heat balance equations $= 5$
> $DF = 11 - (5 + 2) = 11 - 7 = 4$

Mixer

> Number of unknowns $= 15$
> Number of specified relations

$$n_{T_1} = 97.8 \text{ mol/h}, \quad Y_{A_1} = 0.15, \quad Y_{B_1} = 0.75, \quad n_{C_1} = 0,$$
$$T_1 = 323 \text{ K}$$

> so specified relations $= 5$
> Mass and heat balance equations $= 5$
> $DF = 15 - (5 + 5) = 15 - 10 = 5$

Reactor 2

> Number of unknowns $= 11$
> Number of specified relations

$$x_{A_2} = \frac{n_{A_2} - n_{A_3}}{n_{A_2}} = 0.1233$$

> so total specified relations $= 1$
> Mass and heat balance equations $= 5$
> $DF = 11 - (5 + 1) = 11 - 6 = 5$

Exchanger

> Number of unknowns $= 20$
> Number of specified relations

$$T_7 = 698 \text{ K} \quad \text{and} \quad n_{C_6} = 0$$

> so total specified relations $= 2$

Number of mass balance relations $= 8$
Number of heat balance relations $= 1$
DF $= 20 - (2 + 8 + 1) = 20 - 11 = 9$

Cooler

Number of unknowns $= 11$
Number of specified relations

$\quad T_5 = 323$ K

so total specified relations $= 1$
Number of mass and heat balance relations $= 5$
DF $= 11 - (1 + 5) = 11 - 6 = 5$

Separator

Number of unknowns $= 15$
Number of specified relations

$$n_{C_6} = 0, \quad n_{A_6} = 0.9945 n_{A_5}, \quad n_{B_6} = 0.9945 n_{B_5}, \quad T_5 = 323 \text{ K},$$
$$T_6 = T_0, \quad n_{I_6} = 0$$

so total specified relations $= 6$
Number of mass and heat balance relations $= 5$
DF $= 15 - (6 + 5) = 4$

Overall Mass Balance (see Figs. 2.32 and 2.33)

Number of unknowns $= 10 + 2 + 1 = 13$

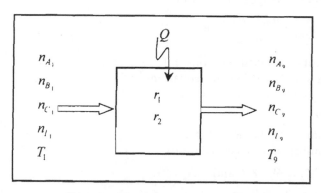

Figure 2.32   Energy flow diagram.

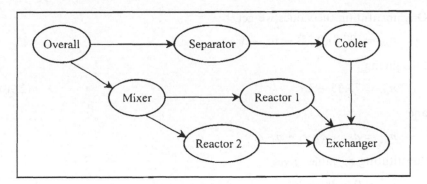

Figure 2.33 Showing solution strategy: where to start and end.

Number of specified relations

$$n_{T_1} = 97.8 \text{ mol/h}, \quad Y_{A_1} = 0.15, \quad Y_{B_1} = 0.75, \quad T_1 = 323 \text{ K},$$
$$n_{I_9} = 0.978$$

so total specified relations = 5
Number of mass and heat balance relations = 5
DF = 13 − (5 + 5) = 13 − 10 = 3

Note that

r_1 = rate of reaction in reactor 1
r_2 = rate of reaction in reactor 2
$r_1 + r_2 = R$

The overall balance for component A is

$$n_{A_9} = n_{A_1} + \sigma_{A_1} r_1 + \sigma_{A_2} r_2$$

which can be written as

$$n_{A_9} = n_{T_1} Y_{A_1} - r_1 - r_2$$

On substituting the values, we get

$$n_{A_9} = (97.8)(0.15) - r_1 - r_2$$

We finally get

$$n_{A_9} = 14.67 - R \tag{2.89}$$

Similarly,

$$n_{B_9} = n_{B_1} + \sigma_{B_1} r_1 + \sigma_{B_2} r_2$$

On substituting the values we get,

$$n_{B_9} = (97.8)(0.75) - 3r_1 - 3r_2$$

thus giving

$$n_{B_9} = 73.35 - 3R \tag{2.90}$$

and

$$n_{C_9} = n_{C_1} + \sigma_{C_1} r_1 + \sigma_{C_2} r_2$$

Substituting of values gives

$$n_{C_9} = 0 + 2r_1 + 2r_2$$

thus giving

$$n_{C_9} = 2R \tag{2.91}$$

and

$$n_{I_9} = n_{I_1}$$

Substitution of values gives

$$n_{I_9} = (97.8)(0.1)$$

and we get

$$n_{I_9} = 9.78 \tag{2.92}$$

Heat balance

$$\left\{ \sum n_{if}(H_{if} - H_{ir}) \right\} + Q = \left\{ \sum n_i(H_i - H_{ir}) \right\} + r_1 \Delta H_r + r_2 \Delta H_r$$

Take the reference as $25°C = 25 + 273 = 298$ K

Thus, we get

$$\left\{ n_{A_1} \int_{298}^{323} C_{P_A}\, dT + n_{B_1} \int_{298}^{323} C_{P_B}\, dT + n_{C_1} \int_{298}^{323} C_{P_C}\, dT + n_{I_1} \int_{298}^{323} C_{P_I}\, dT \right\} + Q$$

$$= \left\{ n_{A_9} \int_{298}^{T_9} C_{P_A}\, dT + n_{B_9} \int_{298}^{T_9} C_{P_B}\, dT + n_{C_9} \int_{298}^{T_9} C_{P_C}\, dT + n_I \int_{298}^{T_9} C_{P_I}\, dT \right\}$$

$$+ R(\Delta H_r)$$

In the above equation, every value is known except R, Q, and T_9. Therefore, the above is a relation between R, Q, and T_9. Thus any of them can be expressed in term of the other two; for example. we can write $T_9 = F_1(R, Q)$. Also, all n_{i_9} (n_{A_9}, n_{B_9}, n_{C_9}) are expressed in terms of R only. After we have expressed five unknowns in the exit stream number 9 (as well as Q), all in terms of R and Q, we can move to the separator as shown in the following step.

Separator (see Fig. 2.34)

   Number of unknowns,

   $n_{i_5}, n_{i_6}, T_5, T_6, R,$ and $Q = 12$ unknowns

   Number of specified relations,

   $n_{C_6} = 0, \quad n_{A_6} = 0.9945 n_{A_5}, \quad n_{B_6} = 0.9945 n_{B_5}, \quad T_5 = 323$ K,
   $T_6 = T_9, \quad n_{I_6} = 0$

   so total specified relations $= 6$
   Number of mass and heat balance relation $= 5$
   DF $= 12 - (6 + 5) = 1$

Mass Balance

   $n_{A_5} = n_{A_6} + n_{A_9}$

On substituting the values, we get

   $n_{A_5} = 0.9945 n_{A_5} + 14.67 - R$

Finally, it gives

   $n_{A_5} = 2667.27 - 181.8\, R$ $\qquad\qquad$ (2.93)

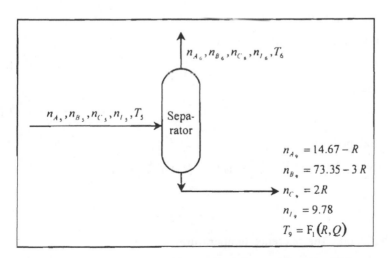

Figure 2.34 Mass and energy flow diagram for separator.

Also, we have
$$n_{A_6} = 0.9945\, n_{A_5}$$
Thus, we get
$$n_{A_6} = 2652.6 - 180.8\, R \tag{2.94}$$
For component B, we get
$$n_{B_5} = n_{B_6} + n_{B_9}$$
On substituting the values, we finally get
$$n_{B_5} = 13336.36 - 545.454R \tag{2.95}$$
Also, we have
$$n_{B_6} = 13263.01 - 542.45R \tag{2.96}$$
For component C, we get
$$n_{C_5} = n_{C_6} + n_{C_9}$$
On substituting the values, we get
$$n_{C_5} = 2R \tag{2.97}$$
Also, we have specified
$$n_{C_6} = 0 \tag{2.98}$$
For component I (the inert), we get
$$n_{I_5} = n_{I_6} + n_{I_9}$$
On substituting the values, we get
$$n_{I_5} = 9.78 \tag{2.99}$$
Also, we get
$$n_{I_6} = 0 \tag{2.100}$$

Heat Balance
$$\sum n_{i_5}(H_{i_5} - H_{ir}) = \sum n_{i_6}(H_{i_6} - H_{ir}) + \sum n_{i_9}(H_{i_9} - H_{ir})$$
We have
$$T_6 = T_9$$
So, let us take the reference temperature
$$T_{\text{ref}} = T_6 = T_9$$

Thus, the right-hand side of the above-mentioned heat balance becomes zero, and we get

$$\sum n_{i_5}(H_{i_5} - H_{i_6}) = 0$$

This can be rewritten as

$$n_{A_5}\int_{T_6}^{323} C_{P_A}\, dT + n_{B_5}\int_{T_6}^{323} C_{P_B}\, dT + n_{C_5}\int_{T_6}^{323} C_{P_C}\, dT + n_{I_5}\int_{T_6}^{323} C_{P_I}\, dT = 0$$

On substituting the values of n_{A_5}, n_{B_5}, n_{C_5}, and n_{I_5} from Eqs. 2.93, 2.95, 2.97, and 2.99, we get

$$(2667.27 - 181.8R)\int_{T_0}^{323} C_{P_A}\, dT + (13336.36 - 545.45R)\int_{T_6}^{323} C_{P_B}\, dT$$

$$+ 2R\int_{T_6}^{323} C_{P_C}\, dT + 9.78\int_{T_6}^{323} dT = 0$$

This gives a direct relation between R and T_6 as follows:

$$T_6 = T_9 = F_2(R)$$

As we had established earlier,

$$T_9 = F_1(R, Q)$$

Therefore, we say

$$F_1(R, Q) = F_2(R)$$

which can be written as

$$Q = F_3(R)$$

Cooler (see Fig. 2.35) Now we can look at the cooler as having only the five unknowns of stream 4 and all other variables are in terms of R.

> Number of unknowns $= 5 + R = 6$
> Number of specified relations $= 0$
> Number of mass and heat balance relations $= 5$
> DF $= 6 - 5 = 1$

Mass Balance

The mass balance for each component can be simply written as

$$n_{A_4} = n_{A_5} = 2667.27 - 181.8\, R$$
$$n_{B_4} = n_{B_5} = 13336.36 - 545.45\, R$$
$$n_{C_4} = n_{C_5} = 2\, R$$
$$n_{I_4} = n_{I_5} = 9.78$$

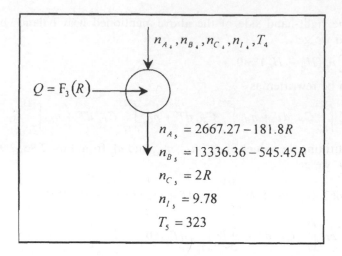

Figure 2.35 Mass and energy flow diagram for the cooler.

Heat Balance

The generalized heat balance equation is

$$\sum n_{i_4}(H_{i_4} - H_{ir}) + Q = \sum n_{i_5}(H_{i_5} - H_{ir})$$

Take the reference temperature as

$$T_{ref} = T_4$$

We get

$$Q = \sum n_{i_5}(H_{i_5} - H_{i_4})$$

Calculations give

$$Q = n_{A_5} \int_{T_4}^{323} C_{P_A} \, dT + n_{B_5} \int_{T_4}^{323} C_{P_B} \, dT + n_{C_5} \int_{T_4}^{323} C_{P_C} \, dT + n_{I_5}$$
$$\int_{T_4}^{323} C_{P_I} \, dT$$

On substituting the values of $n_{A_5}, n_{B_5}, n_{C_5}$, and n_{I_5} from Eqs. (2.93), (2.95), (2.97), and (2.99), we get

$$(2667.27 - 181.8R) \int_{T_4}^{323} C_{P_A} \, dT + (13336.36 - 545.45\,R)$$

$$\int_{T_4}^{323} C_{P_B} \, dT + 2R \int_{T_4}^{323} C_{P_C} \, dT + 9.78 \int_{T_4}^{323} C_{P_I} \, dT = Q$$

As $Q = F_3(R)$ and all the above terms involve either R or T_4, we can write

$$T_4 = F_4(R)$$

Thus, we have $n_{A_5}, n_{B_5}, n_{C_3}$, and n_{I_5}, all functions of R, and $Q = F_3(R)$, together with $T_4 = F_4(R)$. Thus, whenever we compute R, we will get the values of Q and T_4, and back substitution will give us the values of all the variables.

Now we move to the other path as shown in Figure 2.33.

Mixer (see Fig. 2.36)

Mass Balance
Mass balances for all the components give the following:

For component A:

$$n_{A_1} + n_{A_8} = n_{A_2}$$

On substituting the values, we get

$$n_{A_2} = 14.67 + n_{A_8}$$

For component B:

$$n_{B_1} + n_{B_8} = n_{B_2}$$

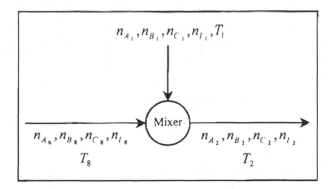

Figure 2.36 Mass and energy flow diagram for the mixer.

On substituting the values, we get

$$n_{B_2} = 73.35 + n_{B_8}$$

For component C:

$$n_{C_1} + n_{C_8} = n_{C_2}$$

On substituting the values, we get

$$n_{C_2} = n_{C_8}$$

For component I:

$$n_{I_1} + n_{I_8} = n_{I_2}$$

On substituting the values, we get

$$n_{I_2} = 9.78 + n_{I_8}$$

Heat Balance

The generalized heat balance equation is

$$\sum n_{i_1}(H_{i_1} - H_{ir}) + \sum n_{i_8}(H_{i_8} - H_{ir}) = \sum n_{i_2}(H_{i_2} - H_{ir})$$

Take the reference temperature as

$$T_{\text{ref}} = T_2$$

We get

$$\sum n_{i_1}(H_{i_1} - H_{i_2}) + \sum n_{i_8}(H_{i_8} - H_{i_2}) = 0$$

Calculations give

$$14.67 \int_{T_2}^{323} C_{P_A}\, dT + 73.35 \int_{T_2}^{323} C_{P_B}\, dT + 0 + 9.78 \int_{T_2}^{323} C_{P_I}\, dT$$

$$+ n_{A_8} \int_{T_2}^{T_8} C_{P_A}\, dT + n_{B_8} \int_{T_2}^{T_8} C_{P_B}\, dT + n_{C_8} \int_{T_2}^{T_8} C_{P_C}\, dT + n_{I_8} \int_{T_2}^{T_8} C_{P_I}\, dT =$$

Thus, we get

$$T_2 = F_5(T_8, n_{i_8}), \quad i = A, B, C, \text{and } I$$

so, we can write

$$T_2 = F_5(T_8, n_{i_8})$$

As n_{i_8} and n_{i_2} are related, we can write

$$T_2 = F_6(T_8, n_{i_2})$$

Reactor 2 (see Fig. 2.37)

For component A, we get

$$n_{A_3} = n_{A_2} + \sigma_{A_2} r_2$$

On substituting the values (note that $\sigma_{A_2} = -1$), we get

$$n_{A_3} = 14.67 + n_{A_8} - r_2 \qquad (2.101)$$

For component B, we get

$$n_{B_3} = n_{B_2} + \sigma_{B_2} r_2$$

On substituting the values (note that $\sigma_{B_2} = -3$), we get

$$n_{B_3} = 73.35 + n_{B_8} - 3r_2 \qquad (2.102)$$

For component C, we get

$$n_{C_3} = n_{C_2} + \sigma_{C_2} r_2$$

On substituting the values (note that $\sigma_{C_2} = 2$), we get

$$n_{C_3} = n_{C_8} + 2r_2 \qquad (2.103)$$

For component I, we get

$$n_{I_3} = n_{I_2} + \sigma_{I_2} r_2$$

On substituting the values (note that $\sigma_{I_2} = 0$), we get

$$n_{I_3} = 9.78 + n_{I_8} \qquad (2.104)$$

Now, we make use of the conversion value given in the problem:

$$x_{A_2} = 0.1233 = \frac{n_{A_2} - n_{A_3}}{n_{A_3}}$$

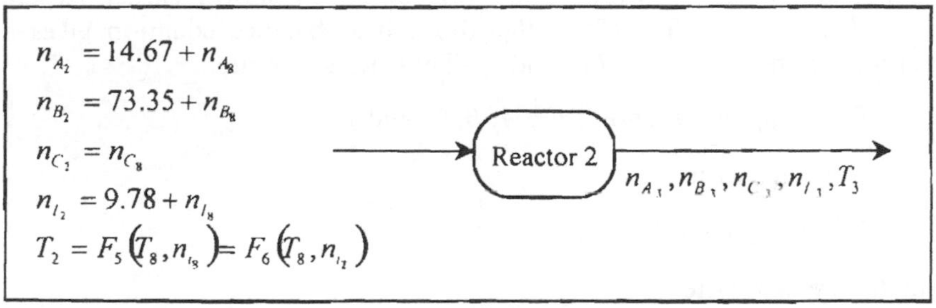

Figure 2.37   Mass and energy flow diagram for reactor 2.

On substituting the values, we get

$$0.1233 = \frac{(14.67 + n_{A_8}) - n_{A_3}}{n_{A_3}}$$

Using the mass balance relations derived above, we get

$$0.1233 = \frac{(14.67 + n_{A_8}) - (14.67 + n_{A_8} - r_2)}{(14.67 + n_{A_8} - r_2)}$$

which can be simplified to

$$r_2 = 1.61 + 0.11 n_{A_8} \tag{2.105}$$

Heat Balance

The generalized heat balance equation is

$$\sum n_{i_2}(H_{i_2} - H_{ir}) + \underbrace{Q}_{\text{zero}} = \sum n_{i_3}(H_{i_3} - H_{ir}) + r_2(\Delta H_r)$$

Take the reference temperature as

$$T_{\text{ref}} = 25°C = 298 \text{ K}$$

We can write

$$n_{A_2} \int_{298}^{T_2} C_{P_A} dT + n_{B_2} \int_{298}^{T_2} C_B dT + n_{C_2} \int_{298}^{T_2} C_{P_C} dT + n_{I_2} \int_{298}^{T_2} C_{P_I} dT$$

$$= n_{A_3} \int_{298}^{T_3} C_{P_A} dT + n_{B_3} \int_{298}^{T_3} C_{P_B} dT + n_{C_3} \int_{298}^{T_3} C_{P_C} dT + n_{I_3}$$

$$\int_{298}^{T_3} C_{P_I} dT + r_2(\Delta H_r)$$

Note that all n_{i_2} can be put in terms of n_{i_8} as shown earlier (where $i = A, B, C$, and I). Therefore, the above heat balance equation gives a relation among $n_{i_8}, n_{i_3}, T_2, T_3$, and r_2. Thus, we can write

$$T_3 = G(n_{i_8}, n_{i_3}, T_2, r_2), \quad i = A, B, C, \text{ and } I$$

We can further write

$$T_3 = G(n_{i_8}, n_{i_3}, F_5(T_8, n_{i_8}), r_2)$$

Finally, we can write

$$T_3 = F_7(n_{i_8}, n_{i_3}, T_8, r_2)$$

However, we already know

$$n_{i_3} = \hat{F}(n_{i_8}, r_2)$$

Thus, we get

$$T_3 = F_8(n_{i_{i}}, T_8, r_2)$$

Then, from Eq. (2.105), we have

$$r_2 = \tilde{F}(n_{A_8})$$

Finally, we get

$$T_3 = F_9(n_{i_8}, T_8)$$

Reactor 1 (See Fig. 2.38)   For component A, we get

$$n_{A_8} = n_{A_7} + \sigma_{A_1} r_1$$

On substituting the values (note that $\sigma_{A_1} = -1$), we get

$$n_{A_8} = n_{A_7} - r_2$$

This can be rewritten as

$$n_{A_7} = n_{A_8} + r_2 \tag{2.106}$$

In the very same way, we can write the mass balance equations for components B, C, and I, we get

$$n_{B_7} = n_{B_8} + 3r_1 \tag{2.107}$$
$$n_{C_7} = n_{C_8} - 2r_1 \tag{2.108}$$
$$n_{I_7} = n_{I_8} \tag{2.109}$$

Now, we make use of the conversion value given in the problem:

$$x_{A_1} = 0.1 = \frac{n_{A_7} - n_{A_8}}{n_{A_7}}$$

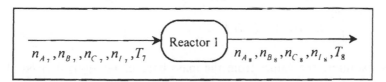

$$n_{A_7}, n_{B_7}, n_{C_7}, n_{I_7}, T_7 \quad \boxed{\text{Reactor 1}} \quad n_{A_8}, n_{B_8}, n_{C_8}, n_{I_8}, T_8$$

Figure 2.38   Mass and energy flow diagram for reactor 1.

Using the mass balance relations derived above, we get

$$0.1 = \frac{(n_{A_8} + r_1) - n_{A_8}}{(n_{A_8} + r_1)}$$

which can be simplified to

$$r_1 = 0.111 n_{A_8} \tag{2.110}$$

As can be seen, r_1 is only a function of n_{A_8}. Moreover, all n_{i_7} involve both r_1 and n_{i_8}, so we can easily say that

$$n_{i_7} = \hat{G}(n_{i_8}) \quad i = A, B, C, \text{ and } I$$

Heat Balance

The generalized heat balance equation is

$$\sum n_{i_7}(H_{i_7} - H_{ir}) + \underbrace{Q}_{\text{zero}} = \sum n_{i_8}(H_{i_8} - H_{ir}) + r_1(\Delta H_r)$$

Take the reference temperature as

$$T_{\text{ref}} = 25°C = 298 \text{ K}$$

It is given that

$$T_7 = 425°C = 698 \text{ K}$$

We can write

$$n_{A_7} \int_{298}^{698} C_{P_A} dT + n_{B_7} \int_{298}^{698} C_{P_B} dT + n_{C_7} \int_{298}^{698} C_{P_C} dT + n_{I_7} \int_{298}^{698} C_{P_I} dT$$

$$= n_{A_8} \int_{298}^{T_8} C_{P_A} dT + n_{B_8} \int_{298}^{T_8} C_{P_B} dT + n_{C_8} \int_{298}^{T_8} C_{P_C} dT + n_{I_8}$$

$$\int_{298}^{T_8} C_{P_I} dT + r_1(\Delta H_r)$$

On substituting the values from the mass balance relations derived above, we get

$$(n_{A_8} + 0.111n_{A_8}) \int_{298}^{698} C_{P_A} dT + [n_{B_8} + 3(0.111n_{A_8})] \int_{298}^{698} C_{P_B} dT$$

$$+ [n_{C_8} - 2(0.111n_{A_8})] \int_{298}^{698} C_{P_C} dT + n_{I_8} \int_{298}^{698} C_{P_I} dT$$

$$= n_{A_8} \int_{298}^{T_8} C_{P_A} dT + n_{B_8} \int_{298}^{T_8} C_{P_B} dT + n_{C_8} \int_{298}^{T_8} C_{P_C} dT$$

$$+ n_{I_8} \int_{298}^{T_8} C_{P_I} dT + (0.111n_{A_8})(\Delta H_r)$$

Therefore, the above heat balance equation gives a relation between n_{i_8} and T_8. Thus, we can write

$$T_8 = G_2(n_{i_8}), \quad i = A, B, C, \text{ and } I$$

Exchanger (See Fig. 2.39)

Mass Balance

The mass balance on streams 3 and 4 gives the following: For component A, we get

$$n_{A_3} = n_{A_4}$$

which can be written as

$$(14.67 - 1.61) + n_{A_8}(1 - 0.11) = F_A''(n_{A_8})$$

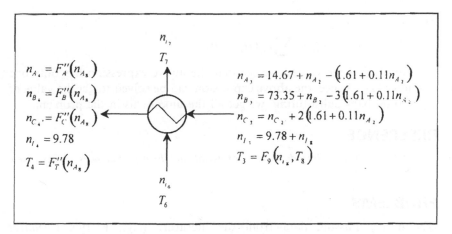

$$n_{A_4} = F_A''(n_{A_8})$$
$$n_{B_4} = F_B''(n_{A_8})$$
$$n_{C_4} = F_C''(n_{A_8})$$
$$n_{I_4} = 9.78$$
$$T_4 = F_T''(n_{A_8})$$

$$n_{A_3} = 14.67 + n_{A_2} - (1.61 + 0.11n_{A_2})$$
$$n_{B_3} = 73.35 + n_{B_2} - 3(1.61 + 0.11n_{A_2})$$
$$n_{C_3} = n_{C_2} + 2(1.61 + 0.11n_{A_2})$$
$$n_{I_3} = 9.78 + n_{I_2}$$
$$T_3 = F_9(n_{i_2}, T_8)$$

Figure 2.39 Mass and energy flow diagram for the exchanger.

As the above relation is in terms of only n_{A_8}, we can solve it to get the value of n_{A_8}.

Similarly, we obtain the values for n_{B_8}, n_{C_8}, and n_{I_8} using the mass balances for components B, C, and I. Note that $n_{i_3} = n_{i_4}$, where $i = A, B, C$, and I. Because the value of n_{A_8} has been calculated, we can get the value of T_4 as $T_4 = F_T''(n_{A_8})$. Also, because n_{A_8} has been calculated, $T_3 = F_9(n_{i_8}, T_8)$ becomes

$$T_3 = F_{10}(T_8)$$

Then from the individual mass balance for the separator, we have

$$n_{i_6} = F'''(R) = F'''(n_{A_8}).$$

Thus, it is known. Also, we get

$$T_6 = \overline{F}'''(R) = \overline{F}'''(n_{A_8}).$$

Thus, it is also known. Using the above two known values, n_{i_7} and T_7 are known.

Heat Balance

The generalized heat balance equation is

$$\sum n_{i_3}(H_{i_3} - H_{ir}) + \sum n_{i_6}(H_{i_6} - H_{ir})$$
$$= \sum n_{i_7}(H_{i_7} - H_{ir}) + \sum n_{i_4}(H_{i_4} - H_{ir})$$

Because

$$n_{i_3} = n_{i_4} \quad \text{and} \quad n_{i_6} = n_{i_7}$$

we get

$$\sum n_{i_3}(H_{i_3} - H_{i_4}) = \sum n_{i_7}(H_{i_7} - H_{i_6})$$

On substituting the values, we see that the above expression has only one unknown (T_8), hence, the above expression can be solved to get the value of T_8, and on back substituting, we get all the unknowns in the problem.

REFERENCE

1. Reklaitis, G. V. *Introduction to Material and Energy Balances*. Wiley, New York, 1983.

PROBLEMS

Obtain any missing data from the literature (e.g., Perry's *Chemical Engineering Handbook* 1999).

Problem 2.1

Superphosphate is produced by reacting calcium phosphate with sulfuric acid according to the reaction

$$Ca_3(PO_4)_2 + 2H_2SO_4 \rightarrow CaH_4(PO_4)_2 + 2CaSO_4$$

If 20,000 kg/day of raw calcium phosphate containing 14% inert impurities is reacted with 15,000 kg/day of H_2SO_4 of 92% concentration, determine the rate of production of superphosphate assuming the reaction is 95% complete (95% conversion). Which is the limiting reactant? ·

Problem 2.2

Carbon disulfide is used in viscose rayon and cellophane manufacture and in the production of carbontetrachloride. In the generally preferred process, vaporized sulfur is reacted with methane according to the reactions

$$CH_4 + 4S \rightarrow CS_2 + 2H_2S$$
$$CH_4 + 2S \rightarrow CS_2 + 2H_2$$
$$CH_4 + 2H_2S \rightarrow CS_2 + 4H_2$$

For a feed containing 4 mol of sulfur per mole methane, calculate the composition of the product if 90% conversion of methane and 70% conversion of sulfur are achieved.

Problem 2.3

Formaldehyde can be made by the partial oxidation of natural gas using pure oxygen made industrially from liquid air.

$$CH_4 + O_2 \rightarrow CH_2O + H_2O$$

The natural gas must be in large excess. The CH_4 is heated to 400°C and O_2 to 300°C and introduced into the reactor. The products leave the reactor at 600°C and have the following analysis: H_2O 1.9%, CH_2O 11.7%, O_2 3.8%, and CH_4 82.6%. How much heat is removed from the reactor by cooling per hour for a production rate of 200 kg/h formaldehyde.

Problem 2.4

Methane and oxygen react in the presence of a catalyst to form formaldehyde. In a parallel side reaction, some of the methane is instead oxidized to carbon dioxide and water:

$$CH_4 + O_2 \rightarrow HCHO + H_2O$$
$$CH_4 + 2O_2 \rightarrow CO_2 + 2H_2O$$

The feed to the reactor contains equimolar amounts of methane and oxygen.

The fractional conversion of methane is 95%, and the fractional yield of formaldehyde is 90%. Calculate the molar composition of the reactor output stream and the selectivity of formaldehyde production relative to carbon dioxide production.

In another case, the reactor output stream contains 45 mol% formaldehyde, 1% carbon dioxide, 4% unreacted methane, and the balance oxygen and water. Calculate the fractional conversion of methane, the fractional yield of formaldehyde and the selectivity of formaldehyde production relative to carbon dioxide production.

Problem 2.5

Ethylene oxide is produced by the catalytic partial oxidation of ethylene:

$$C_2H_4 + \tfrac{1}{2}O_2 \rightarrow C_2H_4O$$

An undesired competing reaction is the combustion of ethylene:

$$C_2H_4 + 3O_2 \rightarrow 2CO_2 + 2H_2O$$

The feed to the reactor (not the fresh feed to the process) contains 3 mol of C_2H_4 per mole of oxygen. The fractional conversion of ethylene in the reactor is 20%, and the yield of ethylene oxide based on ethylene consumed is 80%. A multiple-unit process is used to separate the products (C_2H_4O, CO_2, H_2O) from the output gases leaving the reactor: C_2H_4 and O_2 are recycled back to the reactor, C_2H_4O is sold as a product, and CO_2 and H_2O are discarded.

    (a) Construct a degree-of-freedom table. Is the problem correctly specified?
    (b) Calculate the molar flow of C_2H_4 and O_2 in the fresh feed. What is the mass fraction of C_2H_4 in the feed?
    (c) Calculate the overall conversion and the overall yield based on ethylene fed.
    (d) Calculate the selectively of C_2H_4O based on reactor output.

Problem 2.6

Oxychlorination of hydrocarbons refers to a chemical reaction in which oxygen and hydrogen chloride react with a hydrocarbon in the vapor phase over a supported copper chloride catalyst to produce a chlorinated hydrocarbon and water. The oxychlorination of ethylene to produce 1,2-dichloroethane [commonly called ethylene dichloride (EDC)] is of the greatest commercial importance. EDC is a precursor for poly(vinyl chloride)

(PVC), which is one of the most widely used commercial plastics. The overall oxychlorination reaction of ethane is given as follows:

$$2C_2H_4 + 4HCl + O_2 \rightarrow 2C_2H_4Cl_2 + 2H_2O$$

Determine the mass, moles, and weight percent of the reactant and product streams if 2000 kmol of the limiting reactant HCl is fed with 10% excess air and 5% excess ethane. Ninety-five percent conversion of ethylene occurs in the reactor.

Problem 2.7

A mixture containing 68.4% H_2, 22.6% N_2, and 9% CO_2 react according to the following reaction scheme:

$$N_2 + 3H_2 \Leftrightarrow 2NH_3$$
$$CO_2 + H_2 \Leftrightarrow CO + H_2O$$

The reaction proceeds until the mixture contains 15.5% NH_3 and 5% H_2O. Calculate the mole fraction of N_2, H_2, CO_2, and CO.

Problem 2.8

A gas mixture consisting of NO_2 and air is bubbled through a process into which water is fed at a rate of 55.5 kmol/h (as shown in Fig. P2.8). Water absorbs most of the NO_2 and none of the air. The volumetric flow rate of the feed gas is determined from the pressure drop across the orifice. An equation relating the pressure drop and volumetric flow rate is given by

$$V \ (m^3/h) = 13.2 \, h^{0.515}$$

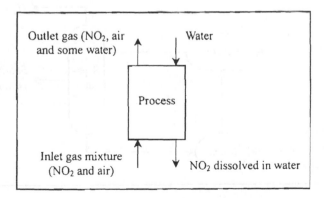

Figure P2.8   Process diagram for Problem 2.8.

where h is in mm Hg. The molar flow rate of the gas mixture is related to volumetric flow rate by the ideal gas law

$$PV = nRT$$

where $R = 0.082$ L atm/g mol K, V is the volumetric flow rate, and n is the molar flow rate.

An instrument was used to measure the composition of the gas mixture, where the mole fraction x of the mixture is determined from the reading M obtained from the instrument using the relation

$$x = 5.00 \times 10^{-4} \, e^{0.06M}$$

The following data are recorded for the inlet gas mixture: $T = 24°C$, $P = 11.2$ atm, h (orificemeter) $= 210$ mm Hg, and $M = 80.4$. For the outlet gas, $M = 11.6$.

Determine the amount of NO_2 absorbed by the water.

Problem 2.9

The recycle process, for producing perchloric acid ($HClO_4$) is shown in Figure P2.9. The reaction

$$Ba(ClO_4)_2 + H_2SO_4 \rightarrow BaSO_4 + 2HClO_4$$

proceeds in the reactor with 85% conversion of $Ba(ClO_4)_2$. The ratio of the number of moles of $Ba(ClO_4)_2$ in feed (fed to the mixer in the figure) to the number of moles of H_2SO_4 fed to the reactor is $1.0:1.1$. The process main feed (fed to the mixer in the figure) consists of 88% (by weight) $Ba(ClO_4)_2$

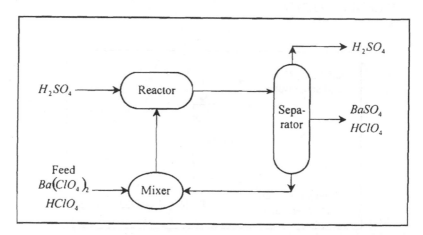

Figure P2.9 Flowsheet of the process for Problem 2.9.

and the rest (12% by weight) is $HClO_4$. The recycle contains only $Ba(ClO_4)_2$. Calculate all flow rates and compositions for all the streams in the process.

Problem 2.10

Refined sugar (sucrose) can be converted to glucose and fructose by the inversion process

$$\underbrace{C_{12}H_{22}O_{11}}_{\text{sucrose}} + H_2O \rightarrow \underbrace{C_6H_{12}O_6}_{d\text{-glucose}} + \underbrace{C_6H_{12}O_6}_{d\text{-fructose}}$$

The combined quantity glucose/fructose is called inversion sugar. If 90% conversion of sucrose occurs on one pass through the reactor, what would be the recycle stream flow per 100 kg/h of sucrose solution entering the process shown in Figure P2.10? What is the concentration of inversion sugar in the recycle stream and in the product stream? The concentrations of the components in the recycle stream and product streams are the same. *Note*: All percentages are weight percentages.

Problem 2.11

Product P is produced from reactant R according to the reaction

$$2R \rightarrow P + W$$

Unfortunately, because both reactant and product decompose to form the by-product B according to the reactions

$$R \rightarrow B + W$$
$$P \rightarrow 2B + W$$

only 50% conversion of R is achieved in the reactor and the fresh feed contains 1 mol of inerts I per 11 mol of R. The unreacted R and inerts I

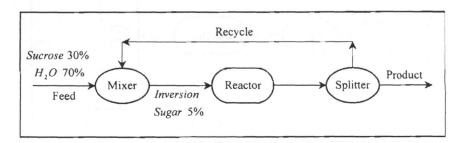

Figure P2.10 Process flow diagram for Problem 2.10.

are separated from the products and recycled. Some of the unreacted R and inerts I must be purged to limit the inerts I in the reactor feed to 12% on a molar basis (see Fig. P2.11). For a product stream produced at a rate of 1000 lb mol/h and analyzing 38% product P on molar basis, calculate the following (see Fig. P2.11):

(a) The composition of the recycle stream on molar basis
(b) The fraction of the recycle which is purged [purge/purge + recycle)]
(c) The fresh feed rate in lb mol/h
(d) The composition of the product stream on a molar basis
(e) The fraction of reactant R which reacts via the reaction $R \rightarrow B + W$.

Problem 2.12

In the recycle system shown in Figure P2.12, a feed of 1000 mol/h consisting of one-third A and two-thirds B is mixed with a recycle stream and reacted following the stoichiometry

$$A + B \rightarrow D$$

In the reactor, 20% of the entering A is converted to product. The resulting stream is separated so that the recycle stream contains 80% of A, 90% of B, and 10% of D fed to the separator.

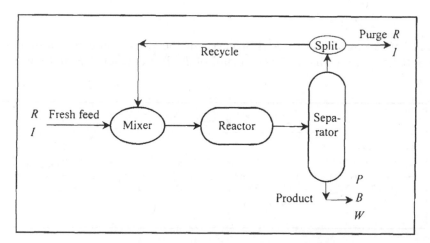

Figure P2.11   Process flow diagram for Problem 2.11.

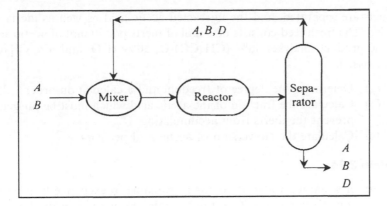

Figure P2.12 Process flow diagram for Problem 2.12.

(a) Show that the process is correctly specified (i.e., mass balance is solvable).
(b) Determine a calculation order for manual solution.
(c) Solve the problem.

Problem 2.13

Acetic anhydride can be produced from acetic acid by catalytic cracking. In the conceptual process shown in Figure P2.13, the acetic acid is reacted, the

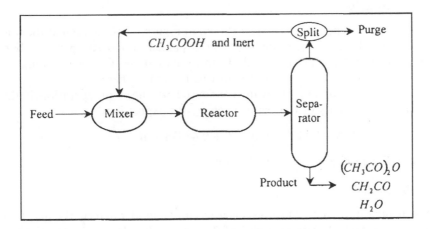

Figure P2.13 Process flow diagram for Problem 2.13.

products are separated, and the unreacted acetic acid as well as inerts are recycled. The fresh feed consists of 1 mol of inerts per 50 mol of acetic acid, and the product analyses 46% $(CH_3CO)_2O$, 50% H_2O, and 4% CH_2CO (molar basis).

(a) Determine the degree of freedom and a calculation order.
(b) Calculate the fraction of the acetic acid which must be purged to prevent the inerts from accumulating.
(c) Calculate the conversion of acetic acid per pass.

Problem 2.14

A stream of saturated water at 10 bar is available to exchange heat with a brine solution at 1 bar and 50°C. If the brine flow is twice the water flow and if the water stream can be cooled to 75°C, what is the temperature to which the brine can be heated? The brine can be assumed to have the properties of water.

Problem 2.15

Given that the net heat of combustion of CH_4 (g) is -191.76 kcal/g mol, calculate its heat of formation. The heats of formation of CO_2 and H_2O are as follows

$$\Delta H^\circ_{f_{CO_2(g)}} = -94.0518 \text{ kcal/g mol}$$

$$\Delta H^\circ_{f_{H_2O(g)}} = -57.7979 \text{ kcal/ g mol}$$

Problem 2.16

In the double-pipe heat exchanger shown in Figure P2.16, oxygen at the rate of 100 kmol/h is being heated by condensing saturated steam at 1.5 bar. Oxygen enters the inner pipe at 25°C and leaves at 200°C. The exchanger is well insulated such that no heat is lost to the atmosphere.

Find the amount of steam condensed per hour of operation if the molar heat capacity of oxygen is given by the expression

$$C_P = 29.88 - 0.11384 \times 10^{-1} T + 0.43378 \times 10^{-4} T^2$$

where T is in K and C_P is in kJ/kmol.

Problem 2.17

The temperature in a CO shift converter can be moderated by the injection of excess steam. Assuming a feed of 30% CO, 20% H_2, and 50% H_2O at 600°F and assuming 90% of CO will be converted. Determine the additional

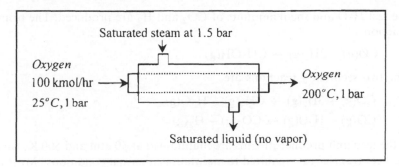

Figure P2.16 Flow diagram for Problem 2.16.

550°F steam required (per mole of feed) to maintain the reactor outlet temperature below 850°F. The reaction stoichiometry is:

$$CO(g) + H_2O(g) \rightarrow CO_2(g) + H_2(g)$$

Problem 2.18

In the production of ethylene oxide by partial oxidation of ethylene, the following reactions take place:

$$2C_2H_4 + O_2 \rightarrow 2C_2H_4O$$
$$C_2H_4 + 3O_2 \rightarrow 2CO_2 + 2H_2O$$

In a given reactor with a feed consisting of 12% C_2H_4 and the rest air, 26% conversion of C_2H_4 and an 78% fractional yield of C_2H_4O from C_2H_4 converted are attained when the reactor is operated at 245°C. Calculate the required heat removal rate from the reactor for a feed rate of 1200 kmol/h if the reactor feed mixture is at 105°C. All species are in the gas phase.

Problem 2.19

Carbon monoxide is completely burned at a pressure of 1 atm with excess air. If reactants enter at 200°F, the products leave at 1800°F and the heat losses are negligible, what percentage of excess air was used?

Problem 2.20

Methanol is synthesized from CO and H_2 at 50 atm and 550 K over a copper-based catalyst. With a feed of 75% H_2 and 25% CO at these conditions, a product stream consisting of 20% CH_3OH, 1% CH_4, 20% CO, and the rest H_2O and small amounts of CO_2 and H_2 are produced. The primary reaction

$$CO(g) + 2H_2(g) \rightarrow CH_3OH(g)$$

The two secondary reactions are

$$CO(g) + 3H_2(g) \rightarrow CH_4(g) + H_2O(g)$$
$$CO(g) + H_2O(g) \rightarrow CO_2(g) + H_2(g)$$

If the feed and product streams are maintained at 50 atm and 500 K, what is the heat transfer rate required to maintain isothermal conditions? Must heat be removed or added to the reactor?

Problem 2.21

For the reaction

$$2A + B \rightarrow C$$

the heat of reaction (at 300 K) is -10000 cal/g mol. The heat capacities of the components are

$$C_{P_A} = 16.0 - \frac{1.5 \times 10^3}{T}$$

$$C_{P_B} = 11.0 - \frac{0.5 \times 10^3}{T}$$

$$C_{P_C} = 25.0 - \frac{1000}{T}$$

In the above relations, T is in K and C_P in cal/g mol K. The heat capacity equations are valid in the range 300 K $\leq T \leq$ 1000 K.

    (a)   Derive an equation for the heat of reaction as a function of temperature.
    (b)   Calculate the temperature at which the reaction changes from exothermic to endothermic.
    (c)   Calculate the heat of reaction at 500 K assuming that substance A undergoes a change of phase at 400 K with $\Delta H_{V L_A} = 928$ cal/g mol after which its heat capacity becomes constant at 10 cal/g mol K.

Problem 2.22

A hot process stream is cooled by heat exchanger with boiler feed water (BFW), thus producing steam. The BFW enters the exchanger at 100°C and 100 bar; the stream is saturated at 100 bar. The process flow stream is 1000 kmol/h, its inlet molar enthalpy is equal to 2000 kJ/kmol, and the exit enthalpy is 800 kJ/kmol. Calculate the required BFW flow rate (see Fig. P2.22).

Problem 2.23

Methanol at 675°C and 1 bar is fed to an adiabatic reactor where 25% of it is dehydrogenated to formaldehyde according to the reaction

$$CH_3OH(g) \rightarrow HCHO(g) + H_2(g)$$

which can be represented by the following reaction for simplification:

$$A \rightarrow B + C$$

Calculate the temperature of the gases leaving the reactor. The given data are as follows:

Average heat capacity of $CH_3OH(g) = C_{P_A} = 17$ kcal/g mol °C
Average heat capacity of $HCHO(g) = C_{P_B} = 12$ kcal/g mol °C
Average heat capacity of $H_2(g) = C_{P_C} = 7$ kcal/g mol °C
Heat of formation of $CH_3OH(g)$ at 25°C and 1 bar = −48.08 kcal/g mol
Heat of formation of $HCHO(g)$ at 25°C and 1 bar − −27.7 kcal/g mol
Heat of formation of $H_2(g)$ at 25°C and 1 bar = 0.0 kcal/g mol

Problem 2.24

An evaporator is a special type of heat exchanger in which steam is used to heat a solution to partially boil off some of the solvent. In the evaporator shown in Figure P2.24, a brine containing 1% (weight) salt in water is fed at

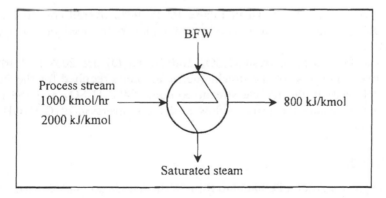

Figure P2.22 Flow diagram for Problem 2.22.

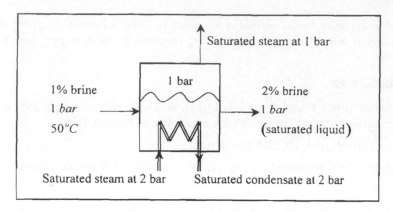

Figure P2.24   Setup for Problem 2.24.

1 bar and 50°C. The exit brine contains 2% (weight) salt and is saturated liquid at 1 bar. The evaporated water is saturated steam at 1 bar. If saturated steam at 2 bar is used as heat source and its condensate is assumed to be saturated liquid at 2 bar, calculate the kilograms of 2-bar steam required per kilogram of evaporated water

Problem 2.25

Acetic acid can be produced via the reaction

$$3C_2H_5OH + 2Na_2Cr_2O_7 + 8H_2SO_4 \rightarrow 3CH_3COOH + 2Cr_2(SO_4)_3$$
$$+ 2Na_2SO_4 + 11H_2O$$

In the recycle stream shown in Figure P2.25, 90% overall conversion of C_2H_5OH is obtained with a recycle flow equal to the feed rate of fresh C_2H_5OH.

The feed rates of fresh H_2SO_4 and $Na_2Cr_2O_7$ are 20% and 10%, respectively, in excess of the stoichiometric amounts required for the fresh C_2H_5OH feed. If the recycle stream contains 94% H_2SO_4 and the rest C_2H_5OH, calculate the product flow and the conversion of C_2H_5OH in the reactor.

Problem 2.26

The dehydrogenation of ethanol to form acetaldehyde takes place according to the stoichiometric relation

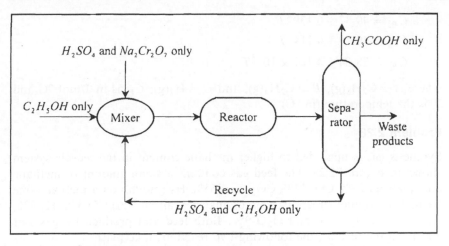

Figure P2.25 Process flow diagram for Problem 2.25.

$$C_2H_5OH(g) \rightarrow CH_3CHO(g) + H_2(g)$$
$$\Delta H_R^\circ = \Delta H_{rR}|_{at\ 25°C} = 68.95 \text{ kJ/mol of ethanol reacted}$$

The reaction is carried out in an adiabatic reactor. Ethanol vapor is fed to the reactor at 300°C and a conversion of 30% is obtained. Calculate the product (outlet) temperature of this adiabatic reactor.

Heat Capacities
[Average values — Assumed constant (independent of temperature)]:

$C_2H_5OH(g)$: $C_P = 0.11$ kJ/mol °C
$C_3CHO(g)$: $C_P = 0.08$ kJ/mol °C
$H_2(g)$: $C_P = 0.029$ kJ/mol °C

Problem 2.27

Ethylene is made commercially by dehydrogenating ethane:

$$C_2H_6(g) \Leftrightarrow C_2H_4(g) + H_2(g)$$
$$\Delta H_R(200°C) = 134.7 \text{ kJ/mol}$$

Ethane is fed to a continuous adiabatic reactor at $T_f = 2000°C$. Calculate the exit temperature that would correspond to 95% conversion ($x_A = 0.95$, where $A \equiv C_2H_6$). Use the following data for the heat capacities in your calculations:

$$C_{P_A} = 49.37 + 0.1392T$$
$$C_{P_B} = 40.75 + 0.1147T$$
$$C_{P_C} = 28.04 + 4.167 \times 10^{-3}T$$

where $A \equiv C_2H_6(g)$, $B \equiv C_2H_4(g)$, and $C \equiv H_2(g)$; C_{P_i} is in J/mol °C and T is the temperature (in °C)

Problem 2.28

Synthesis gas is upgraded to higher methane content in the recycle system shown in Figure P2.28. The feed gas contains a small amount of methane and analyzes 22% CO, 13% CO_2, and 65% H_2 (mol%) on a methane-free basis. The product stream (stream 6) analyzes (all mol %) CO 5%, H_2 9%, CH_4 50%, CO_2 27%, and H_2O 9%. Both feed and product are gases at 200°F. In the reactor, the following two reactions take place:

$$CO + 3H_2 \rightarrow CH_4 + H_2O$$
$$CO + H_2O \rightarrow CO_2 + H_2$$

The reactor effluent is cooled to 500°F in exchanger 1 and then it is further cooled in exchanger 2. Part of the effluent is split off as product. The

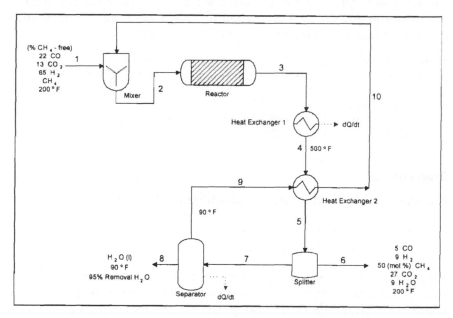

Figure P2.28   Process flow diagram for Problem 2.28.

remainder is sent to a separator in which the stream is cooled to 90°F and 95% of the H_2O is separated as liquid. The remaining gas stream is reheated in exchanger 2, mixed with the fresh feed, and returned to the reactor. Assume that the mixer, splitter, exchanger 2, and reactor operate adiabatically. Use the average C_P values given in Table P2.28.

(a) Check if the process is correctly specified (i.e., process degrees of freedom = 0); if not, assume any missing information and give sound engineering justification for your assumption.
(b) Determine a calculation order of the problem.
(c) Write a computer program for the solution based on subroutines for the standard four mass and heat balance modules.
(d) Calculate the reactor outlet temperature.

Problem 2.29

A vertical lime kiln (Fig. P2.29) is charged with pure limestone ($CaCO_3$) and pure coke (C), both at 25°C. Dry air is blown in at the bottom and provides the necessary heat for decomposition of the carbonate by burning the carbon to carbon dioxide. Lime (CaO) leaves the bottom at 950°C and contains 5% carbon and 1% $CaCO_3$. The kiln gases leave the top at 600°C and consist only of CO_2 and N_2. Assume heat losses are negligible. The two reactions are as follows:

$$CaCO_3(s) \rightarrow CaO(s) + CO_2(g)$$
$$C(s) + O_2(g) \rightarrow CO_2(g)$$

Heats of formation are given in terms of kcal/gmol:

$$\Delta H^{\circ}_{f_{CaCO_3(s)}} = -289.5, \qquad \Delta H^{\circ}_{f_{CaO(s)}} = -151.7,$$
$$\Delta H^{\circ}_{f_{CO_2(g)}} = -289.5$$

Table P2.28 Values of Specific Heats for Problem 2.28

	$C_P(g)$ (Btu/lb mol °F)	$C_P(l)$ (Btu/lb mol °F)
CO	7.3	—
H_2	7.0	—
CH_4	12.0	—
CO_2	11.4	—
H_2O	8.7	18.0

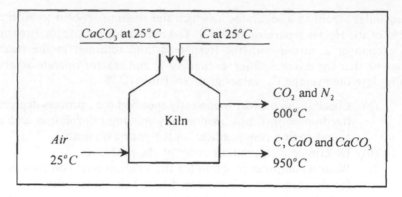

Figure P2.29 Schematic diagram of the kiln for Problem 2.29.

The molar heat capacities (C_P) are given in units of cal/g mol K:

$$C_{P_{CaCO_3(s)}} = 28.6, \qquad C_{P_{C(s)}} = 4.6 \qquad C_{P_{CaO(s)}} = 13.7,$$

$$C_{P_{CO_2(g)}} = 12.2, \qquad C_{P_{N_2(g)}} = 7.5, \qquad C_{P_{O_2(g)}} = 8.0$$

(a) Calculate the heats of reaction at 25°C.
(b) Analyze the degrees of freedom of this process. Can the material and energy balance be coupled?
(c) Calculate the required feed ratio of $CaCO_3$ to C required.

Problem 2.30

In the manufacture of sulfur dioxide by the direct oxidation of sulfur with pure oxygen,

$$S + O_2 \rightarrow SO_2$$

Cool sulfur dioxide must be recycled to the burner to lower the flame temperature below 1000 K so as not to damage the burner. The flowsheet and the temperatures of the different streams are given in Figure P2.30. For per lb mol of SO_2 product, calculate the lb mol SO_2 recycled and the pounds of steam produced.

Given data for sulfur: melting point $= 113°C$, heat of fusion (at melting point) $= 0.3$ kcal/mol, and heat capacity of solid $= 5.8$ cal/mol K. Obtain any other necessary data from appropriate tables.

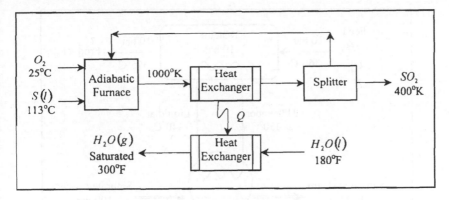

Figure P2.30   Process flow diagram for Problem 2.30.

Problem 2.31

The production of ethylene oxide by partial oxidation of ethylene over a silver catalyst involves the exothermic reaction

$$2C_2H_4(g) + O_2(g) \rightarrow 2C_2H_4O(g)$$

and the secondary, even more exothermic, reaction

$$C_2H_4(g) + 3O_2(g) \rightarrow 2CO_2(g) + 2H_2O(g)$$

Temperature control is essential and is achieved by boiling a hydrocarbon heat transfer fluid on the outside of the reactor tubes. This vaporized fluid is then condensed in a heat exchanger by transferring heat to a stream of saturated liquid water at 100 bar to produce saturated stream at 100 bar. The reactor feed is 11% C_2H_4, 13% O_2 and the rest N_2 at 360°C at 10 bar. The reactor product stream is at 375 C and 10 bar and the conversion of C_2H_4 is 22% with a 83% selectivity for C_2H_4O. At the operating pressure used in this system, the heat transfer fluid boils at 350°C with a heat of vaporization of 500 Btu/lb and has a liquid heat capacity of 0.8 Btu/lb °C and a vapor heat capacity of 0.4 Btu/lb °C. The flow of heat transfer fluid is adjusted so that it enters the reactor as liquid at 340°C and leaves as a two-phase mixture with vapor fraction of 21% (see Fig. P2.31).

   (a) Calculate the mass of steam produced per mole of C_2H_4O produced.

   (b) Calculate the recirculation rate of the heat transfer fluid per mole of C_2H_4O produced.

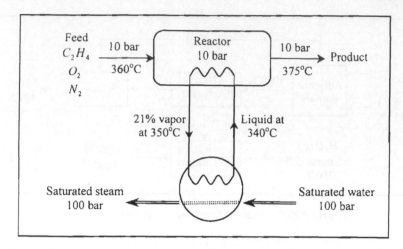

Figure P2.31 Flow diagram for Problem 2.31.

(c) Why is a heat transfer fluid used instead of the direct heat transfer using water?

(d) What pressure would the reactor tubes have to withstand if water were boiled at 350°C on the outside of the reactor tubes?

3

Mathematical Modeling (I): Homogeneous Lumped Systems

This chapter concentrates on the transformation of material and energy balance equations of homogeneous lumped systems into mathematical models (design equations).

3.1 MATHEMATICAL MODELING OF HOMOGENEOUS LUMPED PROCESSES

3.1.1 Basic Concepts for the Mathematical Modeling of Processes

Mathematical modeling of mass transfer, heat transfer, and reaction in the petrochemical, petroleum refining, biochemical, and electronic systems has been a very strong tool for design, simulation, and control as well as research and development (R&D) in these industries. It has led to a more rational approach for design and control in addition to elucidating many important phenomena associated with these systems. It has also led to a higher level of organization for research in these fields. Rigorous highly sophisticated mathematical models of varying degrees of complexity are being developed and used in industrial design and control as well as academic and industrial research (1–3). An important point to be noted with regard to the state of the art in these relatively well-established fields is that steady-state modeling is more advanced than unsteady-state modeling because of the additional complexities associated with unsteady-state behavior and the additional physicochemical information necessary (1,4–8).

Based on these achievements, continuous processing has advanced in most chemical and related industries over the years and became the dominant processing mode. Computerized reliable design packages have been developed and computer control of units and whole plants have been introduced widely (9–13). All of this has been coupled with a considerable increase in the productivity of industrial units and plants with a high reduction in manpower and tight control over the product quality.

On the industrial and academic levels, these advancements have led to innovative designs and configurations that open new and exciting avenues for developing new compact units with very high productivity (14–17).

In this book, a system approach is adopted which treats any process or processing plant as a system consisting of subsystems, with their properties and interactions giving the overall system its characteristics. Before getting into detail, it is important to give a more applied exposition of system theory than that given in Chapter 1. To make the exposition more applied, we will give some details regarding the principles of mathematical models building with some emphasis on fixed-bed catalytic reactors, which represent one of the most important units in the above fields.

3.1.2 Systems and Mathematical Models

In this subsection, the basic concepts of system theory and principles for mathematical modeling of chemical processes are briefly revisited. The basic principles have been discussed in Chapter 1, these basic principles are discussed here once again in a simple and brief manner.

A system is a whole consisting of elements and subsystems. The concept of system–subsystems–elements is relative and depends on the level of analysis. The system has a boundary that distinguishes it from the environment. The system may exchange matter and/or energy with the environment, depending on the type of system from a thermodynamics point of view. A system (or subsystem) is described by its elements (or subsystems), the interaction between the elements, and its relation with the environment. The elements of the system can be material elements distributed topologically within the boundaries of the system and giving the configuration of the system, or they can be processes taking place within the boundaries of the system and defining its functions. They can also be both, together with their complex interactions. An important property of the irreducibility of the complex to the simple or of the whole to its elements is related to the fact that the whole system will possess properties and qualities not found in its constituent elements. This does not mean that certain information about the behavior of the system cannot be deduced from the properties of its elements, but it rather adds to it.

Systems can be classified on different bases. The most fundamental classification is that based on thermodynamic principles, and on this basis, they can be classified into isolated systems, closed systems, and open systems (18,19), as discussed in Chapter 1.

3.1.3 What Are Mathematical Models and Why Do We Need Them?

Because much of the basic information for detailing the answer to this question are discussed in other parts of this book, we present a brief answer to this question here.

For the first part of the question, we present briefly Denn's answer (20):

A mathematical model of a process is a system of equations whose solution, given specified input data, is representative of the response of the process to a corresponding set of inputs.

This brief answer contains the essential components of a true mathematical model. However, for the definition to be more explicitly complete, we have to add that the term process in the above context means the configuration of the unit and all of the parameters related to its dimension and design. We have also to add that the answer given by Denn (20) is phrased for the simulation of existing processes. Of course, mathematical models are basically rigorous design equations which can and must be used for design purposes. In design problems, the output variables are specified as required from the process together with the specifications of the product and their rates of production. Thus, in most design problems, the output variables are predetermined, as well as the input variables, and it is the design parameters which we seek to compute. Therefore, both input and output variables are fed to the model (actually the computer program of the model) in order to compute (usually in an iterative manner) some of the design parameters of the unit itself and also to check that the obtained design does not violate the physically imposed constraints. When the design results violate the physical constraints, the model is further used to suggest solutions for this crucial problem.

The second part of the question has been explicitly and implicitly answered in many places in this book and we can summarize the answer in the following points regarding the advantages of using rigorous mathematical models in design:

1. It gives more precise and optimal design of industrial units.
2. It allows the investigation of new more effective designs.
3. It can be used for simulation of the performance of existing units to ensure their smooth operation.

4. It is usable for the optimization of the performance of existing units and compensation for external disturbances in order to always keep the unit working at its optimum conditions.
5. Dynamic models are used in the design of control loops for different industrial units to compensate for the dynamic effects associated with external disturbances.
6. Dynamic models are also used for stabilization of the desirable unstable steady state by designing the necessary stabilizing control.
7. It is utilized in the organization and rationalization of safety considerations.
8. It is extremely useful for training purposes.
9. It provides an excellent guide regarding environmental impact and protection.

3.1.4 Empirical (Black Box) and Physical (Mathematical) Models

Empirical models are based on input–output relations that do not take into consideration the description of the processes taking place within the system. This is why it is sometimes referred to as the "black-box" approach, because the system is considered as a nontransparent box with attention focused only on input and output variables, without much concern about what is going on inside the box. Many procedures exist for determining the appropriate experimental design and the way in which data are converted to equations. It is usually necessary to make some a priori assumptions about the physical structure of the system as well as the mathematical structure of the equations. Empirical models are highly unreliable and can hardly be extrapolated outside the region where the experiments were conducted. However, they are relatively easy and some empirical models are sometimes used as parts of an overall physical (or mathematical) model. Thus, empirical models are unreliable as overall process models, but sometimes they are used as parts of overall mathematical models.

On the other hand, physical (or mathematical) models are built on the basis of understanding the processes taking place within the boundaries of the system, the laws governing the rates of these processes, the interaction between the processes, as well as the interaction with the input variables and surrounding environment. These models are sometimes called physical models. This is to distinguish them from empirical models, because they describe the physical situation instead of ignoring it. Empirical models, on the other hand, relate input and output through linear and nonlinear regression. Physical models are also called mathematical models because of the mathematical description of the processes taking place within the boundaries of the system and in the description of the overall system.

Of course, no mathematical model is completely rigorous. Each mathematical model will always contain some empirical parts. As the degree of empiricism decreases, the model reliability increases. As an example, in the development of mathematical models for packed beds, the mass and heat transfer resistances are usually computed using the empirical j-factor correlations for mass and heat transfer between the solid surface and the bulk fluid. Having some of the components of the mathematical models as empirical relations does not make the model empirical; it is only that some of the parameters involved in the quantitative description of the processes taking place within the system are determined from empirical relations obtained experimentally.

To illustrate the above points in a simple manner, let us consider a nonisothermal adiabatic fixed-bed reactor packed with nonporous catalyst pellets for a simple exothermic reaction

$$A \to B$$

Assume that the pellets are of large thermal conductivity and that the rate of reaction is first order in the concentration of component A:

$$r = k_0 e^{-E/RT} C_A \qquad \frac{\text{kmol}}{\text{kg catalyst h}} \tag{3.1}$$

A mathematical model for this system will be a heterogeneous model taking into consideration the differences in temperature and concentration between the bulk gas phase and the solid catalytic phase. For the catalyst pellet, the steady state is described by algebraic equations as follows.

Using the symbols in the schematic diagram of the catalyst pellet as shown in Figure 3.1, the mass balance for component A (for a case with negligible intraparticle mass and heat transfer resistances) is given by

$$a_p k_{gA}(C_{AB} - C_A) = W_p r \tag{3.2}$$

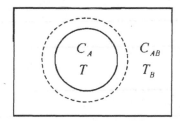

Figure 3.1　Nonporous catalyst pellet.

where a_P is the external surface area of the pellet (in m^2), W_P is the weight of the catalyst pellets (in kg), k_{g_A} is the external mass transfer coefficient of component A (in m/h), r is the rate of reaction per unit mass of the catalyst, C_A is the concentration of component A in the catalyst pellet, and C_{A_B} is the bulk concentration around the pellet. The heat balance for the pellet is given by

$$a_P h(T - T_B) = W_P r(-\Delta H_R) \tag{3.3}$$

where $-\Delta H_R$ is the heat of reaction (in kJ/k mol, h is the external heat transfer coefficient (in kJ/m^2 K h), T is the pellet temperature (in K), and T_B is the bulk temperature (in K). The bulk phase balances are shown in Figure 3.2. The bulk mass balance equation for the variation of C_{AB} along the length of the reactor (details of its derivation will be given in a later section) is

$$q \frac{dC_{AB}}{dl} = k_{gA} A_t a_S \rho_B (C_A - C_{AB}) \tag{3.4}$$

Similarly, the bulk temperature variation along the length of reactor is given by

$$q \rho_{\text{mix}} C_{P\text{mix}} \frac{dT_B}{dl} = h A_t a_S \rho_B (T - T_B) \tag{3.5}$$

with initial conditions (at $l = 0$)

$$C_{AB} = C_{Af} \quad \text{and} \quad T_B = T_f \tag{3.6}$$

where q is the volumetric flow rate (in m^3/h), ρ_{mix} is the density of the gas mixture, ρ_B is the density of the catalyst (in kg/m^3, A_t is the cross-sectional area of the catalyst tube (in m^2), $C_{P_{\text{mix}}}$ is the specific heat of the mixture (in kJ/kg), and a_S is the specific surface area of the catalyst (in m^2/kg catalyst).

To obtain the effect of input variables (e.g., q, C_{Af}, and T_f on the output variables C_B and T_B), Eqs. (3.1)–(3.6) must be solved simultaneously and numerically to obtain necessary results, as shown in Figure 3.3, where L is the length of the reactor.

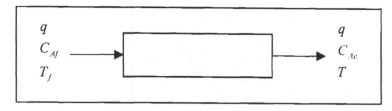

Figure 3.2   Mass and energy flow diagram for the bulk phase.

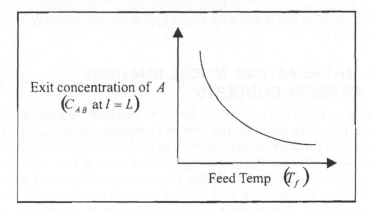

Figure 3.3 Effect of feed temperature on output concentration of reactant A.

In the above simple model, the parameters of the system are obtained from independent experiments to evaluate k_0, E, and $-\Delta H_R$. The heat and mass transfer coefficients (h and k_{gA}, respectively) are obtained from empirical j-factor correlations. For each value of T_f shown in Figure 3.3, the equations are solved to obtain the output variables C_{AB} and T_B. The solution also provides the concentration and temperature profiles along the length of the catalyst bed. It will also provide the temperature of the catalyst pellet T and the concentration C_A at every position along the length of the reactor. This is obviously a physical (mathematical) model, although some of the parameters are obtained using empirical relations (e.g., h and k_{gA} from j-factor correlations).

If the relations in Figure 3.3 are to be obtained by an empirical model, a number of experiments have to be performed, where, for example, T_f is changed and $C_{AB}|_{\text{at } l=L}$ is measured experimentally, and then the results are fitted to a polynomial (or any other suitable function) by nonlinear regression to obtain an empirical model of the form

$$C_{AB}|_{\text{at } l=L} = f(T_f)$$

If other input variables need to be incorporated, then a very large number of experiments must be performed and multiple regression is used to obtain an empirical model of the form

$$C_{AB}|_{\text{at } l=L} = f(T_f, q, C_{Af})$$

The difference in rigor, reliability, and predictability between the two approaches should be obvious from this simple example. It is obvious that such empirical relations have no physical basis and, therefore, it is highly unreliable and it is not very wise (nor necessary nowadays) to try to extrapolate it.

3.2 MATHEMATICAL MODEL BUILDING: GENERAL CONCEPTS

Building a mathematical model for any chemical engineering system depends to a large extent on the level of knowledge regarding the physical and chemical processes taking place within the boundaries of the system and the interaction between the system and the environment. It is, of course, also essential to know the laws governing the processes taking place within the boundaries of the system as well as the laws governing its interaction with the surroundings. For example, if we consider the more general catalytic reacting systems, these will include the diffusion mechanism and rates of diffusion of reacting species to the neighborhood of active centers of reaction, the chemisorption of the reacting species on these active sites, the diffusion of reactants through the pores of the catalyst pellets (intraparticle diffusion), the mechanism and kinetic rates of the reaction of these species, the desorption of products, and the diffusion of products away from the reaction centers. It also includes the thermodynamic limitations that decide the feasibility of the process to start with, and also includes heat production and absorption as well as heat transfer rates. Of course, diffusion rates and heat transfer rates are both dependent to a great extent on the proper description of the fluid flow phenomena in the system. The ideal case is when all of these processes are determined separately and then combined into the system's model in a rigorous manner. However, very often this is quite difficult to achieve; therefore, special experiments need to be devised, coupled with the necessary mathematical modeling, in order to decouple the different processes interacting in the measurements.

Mathematical models (design equations) of different degrees of sophistication and rigor were built in last three to four decades to take their part in directing design procedure as well as directing scientific research in all fields of chemical engineering and its related disciplines. It is important in this respect to stress the fact mentioned a number of times previously that most mathematical models are not completely based on rigorous mathematical formulation of the physical and chemical processes taking place within the boundaries of the system and the interaction between the system and its environment. Every mathematical model contains a certain degree of empiricism. The degree of empiricism, of course, limits the generality of the model, and as our knowledge of the fundamentals of the processes taking place increases, the degree of empiricism decreases and the degree of rigor (generality) of the model increases. The existing models at any stage, with this stage's appropriate level of empiricism, helps greatly in the advancement of the knowledge of the fundamentals and therefore helps to decrease the degrees of empiricism and increase the level of rigor in the mathematical models. In addition, any model will contain simplifying assumptions which are believed, by the model builder, not to affect the predictive nature of the model in any manner that sabotages the purpose of the model.

With a given degree of fundamental knowledge at a certain stage of scientific development, one can build different models with different degrees of sophistication depending on the purpose of the model building and the level of rigor and accuracy required. The choice of the appropriate level of modeling and the degree of sophistication required in the model is an art that needs a high level of experience. Models that are too simplified will not be reliable and will not serve the purpose, whereas models that are too sophisticated will present unnecessary and sometimes expensive overburden. Models that are too sophisticated can be tolerated in academia and may sometimes prove to be useful in discovering new phenomena. However, oversophistication in modeling can hardly be tolerated or justified in industrial practice.

The procedure for model building differs depending on the process itself and the modeling group. However, a reasonably reliable procedure can be summarized in the following steps:

1. The identification of the system configuration, its environment, and the modes of interaction between them.
2. The introduction of the necessary justifiable simplifying assumptions.
3. The identification of the relevant state variables that describe the system.
4. The identification of the processes taking place within the boundaries of the system.
5. The determination of the quantitative laws governing the rates of the processes in terms of the state variables. These quantitative laws can be obtained from the known information in the literature and/or through an experimental research program coupled with the mathematical modeling program.
6. The identification of the input variables acting on the system.
7. The formulation of the model equations based on the principles of mass, energy, and momentum balance appropriate to the type of system.
8. The development of the necessary algorithm for the solution of the model equations.

9. The checking of the model against experimental results to ensure its reliability and re-evaluation of the simplifying assumptions which may result in imposing new simplifying assumptions or relaxing some of them.

It is clear that these steps are interactive in nature and the results of each step should lead to a reconsideration of the results of all previous steps. In many instances, steps 2 and 3 are interchanged in the sequence, depending on the nature of the system and the degree of knowledge regarding the processes taking place, within its boundaries, and its interaction with the environment.

3.2.1 Classification of Models

We revisit the different types of models which are essential in model development for the different chemical/biochemical engineering processes. Batch processes are usually of the isolated- or closed-system type, whereas the continuous processes are almost always of the open-system type.

For continuous processes, a classification from a mathematical point of view is very useful for both model formulation and algorithms for model solution. According to this basis, systems can be classified as follows:

1. *Lumped systems*: These are systems in which the state variables describing the system are lumped in space (invariant in all space dimensions). The simplest chemical engineering example is the perfectly mixed continuous-stirred tank reactor. These systems are described at steady state by algebraic equations, whereas the unsteady state is described by initial-value ordinary differential equations for which time is the independent variable.

2. *Distributed systems*: These are systems in which the state variables are varying in one or more direction of the space coordinates. The simplest chemical engineering example is the plug flow reactor. These systems are described at steady state by either an ordinary differential equation [when the variation of the state variable is only in one direction of the space coordinates, (i.e., one-dimensional models) and the independent variable is the space direction] or partial differential equations [when the variation of the state variables is in more than one direction of the space coordinates (i.e., two-dimensional models or higher), and the independent variables are these space directions]. The ordinary differential equations of the steady state of the one-dimensional distributed parameter models can be either initial-value differential equations (e.g., plug flow models) or two-point boundary-value differential equations (e.g., models with super-

imposed axial dispersion or diffusion through pellets or membranes). The equations describing the unsteady state of distributed models are invariably partial differential equations.

Another classification of systems, which is very important for deciding the algorithm for model solution, is that of linear and nonlinear systems. The equations of linear systems can usually be solved analytically, whereas the equations of nonlinear systems are almost always solved numerically. In this respect, it is important to recognize the important fact that physical systems are almost always nonlinear, and linear systems are either an approximation that should be justified or intentionally linearized in the neighborhood of a certain state of the system and are strictly valid only in this neighborhood.

A third classification, which is very relevant and important to chemical engineers, is the classification based on the number of phases involved within the boundaries of the system. According to this classification systems are divided as follows (21):

1. *Homogeneous systems*: These are systems in which only one phase is involved in the process taking place within the boundaries of the system. The behavior of these systems (for reacting systems) is basically governed by the kinetics of the reaction taking place without the interference of any diffusion processes between phases. Nonreacting homogeneous systems are usually very simple (e.g., mixer, splitter, filling/emptying of a tank, etc.).

2. *Heterogeneous systems*: These are systems in which more than one phase is involved in the processes taking place within the boundaries of the system. The behavior of these systems (for reacting systems) is governed not only by the kinetics of the reactions taking place but also by the complex interaction between the kinetics and the relevant diffusion processes. The modeling and analysis of these systems is obviously much more complicated than for homogeneous systems. It is clear that the systems for fixed-bed catalytic reactors fall into the category of heterogeneous systems and, more specifically into the category of gas–solid systems; therefore, the behavior of the system is dependent on a complex interaction between kinetics and diffusion. Nonreacting heterogeneous systems are not very simple (e.g., absorption, distillation, adsorption, etc.).

The reader interested in more details about system theory and the general concepts of mathematical modeling will find a large number of very good and interesting books (22–24).

3.2.2   Difference Between Modeling and Simulation

Aris (25) in his 1990 Dankwerts Memorial Lecture entitled "Manners Makyth Modelers" distinguishes between modeling and simulation in a special manner that follows the reasoning of Smith (26) in ecological models,

> It is essential quality in a model that it should be capable of having a life of its own. It may not, in practice, need to be sundered from its physical matrix. It may be a poor, an ill-favored thing when it is by itself. But it must be capable of having this independence.

He also adds in the same lecture:

> At the other end of the scale a model can cease to be a model by becoming too large and too detailed a simulation whose natural line of development is to the particular rather than the general. It ceases to have a life of its own by becoming dependent for its vitality on its physical realization.

As discussed in Chapter 1, the basis of the classification given by Aris is very interesting, true, and useful. It is typical of the Minnesota school, which actually revolutionized the field of mathematical modeling in chemical engineering with their extensive work in this field since the mid-1940s until today. Experimental verifications of the arsenal of new and interesting steady-state and dynamic phenomena discovered by the Minnesota school were carried out in other universities, mostly by graduates from Minnesota. The most interesting outcome is the fact that not a single phenomena which was discovered theoretically using mathematical models by the Minnesota group was not experimentally confirmed later. This demonstrates the great power of mathematical modeling discipline as expressed by Aris (25), where the model is stripped of many of its details in order to investigate the most fundamental characteristics of the system. The Minnesota school using this approach has achieved a real revolution in chemical engineering in general and in chemical engineering mathematical modeling in particular.

In this book, a different definition for mathematical modeling and simulation than the one of Aris (25) will be adopted. Our definition will be more pragmatic and will serve our specific purpose in this book. The definition we will adopt is that mathematical modeling will involve the process of building up the models themselves, whereas simulation involves simulating experimental or industrial units using the developed models. Thus, simulation in this sense is actually closely linked to the verification of the model against experimental and industrial units. However, it will also include the use of the verified models to simulate a certain practical situation specific to a unit, a part of a production line, or an entire production line. The main purpose of developing and verifying these models is to use them as rigorous design equations.

3.2.3   Design Equations and Mathematical Models

From the process design point of view, design means deciding on the configuration of the unit, its dimensions, and the optimal input variables. Of course, the guidelines for such a design are the imposed constraints, such as the amount of raw material to be processed in the unit, the properties of the raw material and the products, chemical and physical paths between the raw material and the products, the rate of these paths, the thermochemical properties of the materials and processes (or paths), the maximum temperature that the material of construction and catalyst can withstand, the quality of the products and their rate of production, and so forth.

Design can be carried out based on previous experience with very few calculations and that was the case at the very early stages of the chemical industry before even chemical engineering was established. Certainly, some elements of this highly empirical approach still exists in today's advanced design procedure. A relatively higher degree of rationalization as compared with the highly empirical approach will be based on scaling up from bench scale to pilot plant, then to commercial scale directly or through a semi-commercial-scale unit. This procedure includes extensive experimental work with a limited amount of computation. Another procedure is to use design equations that are of a certain degree of rigor coupled with design guidelines that accumulated over the years.

What are usually called design equations are usually expressed in terms of nonlinear sets of algebraic, differential, or integral equations and can be called mathematical models.

Actually, there is no difference in principle between design equations and mathematical models. It is just that when the process is described by a highly empirical and simplified set of equations, we call them design equations, whereas when the equations are more rigorous and therefore more reliable, they are called a mathematical model.

3.2.4   Simplified Pseudohomogeneous Models
Versus Rigorous Heterogeneous Models

We will use the industrially important fixed bed-catalytic reactor as a tool to illustrate some basic modeling principles. The main steps for reactions taking place in fixed-bed catalytic reactors are as follows:

1. External mass transfer from the bulk gas phase to the external surface of the pellet
2. Intraparticle diffusion of reactants through the pores of the catalyst pellets toward the center of the pellet
3. Chemisorption of reaction molecules onto the surface of the pellet
4. Surface reaction of the adsorbed species
5. Desorption of the product molecules
6. Intraparticle diffusion of product molecules toward the surface of the pellet
7. External mass transfer of the product molecules from the surface of the pellet to the bulk gas phase

This sequence of events makes the concentrations under which the reactions actually take place quite different form those at the measurable bulk conditions. In addition to that, when the heats of reactions are appreciable, there will also be a temperature difference between the bulk gas phase and the solid catalyst phase.

Heterogeneous models (to be discussed in Chapter 6) take into consideration these differences between the bulk gas phase and the catalyst solid phase, whereas pseudohomogeneous models ignore these differences. These differences between the two phases are best expressed through a coefficient called the effectiveness factor η (or sometimes called the efficiency factor) which is defined as the ratio of the actual rate of reaction to the rate of reaction when mass and heat transfer resistances are neglected.

In other words,

$$\eta = \frac{\text{Correct rate of reaction computed from heterogeneous models}}{\text{Rate of reaction computed from homogeneous models}}$$

The effectiveness factor computation involves the solution of nonlinear two-point boundary-value differential equations of varying degrees of complexity depending on the specific reactions taking place in the reactor. The effectiveness factors vary with the change of the bulk-gas-phase conditions and therefore must be computed at every point along the length of the reactor. Efficient numerical techniques have been developed for solving this problem.

From the above simple discussion, it is clear that the pseudo-homogeneous model is simply a heterogeneous model but with $\eta = 1.0$ [or at least $\eta =$ constant (i.e., it is not changing along the length of the reactor)]. Therefore, when η approaches 1.0, the pseudohomogeneous models are valid for design, operation, and optimization of catalytic reactors. However, when η is far from unity and is varying along the length of the reactor, the use of the heterogeneous model is a must.

In industrial fixed-bed catalytic reactors, the catalysts are usually relatively large to avoid a pressure drop, and, therefore, the η's are usually far from unity and heterogeneous models are necessary for the accurate design, operation, and optimization of these reactors. For example, ammonia converters have η's is in the range 0.3–0.7, and an error of 30–70% results from the use of the pseudohomogeneous models. For steam-reforming η is in the range 10^{-2}–10^{-3}; therefore, the use of pseudohomogeneous models will give extremely wrong predictions (10,000–100,000% error).

In some cases of highly exothermic reactions and reactions with nonmonotonic kinetics, η can be much larger than unity and, therefore, the use of pseudohomogeneous models can give very erroneous results.

It is clear that the use of pseudohomogeneous models is not suitable for most industrial fixed-bed catalytic reactors, and heterogeneous models should almost always be used. To the contrary, for fluidized-bed catalytic reactors, the catalyst particles are in the form of fine powder and, therefore, η's are very close to unity. However, in fluidized-bed reactors, the hydrodynamics of the fluidized bed should be taken into consideration especially the bubbles of gas rising through the fluidized powder. Thus, for fluidized beds, although the difference in conditions between the gas surrounding the solid particles and the solid particles in the emulsion phase can be neglected, the difference in concentration and temperature between the emulsion phase and the bubble phase cannot be neglected.

In the petrochemical industry, fluidized beds are used in the production of polyethylene and polypropylene, and in the petroleum refining industry, they are used in fluid catalytic cracking (FCC).

In this chapter, we will concentrate on the development of homogeneous models; and Chapter 6 is devoted to the heterogeneous systems.

3.2.5 Steady-State Models Versus Dynamic Models

Steady-state models are those sets of equations which are time invariant and describe the conditions of the system at rest (i.e., when the states of the system are not changing with time). This will automatically presuppose that the system parameters are also time invariant (i.e., input variables, heat transfer coefficients, catalyst activity, and so forth are not changing with time). Of course, this is a theoretical concept, for no real system can fulfill these requirements perfectly. However, this theoretical concept represents the basis for the design and optimization of almost all chemical/biochemical engineering equipment. The philosophy is that we assume that the system can attain such a time-invariant state and design the system on that basis. Then, we design and implement the control system that always "pushes" the system back to its optimally designed steady state.

For design and optimization purposes, we use steady-state models, whereas for start-up and control of units, we must use dynamic models (unsteady-state models). To illustrate these concepts we use a very simple example of a consecutive reaction,

$$A \xrightarrow{k_1} B \xrightarrow{k_2} C$$

taking place in an isothermal continuous-stirred tank reactor (CSTR), with B being the desired product. For simplicity, we assume that the feed is pure A (contains no B or C):

$$C_{Bf} = C_{Cf} = 0$$

Now, the unsteady-state mass balance equation for component A is:

$$V\frac{dC_A}{dt} = qC_{Af} - qC_A - Vk_1C_A \tag{3.7}$$

and for component B is:

$$V\frac{dC_B}{dt} = -qC_B + V(k_1C_A - k_2C_B) \tag{3.8}$$

with the initial conditions ($t = 0$)

$$C_A = C_{A0} \quad \text{and} \quad C_B = C_{B0}$$

We want to find the size of the reactor V that gives maximum concentration of the desired product B for given q, C_{Af}, k_1, and k_2. To achieve this first task, we use the steady-state equations, which can be simply obtained by setting the time derivatives in Eqs. (3.7) and (3.8) equal to zero, thus giving

$$q(C_{Af} - C_A) = Vk_1C_A \tag{3.9}$$

and

$$qC_B = V(k_1C_A - k_2C_B) \tag{3.10}$$

Some simple manipulations of Eqs. (3.9) and (3.10) give

$$C_B = \frac{qk_1C_{Af}V}{(q + Vk_1)(q + Vk_2)} \tag{3.11}$$

To obtain $V_{optimum}$, we differentiate Eq. (3.11) with respect to V to get

$$\frac{dC_B}{dV} = qk_1C_{Af}\left[\frac{q^2 - k_1k_2V^2}{(q + Vk_1)^2(q + Vk_2)^2}\right] \tag{3.12}$$

Then, putting $dC_B/dV = 0$ gives V_{optimum} (denoted as V_{opt}),

$$V_{\text{opt}} = \frac{q}{\sqrt{k_1 k_2}} \tag{3.13}$$

Therefore,

$$C_{B_{\text{max}}} = \frac{q k_1 C_{Af} V_{\text{opt}}}{(q + V_{\text{opt}} k_1)(q + V_{\text{opt}} k_2)} \tag{3.14}$$

If the reactor is operating at this output concentration, disturbances will cause it to deviate from it. The deviation can be dynamic, (varying with time) or static (very slow disturbance) so that the system settles down to a new steady state.

If we want to follow the change of the state variables with time due to a disturbance, say a change in q, we have to solve the unsteady-state equations. Suppose that the system is at its optimum steady-state conditions; then, the initial condition ($t = 0$) is given by

$$C_A = C_{A_{\text{opt}}} \quad \text{and} \quad C_B = C_{B_{\text{opt}}}$$

where

$$C_{A_{\text{opt}}} = \frac{q C_{Af}}{q + V_{\text{opt}} k_1} \tag{3.15}$$

and $C_{B_{\text{opt}}}$ is equal to $C_{B_{\text{max}}}$ in Eq. (3.14).

If q changes to q', then we insert q' in the dynamic equations instead of q and solve the differential equations from $t = 0$ to higher values of time in order to follow the change with time. At large values of t, the system will settle to new steady-state values corresponding to the new q'. These new values of C_A and C_B will be C'_A and C'_B:

$$C'_A = \frac{q' C_{Af}}{(q' + V_{\text{opt}} k_1)} \tag{3.16}$$

$$C'_B = \frac{q' k_1 C_{Af} V_{\text{opt}}}{(q' + V_{\text{opt}} k_1)(q' + V_{\text{opt}} k_2)} \tag{3.17}$$

V_{opt} is used here as a symbol to indicate the value of V chosen to give $C'_{B_{\text{max}}}$ for q. However, for q', the volume V_{opt} is no longer optimum.

This steady-state deviation from $C_{B_{\text{max}}}$ can be compensated for by the change of another variable such as C_{Af} and/or the operating temperature that will change k_1 and k_2. The requirement for this compensation (say, in feed concentration) will be that

$$\frac{q'C'_{Af}}{q' + V_{opt}k'_1} = \frac{qC_{Af}}{q + V_{opt}k_1} \tag{3.18}$$

and

$$\frac{q'k'_1 C'_{Af} V_{opt}}{(q' + V_{opt}k'_1)(q' + V_{opt}k'_2)} = \frac{qk_1 C_{Af} V_{opt}}{(q + V_{opt}k_1)(q + V_{opt}k_2)} \tag{3.19}$$

These are two equations in three unknowns (C'_{Af}, k'_1 and k'_2). However, note that both k'_1 and k'_2 are functions of the new temperature T'. Equations (3.18) and (3.19) can be solved for C'_{Af} and T' (gives k'_1 and k'_2) to find the new feed concentration and temperature needed to maintain the system at its original optimum state despite the change of q to q'.

However, the question arises with regard to the dynamic variation due to change in input. What is to be done when the variation in input parameters is continuous (i.e., when the system does not have enough time to settle to any new steady state)? In this case, the dynamic model equations must be used to design a controller that introduces compensation which is changing with time. It is also important in this respect to make clear that when the disturbances are very slow (e.g., slowly deactivating catalyst), then the quasi-steady-state approximation can be used, where the change with time is considered as a series of steady states each corresponding to the value of the changing variable at the sequence of time intervals.

3.2.6 A Simple Feedback Control Example

A simple example can be used to illustrate the concept of the use of dynamic models in simulation and control. Consider the water tank shown in Figure 3.4, where the valve at the bottom discharges water at a rate proportional to the head h. It is well known that the discharge is proportional to $\sqrt{h}$; however, we use the assumption that it is proportional to h in order to make the equations linear and, therefore, illustrate the ideas in a simple manner.

The mass balance on water gives

$$q_{in} = q_0 + A\frac{dh}{dt} = Ch + A\frac{dh}{dt} \tag{3.20}$$

where C is the valve coefficient and A is the cross-sectional area of the cylindrical tank. This is a simple linear equation. Suppose that the tank was originally empty; then, at $t = 0$, $h = 0$. Also suppose that q_{in} is constant; then, Eq. (3.20) can be solved analytically and we get the change of h with time:

$$h(t) = \frac{q_{in}}{A\alpha}(1 - e^{-\alpha t}) \tag{3.21}$$

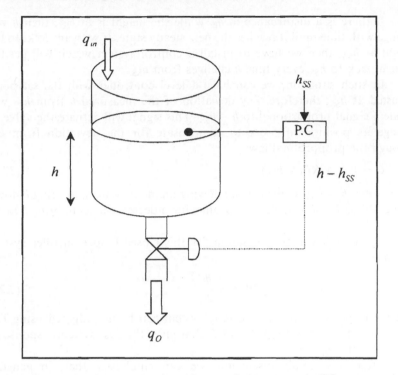

Figure 3.4 Schematic diagram for the controlled water tank (PC = proportional controller).

where

$$\alpha = \frac{C}{A}$$

The steady-state height can be obtained from the steady-state equation

$$q_{in} = Ch_{SS} \tag{3.22}$$

Thus,

$$h_{SS} = \frac{q_{in}}{C}$$

or from the solution of Eq. (3.21) by putting $t \to \infty$,

$$h_{SS} = \frac{q_{in}}{C} \tag{3.23}$$

The change of h with time from $h = 0$ to $h = h_{SS}$ follows Eq. (3.21).

If there is a disturbance in q_{in} while the height is at h_{SS}, then h will change with time until it reaches the new steady state. If we want to keep the height at h_{SS}, then we have to install a control system which will get the system back to h_{SS} every time h changes from h_{SS}.

In such situation, we can have a level controller with the set point adjusted at h_{SS}; therefore, any deviation of the measured h from h_{SS} will create a signal proportional to $h - h_{SS}$. This signal will actuate the valve to change its position in order to compensate for the deviation from h_{SS} through the proportional law:

$$C = C_{SS} + K(h - h_{SS}) \tag{3.24}$$

Thus, when $h > h_{SS}$, C increases, allowing an increase in q_0 to bring h down to h_{SS}. When $h < h_{SS}$, C decreases, allowing a decrease in q_0 to bring h up to h_{SS}.

The dynamic model equation for this closed-loop controller system becomes

$$q_{in} = [C_{SS} + K(h - h_{SS})]h + A\frac{dh}{dt} \tag{3.25}$$

and the response of this closed-loop system can be investigated using Eq. (3.25) in order to find the value of K that gives the best system response to external disturbances.

After this simple discussion, we can emphasize that, in general, "steady-state models are used for design and optimization, whereas dynamic models are used for start-up, shutdown, and process control."

3.3 GENERIC AND CUSTOMIZED MODELS

Generic models is a term usually used to describe models which are not developed for a specific unit but are simple models which have the qualitative behavior of the unit rather than the exact quantitative behavior. It gives trends for the behavior of the unit and is therefore not verified against specific units. They are usually simple and formed from a collection of semiempirical relations that are supposed to give the qualitative behavior of the unit. These models are quite useful in many cases, especially for training purposes. However, it is our experience that in some situations, these generic models do not even give the correct qualitative behavior of the unit.

In this section, we will always refer to heterogeneous catalytic chemical reactors as our industrial example. This is because many other important chemical engineering processes can be treated as special cases of these complex units, as will be shown later. The mathematical modeling of homogeneous reactors is a special case of heterogeneous catalytic reactors when the diffusional resistances are neglected; the modeling of a nonreacting system is a special case when the rates of reactions are set equal to zero.

For catalytic reactors (which represent some of the most complex chemical engineering units), generic models are less widespread than in separation and heat transfer processes. This is because catalytic systems have their own specific characteristics that depend on the intrinsic kinetics, the diffusional processes, as well as the reactor's configuration and thermal characteristics. The term *generic* in the field of chemical reactors is usually and loosely applied to reactors models based on thermodynamic equilibrium (Gibbs free-energy reactor models) or at their highest level of "sophistication", they refer to plug flow and perfectly mixed homogeneous steady-state models. These models hardly represent actual catalytic reactors. Thermodynamic equilibrium models do not represent even homogeneous reactors because they neglect the kinetics of reactions altogether, whereas pseudohomogeneous plug flow and perfectly mixed models, although they do not neglect the kinetics of the reaction, they actually completely neglect diffusional processes. In heterogeneous reactors, the concept of the effectiveness factor η is extremely important. It expresses the effect of diffusional resistance by one number. The pseudohomogeneous models, which neglect diffusional resistances, assume that the effectiveness factors η's are equal to unity (or any other constant number). Taking into consideration the fact that diffusional processes expressed in terms of effectiveness factors can be very different from unity (or constancy) and depend on intrinsic kinetics, operating variables, particle size, and so forth, one can easily realize that such models cannot seriously be a representation of industrial catalytic reactors. For example, η for steam reforming of natural gas is in the range $10^{-2}-10^{-3}$(and varies strongly along the length of the reactor), whereas for some highly exothermic reactions, η may reach values of several hundreds and sometimes several thousands. These simple and clear facts show very clearly that generic models are actually not suitable for serious consideration when we are dealing with industrial catalytic reactors as well as other industrial chemical engineering processes involving diffusional limitations (e.g., adsorption column).

On the other hand, customized models for catalytic reactors include all of the main processes taking place inside the catalytic reactors. The most important of these processes for catalytic reactors are those associated with the catalyst pellets, namely intrinsic kinetics (which includes chemisorption and surface reaction), intraparticle diffusion of mass and heat, external mass and heat transfer resistances between the catalyst surface and the bulk of the fluid, as well as all the heat production and heat consumption accompanying the catalytic reaction.

For fluidized-bed reactors, the diffusional processes associated with the catalyst pellets are not very important because the particles are quite small; however, the intrinsic kinetics is still of prime importance, as well as the hydrodynamics of the fluidization process, especially those associated with the bubbles and the mass and heat transfer between the bubble and dense phases, as will be shown in Chapter 6.

Such customized models for industrial catalytic reactors (or any other chemical engineering units) should be verified against the performance of industrial units. After some verification work and identification of the basic physicochemical parameters of the industrial units, the model should be able to predict the behavior of the unit without adjustable parameters. The introduction of too many adjustable parameters reduces the predictive power of the model.

Such kinetic-diffusion highly rigorous models should not be verified against inefficiently designed units operating very close to thermodynamic equilibrium for in this case, the verification will not be critical enough because many models without the required degree of precision will predict correct results. Units operating near thermodynamic equilibrium are not good for testing models based on rate processes and which are built for the correct prediction of the behavior of these units away from thermodynamic equilibrium.

3.3.1 Practical Uses of Different Types of Models

Mathematical models are very useful and efficient tools for the design, optimization, and control of different chemical engineering units. Steady-state models are suitable for design and optimization, whereas unsteady-state models are necessary for dynamic analysis, start-up, shutdown, and control.

Rigorous and reliable models are excellent tools during the design stage and also during the entire operational life period of units and plants. When correctly developed, these models can eliminate the stage of pilot plant and semicommercial units and, therefore, achieve a great economical saving. When a pilot plant is still needed, rigorous mathematical models help to exploit the pilot plant to the maximum level and will also make the pilot plant stage shorter, less expensive, and more fruitful.

We will still use heterogeneous catalytic reactors as our illustration example, for the reasons described earlier (i.e., mathematical models of most other important chemical engineering units will be shown to be special cases of the heterogeneous catalytic reactors). Our descriptive discussion here will include steady states as well as dynamic models and their applications.

3.3.2 Steady-State Models

Models for industrial catalytic reactors, like all other units, can be built with different degrees of sophistication and rigor. The level of sophistication and rigor is strongly linked to the intended use of the model. Obviously, the more sophisticated and rigorous the model, the more expensive it is to buy or develop because highly sophisticated models require a higher level of expertise and more time and effort for their development.

The most beneficial practical use of these models is in the design and optimization of these industrial units. In fact, most industrial unit's design and optimization is based on steady-state design, with the dynamic models developed in a later stage for the design of the proper control loops in order to keep the reactor dynamically operating near its optimum steady-state design in the face of external disturbances.

Models intended for design, optimization, and on-line steady-state adjustment of variables to keep the reactor operating at its optimum conditions should be sophisticated, rigorous, high-fidelity models. However, if the models are intended for operator training or rough design calculations, then models of lower level of rigor and fidelity can be used.

For high-fidelity models, there is also an optimum degree of sophistication to be used, for if one tries to describe all of the processes taking place in the unit to the highest degree of rigor, the model can become too complicated and too expensive. The expert developer of the model should emphasize the most influential processes in the model's development and try to formulate it to the highest degree of rigor. However, less influential processes should be treated in a more approximate and less rigorous manner. There is always a trade-off between the accuracy of the model and its cost; a critical point exists beyond which added accuracy is too small to justify the extra cost involved in the extra development of the model. This critical point is not easily distinguished and depends to a high degree on the experience of the model developers. We can call this side of the problem "the optimum degree of model sophistication".

The practical advantages gained from the use of steady-state models in design, optimization, and operation of catalytic reactors are tremendous. It is estimated that about 80–85% of the success of the process depends on the steady-state design and the remaining 15–20% depends on the successful dynamic control of the optimum steady state. These estimates are, of course, made for a process operating smoothly with conventional control which is not model based. However, in certain cases, inefficient dynamic control may cause temperature runaway or a complete shutdown of the process.

3.3.3 Dynamic Models

For most continuous processes operating under steady-state conditions, dynamic models are used to design appropriate control loops that minimize the deviation of the process dynamically from the optimum steady-state operation. This is the classical control objective whether we use analog or digital computer control. However, a more crucial but less common objective is when the process is intrinsically unstable and the control objective is to stabilize this intrinsically unstable process. This objective does not cancel the first one, but rather adds to it, for either the stabilizing controller is also able to achieve the first objective or extra control loops are added to achieve that.

Dynamic models are essential for discovering intrinsic instability and for choosing the proper control action to stabilize the unstable process. With regard to the classical control objective, although it may be achieved without the need of a model for the process, this, in fact, can be quite dangerous because the closed-loop dynamics of a process can be unstable, even when the process is intrinsically stable. Also, the model-based control is almost always more efficient and robust than control not based on reliable dynamic models.

In industrial catalytic reactors with their heterogeneous and distributed nature (variation of the state variables with respect to the space coordinates), dynamic temperature runaways may occur, especially for highly exothermic reactions. A reliable dynamic model is one of the best ways to discover and monitor these temperature runaways, which may cause explosions or, at the least, emergency shutdowns, which are quite expensive, especially with today's large-capacity production lines.

Both steady-state and dynamic models are essential tools for the full exploitation of modern digital computers used in the design, optimization, and control of petrochemical and petroleum refining plants, as well as biochemical systems. They allow the on-line computation of optimum operating conditions and optimum actions to be taken in order to keep the plant at its maximum production capacity. These models are also essential for the optimum performance of adaptive control, where the set points obtained from steady-state design need to be changed in the case of long-term changes in feed conditions (especially changes in feedstock composition).

3.3.4 Measures for the Reliability of Models and Model Verification

"A theory is just a model of the universe, or a restricted part of it, and a set of rules that relate quantities in the model to observations that we make. It exists only in our minds and does not have any other reality (whatever that might mean). A theory is a good theory if it satisfies two requirements: it must accurately describe a large class of observations on the basis of a model that contains only a few arbitrary elements, and it must make definitive predictions about the results of future observations." (27)

The reliability of a mathematical model for any chemical engineering process depends on many integrated factors which are inherent in the structure of the model itself. Although the final test of the model is through its comparison with experimental or industrial results, this, in fact, is not enough; more scientific and intellectual logical insight into the model is necessary. Blind reliance upon comparison with experimental and industrial data, although of great value, can sometimes be extremely misleading. Therefore, before the model is put to the test through a comparison with experimental data, physical and chemical information should be put into the model. The model should stand the test of logical and scientific theoretical analysis before it is put to the test of a comparison with experimental data. Before we proceed further, it is important to make the above statement clearer, for it may sound strange to many practitioners who put great faith on the comparison of the model with experimental results. To that end, we recall that "a mathematical model of a process is a system of equations whose solution, given specific input data, is representative of the response to a corresponding set of data." We also recall a simple theoretical fact: "One never proves that a theory is correct (validate a model), for this would require an infinite number of experiments." In other words, the model must have a certain logic in order to be reliable without the need to use a large number of experimental runs for comparison.

In this sense, the situation changes drastically; it becomes the following. The model should contain, in a rational and scientifically sound basis, the important processes taking place within the system. These processes are included in the model through the use of scientifically sound laws and through the correct estimation of the parameters. Such a model should be a good representation of the real system; it should match the results of the real system. A few experiments need to be compared with the model predictions to confirm these facts and/or fine-tune the model to represent reality more accurately.

To illustrate the above argument by a simple industrial example, consider the industrial steam reformer, where three reversible reactions are taking place catalytically in the reactor. If the industrial or experimental reactor is operating under conditions such that the exit conditions are close to the thermodynamic equilibrium of the mixture and the comparison between model predictions and industrial performance is based on exit conditions (which is actually the case, because it is quite difficult to measure profiles of variables along the length of an industrial steam reformer), then this comparison is not a valid check for the accuracy of the model with regard to its correct prediction of rates of reaction and mass transfer. This is because the results do not depend on the rates because the reactions are close to thermodynamic equilibrium. Therefore, if a diffusion-reaction model which is not rigorously built is compared with a large number of industrial cases, all close to thermodynamic equilibrium, it does not mean that the model is reliable, because a thermodynamic equilibrium model (which is much simpler) will also give good results. Such a model, although checked successfully against a large number of industrial cases, may prove to be a failure when compared with a case operating far from thermodynamic equilibrium.

Another example, it is our experience that models using the simplified Fickian diffusion for the catalyst pellet give good results for cases not very far from thermodynamic equilibrium. However, for cases far from thermodynamic equilibrium, such models fail and the more rigorous dusty gas model must be used for the modeling of the catalyst pellets.

Denn (20) discusses a similar case for coal gasifiers, where the kinetic-free model (thermodynamic equilibrium model) fails in certain regions of parameters, necessitating the use of a kinetic model.

In a nutshell, the reliability of the model depends on the following:

1.   The structure of the model as a combination of accurate description of the processes taking place within the system, together with the interaction between the processes themselves and between them and the environment
2.   The verification of the model against experimental and industrial units, provided that these tests are for critical cases that actually test the rates of the processes in the system

Verification of the model does not mean the use of a number of adjustable parameters. The main role of the model is prediction, and adjustable parameters reduce the predictive nature of the model; when overused, they can reduce it to zero. It is best when the model is verified without any adjustable parameters. When this is too difficult, one (or more) set(s) of data can be used to adjust some parameters; however, there should be enough data to test the model successfully against other sets of data without any adjustable parameters.

3.4   ECONOMIC BENEFITS OF USING HIGH-FIDELITY CUSTOMIZED MODELS

These benefits can be classified into the following two broad categories: design and operation, and control.

3.4.1 Design and Operation

Chemical reactors (as well as many other chemical/biochemical process units) design differs fundamentally from other engineering design in the fact that "Design Safety Factors" which are widely and comfortably used in other engineering designs are not applicable in the case of many chemical/biochemical engineering processes. This is due to the complexity of these units, in which a large number of processes are taking place. However, the core of the matter is the fact that, for chemical reactors, reaction networks are usually formed of complex consecutive and parallel steps which give rise to selectivity and yield problems. For a large class of reactions, the dependence of the yield of desired product(s) on design variables is nonmonotonic. This nonmonotonic dependence implies that the use of "Design Safety Factors" can be rather catastrophic for the process.

To illustrate this point in the simplest possible fashion, consider the consecutive reaction

$$A \rightarrow B \rightarrow C$$

taking place in an isothermal CSTR with component B being the desired product. For this simple system, the yield of the desired component B, Y_B, will depend on the reactor volume V_R or the volumetric flow rate to the reactor q_R in the manner shown schematically in Figure 3.5. It is clear from

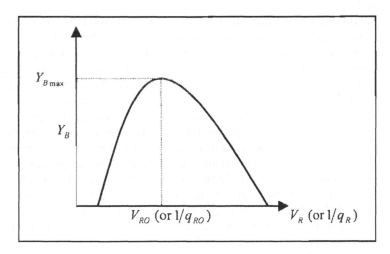

Figure 3.5 Schematic representation of Y_B versus V_R (for certain values of q_R and other parameters), or Y_B versus $1/q_R$ (for certain values of V_R and other parameters).

Figure 3.5 that there is an optimum value of V_R (or $1/q_R$), V_{RO}, which gives the maximum yield of B, $Y_{B_{max}}$. It is obvious that the celebrated engineering "Design Safety Factors" used in many engineering designs (including separation processes when the number of trays are increased over the design calculated value by multiplying it by a safety factor or dividing by what is called tray efficiency) can be disastrous in the case of certain chemical/biochemical units.

Although the above simple illustration of the concept for a homogeneous isothermal lumped system is applicable to other more complicated systems, the situation for catalytic reactors is much more involved because of the complexity of the intrinsic kinetics as well as the complex interaction among reactions, heat release (or absorption), and mass and heat diffusion inside the reactor. For example, many catalytic and biocatalytic reactions show nonmonotonic dependence of the rate of reaction on reactant concentrations. For example, the hydrogenation of benzene to cyclohexane over different types of nickel catalyst has an intrinsic rate of reaction of the form

$$r = \frac{kC_A C_B}{(1 + K_A C_A + K_B C_B)^2}$$

where A is hydrogen and B is benzene. This rate equation has nonmonotonic dependence on reactant concentrations, and, therefore, in some regions of parameters, the rate of reaction may decrease when the concentration increases!

Furthermore, the interaction between diffusion and reaction in porous catalyst pellets gives rise to an arsenal of complicated behaviors. The simplest is the reduction of the rate of reaction due to diffusional resistances, giving rise to effectiveness factors less that 1.0. However, effectiveness factors greater than 1.0 are also widespread in catalytic and biocatalytic systems for nonmonotonic kinetics and/or nonisothermal pellets. In addition, negative effectiveness factors (for reversible reactions) indicating that the reactions reverse their direction due to diffusion are also possible for important industrial catalytic reactors with multiple reactions [e.g., steam reforming of natural gas, partial oxidation of o-xylene to phthalic anhydride, methanation, etc. (28)]. Nonmonotonic change of the effectiveness factor η along the length of the reactor is also possible for reversible single reactions (29, 30). Fluctuation in the value of the effectiveness factor along the reactor length is possible in many industrial reactors (e.g., steam reforming of natural gas and methanol synthesis).

From the above, it is clear that the widespread engineering design procedure of using simple design equations coupled with a "Design Safety Factor" is not generally applicable in the design of catalytic and noncatalytic chemical and biochemical reactors. Therefore, rigorous high-fidelity, steady-state models (design equations) should be used for the precise design of these units. This does not mean that it is not possible to design such units without such models. This false statement is contradictory to the history of the chemical and biochemical industries because different units were designed and operated before chemistry and chemical engineering principles were even discovered. Units can be designed and operated using empirical techniques and accumulation of practical experience and even using trial and error and sequential scaling up. However, such designs will have a very high cost and will not be the most efficient and safe designs. With today's large continuous production lines, design cost, productivity, and safety are of the highest concern.

The first step in the design of any unit is the sizing of the unit for a given configuration and operating conditions. The high-fidelity mathematical models allow optimizing the design not only with respect to operating and design parameters but also with regard to the configuration of the unit itself. Design and optimization on this basis can have a very strong impact on the process. To take a very well-known example from an important chemical industry, consider an ammonia production line of 1200 MTPD capacity. For this line, an improvement of 1% in the production line will mean about 1.5–2 million \$/year in extra revenues. Optimization of an existing plant using high-fidelity models can lead to up to 10% improvement in the ammonia production (equivalent to 15–20 million \$/year in added revenues). The opportunities for improvement are much higher when the high-fidelity models and optimization algorithms are used in the design stage.

Despite the great importance of efficient control of processes and production lines, it is now well established in industry that the overall value of the project is not dominated by controller's action but by the ability to predict the best operating point in terms of process variables. For example, 0.03–0.05 \$/barrel might be achieved by holding an FCC plant under tight control near a desired point or constraint. Complicated control loops will help to achieve smaller excursions and faster responses to set-point changes and load disturbances by moving the operation to the desired point faster and with less variation in product yields. However, 0.3–0.5 \$/barrel might be realized by moving the steady state of the plant from a steady state which was not properly optimized to new and near-optimal steady-state operating conditions.

The above brief discussion gives an elementary idea about the economic benefits of using rigorous high-fidelity, steady-state models in the design and operation of industrial catalytic reactors. The same principle applies to other industrial chemical and biochemical units also.

The follow-up of the operation of the plant using such models allows an instant and deep insight into what is going on inside the units. This allows smooth operation and early warning of troubles that may cause an expensive shutdown or an even more expensive and catastrophic accident.

3.4.2 Control

Steady-state models are invaluable in what we may call steady-state control. The term *steady-state control* refers to situations in which the operator is not highly concerned about the dynamic behavior of the system, but is mostly concerned about operating the unit at its optimum steady state in the face of long-term external disturbances. This is a situation in which one of the operating parameters changes (usually feedstock composition), and the steady-state control question is: What are the input variables that need to be changed in order to keep the unit at the same steady state and producing the desired yield and production rates? Both off-line and on-line use of high-fidelity mathematical models can be utilized here. Of course, the on-line option is more efficient and reliable, however, it has a higher cost with regard to fixed investment (computer, measurement, and manipulation hardware).

On the other hand, dynamic control is when it is desired to avoid the losses associated with dynamic excursions of the process from steady-state optimum operation resulting from transient disturbances and the corresponding process responses.

Economic incentives for advanced control (which is usually model based) can be considered to fall into three categories:

1. The major benefit results from moving the steady-state operation to a better operating point, which has been discussed earlier and which mainly depends on high-fidelity, steady-state models.
2. Improved control allows operation closer to any limiting constraint, thus resulting in additional benefits. This relies upon steady-state as well as dynamic models.
3. Smaller excursions and faster responses to setpoint changes and load disturbances move the operation back to the desired steady state faster and with less variation in the yield and productivity (and, of course, the product quality). This relies mainly on steady-state and dynamic models.

Of course, all of the above is for intrinsically stable processes. For intrinsically unstable processes, stabilization of the unstable process must precede all of the above-discussed points. We should note that in some cases, unstable operation may give higher yield and productivity than stable states. However, this is well beyond the scope of this undergraduate book.

Dynamic models are also essential for the smooth and fast start-up of the process and for shutdown procedures. These two crucial stages in the plant operation are intrinsically unsteady state and therefore dynamic models are the ones applicable to these stages. In the start-up and shutdown phases, the controller settings are very different from the settings at steady state operation because the objectives are completely different.

It was estimated that the close control over an ammonia production line based on objective 3 given above, can achieve a 2–3% improvement in ammonia productivity. For an ammonia plant with a capacity of 1200 MTPD, this amounts to about 2.4–6.0 million $/year.

The use of rigorous, high-fidelity, steady-state mathematical models in the design stage allows integration between steady-state design and control considerations. This is due to the fact that changes in the process design could influence profoundly the process dynamics. Thus, for the optimal overall design of the process, control configurations should be considered during the design of the process itself. This crucial objective is best achieved using rigorous, high-fidelity, steady-state and dynamic models.

3.5 INCORPORATION OF RIGOROUS MODELS INTO FLOWSHEET SIMULATORS AND PUTTING MATHEMATICAL MODELS INTO USER-FRIENDLY SOFTWARE PACKAGES

Commercial steady-state flowsheet simulators are usually computer packages equipped with an extensive physical properties database and modular steady-state mass and heat balance calculations facilities. These simulators usually also contain modules for the process design of multistage operations such as distillation, absorption, extraction, and so forth. Most of these simulators are based on equilibrium stage calculations coupled with empirical formulas for calculation of stage efficiencies. Very few packages contain design procedures based on rates of mass and heat transfer. Therefore, most of these packages are not completely suitable for the design of continuous contact separation processes such as packed and spray columns. However, most of them contain some highly empirical correlations for the rough design of such equipment.

With regard to chemical reactors, the most sophisticated simulators usually contain modules for idealized plug flow and perfectly mixed continuous-stirred tank reactors. The less sophisticated simulators usually contain only procedures for reactor design based on thermodynamic equilibrium (Gibbs free energy).

However, with regard to catalytic reactors, these simulators usually contain nothing. This state of affairs is due to the fact that each catalytic reactor has its own characteristics. The development of catalytic reactor models needs a deep understanding of catalysis, kinetics, kinetic modeling, mass and heat transfer processes, reactor modeling, and advanced numerical techniques. The development of models for these catalytic reactors needs a modeler with long experience in a number of fields related to catalytic processes and it is also a time-consuming process.

High-fidelity models for industrial catalytic reactors can be incorporated into these flowsheet simulators via a number of techniques depending on the characteristics of the flowsheet simulator. In some steady-state flowsheet simulators, there is a built-in facility for the user to add his own module. In such cases, the high-fidelity reactor model can be added as one of these special modules. The reactor model in this case can make full use of the physical properties database available in the simulator. However, in most flowsheet simulators, this facility is not available; in such cases, the module can be used with the flowsheet simulator on the basis that the output variables from the last unit preceding the catalytic reactor is used as an input file to the catalytic reactor model, and when the calculations for the reactor are complete, the output file from the reactor model is used as an input file to the unit next to the reactor in the process flowsheet. Clearly, this last technique is less efficient and takes more effort to implement than in the first case.

In many cases, the catalytic reactor model is used as a stand-alone unit in the design, simulation, and optimization of catalytic reactors. There are some typical cases in the petrochemical industry where the catalytic reactors dominate the production lines (e.g., the ammonia production line usually contains about six catalytic reactors representing almost 90% of the production line). In these cases, an evaluation has to be made in order to decide whether the reactor modules should be added to the flowsheet simulator or the few noncatalytic processes should be borrowed from the flowsheet simulator and added to a specially formulated flowsheet simulator of catalytic reactors forming the production line.

With regard to dynamic simulation, very few simulators exist and they usually give the trends of the dynamic behavior rather than the precise simulation of the dynamics. High-fidelity dynamic models for catalytic reactors are quite rare and they are usually developed through special orders and agreements. The incorporation of these models into commercial dynamic simulators follows basically the same general principles outlined for the steady-state models.

Putting Mathematical Models into User-Friendly Software Packages

Great efforts are expended in the development and verification of rigorous, high-fidelity mathematical models for any chemical and biochemical engineering unit(s). Therefore, it is quite important to maximize the benefits from these models and to facilitate their use to a wider spectrum of users. This aim is achieved through putting the models into user-friendly software packages with advanced graphic capabilities. This facilitates the use of the model by many people in the plant or design office without much background in mathematical modeling, numerical analysis, or computer programming. Also, the results are usually presented in both graphical and tabulated forms. The user can have a summary screen with all of the most important input–output variables or the user can zoom to any set of variables and get more details about them. The user can also obtain a hardcopy of the results independently or use an on-line insertion of the tabulated or graphical results into a report. In addition to these facilities, the user can follow the progress of the solution through interesting visual means.

Of course, it is beyond the scope of this undergraduate book to aim at teaching and training the reader on how to put the models into these software packages. However, we can give a general outline for the procedures of developing such computer packages. An overall introduction for undergraduate chemical and biochemical engineers can be summarized in the following very general and brief points:

I. **What is a software package?**
    A software package is a collection of modules interlinked to achieve certain purpose(s).
II. **Components of software packages**
    1. Input section for acquiring necessary information
    2. Database for retrieving basic data necessary for computations
    3. Calculation section for creating new data from input, a database, and a mathematical model (a numerical algorithm is, of course, developed and incorporated into the software package for the accurate solution of the model equations)
    4. Output section for presenting computed information
III. **Basic decisions for software design**
    1. Language selection: Fortran, C, C++, and so forth.
    2. Operating system selection: DOS, Unix, VMS, windows, and so forth
IV. **Design steps for software packages**
    A. Preliminary steps:

    1. Select (or develop) the proper mathematical model with the appropriate state variables
    2. Identify input parameters
    3. Identify calculation procedures
    4. Identify output variables (the state variables at the exit of the unit)
    5. Test run with I/O from files

B. Input and Output:
    1. Design input screen layout
    2. Test run readings from screen
    3. Select output variables
    4. Design output selection screen
    5. Decide output variable format

C. Alert the user:
    1. Select indicative intermediate variables
    2. Decide on intermediate displays
    3. Test displays of intermediate results
    4. Add as many variables and indicators as necessary to keep the user alert

D. Error trapping (input):
    1. Include limits on input parameters
    2. Check input parameters for type and value
    3. Check other related values

Warning: Take nothing for granted, you have to check everything and test run it.

E. Error trapping (calculations):
    1. Check for overflow and underflow
    2. Give user a proper warning when there is a problem in calculations
    3. Advise user to possible action
    4. Always make an option to quit gracefully

F. Error trapping (files):
    1. Check for existence
    2. Never write over a file
    3. Never erase a file
    4. Avoid end of file trap

G. Fine-tuning:
    1. Avoid redundant inputs
    2. Keep a consistent use of keys
    3. Keep messages in one area
    4. Use short indicative messages

V. Ultimate package
  1. General in nature
  2. Based on sound mathematical model and an efficient solution algorithm
  3. Easy to use by nonspecialists
  4. User friendly with error trapping
  5. Flexibility in data entry
  6. Provides intermediate, summary, and detailed results
  7. Industrially verified against industrial units

3.6 FROM MATERIAL AND ENERGY BALANCES TO STEADY-STATE DESIGN EQUATIONS (STEADY-STATE MATHEMATICAL MODELS)

Let us first review the most general material and heat balance equations and all of the special cases which can be easily obtained from these equations. This will be followed by the basic idea of how to transform these material and energy balance equations into design equations, first for lumped systems and then followed by the same for distributed systems. We will use homogeneous chemical reactors with multiple inputs, multiple outputs, and multiple reactions. It will be shown in Chapter 6 how to apply the same principles to heterogeneous system and how other rates (e.g., rates of mass transfer) can systematically replace (or is added to) the rates of reactions.

This approach will put the mathematical modeling of all chemical/biochemical engineering systems into one unified and very easy-to-use framework. We will also extend this unified framework to dynamic models.

3.6.1 Generalized Mass Balance Equation (Fig. 3.6)

This is the most general one-phase (homogeneous) chemical/biochemical engineering lumped system. It has L input streams $(n_{if_1}, n_{if_2}, \ldots, n_{if_l}, \ldots, n_{if_L})$ and K output streams $(n_{i_1}, n_{i_2}, \ldots, n_{i_k}, \ldots, n_{i_K})$ with N reactions $(r_1, r_2, \ldots, r_j, \ldots, r_N)$ and the number of species is $M(i = 1, 2, \ldots, M)$.

The generalized material balance equation is

$$\sum_{k=1}^{K} n_{i_k} = \sum_{l=1}^{L} n_{if_l} + \sum_{j=1}^{N} \sigma_{ij} r_j, \quad i = 1, 2, \ldots, M \tag{3.26}$$

These are M equations having N reactions $(r_j, j = 1, 2, \ldots, N)$ and σ_{ij} is the stoichiometric number of component i in reaction j. The generalized rate of reaction for reaction j is given by r_j. Note that r_j is the overall rate of reaction for the whole unit (it is not per unit volume, or per unit

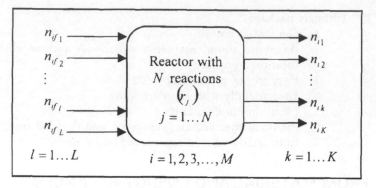

Figure 3.6 Mass flow diagram.

mass of catalyst, and so forth, and that is why our equations are mass balance equations, not design equations, as shown later).

This general system can also be represented as three systems: a mixer, a single input–single output reactor, and a splitter, as shown in Figure 3.7.

Equation (3.26) can be put in an equivalent form consisting of three sets of equations:

$$\bar{n}_i = \sum_{l=1}^{L} n_{if_l} \quad \text{and} \quad \bar{n}_i = n_{if} \quad \text{for } i = 1, 2, \ldots, M \tag{3.27}$$

$$n_i = \bar{n}_i + \sum_{j=1}^{N} \sigma_{ij} r_j \quad \text{and} \quad n_i = \bar{n}_{if} \quad \text{for } i = 1, 2, \ldots, M \tag{3.28}$$

$$\sum_{k=1}^{K} n_{ik} = \bar{n}_{if} = n_i \quad \text{for } i = 1, 2, \ldots, M \tag{3.29}$$

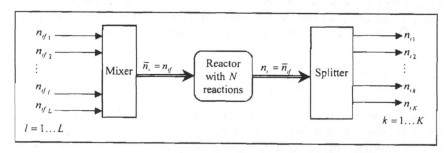

Figure 3.7 Another representation of the system.

Note also that Eq. (3.26) is the general form for all kinds of special cases:

1. For a single reaction ($N = 1$), the equation becomes

$$\sum_{k=1}^{K} n_{ik} = \sum_{l=1}^{L} n_{if_l} + \sigma_i r \quad \text{for } i = 1, 2, \ldots, M$$

2. For a single input ($L = 1$), the equation becomes

$$\sum_{k=1}^{K} n_{ik} = n_{if} + \sum_{j=1}^{N} \sigma_{ij} r_j \quad \text{for } i = 1, 2, \ldots, M$$

3. For a single output ($K = 1$), the equation becomes

$$n_i = \sum_{l=1}^{L} n_{if_l} + \sum_{j=1}^{N} \sigma_{ij} r_j \quad \text{for } i = 1, 2, \ldots, M$$

4. For no reactions ($N = 0$), the equation becomes

$$\sum_{k=1}^{K} n_{ik} = \sum_{l=1}^{L} n_{if_l} \quad \text{for } i = 1, 2, \ldots, M$$

5. For $N = 0$ and $L = 1$, we have thus the equation for a single input splitter:

$$\sum_{k=1}^{K} n_{ik} = n_{if}$$

6. For $N = 0$ and $K = 1$, we have the equation of a mixer:

$$n_i = \sum_{l=1}^{L} n_{if_l}$$

and so on.

Now, can we change this general mass balance equation to a design equation? Very simply, we use r' instead of r, where r' is the rate of reaction per unit volume of the reactor, and the equation becomes

$$\sum_{k=1}^{K} n_{i_k} = \sum_{l=1}^{L} n_{if_l} + \sum_{j=1}^{N} \sigma_{ij} r'_j V_R$$

Now, this is not a mass balance equation; it is actually a design equation which can be used for the design of a lumped system (e.g., an isothermal CSTR). Let us illustrate this for a very simple case of a first-order irreversible liquid-phase reaction, where

$$A \rightarrow B$$

and the rate of reaction is given by $r' = kC_A$, where k is the reaction rate constant (in s^{-1}) and C_A is the concentration of component A (in gmol/L). Later in this book, we will show how the same principles can be applied to distributed system and also for other rates like the rate of mass transfer for heterogeneous systems (Chapter 6).

Let us consider the single–input, single–output case, thus,

$$n_i = n_{if} + \sigma_i r' V_R$$

which, when applied to component A, gives

$$n_A = n_{Af} - kC_A V_R \tag{3.30}$$

We have a simple problem; that is, the existence of C_A and n_A in the same equation. For liquid-phase systems when there is no change in volume, this problem is trivial (for a gas-phase system with a change in number of moles accompanying the reaction, although it is simple, it is not as trivial, as will be shown later).

For this case of the liquid-phase system, we can easily write

$$n_A = qC_A \quad \text{and} \quad n_{Af} = qC_{Af}$$

Thus, Eq. (3) becomes

$$qC_A = qC_{Af} - kC_A V_R \tag{3.31}$$

Thus, for a specific volumetric flow rate q and the rate of reaction constant k, we can obtain the volume of the reactor that is needed to achieve a certain conversion.

Rearranging Eq. (3.31) gives

$$V_R = \frac{qC_{Af}}{(q+k)C_A}$$

Thus, for example, if we want 90% conversion, then $C_A = 0.1C_{Af}$ and V_R can be obtained as

$$V_R = \frac{q}{0.1(q+k)}$$

This is how simply the mass balance equation has been turned into a design equation (a mathematical model for the lumped isothermal system).

The same very simple principles apply to the heat balance equations for a nonisothermal system and also for distributed systems as shown in the following section. It also applies to the heterogeneous system, as shown in Chapter 6.

Now, we will apply these simple principles to batch reactors, CSTRs, and tubular reactors (distributed systems), starting with isothermal systems followed by nonisothermal systems.

3.6.2 Isothermal Reactors (Temperature Is Constant)

We will illustrate the design problem using the consecutive reactions network problem for CSTR, batch and tubular reactors. We consider the following reaction network:

$$3A \xrightarrow{k_1} 2B \xrightarrow{k_2} 5C$$

The Rates of Reaction

- This network is a three-component, two-reaction network. The two reactions are obviously independent.
- Accordingly, the system can be described in terms of two state variables, for example, n_A and n_B or x_A and Y_B. [n_A and n_B are the molar flow rates (or number of moles of A and B respectively, and $x_A = (n_{Af} - n_A)/n_{Af}$ and $Y_B = (n_B - n_{Bf})/n_{Af}$ are the conversion of A and yield of B, respectively. n_{Af} and n_{Bf} are the molar flow rates at the entrance (feed point) of the reactor for the continuous-flow reactors (CSTR and tubular) or the initial number of moles in a batch reactor].

The rates of reaction can be given for the production or consumption of the components or they can be given for the reactions that lead to the generalized rates of reaction studied earlier.

The network can be written as

$$3A \xrightarrow{k_1} 2B \qquad (3.32)$$

$$2B \xrightarrow{k_2} 5C \qquad (3.33)$$

Define all rates for components as production rates:

$$R_A = \text{Rate of production of } A = -K_1 C_A \; \frac{\text{moles of } A \text{ converted}}{\text{L s}}$$

$$R_B = \text{Rate of production of } B = \tfrac{2}{3}K_1 C_A - K_2 C_B \; \frac{\text{moles of } B \text{ produced}}{\text{L s}}$$

$$R_C = \text{Rate of production of } C = \tfrac{5}{2}K_2 C_B \; \frac{\text{moles of } C \text{ produced}}{\text{L s}}$$

In a more detailed fashion, we write

R_{A1} = Rate of production of A in reaction (3.32)

$$= -K_1 C_A \frac{\text{moles of } A \text{ converted}}{\text{L s}}$$

R_{B1} = Rate of production of B in reaction (3.32)

$$= \frac{\sigma_{B1}}{\sigma_{A1}}(-K_1 C_A) = \left(\frac{2}{-3}\right)(-K_1 C_A) = \left(\frac{2}{3} K_1 C_A\right) \frac{\text{moles of } B \text{ produced}}{\text{L s}}$$

R_{B2} = Rate of production of B in reaction (3.33)

$$= -K_2 C_B \frac{\text{moles of B produced}}{\text{L s}}$$

R_{C2} = Rate of production of C in reaction (3.33)

$$= \frac{\sigma_{C2}}{\sigma_{B2}}(-K_2 C_B) = \left(\frac{5}{-2}\right)(-K_2 C_B) = \left(\frac{5}{2} K_2 C_B\right) \frac{\text{moles of } C \text{ produced}}{\text{L s}}$$

As explained earlier, it is much more convenient to define the rates for the reaction rather than the components:

$$3A \xrightarrow{k_1} 2B \quad r_1' \text{ is the generalized rate of this first reaction}$$

$$2B \xrightarrow{k_2} 5C \quad r_2' \text{ is the generalized rate of this second reaction}$$

The generalized rates of reactions r_1' and r_2' (moles per unit volume per unit time) can be defined as explained in the following very simple steps.

The generalized rate r_j' for any reaction j is defined as the rate of production of any component in reaction j divided by the stoichiometric number of the said component; therefore,

$$r_1' = \frac{R_{A1}}{\sigma_{A1}} = \frac{-K_1 C_A}{-3} = \frac{K_1}{3} C_A = k_1 C_A \frac{\text{gmol}}{\text{L s}}$$

where

$$k_1 = \frac{K_1}{3}$$

or (giving exactly the same result),

$$r_1' = \frac{R_{B1}}{\sigma_{B1}} = \frac{2/3 K_1 C_A}{2} = \frac{K_1}{3} C_A = k_1 C_A \frac{\text{gmol}}{\text{L s}}$$

where

$$k_1 = \frac{K_1}{3}$$

and

$$r_2' = \frac{R_{B2}}{\sigma_{B2}} = \frac{-K_2 C_B}{-2} = \frac{K_2}{2} C_B = k_2 C_B \frac{\text{gmol}}{\text{L s}}$$

where

$$k_2 = \frac{K_2}{2}$$

or (giving exactly the same result)

$$r_2' = \frac{R_{C2}}{\sigma_{C2}} = \frac{5/2 K_2 C_B}{5} = \frac{K_2}{2} C_B = k_2 C_B \frac{\text{gmol}}{\text{L s}}$$

where

$$k_2 = \frac{k_2}{2}$$

Note: Notice that we called the generalized rates used above r_1' and r_2' as per unit volume of the reactor, whereas earlier we called them r_1 and r_2, as the overall rates. In other words, r_1' and r_2' are in moles per unit volume per unit time, whereas r_1 and r_2 were in moles per unit time for the whole unit. As explained in the previous few pages, this is the only real difference between the mass balance (inventory) equations and the design equations; for example,

1. For a CSTR,

$$r_j = V_{\text{CSTR}} r_j'$$

2. For a tubular reactor,

$$r_j = \Delta V_t r_j'$$

where ΔV_t is a difference element of the volume of the tubular reactor.

3. For a batch reactor,

$$r_j = V \Delta t r_j'$$

where Δt is a difference element of time.

With this simple and obvious background material, we can tackle the design problem for different configurations of isothermal reactors.

The general mass balance equation for the reacting system with single input–single output (SISO) and N multiple reactions is given by

$$n_i = n_{if} + \sum_{j=1}^{N} \sigma_{ij} r_j \quad i = 1, 2, \ldots, M$$

where M is the number of components, N is the number of reactions, $r_j = \text{mol/time} = V r_j'$ and $r_j' = \text{mol/(volume)(time)}$

I. The Continuous-Stirred Tank Reactor (CSTR)

The design equation for component A is

$$n_A = n_{Af} + \sigma_{A1} V r_1' + \sigma_{A2} V r_2'$$

It can be rearranged in the form

$$n_A = n_{Af} - 3 V k_1 C_A \tag{3.34}$$

For component B,

$$n_B = n_{Bf} + \sigma_{B1} V r_1' + \sigma_{B2} V r_2'$$

On rearrangement, we get

$$n_B = n_{Bf} + 2 V k_1 C_A - 2 V k_2 C_B \tag{3.35}$$

For component C,

$$n_C = n_{Cf} + \sigma_{C1} V r_1' + \sigma_{C2} V r_2'$$

On rearrangement,

$$n_C = n_{Cf} + 5 V k_2 C_B \tag{3.36}$$

From Eqs. (3.34)–(3.36), by addition we get

$$\frac{n_A}{3} + \frac{n_B}{2} + \frac{n_C}{5} = \frac{n_{Af}}{3} + \frac{n_{Bf}}{2} + \frac{n_{Cf}}{5} \tag{3.37}$$

Equation (3.37) is essentially the same overall mass balance relation obtained for any kind of reactor:

$$\sum_{i=1}^{M} \frac{n_i}{|\sigma_i|} = \sum_{i=1}^{M} \frac{n_{if}}{|\sigma_i|}$$

It is obvious that we need only Eqs (3.34) and (3.35) to solve the system, because n_C can be obtained from Eq. (3.37) when n_A and n_B are computed. Design equations (3.34) and (3.35) can be formulated in term of n_A and n_B or in terms of C_A and C_B or in terms of x_A (conversion of A) and Y_B (yield of B) as discussed earlier.

We will illustrate this for the more difficult gas-phase case (isothermal and constant pressure). Formulation of the design equations in terms of n_A and n_B can be done as follows:

$$C_A = \frac{n_A}{q} \quad \text{and} \quad C_B = \frac{n_B}{q}$$

where q is the volumetric flow rate.

Also, we know that

$$Pq = n_T RT$$

(for nonideal gases we have to use the compressibility factor Z).

where n_T is the total molar flow rate. Thus, the relation between the volumetric flow rate and n_T (the total molar flow rate), T (temperature), and P (pressure) will be

$$q = \frac{n_T RT}{P}$$

which can be rewritten as

$$q = \alpha n_T$$

where

$$n_T = n_A + n_B + n_C \quad \text{and} \quad \alpha = \frac{RT}{P}$$

From Eq. (3.37),

$$n_C = 5\left(\frac{\sum n_{if}}{|\sigma_i|} - \frac{n_A}{3} - \frac{n_B}{2}\right) = f_1(n_A, n_B)$$

and

$$q = \alpha[n_A + n_B + f_1(n_A, n_B)] = f_2(n_A, n_B)$$

where f_1 and f_2 are functions of n_A and n_B only. Thus,

$$C_A = \frac{n_A}{q} = \frac{n_A}{f(n_A, n_B)} = g_A(n_A, n_B) \tag{3.38}$$

$$C_B = \frac{n_B}{q} = \frac{n_B}{f(n_A, n_B)} = g_B(n_A, n_B) \tag{3.39}$$

where g_A and g_B are functions of n_A and n_B only.

Using Eqs. (3.38) and (3.39) in Eqs. (3.34) and (3.35), we get

$$n_A = n_{Af} - 3Vk_1 g_A(n_A, n_B) \tag{3.40}$$

$$n_B = n_{Bf} + 2Vk_1 g_A(n_A, n_B) - 2Vk_2 g_B(n_A, n_B) \tag{3.41}$$

Equations (3.40) and (3.41) are two equations in two variables n_A and n_B; they can be solved to obtain n_A and n_B. When n_A and n_B are computed, we can easily compute q (is a function of n_A and n_B), C_A, C_B, x_A, and Y_B, where

$$x_A = \frac{n_{Af} - n_A}{n_{Af}} \quad \text{and} \quad Y_B = \frac{n_B - n_{Bf}}{n_{Af}}$$

and C_A and C_B in terms of n_A, n_B, and q as shown above.

Exercise

Formulate the design equations for the above-discussed CSTR in terms of C_A and C_B and in terms of x_A and Y_B.

II. Batch Reactors

This is a system which is "distributed" with respect to time.

For the batch reactor with variable volume (constant pressure), the gas-phase reaction is

$$3A \xrightarrow{k_1} 2B$$

$$2B \xrightarrow{k_2} 5C$$

The design equation for component A (carried out over an element of time, Δt) is

$$n_A(t + \Delta t) = n_A(t) + \sigma_{A1} V r'_1 \Delta t$$

This difference form can be written as

$$n_A + \Delta n_A = n_A + \sigma_{A1} V r'_1 \Delta t$$

which, after rearrangement and taking the limit as $\Delta t \to 0$, gives

$$\frac{dn_A}{dt} = \sigma_{A1} V r'_1 = \sigma_{A1} V k_1 C_A = -3V k_1 C_A \tag{3.42}$$

Similarly,

$$\frac{dn_B}{dt} = \sigma_{B1} V r'_1 + \sigma_{B2} V r'_2 = V(\sigma_{B1} r'_1 + \sigma_{B2} r'_2)$$

Thus,

$$\frac{dn_B}{dt} = V[2k_1 C_A - 2k_2 C_B] = 2V(k_1 C_A - k_2 C_B) \tag{3.43}$$

Similarly,

$$\frac{dn_C}{dt} = \sigma_{C1} V r'_2 = 5V k_2 C_B \tag{3.44}$$

Here, we will demonstrate some of the possible simple manipulations that help to reduce the dimensionality of many problems.

We can easily obtain a relation among n_A, n_B, and n_C, and therefore eliminate one of the three differential equations. Write Eqs. (3.42)–(3.44) as follows:

$$\frac{1}{3}\frac{dn_A}{dt} = -Vk_1C_A \tag{3.45}$$

$$\frac{1}{2}\frac{dn_B}{dt} = V(k_1C_A - k_2C_B) \tag{3.46}$$

$$\frac{1}{5}\frac{dn_C}{dt} = Vk_2C_B \tag{3.47}$$

By addition of equations (3.45)–(3.47), we get

$$\frac{d}{dt}\left(\frac{n_A}{3} + \frac{n_B}{2} + \frac{n_C}{5}\right) = 0$$

Thus, by integration, we get

$$\frac{n_A}{3} + \frac{n_B}{2} + \frac{n_C}{5} = C_1$$

where C_1 is a constant and is evaluated from the initial conditions, at $t = 0$, $n_A = n_{Af}$, $n_B = n_{Bf}$, and $n_C = n_{Cf}$. Therefore,

$$\frac{n_A}{3} + \frac{n_B}{2} + \frac{n_C}{5} = \frac{n_{Af}}{3} + \frac{n_{Bf}}{2} + \frac{n_{Cf}}{5} \tag{3.48}$$

Equation (3.48) can be rewritten in the generalized form

$$\sum_{i=1}^{M} \frac{n_i}{|\sigma_i|} = \sum_{i=1}^{M} \frac{n_{if}}{|\sigma_i|} \tag{3.49}$$

where M is the number of components. Therefore, it is clear that Eq. (3.48) can be used to compute n_C without the need to solve the differential equation (3.47).

Now, we are left with Eqs. (3.45) and (3.46), which have the state variables n_A, n_B, C_A, and C_B; however, it is very easy to note that (for the case of constant reactor volume)

$$C_j V = n_j \tag{3.50}$$

Therefore, Eqs. (3.45) and (3.46) can be rewritten as

$$\frac{dn_A}{dt} = -3k_1n_A \tag{3.51}$$

and

$$\frac{dn_B}{dt} = 2(k_1 n_A - k_2 n_B) \tag{3.52}$$

with the initial conditions, at $t = 0$, $n_A = n_{Af}$ and $n_B = n_{Bf}$. The two equations are linear and can be solved analytically to obtain the change of n_A and n_B with time as follows. Note that Eq. (3.51) does not include n_B and can thus be solved independently of Eq. (3.52) as follows:

$$\frac{dn_A}{n_A} = -3k_1 dt$$

Integration gives

$$\ln n_A = -3k_1 t + C_1$$

Because, at $t = 0$, $n_A = n_{Af}$, we have,

$$\ln n_{Af} = C_1$$

and

$$\ln\left(\frac{n_A}{n_{Af}}\right) = -3k_1 t$$

Thus,

$$n_A = n_{Af} e^{-3k_1 t} \tag{3.53}$$

Substituting the value of n_A from Eq. (3.53) into Eq. (3.52) gives

$$\frac{dn_B}{dt} = 2k_1 n_{Af} e^{-3k_1 t} - 2k_2 n_B$$

which can be arranged in the following form in order to use the integration factor method to solve it:

$$\frac{dn_B}{dt} + 2k_2 n_B = 2k_1 n_{Af} e^{-3k_1 t}$$

Multiplying by the integration factor $(e^{2k_2 t})$, we get

$$e^{2k_2 t}\frac{dn_B}{dt} + 2e^{2k_2 t}k_2 n_B = 2k_1 n_{Af} e^{-3k_1 t} e^{2k_2 t}$$

We can put this in the following form:

$$\frac{d\left(n_B e^{2k_2 t}\right)}{dt} = 2k_1 n_{Af} e^{(2k_2 - 3k_1)t}$$

By integration we obtain

$$e^{2k_2 t} n_B = \frac{2k_1 n_{Af} e^{(2k_2 - 3k_1)t}}{2k_2 - 3k_1} + \overline{C}_1 \tag{3.54}$$

At $t = 0$, $n_B = n_{Bf}$; thus we get:

$$n_{Bf} = \frac{2k_1 n_{Af}}{2k_2 - 3k_1} + \overline{C}_1$$

Therefore,

$$\overline{C}_1 = n_{Bf} - \frac{2k_1 n_{Af}}{2k_2 - 3k_1} \tag{3.55}$$

Substituting the value of $\overline{C}_1$ from Eq. (3.55) into Eq. (3.54) gives

$$e^{2k_2 t} n_B = \frac{2k_1 n_{Af} e^{(2k_2 - 3k_1)t}}{2k_2 - 3k_1} + n_{Bf} - \frac{2k_1 n_{Af}}{2k_2 - 3k_1}$$

which can be rearranged as

$$e^{2k_2 t} n_B = \frac{2k_1 n_{Af} \left(e^{(2k_2 - 3k_1)t} - 1 \right)}{2k_2 - 3k_1} + n_{Bf}$$

Thus,

$$n_B = \frac{2k_1 n_{Af} \left(e^{-3k_1 t} - e^{-2k_2 t} \right)}{2k_2 - 3k_1} + n_{Bf} e^{-2k_2 t} \tag{3.56}$$

Equations (3.53) and (3.56) give the change of n_A and n_B with time; if n_C is required, it can be easily computed using Eq. (3.48), as the values of n_A and n_B are known from Eqs. (3.53) and (3.56).

   If the volume change is also required, then from the following relation for ideal gases

$$PV = n_T RT \quad \text{(for nonideal gases we have use compressibility factor } Z)$$

we get

$$V = (n_A + n_B + n_C)\frac{RT}{P} = \alpha(n_A + n_B + n_C)$$

Thus,

$$V = \alpha \left(n_{Af} e^{-3k_1 t} + \frac{2k_1 n_{Af} \left(e^{-3k_1 t} - e^{-2k_2 t} \right)}{2k_2 - 3k_1} + n_{Bf} e^{-2k_2 t} \right.$$

$$+ \sum_{i=1}^{M} \frac{n_{if}}{|\sigma_i|} - \frac{n_{Af} e^{-3k_1 t}}{3} - \frac{1}{2} \frac{2k_1 n_{Af} \left(e^{-3k_1 t} - e^{-2k_2 t} \right)}{2k_2 - 3k_1}$$

$$\left. - \frac{1}{2} n_{Bf} e^{-2k_2 t} \right)$$

which simplifies to give

$$V = \alpha \left(\sum_{i=1}^{M} \frac{n_{if}}{|\sigma_i|} + \frac{2}{3} n_{Af} e^{-3k_1 t} + \frac{k_1 n_{Af} \left(e^{-3k_1 t} - e^{-2k_2 t} \right)}{2k_2 - 3k_1} + \frac{1}{2} n_{Bf} e^{-2k_2 t} \right)$$

(3.57)

If the conversion and yield are required, then we obtain the following:

Conversion of A,

$$x_A = \frac{n_{Af} - n_A}{n_{Af}}$$

Using Eq. (3.53), we get

$$x_A = 1 - \left(\frac{n_A}{n_{Af}} \right) \Rightarrow 1 - e^{-3k_1 t}$$

Yield of B,

$$Y_B = \frac{n_B - n_{Bf}}{n_{Af}}$$

Using Eq. (3.56), we get

$$Y_B = \left(\frac{n_B}{n_{Af}} - \frac{n_{Bf}}{n_{Af}} \right) = \frac{2k_1 \left(e^{-3k_1 t} - e^{-2k_2 t} \right)}{2k_2 - 3k_1} + \left(\frac{n_{Bf}}{n_{Af}} \right) e^{-2k_2 t} - \left(\frac{n_{Bf}}{n_{Af}} \right)$$

This can be further simplified to get

$$Y_B = \frac{2k_1 \left(e^{-3k_1 t} - e^{-2k_2 t} \right)}{2k_2 - 3k_1} + \left(\frac{n_{Bf}}{n_{Af}} \right) \left(e^{-2k_2 t} - 1 \right)$$

From this relation, we can obtain the optimum time for obtaining the maximum yield of B ($Y_{B_{max}}$) as depicted in Figure 3.8:

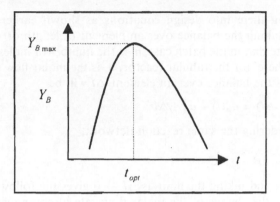

Figure 3.8   Maximum yield.

$$\frac{dY_B}{dt} = \frac{2k_1}{2k_2 - 3k_1}\left(-3k_1 e^{-3k_1 t} + 2k_2 e^{-2k_2 t}\right) + \left(\frac{n_{Bf}}{n_{Af}}\right)\left(-2k_2 e^{-2k_2 t}\right)$$

Now, we obtain the optimum time (t_{opt}) by equating dY_B/dt to zero:

$$0 = \frac{2k_1}{2k_2 - 3k_1}\left(-3k_1 e^{-3k_1 t} + 2k_2 e^{-2k_2 t}\right) + \left(\frac{n_{Bf}}{n_{Af}}\right)\left(-2k_2 e^{-2k_2 t}\right)$$

Solving the above equation to obtain t_{opt}, gives $Y_{B_{max}}$.
   For the special case of $n_{Bf} = 0$, we get

$$2k_2 e^{-2k_2 t_{opt}} = 3k_1 e^{-3k_1 t_{opt}}$$

This gives

$$e^{(3k_1 - 2k_2)t_{opt}} = \frac{3k_1}{2k_2}$$

Thus,

$$t_{opt} = \frac{\ln(3k_1/2k_2)}{3k_1 - 2k_2}$$

Thus, for such a reaction and for $n_{Bf} = 0$, the optimum time (t_{opt}) for obtaining the maximum yield of B $(Y_{B_{max}})$, which is only a function of k_1 and k_2 as shown above.

III. Tubular Reactors

For the homogeneous tubular reactor (the simplest model is the plug flow), the design equations are again obtained from the mass balance equations after turning them into design equations as shown earlier for the batch reactor by taking the balance over an element of length or volume instead of time. Note that in the batch case, n_i is the number of moles of component i, whereas, here, for the tubular reactor, n_i is the molar flow rate of component i. Thus the balance over an element Δl will be

$$n_A(l + \Delta l) = n_A(l) + \sigma_A A_t \Delta lr'$$

Again considering the same reaction network,

$$3A \xrightarrow{k_1} 2B \xrightarrow{k_2} 5C$$

Rearranging and taking the limit as $\Delta l \to 0$ gives the following differential equation for the change of the molar flow rate of component A along the length of the tubular reactor:

$$\frac{dn_A}{dl} = -3A_t k_1 C_A \tag{3.58}$$

The above equation can also be written in terms of dV:

$$\frac{dn_A}{dV} = -3k_1 C_A$$

where

$$dV = A_t \, dl$$

and A_t is the cross-sectional area of the tubular reactor.
   With the initial condition at

$$l = 0 \quad \text{(or } V = 0\text{)},$$

$$n_A = n_{Af} \quad \text{(or } C_A = C_{Af}\text{)}$$

Important notes: Note that in the batch reactor, n_A is the number of moles of component A, whereas, here, in the continuous reactor, n_A is the molar flow rate of A (mol/time). Also in the batch case, C_A is the concentration of the component A inside the batch reactor (n_A/V = number of moles in reactor divided by the volume of reactor). In the present continuous case, this in not the case; in fact, $C_A = n_A/q$ is molar flow rate of component A divided by the total volumetric flow rate. These facts will slightly complicate the formulation of the design equation as will be shown in this section.

   In this continuous-flow system (gas flow system), the change in number of moles causes a change in the volumetric flow rate (q) and no change in pressure. The balance on component B gives

$$\frac{dn_B}{dl} = A_t(2k_1 C_A - 2k_2 C_B) \tag{3.59}$$

Similarly,

$$\frac{dn_B}{dV} = (2k_1 C_A - 2k_2 C_B)$$

For component C,

$$\frac{dn_C}{dl} = 5A_t k_2 C_B \tag{3.60}$$

From Eqs. (3.58)–(3.60), it is clear that the same relation among n_A, n_B, and n_C obtained earlier for the CSTR and batch reactors can be obtained:

$$\frac{n_A}{3} + \frac{n_B}{2} + \frac{n_C}{5} = \frac{n_{Af}}{3} + \frac{n_{Bf}}{2} + \frac{n_{Cf}}{5} \tag{3.61}$$

Therefore, only Eqs. (3.58) and (3.59) need to be solved [because n_C can be computed from Eq. (3.61) when n_A and n_B have been computed). However, these equations contain the expression for components A and B in two different forms; we have n_A and n_B as well as C_A and C_B. We can handle this situation by writing everything in terms of n_A and n_B, or C_A and C_B, or x_A and Y_B, as discussed earlier.

Formulation in terms of n_A and n_B yields

$$C_A = \frac{n_A}{q} \quad \text{and} \quad C_B = \frac{n_B}{q}$$

Also,

$$Pq = n_T RT$$

$$q = \frac{n_T RT}{P}$$

Therefore,

$$q = \alpha n_T$$

where

$$n_T = n_A + n_B + n_C$$

From Eq. (3.61),

$$n_C = 5\left(\frac{\sum n_{if}}{|\sigma_i|} - \frac{n_A}{3} - \frac{n_B}{2}\right)$$

Therefore, we can write

$$n_T = n_A + n_B + 5\left(\frac{\sum n_{if}}{|\sigma_i|} - \frac{n_A}{3} - \frac{n_B}{2}\right)$$

which gives

$$n_T = 5\left(\frac{\sum n_{if}}{|\sigma_i|}\right) - \frac{2}{3}n_A - \frac{3}{2}n_B$$

This can be rewritten simply as

$$n_T = 5\frac{n_{if}}{3} + 5\frac{n_{Bf}}{2} + n_{Cf} - \frac{2}{3}n_A - \frac{3}{2}n_B$$

On reorganizing, we have

$$n_T = n_{Cf} + \tfrac{1}{3}\left(5n_{Af} - 2n_A\right) + \tfrac{1}{2}\left(5n_{Bf} - 3n_B\right)$$

Thus, n_T can now be written in terms of n_A and n_B:

$$n_T = f_1(n_A, n_B)$$

Thus,

$$q = \alpha f_1(n_A, n_B)$$

So we get

$$C_A = \frac{n_A}{\alpha f_1(n_A, n_B)} = g_A(n_A, n_B) \tag{3.62}$$

and

$$C_B = \frac{n_B}{\alpha f_1(n_A, n_B)} = g_B(n_A, n_B) \tag{3.63}$$

Substituting Eqs. (3.62) and (3.63) into Eqs. (3.58) and (3.59) gives

$$\frac{dn_A}{dl} = -3A_t k_1 g_A(n_A, n_B) \tag{3.64}$$

$$\frac{dn_B}{dl} = A_t[2k_1 g_A(n_A, n_B) - 2k_2 g_B(n_A, n_B)] \tag{3.65}$$

Equations (3.64) and (3.65) are two differential equations that can be solved simultaneously using one of the standard subroutines and the initial conditions at $l = 0$ (or $V = 0$), $n_A = n_{Af}$ and $n_B = n_{Bf}$. Of course, by computing n_A and n_B at any position along the length of the reactor, all other variables can be computed, including x_A, Y_B, C_A, C_B, q, and so forth.

Formulation in terms of C_A and C_B (instead of in terms of n_A and n_B) is as follows: For this case, we write Eq. (3.58) as

$$\frac{d(qC_A)}{dl} = -3A_t k_1 C_A$$

Thus,

$$q\frac{dC_A}{dl} + C_A \frac{dq}{dl} = -3A_t k_1 C_A \tag{3.66}$$

From the formulation in terms of n_A and n_B we have

$$q = \alpha(n_A + n_B + n_C)$$

Thus,

$$\frac{dq}{dl} = \alpha\left(\frac{dn_A}{dl} + \frac{dn_B}{dl} + \frac{dn_C}{dl}\right) \tag{3.67}$$

Substituting Eqs. (3.58)–(3.60) into Eq. (3.67), we get

$$\frac{dq}{dl} = \alpha A_t(-3k_1 C_A + 2k_1 C_A - 2k_2 C_B + 5k_2 C_B)$$

which reduces to

$$\frac{dq}{dl} = \alpha A_t[-k_1 C_A + 3k_2 C_B] \tag{3.68}$$

Substituting the value of dq/dl from Eq. (3.68) into Eq. (3.66), we get

$$q\frac{dC_A}{dl} + C_A \alpha A_t(-k_1 C_A + 3k_2 C_B) = -3A_t k_1 C_A$$

On rearrangement, we get

$$\frac{dC_A}{dl} = \frac{A_t}{q}[-\alpha C_A(-k_1 C_A + 3k_2 C_B) - 3k_1 C_A] \tag{3.69}$$

Equation (3.59) can also be rearranged in term of C_B and q, and the resulting equation is solved simultaneously with Eqs. (3.68) and (3.69).

It is clear that the formulation in term of n_A and n_B is more straightforward than the formulation in terms of C_A and C_B.

Exercise

Formulate the design equations in terms of x_A and Y_B, as defined earlier.

The Simple Special Case of No Change in Number of Moles (or Liquid Phase) Systems

The simple procedure given here is applicable to gas phase systems with no change in the number of moles and also to liquid phase in general, where the change of number of moles accompanying the reaction has negligible effect on volumes and volumetric flow rates.

Consider the reaction

$$A \xrightarrow{k_1} B \xrightarrow{k_2} C$$

This is a much simpler case and the design equations can be formulated easily in terms of C_A and C_B (or any other form: n_A and n_B or x_A and Y_B).

CSTR

The balance for component A (design equation) is given by

$$n_A = n_{Af} - V k_1 C_A$$

In this case,

$$n_A = q C_A$$

where q is a constant. So we can also write

$$n_{Af} = q C_{Af}$$

Thus, the above equation can be written as

$$q C_A = q C_{Af} - V k_1 C_A$$

Similarly,

$$q C_B = q C_{Bf} + V k_1 C_A - V k_2 C_B$$
$$q C_C = q C_{Cf} + V k_2 C_B$$

and

$$C_A + C_B + C_C = C_{Af} + C_{Bf} + C_{Cf}$$

Clearly, the resulting design equations are much simpler than the gas-phase case, which is accompanied by a change in the number of moles.

Batch Reactor

The balance for component A is

$$\frac{dn_A}{dt} = -V k_1 C_A$$

where

$$n_A = V C_A \quad \text{and} \quad V = \text{constant}$$

Thus,

$$\frac{d(VC_A)}{dt} = -Vk_1C_A$$

Because V is constant, we can write,

$$V\frac{dC_A}{dt} = -Vk_1C_A$$

which gives

$$\frac{dC_A}{dt} = -k_1C_A$$

Similarly,

$$\frac{dC_B}{dt} = k_1C_A - k_2C_B$$

and

$$\frac{dC_C}{dt} = k_2C_B$$

with initial conditions at $t = 0$, $C_A = C_{Af}$, $C_B = C_{Bf}$, and $C_C = C_{Cf}$. Also,

$$C_A + C_B + C_C = C_{Af} + C_{Bf} + C_{Cf}$$

Again, the design equations are much simpler than the gas-phase case, which is accompanied by change in the number of moles.

Plug Flow (Tubular)

The balance for component A is

$$\frac{dn_A}{dl} = -A_t k_1 C_A$$

Here,

$$n_A = qC_A \quad \text{and} \quad q = \text{constant}$$

Therefore,

$$q\frac{dC_A}{dl} = -A_t k_1 C_A$$

Similarly,

$$q\frac{dC_B}{dl} = A_t(k_1 C_A - k_2 C_B)$$

and,

$$q\frac{dC_C}{dl} = A_t k_2 C_B$$

With initial conditions at $l = 0$ (or $V = 0$), $C_A = C_{Af}$, $C_B = C_{Bf}$, and $C_C = C_{Cf}$ and

$$C_A + C_B + C_C = C_{Af} + C_{Bf} + C_{Cf}$$

As seen for the previous two cases, these design equations are much simpler than the gas-phase system accompanied by a change in the number of moles.

3.6.3 Nonisothermal Reactors

The generalized heat balance equation for L inputs, K outputs, M components, and N reactions is the following single equation:

$$\sum_{l=1}^{L} \sum_{i=1}^{M} n_{if_l}(H_{if_l} - H_{ir}) + Q = \sum_{k=1}^{K} \sum_{i=1}^{M} n_{ik}(H_{ik} - H_{ir}) + \sum_{j=1}^{N} r_j \Delta H_j$$

Similar to what we have shown earlier for the mass balance, the above equation is the very general heat balance equation for homogeneous systems. All other cases are special cases of this generalized form:

1. For no reactions, we just put all r_j's equal to zero.
2. For adiabatic operation, we simply put $Q = 0$.
3. For single-input, single-output, and single reaction, the general equation becomes

$$\sum_{i=1}^{M} n_{if}(H_{if} - H_{ir}) + Q = \sum_{i=1}^{M} n_i(H_i - H_{ir}) + r\Delta H$$

and so on.

Now, we start with some specific examples combining the mass and heat balance equations. In order to turn this heat balance equation into a part of the design equations, we simply replace r_j with Vr'_j, as we did earlier for mass balance.

I. Nonisothermal CSTR

Mass Balance

For a case of single-input, single-output, and N reactions, the mass balance equations for any component is given by

$$n_i = n_{if} + \sum_{j=1}^{N} \sigma_{ij} r_j \tag{3.70}$$

where $i = 1, 2, 3, \ldots, M$, M is the number of components, N is the number of reactions, and r_j is the generalized overall rate of reaction for reaction j [in mol/time (rate of reaction for the entire unit)].

As we have discussed earlier, in order to turn the mass balance [Eq. (3.70)] into a design equation, we write it as

$$n_i = n_{if} + V \sum_{j=1}^{N} \sigma_{ij} r_j' \qquad (3.71)$$

Here, r_j' is the generalized rate of reaction for reaction j [in mol/(time $\times$ volume)] (it is equal to r_j for a unit volume of reactor). For a single reaction, the design equation simply becomes

$$n_i = n_{if} + V \sigma_i r' \qquad (3.72)$$

Heat Balance

The heat balance for a single-input, single-output and N reactions can be written as

$$\sum_{i=1}^{M} n_{if} \left(H_{if} - H_{ir} \right) + Q = \sum_{i=1}^{M} n_i (H_i - H_{ir}) + \sum_{j=1}^{N} (\Delta H_j)_r r_j \qquad (3.73)$$

where $(\Delta H_j)_r$ is the heat of reaction for reaction j and Q is the heat added to the system.

Derivation of the Heat Balance Equation (3.73) for a Single Reaction

To remind the reader of the derivation of this equation, which is quite important, we derive it again here. The heat balance equation is as follows:

$$\sum_{i=1}^{M} n_{if} H_{if} + Q = \sum_{i=1}^{M} n_i H_i \qquad (3.74)$$

However, now we have to put it in the form of enthalpy difference, this can be done as follows:

1. Subtract $\sum_{i=1}^{M} n_{if} H_{ir}$ from the left-hand side (therefore, subtract it from the right-hand side so as not to change the equation).
2. Subtract $\sum_{i=1}^{M} n_i H_{ir}$ from the right-hand side (therefore, subtract it from the left-hand side so as not to change the equation).

Thus, we get

$$\sum_{i=1}^{M} n_{if} H_{if} - \sum_{i=1}^{M} n_{if} H_{ir} - \sum_{i=1}^{M} n_i H_{ir} + Q = \sum_{i=1}^{M} n_i H_i - \sum_{i=1}^{M} n_{if} H_{ir} - \sum_{i=1}^{M} n_i H_{ir}$$

(3.75)

Rearranging gives

$$\sum_{i=1}^{M} n_{if} (H_{if} - H_{ir}) + Q = \sum_{i=1}^{M} n_i (H_i - H_{ir}) + \sum_{i=1}^{M} (n_i - n_{if}) H_{ir}$$

(3.76)

However, from the mass balance for a single reaction, we have

$$n_i = n_{if} + \sigma_i r$$

(3.77)

Using Eq. (3.77) in Eq. (3.76), we get

$$\sum_{i=1}^{M} n_{if} (H_{if} - H_{ir}) + Q = \sum_{i=1}^{M} n_i (H_i - H_{ir}) + \sum_{i=1}^{M} \sigma_i r H_{ir}$$

(3.78)

From the definition of the heat of reaction, we have

$$\sum_{i=1}^{M} \sigma_i H_{ir} = \Delta H_r$$

Therefore, Eq. (3.78) becomes

$$\sum_{i=1}^{M} n_{if} (H_{if} - H_{ir}) + Q = \sum_{i=1}^{M} n_i (H_i - H_{ir}) + r(\Delta H_r)$$

(3.79)

where ΔH_r is the heat of reaction at reference conditions r for the reaction as stoichiometrically written.

What do we mean by "the heat of reaction for the reaction as stoichiometrically written"? Suppose we have

$$3A + 5B \rightarrow 6C + 8D$$

Because we define the heat of reaction as

$$\Delta H_r = 8H_D + 6H_C - 3H_A - 5H_B$$

which is

$$\Delta H_r = \sum_{i=1}^{M} \sigma_i H_i$$

this ΔH_r is not per mole of A, B, C, or D, but it is actually per 3 mol of A reacted, 5 mol of B reacted, 6 mol of C produced, or 8 mol of D produced. If we want it per mole of any component, we should divide by $|\sigma_i|$, where i is that particular component for which we want to make the heat of reaction per mole of it.

In most cases, ΔH_r is computed from the heat of formation or heats of combustion as follows:

$$\Delta H_r = \sum \sigma_i \Delta H_{fi}$$

or

$$\Delta H_r = -\sum \sigma_i \Delta H_{ci}$$

These heats of reaction are the correct ones to be used in Eq. (46), because they are for the reaction as stoichiometrically written.

However, if ΔH_r is obtained experimentally and is defined as $\Delta \overline{H}_r$, which is in calories per mole of A, then the correct ΔH_r to be used in Eq. (3.79) is $\Delta \overline{H}_r |\sigma_A|$; in general,

$$\Delta H_r = (\Delta \overline{H}_{ri} |\sigma_i|)$$

For reactor design purposes, of course r should be written as Vr' (for CSTR), or $\Delta Vr'$ (for tubular), or $V\Delta tr'$ (for batch). Thus, Eq. (3.79) for a CSTR should be written as

$$\sum_{i=1}^{M} n_{if}\left(H_{if} - H_{ir}\right) + Q = \sum_{i=1}^{M} n_i(H_i - H_{ir}) + Vr'(\Delta H_r)$$

For multiple reactions, it will be

$$\sum_{i=1}^{M} n_{if}(H_{if} - H_{ir}) + Q = \sum_{i=1}^{M} n_i(H_i - H_{ir}) + V\sum_{j=1}^{N} r'_j(\Delta H_{rj}) \qquad (3.80)$$

where M is total number of components involved (i.e., reactants + products + inerts) and N is the number of reactions.

For multiple inputs and multiple outputs, we have

$$\sum_{l=1}^{L}\sum_{i=1}^{M} n_{if_l}\left(H_{if_l} - H_{ir}\right) + Q = \sum_{k=1}^{K}\sum_{i=1}^{M} n_{ik}(H_{ik} - H_{ir}) + V\sum_{j=1}^{N} r'_j \Delta H_j$$

$$(3.81)$$

Because our entire book prior to heterogeneous systems deals with homogeneous systems, in this part of the book no change of phase is involved and therefore the change of enthalpies are changes in sensible heats only. Therefore, Eq. (3.81) can be rewritten as

$$\sum_{i=1}^{M} n_{if} \int_{T_r}^{T_f} C_{pi} \, dT + Q = \sum_{i=1}^{M} n_i \int_{T_r}^{T} C_{pi} \, dT + V \sum_{j=1}^{N} r_j'(\Delta H_{rj}) \qquad (3.82)$$

Equations (3.71) and (3.82) are the basic design equations for the design of nonisothermal CSTR. For tubular and batch reactors, the same equations are slightly modified to put them in a suitable differential equation form.

II. Nonisothermal Tubular Reactors

This case, of course, belongs to the distributed system type, which will be covered in more detail in Chapter 4. For this distributed case, Eq. (3.71) becomes

$$n_{i(l+\Delta l)} = n_{i(l)} + A_t \Delta l \sum_{j=1}^{N} \sigma_{ij} r_j'$$

which can be written as

$$n_i + \Delta n_i = n_i + A_t \Delta l \sum_{j=1}^{N} \sigma_{ij} r_j'$$

After rearrangement and taking the limit as $\Delta l \to 0$, we get the differential equation

$$\frac{dn_i}{dl} = A_t \sum_{j=1}^{N} \sigma_{ij} r_j' \qquad (3.83)$$

For a single reaction, the above equation becomes

$$\frac{dn_i}{dl} = A_t \sigma_i r' \qquad (3.84)$$

To get the heat balance design equation, an enthalpy balance over an element Δl for a single reaction gives

$$\sum_{i=1}^{M} (n_i H_i)_l + Q' \Delta l = \sum_{i=1}^{M} (n_i H_i)_{l+\Delta l}$$

where M is the total number of components involved (reactants + products + inerts), including components not in the feed but created during the reaction (they will have $n_{if} = 0$) and Q' is the rate of heat removed or added per unit length of reactor. Rearranging gives

$$\sum_{i=1}^{M} \frac{\Delta n_i H_i}{\Delta l} - Q' = 0$$

Taking the limit as $\Delta l \rightarrow 0$, we obtain the differential equation

$$\sum_{i=1}^{M} \frac{d(n_i H_i)}{dl} - Q' = 0$$

which gives

$$\sum_{i=1}^{M} \left(n_i \frac{dH_i}{dl} + H_i \frac{dn_i}{dl} \right) - Q' = 0 \qquad (3.85)$$

Using Eqs. (3.84) and (3.85), we get

$$\sum_{i=1}^{M} \left(n_i \frac{dH_i}{dl} + H_i A_t \sigma_i r' \right) - Q' = 0$$

Rearrangement gives

$$\sum_{i=1}^{M} n_i \frac{dH_i}{dl} + \left(A_t r' \sum_{i=1}^{M} H_i \sigma_i \right) - Q' = 0$$

Because

$$\sum_{i=1}^{M} \sigma_i H_i = (\Delta H) \equiv \text{heat of reaction}$$

we get

$$\sum_{i=1}^{M} n_i \frac{dH_i}{dl} + A_t r'(\Delta H) - Q' = 0$$

Finally the above equation can be written as

$$\sum_{i=1}^{M} n_i \frac{dH_i}{dl} = A_t r'(-\Delta H) + Q' \qquad (3.86)$$

For multiple reactions, it will be

$$\sum_{i=1}^{M} n_i \frac{dH_i}{dl} = \left(\sum_{j=1}^{N} A_t r_j'(-\Delta H_j) \right) + Q' \qquad (3.87)$$

The design equations for the multiple reactions case are Eqs. (3.83) and (3.87), whereas for the single-reaction case, the design equations are Eqs. (3.84) and (3.86).

III. Nonisothermal Batch Reactor

For this case, mass balance design equations are obtained as follows (balance over a time element Δt):

$$n_i(t + \Delta t) = n_i(t) + V\Delta t \sum_{j=1}^{N} \sigma_{ij} r'_j$$

Rearranging and taking the limit as $\Delta t \to 0$ gives

$$\frac{dn_i}{dt} = V \sum_{j=1}^{N} \sigma_{ij} r'_j \tag{3.88}$$

For single reaction, it will be

$$\frac{dn_i}{dt} = V\sigma_i r' \tag{3.89}$$

The heat balance (for a single reaction) is

$$\sum_{i=1}^{M} (n_i H_i)_t + Q'\Delta t = \sum_{i=1}^{M} (n_i H_i)_{t+\Delta t}$$

After rearranging and taking the limit as $\Delta t \to 0$, we obtain the following differential equation:

$$\sum_{i=1}^{M} \frac{d(n_i H_i)}{dt} - Q' = 0$$

After differentiation, we obtain

$$\left[\sum_{i=1}^{M} \left(n_i \frac{dH_i}{dt} + H_i \frac{dn_i}{dt} \right) \right] - Q' = 0 \tag{3.90}$$

Using Eqs. (3.89) and (3.90), we get

$$\left[\sum_{i=1}^{M} \left(n_i \frac{dH_i}{dt} + H_i V\sigma_i r' \right) \right] - Q' = 0$$

Rearranging gives

$$\sum_{i=1}^{M} n_i \frac{dH_i}{dt} + \left(Vr' \sum_{i=1}^{M} H_i \sigma_i \right) - Q' = 0$$

Because

$$\sum \sigma_i H_i = (\Delta H) \equiv \text{heat of reaction}$$

we obtain

$$\sum_{i=1}^{M} n_i \frac{dH_i}{dt} + Vr'(\Delta H) - Q' = 0$$

The above equation can be rearranged and written as

$$\sum_{i=1}^{M} n_i \frac{dH_i}{dt} = Vr'(-\Delta H) + Q' \tag{3.91}$$

For multiple reactions,

$$\sum_{i=1}^{M} n_i \frac{dH_i}{dt} = \left(V \sum_{j=1}^{N} r_j'(-\Delta H_j) \right) + Q' \tag{3.92}$$

Equations (3.88) and (3.92) are the design equations for the multiple-reactions case and equations (3.89) and (3.91) are the design equations for the single-reaction case.

3.7 SIMPLE EXAMPLES FOR THE GENERAL EQUATIONS

In this section, we show the design (model) equations for the three types of reactors. We consider only the one-phase (homogeneous) case.

I. Nonisothermal CSTR

Consider the simplest case of a single reaction with no change in the number of moles:

$$A \xrightarrow{k} B, \quad \text{with } r' = kC_A$$

Equation (3.72) becomes

$$n_A = n_{Af} + V(-1)kC_A \tag{3.93}$$

Because there is no change in the number of moles, q (the volumetric flow rate) is constant; therefore,

$$n_A = qC_A \quad \text{and} \quad n_{Af} = qC_{Af}$$

Thus, Eq. (3.93) becomes

$$qC_A = qC_{Af} - VkC_A$$

and k depends on the temperature according to the Arrhenius relation

$$k = k_0 e^{(-E/RT)}$$

The mass balance design equation is thus

$$qC_A = qC_{Af} - Vk_0 e^{(-E/RT)} C_A \tag{3.94}$$

Equation (3.94) has two variables, C_A and T; therefore, we must use the heat balance equations also.

We assume that there is no change in phase and use Eq. (3.82) and after putting $N = 1$ (single reaction), we get

$$\left(\sum_{i=1}^{M} n_{if} \int_{T_r}^{T_f} C_{pi}\, dT \right) + Q = \left(\sum_{i=1}^{M} n_i \int_{T_r}^{T} C_{pi}\, dT \right) + VkC_A(\Delta H_r)$$

We further assume adiabatic operation ($Q = 0$) and also assume that all of the C_{pi}'s are constant (or using constant average values) to obtain

$$\left(\sum_{i=1}^{M} n_{if} C_{pi}(T_f - T_r) \right) = \left(\sum_{i=1}^{M} n_i C_{pi}(T - T_r) \right) + VkC_A(\Delta H_r)$$

We further assume an average $C_{p_{mix}}$, which is the same for feed and products, to obtain

$$\left(C_{p_{mix}}(T_f - T_r) \sum_{i=1}^{M} n_{if} \right) = \left(C_{p_{mix}}(T - T_r) \sum_{i=1}^{M} n_i \right) + VkC_A(\Delta H_r)$$

We further assume that $n_{Tf} = n_T$ (actually, there is no change in the number of moles for this simple reaction; therefore, for this specific, case this is a physical fact not an assumption) and therefore the equation reduces to

$$n_T C_p' T_f = n_T C_p' T + VkC_A(\Delta H_r)$$

$C_p' = C_{p_{mix}}$, where C_p' is the molar specific heat and n_T is the molar flow rate. We can also write it in terms of specific heat C_p(per unit mass as):

$$q\rho C_p T_f = q\rho C_p T + Vk_0 e^{(-E/RT)} C_A(\Delta H_r) \tag{3.95}$$

where q is the volumetric flow rate and ρ is the average constant density of the mixture. Equations (3.94) and (3.95) are the design equations for this nonisothermal case; we can put them in a dimensionless form as follows:

$$X_A = X_{Af} - \alpha' e^{(-\gamma/Y)} X_A \tag{3.96}$$

where

$$X_A = \frac{C_A}{C_{A_{ref}}}, \quad X_{Af} = \frac{C_{Af}}{C_{A_{ref}}}, \quad \alpha' = \frac{Vk_0}{q}, \quad Y = \frac{T}{T_{ref}}$$

$$\text{and} \quad \gamma = \frac{E}{RT_{ref}}$$

Similarly, the heat balance equation can be written in a dimensionless form as

$$Y - Y_f = \alpha' e^{-(\gamma/Y)} X_A \beta \tag{3.97}$$

where

$$\beta = \frac{(-\Delta H_r) C_{ref}}{\rho C_p T_{ref}}$$

When analyzed, eqs. (3.96) and (3.97) show the complex behavior of this system and will be discussed later.

If we take

$$C_{A_{ref}} = C_{Af} \quad \text{and} \quad T_{ref} = T_f$$

Then the dimensionless heat balance equation becomes

$$Y - 1 = \alpha' e^{-(\gamma/Y)} \beta X_A \tag{3.98}$$

and the dimensionless mass balance equation becomes

$$(X_A - 1) = -\alpha' e^{-(\gamma/Y)} X_A \tag{3.99}$$

The two nonlinear algebraic equations [Eqs. (3.98) and (3.99)] have two variables (X_A and Y) and can be solved simultaneously for given values of α', β, and γ.

Note: The nine physical parameters (also design and operating parameters) are $q, \rho, C'_p, T_f, V, k_0, E, C_{Af}$, and $-\Delta H_r$. They all are lumped into three parameters: α', β, and γ

Can we get some insight into the characteristics of this system and its dependence on the parameters before we solve its design equations (3.98) and (3.99)? Let us use some chemical engineering intelligence together with Eqs. (3.98) and (3.99):

$$(Y - 1) = \alpha' e^{-(\gamma/Y)} \beta X_A \tag{3.98}$$

$$(X_A - 1) = -\alpha' e^{-(\gamma/Y)} X_A \tag{3.99}$$

Multiplying Eq. (3.99) by β and adding it to Eq. (3.98) gives

$$Y - 1 + \beta(X_A - 1) = 0$$

which can be rewritten as

$$Y = 1 + \beta(1 - X_A) \tag{3.100}$$

What is the maximum possible value of Y? It is when $X_A \to 0$ (i.e., at 100% conversion):

$$Y_{max} = 1 + \beta$$

For the exothermic reaction, β is called as the maximum dimensionless adiabatic temperature rise.

Can we reduce the two nonlinear coupled equations to one equation in one variable and the other is a simple explicit linear equation? Yes, we can do that through the two alternative equivalent route shown in Table 3.1. From either of the two routes, we get one equation in terms of Y (one variable); we can solve it and get the value of Y. Then, the value of x_A can be obtained easily from Eq. (3.101) or (3.102).

Take, for example,

$$\frac{1}{\alpha'}(Y - 1) = e^{-(\gamma/Y)}(1 + \beta - Y)$$

Table 3.1   Alternate Routes to Solve Eqs. (3.98) and (3.99)

Route 1	Route 2
Equation (3.99) can be rearranged as $$X_A + \alpha' e^{-(\gamma/Y)} X_A = 1$$ which can be written as $$X_A = \frac{1}{1 + \alpha' e^{-(\gamma/Y))}} \quad (3.101)$$ Substitute Eq. (3.101) into Eq. (3.98) to eliminate X_A and obtain $$Y - 1 = \frac{\alpha' e^{-(\gamma/Y)} \beta}{1 + \alpha' e^{-(\gamma/Y)}}$$ We can rearrange it in the following form: $$\underbrace{\frac{1}{\alpha'}(Y - 1)}_{R(Y)} = \underbrace{\frac{\beta e^{-(\gamma/Y)}}{1 + \alpha' e - (\gamma/Y)}}_{G(Y)}$$ which gives $$R(Y) = G(Y)$$ where $R(Y)$ is a heat removal function and $G(Y)$ is a heat generation function.	Equation (3.100) can be rearranged to give $$X_A = \frac{1 + \beta - Y}{\beta} \quad (3.102)$$ Use Eq. (3.102) in Eq. (3.98) to eliminate X_A and obtain $$Y - 1 = \alpha' e^{-(\gamma/Y)} \beta \frac{(1 + \beta - Y)}{\beta}$$ which can be written as $$\underbrace{\frac{1}{\alpha'}(Y - 1)}_{R(Y)} = \underbrace{e^{-(\gamma/Y)}(1 + \beta - Y)}_{\overline{G}(Y)}$$ where $R(Y)$ is a heat removal function and $\overline{G}(Y)$ is a heat generation function

It can be solved numerically (using Newton–Raphson, bisectional methods, etc.; see Appendix B), but for the sake of illustration, we solve it graphically (as shown in Fig. 3.9).

$$\underbrace{\frac{1}{\alpha'}(Y-1)}_{R(Y)} = \underbrace{e^{-(\gamma/Y)}(1+\beta-Y)}_{\overline{G}(Y)}$$

where, $\overline{G}(Y)$ is the heat generation function and $R(Y)$ is the heat removal function.

It is clear from Figure 3.9 that for an exothermic reaction, there is a possibility of three steady states (stationery nonequilibrium states), whereas for an endothermic reaction, there is only one steady state. For more information regarding this very important phenomenon (multiplicity phenomenon or bifurcation behavior), see Chapter 7.

II. Nonisothermal Tubular Reactor

Consider the simple reaction

$$A \xrightarrow{k} B$$

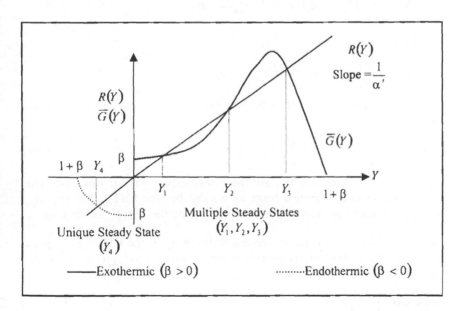

Figure 3.9 Plot of heat generation and heat removal functions (points of intersection are the steady states).

where $r' = kC_A$. The mass balance design equation [Eq. (3.84)] for component A is

$$\frac{dn_A}{dl} = -A_t k C_A$$

Because q is constant, $n_A = qC_A$ and $dn_A/dl = qdC_A/dl$; thus,

$$q\frac{dC_A}{dl} = -A_t k_0 e^{-(E/RT)}C_A \tag{3.103}$$

The heat balance design equation is

$$\sum_{i=1}^{M} n_i \frac{dH_i}{dl} = A_t k_0 e^{-(E/RT)}C_A(-\Delta H) + Q$$

For adiabatic operation, $Q = 0$, and for no change in phase, it becomes

$$\sum_{i=1}^{M} n_i \frac{d(C_{pi}T)}{dl} = A_t k_0 e^{-(E/RT)}C_A(-\Delta H)$$

For constant C_{pi}, we can write

$$\sum_{i=1}^{M} (n_i C_{pi}) \frac{dT}{dl} = A_t k_0 e^{-(E/RT)}C_A(-\Delta H)$$

Using constant molar $C_{p_{\text{mix}}}$ gives,

$$n_T C_{p_{\text{mix}}} \frac{dT}{dl} = A_t k_0 e^{-(E/RT)}C_A(-\Delta H)$$

If we use $C_{p_{\text{mix}}}$ per unit mass instead of per unit mole and call it C_p, we can write

$$q\rho C_p \frac{dT}{dl} = A_t k_0 e^{-(E/RT)}C_A(-\Delta H) \tag{3.104}$$

Equations (3.103) and (3.104) are the design equations for this case. They can be put in dimensionless form as we did for the CSTR. On the same basis, the reader can develop these equation for the batch reactor case.

Note: ΔH here is not at reference condition as in the CSTR case, it can be assumed constant or its variation with temperature can be taken into consideration.

Questions

1. Derive the simplified design equations for the batch nonisothermal reactor.

2. Write the tubular reactor and the batch reactor equations in dimensionless form.
3. Compare your dimensionless groups for the batch and the tubular reactors with those for the CSTR and comment.

3.8 MODELING OF BIOCHEMICAL SYSTEMS

Biochemical systems can be classified into two main categories: The first deals with enzyme systems and the second deals with processes catalyzed by whole-cell micro-organisms.

3.8.1 Modeling of Enzyme Systems

Biochemical systems can be handled in a very similar manner as was done so far for chemical systems in the previous chapters and sections. The few simple differences will be clarified in this subsection through simple and illustrative examples. Also, some of the most important terminology will be introduced.

Microbial systems. These are systems in which there is a biochemical reaction which is catalyzed by micro-organisms.

Micro-organisms. They contain large number of enzymes and they exist in a tissue structure. They not only catalyze the biochemical reaction, but they themselves grow also.

Micro-organism operation (Fig. 3.10):
1. Micro-organisms behave as a catalyst when fed with the substrate (reactant).
2. They grow and accumulate.
3. They release product(s)
4. They are responsive to surroundings (pH, temperature, etc.).

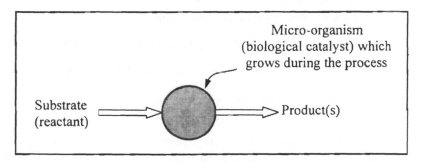

Figure 3.10 Micro-organism operation.

Microbial reactions

Substrates + Micro-organisms → Products + More micro-organisms

The process is autocatalytic, because the micro-organism is not only catalyzing the reactions, but it also grows and reproduces during the reaction.

The parameter used to describe the fraction of substrate (reactant) converted to micro-organism is called the "yield factor" (Y_S):

$$Y_S = \frac{\text{gram of micro-organism produced}}{\text{gram of substrate consumed}}$$

Characteristics of microbial systems

1. Micro-organisms grow and accumulate.
2. The reaction network is quite complex.
3. Historical experience from many processes such as fermentation are important for process development.

Enzyme Systems

There are more than 2000 different kinds of known enzyme (e.g., trypsin, pepsin, etc.).

Example

$$\text{Glucose} \xrightarrow{\text{Glucose isomerase}} \text{Fructose}$$

Enzymes also act as catalysts, but they differ from micro-organisms, as shown in the few simple points given below.

Enzymes

1. They do not grow like microorganisms.
2. Very selective in nature.
3. Soluble in some solvent to form a solution.

Micro-organisms, in contradiction, are characterized by the following:

1. They grow and accumulate in the presence of substrate and nutrients.
2. They contain a large number of enzymes in their network tissues.
3. They are not soluble in solution, they remain suspended.

Both enzymes and micro-organisms are expensive and they have to be protected from washing out with the product. *Thus, enzyme and micro-organism immobilization is used to prevent their washout.*

Immobilization

There are two main types of immobilization techniques:

1. *Attachment to solid surface*: This technique is usually used for microorganisms through attaching the micro-organism onto a solid surface either by physical or chemical bonding in order to prevent its mobility. The advantage of this technique is that there is very little or no mass transfer resistance. The disadvantage of this method is that if micro-organisms are used in a reactor where the flow rate is high, then the shear stress will remove the micro-organisms from the solid's surface.

2. *Microencapsulation*: This technique is suitable for both enzymes and micro-organisms where they are entrapped in capsules made of materials which are permeable only to substrates and products, but are impermeable to enzyme or micro-organism. The advantage of this technique is that it keeps the enzymes or micro-organisms inside the capsules even under high-flow-rate turbulent conditions. The main disadvantage of this method is that it introduces additional mass transfer resistances.

Enzyme Kinetics

A simple enzyme kinetic can be expresssed as,

$$S \xrightarrow{E} P$$

where S is the substrate and P is the product. Concentration and pH are the most important variables affecting the rates of enzyme reactions, and that is why they are the main state variables for enzyme systems.

Michaelis-Menten Kinetics

This is the simplest type of enzyme reaction giving rise to monotonic enzyme kinetics. The mechanism can be simply expressed by the following steps:

$$E + S \underset{k_{-1}}{\overset{k_1}{\Longleftrightarrow}} ES \qquad\qquad (3.105)$$

where E is an enzyme, S is a substrate, and ES is a enzyme–substrate complex. The equilibrium constant for this reversible reaction is K_S, which is given by

$$K_S = \frac{[E][S]}{[ES]} = \frac{k_{-1}}{k_1}$$

The enzyme–substrate complex gives the product P and reproduces the enzyme E according to the reaction

$$ES \xrightarrow{K} P + E \tag{3.106}$$

The rate of product production through this irreversible reaction is given by

$$r = K[ES] \tag{3.107}$$

We cannot measure the intermediate concentration of ES ($[ES]$), so we try to describe the rate of reaction in terms of the substrate concentration $[S]$. E reacts with S and at steady state; thus, we have the relation

$$K[ES] = k_1[S][E] - k_{-1}[ES] \tag{3.108}$$

where $[E]$ is the amount of enzyme available for reaction at any time. This is equal to the difference between the total amount of enzyme and the amount of enzyme used for the formation of ES. Mathematically, it can be written as

$$[E] = [E_T] - [ES]$$

where $[E_T]$ is the total amount of enzyme. Substituting this relation of $[E]$ in Eq. (3.108) gives

$$K[ES] = k_1[S]\{[E_T] - [ES]\} - k_{-1}[ES]$$

which can be rearranged to give

$$[ES]\{K + k_1[S] + k_{-1}\} = k_1[E_T][S]$$

Thus, we can obtain $[ES]$ as a function of $[E_T]$ and $[S]$ as follows:

$$[ES] = \frac{k_1[E_T][S]}{K + k_1[S] + k_{-1}} \tag{3.109}$$

Equation (3.109) can be manipulated into the following form:

$$[ES] = \left(\frac{k_1}{k_1}\right)[E_T][S] \Big/ \left(\frac{K}{k_1} + \frac{k_{-1}}{k_1} + \frac{k_1}{k_1}[S]\right)$$

Because

$$\frac{k_{-1}}{k_1} = K_S$$

we can rewrite the equation as

$$[ES] = \frac{[E_T][S]}{(K/k_1) + K_S + [S]} \tag{3.110}$$

From Eq. (3.107), the reaction rate is given as

$$r = K[ES]$$

Substituting the expression for $[ES]$ from Eq. (3.110) into Eq. (3.106) gives

$$r = \frac{K[E_T][S]}{(K/k_1) + K_S + [S]} \tag{3.111}$$

Let

$$K[E_T] = V_{max}$$

Thus, Eq. (3.111) becomes

$$r = \frac{V_{max}[S]}{(K/k_1) + K_S + [S]} \tag{3.112}$$

If k_1 is large, then the term K/k_1 is negligible, but $K_S(= k_{-1}/k_1)$ is not negligible because it is a division of two large numbers. If we assume that $K/k_1 = 0$, it means that the reaction given by Eq. (3.105) is very fast as compared to the reaction given by Eq. (3.107). Equation (3.112) becomes

$$r = \frac{V_{max}[S]}{K_S + [S]} \tag{3.113}$$

Equation (3.113) is called the Michaelis–Menten equation, and the following can be deduced from the equation:

1. If $[S]$ is very small as compared with K_S, a first-order reaction rate r is dominating, which can be described graphically by a straight line with slope V_{max}/K_S and the rate is $r = V_{max}[S]/K_S$.
2. If K_S is small as compared to $[S]$ (at high concentration), then the reaction rate r is a horizontal line having one value, $r = V_{max}$, and that will be described graphically by a straight line parallel to the x axis. The Michaelis–Menten kinetics is graphically shown in Figure 3.11.

Important Notes

1. The actual differential equations which describe the reaction and the relation between S and ES are

    $$\frac{d[S]}{dt} = -k_1[S][E] + k_{-1}[ES] \tag{3.114}$$

    and

    $$\frac{d[ES]}{dt} = k_1[S][E] - k_{-1}[ES] - K[ES] \tag{3.115}$$

    where

    $$[E] = [E_T] - [ES]$$

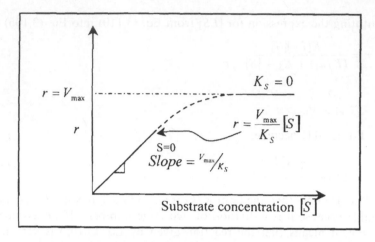

Figure 3.11 Graphical representation of Michaelis–Menten kinetics.

2. If $d[S]/dt$ and $d[ES]/dt$ in Eqs. (3.114) and (3.115) are set equal to zero, then Eq. (3.112) is satisfied and that means that the dynamics of the system is fast (equilibrium).

3. When using Eq. (3.113) to develop unsteady-state models, we should notice that there are many assumptions, including that the operation is quasi-steady-state. Also, we assume that ES produced from reaction (3.105) is the same as that reacted in reaction (3.107). Thus, Michaelis–Menten Eq. (3.113) is not rigorous enough for unsteady-state models.

Substrate-Inhibited Enzyme Kinetics

For noncompetitive substrate inhibition, the overall reaction is given by

$$S \rightarrow P \tag{3.116}$$

The formation of the active enzyme complex can be shown as

$$S + E \Leftrightarrow ES \tag{3.117}$$

The equilibrium constant for reaction (3.117) can be written as

$$K_S = \frac{[S][E]}{[ES]} = \frac{k_{-1}}{k_1}$$

The formation of inactive enzyme complex follows the pathway

$$ES + S \Leftrightarrow ES_2 \tag{3.118}$$

The equilibrium constant for reaction (3.118) is

$$K_S' = \frac{[ES][S]}{[ES_2]}$$

The active enzyme complex produces the product and enzyme again according to the pathway

$$ES \xrightarrow{K} P + E$$

The overall rate can be written as

$$r = K[ES] = \frac{K[S][E]}{K_S}$$

Also,

$$[E_T] = [E] + [ES] + [ES_2]$$

By using the equilibrium relations of reactions (3.116)–(3.118) in the definition of $[E_T]$, we get

$$[E_T] = [E] + \frac{[S][E]}{K_S} + \frac{[ES][S]}{K_S'}$$

By taking $[E]$ as a common part, where $[ES] = [S][E]/K_S$, we get

$$[E_T] = [E]\left(1 + \frac{[S]}{K_S} + \frac{[S]^2}{K_S K_S'}\right)$$

from which we obtain

$$[E] = [E_T]\bigg/\left(1 + \frac{[S]}{K_S} + \frac{[S]^2}{K_S K_S'}\right)$$

From the definition of the reaction rate,

$$r = \frac{K[S][E]}{K_S}$$

We substitute the relation for $[E]$ to get

$$r = \left(\frac{K[S]}{K_S}\right)[E_T]\bigg/\left(1 + \frac{[S]}{K_S} + \frac{[S]^2}{K_S K_S'}\right) = K[S]\left(\frac{[E_T]}{K_S + [S] + [S]^2/K_S'}\right)$$

or we can write

$$r = \frac{V_{\max}[S]}{K_S + [S] + [S]^2/K_S'} \tag{3.119}$$

The difference between Eqs. (3.113) (Michaelis–Menten) and (3.119) is that in Eq. (3.119) when $[S]^2$ is large, it will dominate the equation leading to a nonmonotonic behavior.

We can see from the graph on the right-hand side of Figure 3.12 the existence of a negative slope region (after the dashed line), where the rate of reaction decreases with the increase in concentration of substrate (inhibition). The negative slope region means that as the substrate concentration increases, the reaction rate decreases ($r = K/[S]$).

Equation (3.119) represents what we call nonmonotonic kinetics. This has a very important implication on the behavior of the bioreactors when such bioreactions are conducted.

Explanation of non-monotonic kinetics (Fig. 3.13)

Positive reaction order (region I). It means that the rate of [ES] produced from reaction (3.117) is higher than that consumed in reaction (3.118).

Zero reaction order (region II). It means that the rate of [ES] produced from reaction (3.117) is equal to that consumed by reaction (3.118).

Negative reaction order (region III). It means that the rate of [ES] produced from reaction (3.117) is less than that consumed by reaction (3.118).

Competitive Substrate Inhibition

The above substrate inhibition mechanism is called consecutive substrate inhibition. Another mechanism is the competitive substrate inhibition as shown in this subsection.

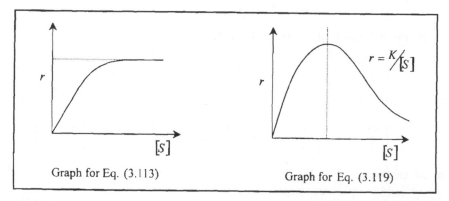

Graph for Eq. (3.113)

Graph for Eq. (3.119)

Figure 3.12 Plots showing the difference between the behavior of rates expressed by Eqs. (3.113) and (3.119).

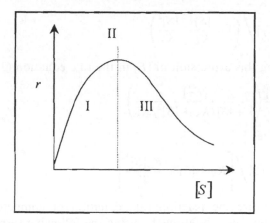

Figure 3.13 Different kinetics regions.

$$E + S \Leftrightarrow ES \qquad \text{Equilibrium constant } = K_S \qquad\qquad (3.120)$$

$$2S + E \Leftrightarrow ES_2 \quad \text{Equilibrium constant } = K_S' \qquad\qquad (3.121)$$

$$ES \xrightarrow{K} P + E \quad \text{Rate of reaction, } r = K[ES]$$

We know that

$$K_S = \frac{[E][S]}{[ES]} \quad \text{and} \quad K_S' = \frac{[S]^2[E]}{[ES_2]}$$

The reaction rate can be written as

$$r = K\frac{[S][E]}{K_S} \qquad\qquad (3.122)$$

Now the total enzyme balance gives

$$[E_T] = [E] + [ES] + [ES_2]$$

By substituting the relations for $[ES]$ and $[ES_2]$ in terms of K_S and K_S', we get

$$[E_T] = [E] + \frac{[E][S]}{K_S} + \frac{[E][S]^2}{K_S'}$$

which can be rearranged to give

$$[E_T] = [E]\left(1 + \frac{[S]}{K_S} + \frac{[S]^2}{K_S'}\right)$$

We can obtain $[E]$ as

$$[E] = [E_T] \bigg/ \left(1 + \frac{[S]}{K_S} + \frac{[S]^2}{K_S'} \right)$$

By substituting this expression of $[E]$ in the rate equation (3.123), we get

$$r = \frac{K[S]}{K_S} \left(\frac{[E_T]}{1 + [S]/K_S + [S]^2/K_S'} \right)$$

or we can write

$$r = V_{max}[S] \bigg/ \left[K_S + [S] + \frac{K_S[S]^2}{K_S'} \right] \tag{3.124}$$

We can generalize the reaction rate equation of competitive and non-competitive substrate inhibitions by using the following formula:

$$r = \frac{V_{max}[S]}{\left[K_S + [S] + \alpha_I[S]^2 \right]} \tag{3.125}$$

If $\alpha_I = 1/K_S'$, we have noncompetitive (consecutive) inhibition; if $\alpha_I = K_S/K_S'$ we have competitive inhibition.

Continuous-Stirred Tank Enzyme Reactor (or Enzyme CSTR) (Fig. 3.14)

For simplicity, assume that the enzymes are immobilized inside of the reactor (i.e., no washout) but, at the same time, there is no mass transfer resistance, then the equation is simply

$$qS_f = qS + Vr \tag{3.126}$$

Note: If r is per unit mass, then multiply by the enzyme concentration C_E.

By dividing Eq. (3.126) by V, we get

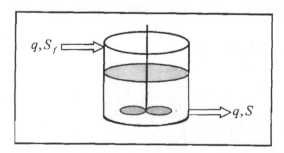

Figure 3.14 Enzyme CSTR.

$$\left(\frac{q}{V}\right)S_f = \left(\frac{q}{V}\right)S + r \tag{3.127}$$

where (q/V) is the dilution rate D (time^{-1}), which is the inverse of the residence time, $\tau = V/q$ Therefore,

$$D(S_f - S) = r \tag{3.128}$$

Equation (3.128) can be easily solved; the nature of the solution(s) will depend on the rate of the bioreaction function r. We will call it the consumption function $[C(S)]$; the left-hand side $\lfloor D(S_f - S) \rfloor$ will be called the supply function $[S(S)]$. Let us solve them graphically as shown in Figure 3.15.

It is clear that for Michaelis–Menten kinetics, we have only one steady state for the whole range of D, whereas for the substrate-inhibited kinetics, we can have more than one steady state over a certain range of D values (this can be easily visualized by Fig. 3.16).

More details regarding multiplicity of the steady states are shown in Figure 3.16. The stability of the different steady states is stated below.

The simplest definition of the stability of steady states is as follows:

> *Stable steady state.* If disturbance is made and then removed, the system will return back to its initial steady state.

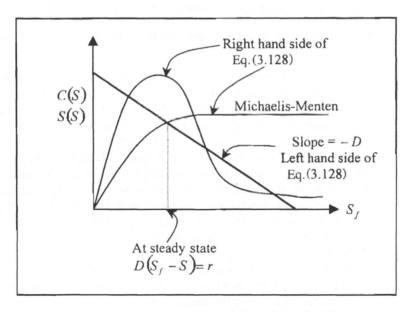

Figure 3.15 Graphical solution of Eq. (3.128).

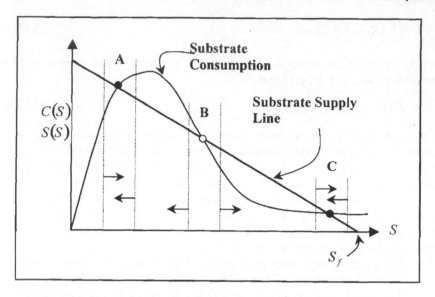

Figure 3.16   Multiple steady states and their stability characteristics.

Unstable steady state. If a disturbance is made and then removed, the system will not return to its initial steady state.

Corresponding to Figure 3.16, we can describe the stability behavior of the three steady states (denoted by points A, B, and C in Fig. 3.16).

Steady-state point A. From a static point of view, it is stable because slightly to the right of it, the rate of consumption of the substrate is greater than the rate of supply. This will give rise to a decrease in the concentration and the reactor goes back to its steady state at A. Also, for concentrations slightly to the left of steady state A, the rate of consumption is smaller than the rate of supply. Therefore, the concentration will increase and the reactor will go back to steady state A.

Steady-state point B. From the static point of view, it is unstable because for any concentration change slightly to the right of this point, the rate of substrate consumption is smaller than the rate of substrate supply. Therefore, the concentration continues to increase and the system does not go back to steady state B. Also, as the concentration is changed slightly to the left of steady state point B, the rate of substrate supply is lower than the rate of consumption and, therefore, the concentration continues to decrease and never returns to steady state B.

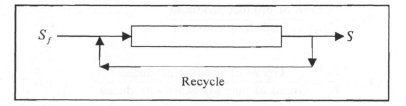

Figure 3.17 Recycle is a source of multiplicity.

Steady-state point C. This steady state is stable for the same reasons described for steady-state point A. (The reader is advised to perform the tests described for steady state points *A* and *B* to verify that steady-state point C is indeed stable).

At which stable steady state will the reactor operate? This is not known and the behavior of the reactor is described by its previous history (non-physical, nonchemical condition).

Sources of multiplicity

1. Nonmonotonic dependence of a rate process (or more) on at least one state variable
2. Feedback of information (a recycle in a tubular reactor or axial dispersion, which will be discussed later) (see Fig. 3.17).

More details about multiplicity of steady states and bifurcation behavior are given in Chapter 7.

3.8.2 Modeling of Microbial Systems

Microbial bioreactors differ from enzyme reactors with regard to the fact that the biocatalyst (micro-organism) grows with the progress of the reaction (i.e., it is an autocatalytic process).

$$\text{Micro-organism concentration } = X \left(\frac{\text{g}}{\text{L}}\right)$$

and the reaction is as follows:

$$S + X \rightarrow P + X$$

The following biochemical yield factors replace the stoichiometric numbers used in reactions:

$$Y_S = \frac{\text{Grams of micro-organism produced}}{\text{Grams of substrate consumed}}$$

$$Y_P = \frac{\text{Grams of product produced}}{\text{Grams of substrate consumed}}$$

or

$$Y'_P = \frac{Y_P}{Y_S} = \frac{\text{Grams of product produced}}{\text{Grams of micro-organism produced}}$$

Monod Equation and Substrate Inhibition

This is the most common reaction rate equation used. It resembles the enzyme rate equation, but it is more empirical:

$$\mu = \frac{\text{Grams of micro-organism produced}}{\text{Grams of micro-organisms existing} \times \text{Time}} = \text{time}^{-1}$$

Mathematically, it can be written as

$$\mu = \frac{\mu_{max}[S]}{K_S + [S]}$$

For the case of substrate inhibition,

$$\mu = \frac{\mu_{max}[S]}{K_S + [S] + K[S]^2} \tag{3.129}$$

- There can be product inhibition also.
- There can be cell inhibition also, where, after a time, the cells stop the reaction.

If the reaction is taking place in a batch reactor, then

$$\frac{d(VX)}{dt} = \mu(VX)$$

where μ is the rate of production of the micro-organism and VX is the total weight of the micro-organism at any time t.

Simplifying Assumptions

1. Assume V is constant.
2. Assume dilute solution where the micro-organism is small and is accumulating slowly.

With these simplifying assumptions, the equation can be modified to give

$$V\frac{dS}{dt} = -\frac{\mu VX}{Y_S} \quad \text{(consumption rate of substrate)}$$

This gives

$$\frac{dS}{dt} = -\frac{\mu X}{Y_S} \tag{3.130}$$

Also, we get

$$\frac{dX}{dt} = \mu X \tag{3.131}$$

with initial condition at $t = 0$, $X = X_f$ and $S = S_f$. By rearranging Eqs. (3.130) and (3.131) and adding them, we get

$$Y_S \frac{dS}{dt} + \frac{dX}{dt} = -\mu X + \mu X = 0$$

which can be rewritten as

$$\frac{d(Y_S S + X)}{dt} = 0$$

To solve this differential equation, we integrate with the initial conditions at $t = 0$, $X = X_f$ and $S = S_f$:

$$\int \frac{d(Y_S S + X)}{dt} = \int 0$$

$$\int d(Y_S S + X) = C = Y_S S_f + X_f$$

We get

$$Y_S S + X = Y_S S_f + X_f$$

On rearranging, we get

$$X = X_f + Y_S(S_f - S)$$

By substituting the value of X in Eq. (3.130) and after some rearrangement, we get the design equation in terms of one state variable S:

$$Y_S \frac{dS}{dt} = -\frac{\mu_{max}[S]}{K_S + [S]} [X_f + Y_S(S_f - S)]$$

where

$$\mu = \frac{\mu_{max}[S]}{K_S + [S]}$$

For other cases, another mechanism can also decrease (eat) the micro-organism through "death". Therefore, we introduce a term k_d to incorporate this decrease in micro-organism number,

$$\frac{dX}{dt} = \mu X - k_d X = X(\mu - k_d) \tag{3.132}$$

and for the substrate, we can write

$$\frac{dS}{dt} = -\frac{\mu X}{Y_S} \tag{3.133}$$

The two differential equations (3.132 and 3.133) can to be solved simultaneously with suitable initial conditions.

Continuous-Stirred Tank Fermentor (Without Micro-organism Death) (Fig. 3.18)

For this case and with the assumption of constant volume, the unsteady-state equations are given as follows:

For micro-organisms

$$FX_f + \mu XV = FX + V\frac{dX}{dt}$$

For substrate

$$FS_f = FS + \frac{\mu XV}{Y_S} + V\frac{dS}{dt}$$

With initial conditions at $t = 0, S = S_f$ and $X = X_f$. The steady-state equations can be obtained by simply setting

$$\frac{dX}{dt} = \frac{dS}{dt} = 0$$

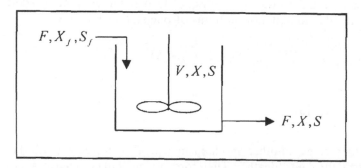

Figure 3.18 Continuous-stirred tank fermentor.

A much more detailed and involved heterogeneous modeling of microbial reaction systems is discussed in Chapter 6.

Solved Example

The continuous anaerobic digester used for treating municipal waste sludge is essentially like any other CSTR except that it involves multiphase biological reactions. Material essentially consisting of organisms of concentration C_{XO} mol/m^3 and a substrate of concentration C_{SO} mol/m^3 is fed to a certain anaerobic digester at F m^3/day. Use the following additional assumptions:

1. The reactions of interest take place only in the liquid phase and the reactor (liquid) volume V (m^3) is constant.
2. The organism concentration in the digester C_X is uniform, as is the substrate concentration C_S.

Obtain a model for this digester process by carrying out organism and substrate balances and using the following constitutive relationships:

1. The organism growth rate (in the reactor) is given by

$$r_X = \frac{dC_X}{dt} = \mu C_X$$

where μ, the specific growth rate, is itself given by the Monod function

$$\mu = \mu_0 \left(\frac{C_S}{K_S + C_S} \right)$$

Here, μ_0 is the maximum specific growth rate and K_S is the saturation constant.

2. The yield (the rate of growth of organism ratioed to the rate of consumption of substrate) in the reactor is given by

$$r_X = -Y_{XS} \, r_S$$

Solution

The process can be shown schematically as in Figure 3.19. The microbial reaction is given by

$$S \xrightarrow{X} P$$

The rate of production of the micro-organism is given by

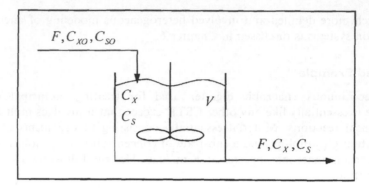

Figure 3.19 Schematic diagram of the process.

$$r_X = \mu C_X = \mu_0 \left(\frac{C_S}{K_S + C_S} \right) C_X$$

The rate of consumption of the substrate is given by (note the negative sign in the rate as the substrate concentration decreases due to its consumption by the micro-organisms)

$$r_S = \frac{dC_S}{dt} = -\frac{1}{Y_{XS}} \mu_0 \left(\frac{C_S}{K_S + C_S} \right) C_X$$

An appropriate model may be obtained from a micro-organism material balance and a substrate mass balance:

Substrate balance

$$V \frac{dC_S}{dt} = FC_{SO} - FC_S - \frac{1}{Y_{XS}} \mu_0 \left(\frac{C_S}{K_S + C_S} \right) C_X \qquad (3.134)$$

Micro-organism balance

$$V \frac{dC_X}{dt} = FC_{XO} - FC_X + \mu_0 \left(\frac{C_S}{K_S + C_S} \right) C_X \qquad (3.135)$$

With initial conditions at $t = 0$, $C_S = C_S(0)$ and $C_X = C_X(0)$. Equation (3.134) and (3.135) with initial conditions formulate the dynamic model for this microbial system.

Note: Volume V is assumed to be constant.

The steady state is obtained by setting the differential terms equal to zero in Eqs. (3.134) and (3.135):

$$\frac{dC_S}{dt} = \frac{dC_X}{dt} = 0$$

The steady-state equations are

$$FC_{SO} - FC_{S_{ss}} - \frac{1}{Y_{XS}}\mu_0\left(\frac{C_{S_{ss}}}{K_S + C_{S_{ss}}}\right)C_{X_{ss}} = 0$$

and

$$FC_{XO} - FC_{X_{ss}} + \mu_0\left(\frac{C_{S_{ss}}}{K_S + C_{S_{ss}}}\right)C_{X_{ss}} = 0$$

where $C_{S_{ss}}$ is the steady-state concentration of the substrate and $C_{X_{ss}}$ is the steady-state concentration of the micro-organism.

REFERENCES

1. Eigenberger, G. *Modelling and Simulation in Industrial Chemical Reaction Engineering.* Springer Series in Chemistry and Physics Vol. 18. Springer-Verlag, New York, 1981, pp. 284–304.
2. Salmi, T. Modeling and simulation of transient states of ideal heterogeneous catalytic reactors. *Chem. Eng. Sci,* 43(3), 503–511, 1988.
3. Aube, F., and Sapoundjiev, H. Mathematical model and numerical simulations of catalytic flow reversal reactors for industrial applications. *Computers Chem. Eng.,* 24(12), 2623-2632, 2000.
4. Elnashaie, S. S., Gaber, A. H., and El-Rifai, M. A. Dynamic analysis of enzyme reactors exhibiting substrate Iinhibition multiplicity. *Chem. Eng. Sci.* 32(5), 557–559, 1977.
5. Baiker, A., and Bergougnan, M. Investigation of a fixed-bed pilot plant reactor by dynamic experimentation. Part 1. Apparatus and experimental results. *Can. J. Chem. Eng.* 63(1), 138–145, 1985.
6. Baiker, A., and Bergougnan, M. Investigation of a fixed-bed pilot plant reactor by dynamic experimentation. Part 2. Simulation of reactor behavior. *Can. J. Chem. Eng.* 63(1), 146–154, 1985.
7. Elnashaie, S. S. E. H., Moustafa, T., Alsoudani, T., and Elshishini, S. S. Modeling and basic characteristics of novel integrated dehydrogenation-hydrogenation membrane catalytic reactors. *Computers Chem. Eng.,* 24(2–7), 1293–1300, 2000.
8. Ajbar, A., Alhumazi, K., and Elnashaie, S. S. E. H. Classification of static and dynamic behavior in a fluidized-bed catalytic reactor. *Chem. Eng. J.,* 84(3), 503–516, 2001.
9. Elnashaie, S. S. E. H., Alhabdan, F. M., and Adris, A. M. A Software *Computer Package for the Simulation of Industrial Steam Reformers for*

Natural Gas. Institute of Chemical Engineering Symposium Series No. 114. 1989, pp. 59–67.

10. Hoelzer, D., and Hahn, H. H. Development of a software package for the dimensioning and operation of activated sludge plants. *Korresp. Abwasser* 42(10), 1836–1838, 1841–1846, 1995.

11. Li, Ke, Crittenden, J. C., and Hand, D. W. A software package for modeling of the H2O2/UV advanced oxidation process. Proceedings of the Annual Conference of the American Water Works Association, 2001, pp. 1421–1424.

12. Schnelle, P. D., Jr., and Richards, J. R. A review of industrial reactor control: Difficult problems and workable solutions. Chemical Process Control (CPC3), Proceedings of the 3rd International Conference on Chemical Process Control, 1986, pp. 749–802.

13. Richards, J. R., and Schnelle, P. D., Jr. Perspectives on industrial reactor control. *Chem. Eng. Prog.* 84(10), 32–36, 1988.

14. Boreskov, G. K., and Matros, Yu. Sh. Fixed catalyst bed reactors operated in steady- and unsteady-state conditions. *Recent Adv. Eng. Anal. Chem. React. Syst.*, 142–155, Wiley, New York, 1984.

15. Adris, A., Grace, J., Lim, C., and Elnashaie, S. S. E. H. Fluidized bed reaction system for steam/hydrocarbon gas reforming to produce hydrogen, U.S. Patent 5,326,550, June 5, 1994.

16. Makel, D. Low cost microchannel reformer for hydrogen production from natural gas. California Energy Commission (CEG), Energy Innovations Small Grant (EISG) Program, 1999.

17. Sammells, A. F., Barton, A. F., Peterson, D. R., Harford, S. T., Mackey, R., Van Calcar, P. H., Mundshau, M. V., and Schultz, J. B. Methane conversion to syngas in mixed conducting membrane reactors. 4th International Conference on Catalysis in Membrane Reactors, 2000.

18. Prigogine, I., George, C., Henin, F., and Rosenfeld, L. Unified formulation of dynamics and thermodynamics. With special reference to non-equilibrium statistical thermodynamics. *Chem. Scr.*, 4(1), 5–32, 1973.

19. Nicolis, G., and Prigogine, I. Self-organization in nonequilibrium systems: From dissipative structures to order through fluctuations. Wiley, New York, 1977.

20. Denn, M. M. *Process Modeling*. Longman Science and Technology, London, 1986.

21. Froment, G. F., and Bischoff, K. B. *Chemical Reactor Analysis and Design*. Wiley, New York, 1990.

22. Dransfield, P. *Engineering Systems and Automatic Control*. Prentice-Hall, Englewood Cliffs, NJ, 1968.

23. Von Bertalanffy, L. *General Systems Theory*. George Braziller, New York, 1968.

24. Robert Shaw, J. E., Mecca, S. J., and Rerick, M. N. *Problem Solving. A System Approach*. Academic Press, New York, 1964.

25. Aris, R. Manners makyth modellers. *Chem. Eng. Res. Des.*, 69(A2), 165–174, 1991.

26. Smith, J. M. *Models in Ecology*. Cambridge University Press, Cambridge, 1974.

27. Hawking, S. W. *A Brief History of Time: From the Big Bang to Black Holes*. Bantam Press, London, 1989.

28. Soliman, M. A. and Elnashaie, S. S. E. H. Negative effectiveness factors for cyclic reversible reactions, *Chem. Eng. Sci.* 45(7), 1925–1928, 1990.
29. Elnashaie, S. S. E. H., Abashar, M. E., and Alubaid, A. S. Simulation and optimization of an industrial ammonia reactor, *Ind. Eng. Chem. Res.* 27(11), 2015–2022, 1988.
30. Alhabdan, F. M. and Elnashaie, S. S. E. H. Simulation of an ammonia plant accident using rigorous heterogeneous models—Effect of shift converters disturbances on methanator. *Math. Computer Model.* 21(4), 85–106, 1995.

PROBLEMS

Problem 3.1

Derive the steady-state mass balance equations for an isothermal continuous-stirred tank reactor in which a consecutive homogeneous reaction

$$A \xrightarrow{k_1} B \xrightarrow{k_2} C$$

is taking place. Put the resulting equation in a matrix form and suggest the sequence of solution procedures to obtain the output of the three components.

Problem 3.2

A semibatch reactor is run at constant temperature by varying the rate of addition of one of the reactants A. The irreversible exothermic reaction taking place in the reactor is first order in reactants A and B:

$$A + B \rightarrow C$$

The tank is initially filled to its 40% level with the pure reactant B at a concentration C_{BO}. Cooling water is passed through a cooling coil immersed in the reactor and reactant A is slowly added to the perfectly stirred vessel. Write the equations describing the system. Without solving the equations, try to sketch the profiles of F_A (feed flow rate) and C_A (concentration of A in the reactor), and C_B (concentration of B in the reactor) with time.

Problem 3.3

A perfectly mixed nonisothermal adiabatic reactor carries out a simple first-order exothermic reaction in the liquid phase:

$$A \rightarrow B$$

The product from the reactor is cooled from the output temperature T to a temperature T_C and is then introduced into a separation unit where the unreacted A is separated from the product B. The feed to the separation unit is split into two equal parts: top product and bottom product; the bottom product from the separation unit contains 95% of the unreacted A in the effluent of the reactor and 1% of the B in the same stream. This bottom product, which is at temperature T_C (i.e., the separation unit is isothermal), is recycled and mixed with the fresh feed to the reactor and the mixed stream is heated to the reactor feed temperature T_f before being introduced into the reactor. Write the steady-state mass and heat balance equations for this system assuming constant physical properties and heat of reaction. (*Note:* concentrate your modeling attention on the nonisothermal, adiabatic reactor, and for the rest of units, carry out a simple mass and heat balance in order to define the feed conditions to the reactor.)

Problem 3.4

Consider a system that initially consists of 1 mol of CO and 3 mol of H_2 at 1000 K. The system pressure is 25 atm. The following reactions are to be considered:

$$2CO + 2H_2 \Leftrightarrow CH_4 + CO_2 \tag{A}$$

$$CO + 3H_2 \Leftrightarrow CH_4 + H_2O \tag{B}$$

$$CO_2 + H_2 \Leftrightarrow H_2O + CO \tag{C}$$

When the equilibrium constants for reactions A and B are expressed in terms of the partial pressures of the various species (in atm), the equilibrium constants for these reactions have the values $K_{P_A} = 0.046$ and $K_{P_B} = 0.034$. Determine the number of independent reactions, and then determine the equilibrium composition of the mixture.

Problem 3.5

A perfect gas with molecular weight M flows at a rate W_0 (kg/h) into a cylinder through a restriction. The flow rate is proportional to the square root of the pressure drop over the restriction:

$$W_0 = K_0 \sqrt{P_0 - P}$$

where P is the pressure (N/m^2 absolute) in the cylinder, P_0 is constant upstream pressure. The system is isothermal. Inside the cylinder a piston is forced to the right as the pressure P builds up. A spring resists the movement of the piston with a force that is proportional to axial displacement X of the piston:

$$F_S = K_S X \text{ (N)}$$

The pressure on the spring side is atmospheric and is constant. The piston is initially at $X = 0$ when the pressure in the cylinder is zero. The cross-sectional area of the cylinder is A (m^2). Assume that the piston has negligible mass and friction.

(a) Derive the equations describing the system, (i.e., the dynamic behavior of the system).
(b) What will the steady-state piston displacement be?
(c) What is the effect of relaxing the assumption of negligible mass of the piston on both the dynamic behavior and steady-state position of the piston?
(d) What is the effect of relaxing the assumption of constant atmospheric pressure in the spring side (behind the piston)?

Problem 3.6

The reaction of ethylene and chlorine in liquid ethylene dichloride solution is taking place in a CSTR. The stoichiometry of the reaction is,

$$C_2H_4 + Cl_2 \rightarrow C_2H_4Cl_2$$

Equimolar flow rates of ethylene and chlorine are used in the following experiment, which is carried out at 36°C. The results of the experiment are tabulated in Table P3.6.

(a) Determine the overall order of the reaction and the reaction rate constant.
(b) Determine the space time (space time is the residence time which is the ratio between the volume of the reactor and the volumetric flow rate) necessary for 65% conversion in a CSTR.
(c) What would be the conversion in a PFR having the space time determined in part (b)?

Table P3.6 Experimental Data

Space time (s)	Effluent chlorine concentration (mol/cm^3)
0	0.0116
300	0.0094
600	0.0081
900	0.0071
1200	0.0064
1500	0.0060
1800	0.0058

In parts (b) and (c), assume that the operating temperature and the initial concentrations are the same as in part (a).

Problem 3.7

Some of the condensation reactions that take place when formaldehyde (F) is added to sodium paraphenolsulfonate (M) in an alkaline-aqueous solution have been studied. It was found that the reactions could be represented by the following equations:

$$F + M \rightarrow MA, \qquad k_1 = 0.15 \text{ L/gmol min}$$
$$F + MA \rightarrow MDA, \qquad k_2 = 0.49$$
$$MA + MDA \rightarrow DDA, \qquad k_3 = 0.14$$
$$M + MDA \rightarrow DA, \qquad k_4 = 0.14$$
$$MA + MA \rightarrow DA, \qquad k_5 = 0.04$$
$$MA + M \rightarrow D, \qquad k_6 = 0.056$$
$$F + D \rightarrow DA, \qquad k_7 = 0.50$$
$$F + DA \rightarrow DDA, \qquad k_8 = 0.50$$

where M, MA, and MDA are monomers and D, DA, and DDA are dimers. The process continues to form trimers. The rate constants were evaluated using the assumption that the molecularity of each reaction was identical to its stoichiometry.

Derive a dynamic model for these reactions taking place in a single, isothermal CSTR. Carefully define your terms and list your assumptions.

Problem 3.8

The conversion of glucose to gluconic acid is a relatively simple oxidation of the aldehyde group of the sugar to a carboxyl group. A micro-organism in a batch or continuous fermentation process can achieve this transformation. The enzyme (a biocatalyst) glucose oxidase, present in the micro-organism, converts glucose to gluconolactone. In turn, the gluconolactone hydrolyzes to form the gluconic acid. The overall mechanism of the fermentation process that performs this transformation can be described as follows:

Cell growth

$$\text{Glucose} + \text{Cells} \rightarrow \text{Cells}$$

Rate of cell growth

$$R_X = k_1 C_X \left(1 - \frac{C_X}{K_1} \right)$$

where, C_X is the concentration of cells and k_1 and K_1 are rate constants.

Glucose oxidation

$$\text{Glucose} + O_2 \xrightarrow{\text{Glucose oxidase}} \text{Gluconolactone} + H_2O_2$$

and

$$H_2O_2 \rightarrow \tfrac{1}{2}O_2 + H_2O$$

Rate of gluconolactone formation

$$R_{GI} = \frac{k_2 C_X C_S}{K_2 + C_S}$$

where C_S is the substrate (glucose) concentration and k_2 and K_2 are rate constants.

Gluconolactone hydrolysis

$$\text{Gluconolactone} + H_2O \rightarrow \text{Gluconic acid}$$

Rate of gluconolactone hydrolysis

$$R_{GH} = 0.9 k_3 C_{GI}$$

Rate of gluconic acid formation

$$R_G = k_3 C_{GI}$$

where C_{GI} is the gluconolactone concentration and k_3 is a rate constant.

Overall rate of substrate (glucose) consumption

$$R_S = 1.011 \left(\frac{k_2 C_X C_S}{K_2 + C_S} \right)$$

At operating conditions of 32°C and pH $= 6.8$, the values of the parameters are as follows:

$k_1 = 0.949$

$K_1 = 3.439$

$k_2 = 18.72$

$K_2 = 37.51$

$k_3 = 1.169$

(a) Develop a model for the batch fermentor and calculate the change of the concentration of the species with time for the following initial conditions (demonstrate the calculations for a few steps using the simple Euler's method):

$C_X(0) = 0.05$ mg/mL

$C_{GI}(0) = 0.0$ mg/mL

$C_{GA}(0) = 0.05$ mg/mL (gluconic acid concentration)

$C_S(0) = 50.0$ mg/mL

(b) Develop the unsteady-state model for the continuous fermentor. Find the feed flow rate and fermentor active volume, which gives 60% substrate (glucose) conversion at steady state and 50 kg/h of gluconic acid. Show, using the unsteady-state model, the dynamic response of the fermentor to disturbances in the input feed flow rate.

Problem 3.9

A perfectly mixed, isothermal CSTR has an outlet weir as shown in Figure P3.9. The flow rate over the weir is proportional to the height of liquid over the weir, h_{ow}, to the 1.5 power. The weir height is h_w. The cross-sectional area of the tank is A. Assume constant density.

A first-order chemical reaction takes place in the tank:

$$A \xrightarrow{k} B$$

Derive the equations describing the system.

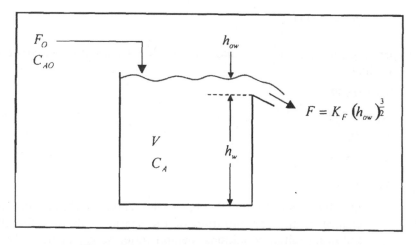

Figure P3.9 Schematic diagram of Problem 3.9.

Problem 3.10

The liquid in a jacketed, nonisothermal CSTR is stirred by an agitator whose mass is significant compared with the reaction mass. The mass of the reactor wall and the mass of the jacket wall are also significant. Write the energy equations for the system. Neglect radial temperature gradients in the agitator, reactor wall, and jacket wall.

Problem 3.11

The reaction

$$3A \rightarrow 2B + C$$

is carried out in an isothermal semibatch reactor as shown in Figure P3.11. Product B is the desired product. Product C is a very volatile by-product that must be vented off to prevent a pressure buildup in the reactor. Gaseous C is vented off through a condenser to force any A and B back into the reactor to prevent loss of reactant and product.

Assume F_V is pure C. The reaction is first-order in C_A. The relative volatilities of A and C to B are $\alpha_{AB} = 1.2$ and $\alpha_{CB} = 10$. Assume perfect gases and constant pressure. Write the equations describing the system. Carefully list all assumptions.

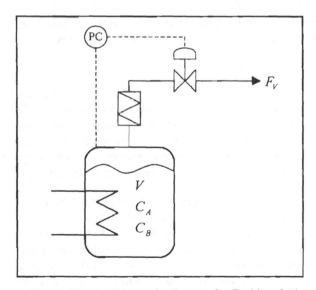

Figure P3.11   Schematic diagram for Problem 3.11.

Problem 3.12

A vertical, cylindrical tank is filled with well water at 65°F. The tank is insulated at the top and bottom but is exposed on its vertical sides to cold night air at 10°F. The diameter of the tank is 2 ft and its height is 3 ft. The overall heat transfer coefficient is 20 Btu/h °F ft^2. Neglect the metal wall of the tank and assume that the water in the tank is perfectly mixed.

(a) Calculate how many minutes it will be until the first crystal of ice is formed.
(b) How long will it take to completely freeze the water in the tank? The heat of fusion of water is 144 Btu/lb$_m$.

Problem 3.13

An isothermal, first-order, liquid-phase, irreversible reaction is conducted in a constant volume batch reactor:

$$A \xrightarrow{k} B$$

The initial concentration of reactant A at the beginning of the batch is C_{AO}. The specific reaction rate decreases with time because of catalyst deactivation as $k = k_0 e^{-\beta t}$.

(a) Solve for $C_A(t)$
(b) Show that in the limit as $\beta \to 0$, $C_A(t) = C_{AO} e^{-k_0 t}$
(c) Show that in the limit as $\beta \to \infty$, $C_A(t) = C_{AO}$

Problem 3.14

There are 3580 lbs of water in the jacket of a reactor that are initially at 145°F. At time equal to zero, 70°F cooling water is added to the jacket at a constant rate of 418 lbs/min. The holdup of water in the jacket is constant because the jacket is completely filled with water and excess water is removed from the system on pressure control as cold water is added. Water in the jacket can be assumed to be perfectly mixed.

(a) How many minutes does it take the jacket water to reach 99°F if no heat is transferred to the jacket?
(b) Suppose a constant 358,000 Btu/h of heat is transferred into the jacket from the reactor, starting at time equal zero when the jacket is at 145°F. How long will it take the jacket water to reach 99°F if the cold water addition rate is constant at 416 lbs/min?

Problem 3.15

Consider an enzymatic reaction being carried out in a CSTR. The enzyme is subject to substrate inhibition, and the rate equation is given by Eq. (3.119):

$$r = \frac{V_{max}[S]}{K_S + [S] + [S]^2/K'_S}$$

(a) Sketch a plot of reaction rate versus the substrate concentration for this system. Use this plot to show that for sufficiently high inlet substrate concentration, there is a range of dilution rates for which multiple steady states are possible.

(b) Explain how would you calculate the highest and lowest dilution rates for which multiple steady states are possible.

4

Mathematical Modeling (II): Homogeneous Distributed Systems and Unsteady-State Behavior

In Chapter 3, we covered the mathematical modeling of lumped systems as well as some preliminary examples of distributed systems using a systematic, generalized approach. The examples for distributed systems were preliminary and they were not sufficiently generalized. In this chapter, we introduce sufficient generalization for distributed systems and give more fundamentally and practically important examples, such as the axial dispersion model resulting in two-point boundary-value differential equations. These types of model equations are much more difficult to solve than models described by initial-value differential equations, specially for nonlinear cases, which are solved numerically and iteratively. Also, examples of diffusion (with and without chemical reaction) in porous structures of different shapes will be presented, explained, and solved for both linear and nonlinear cases in Chapter 6.

In this chapter, the unsteady-state term(s) to be introduced into the models will be explained and discussed for homogeneous as well as heterogeneous systems, approximated using pseudohomogeneous model. Rigorous steady-state and dynamic heterogeneous models will be covered in Chapter 6. The approach adopted in this chapter regarding dynamic modeling, like the rest of the book, is a very general approach which can be applied in a very systematic and easy-to-understand manner. It consists of explaining how to formulate the unsteady-state term(s) in any situation and where to add it to the equations of the steady-state model in order to make the model an unsteady state (dynamic) model. These dynamic models are suitable for start-up, shutdown, process control, and investigation of the dynamic stability characteristics of the system, as will be shown in Chapter 5.

4.1 MODELING OF DISTRIBUTED SYSTEMS

A distributed system is characterized by variation of the state variable(s) along one or more of the space coordinates. When the state variable(s) are varying along only one of the space coordinates, then the distributed model is called a one-dimensional model, and is described by ordinary differential equations (ODEs). When the variation is along two of the space coordinates, the model is called a two-dimensional model, and is described by partial differential equations (PDEs). When the variation is along three space coordinates, it is called a three-dimensional model (PDEs). In this book, only one-dimensional distributed models will be covered. The reason for that is that one-dimensional distributed models are sufficient for most practical purposes, in addition to the fact that the mathematical formulation and solution of higher-dimensional models is beyond the scope of this undergraduate book.

4.1.1 Isothermal Distributed Systems

Here, we present the basic idea for the formulation of one-dimensional distributed models for a single-input, single-output, and single-reaction tubular reactor. Axial dispersion is neglected in this preliminary stage; then, it will be introduced and discussed later in this chapter.

Figure 4.1 illustrates a single reaction in a tubular reactor. The overall steady-state mass balance is

$$\bar{n}_i = n_{if} + \sigma_i r$$

where r is the overall rate of reaction for the whole reactor. This means that it is not a design (model) equation; it is a mass balance equation. The technique used for lumped system [i.e., replacing r by r' (rate per unit

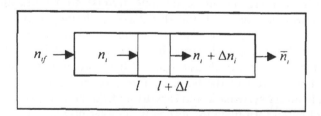

Figure 4.1 Mass flow along a tubular reactor.

volume) multiplied by the reactor volume (V)] is not directly applicable here. Why? Because r' changes along the length of the reactor.

The solution is to carry out the balance over an element of length Δl (or volume $\Delta V = A_t \Delta l$, where A_t is the cross-sectional area of the tubular reactor). Thus the mass balance over this element becomes

$$(n_i + \Delta n_i) = n_i + \sigma_i r' A_t \Delta l$$

Rearranging and dividing the equation by Δl gives

$$\frac{\Delta n_i}{\Delta l} = \sigma_i A_t r'$$

Taking the limit as $\Delta l \to 0$ gives the differential mass balance (design) equation for this simple distributed model:

$$\frac{dn_i}{dl} = \sigma_i A_t r'$$

or (because $\Delta V = A_t \Delta l$)

$$\frac{dn_i}{dV} = \sigma_i r'$$

where, $i = 1, 2, 3, \ldots, M$ and r' is the rate of reaction per unit volume of the reactor.

For multiple reactions, it becomes

$$\frac{dn_i}{dV} = \sum_{j=1}^{N} \sigma_{ij} r'_j$$

where M is the number of components and N is the number of reactions with the initial condition at $l = 0$ (or $V = 0$), $n_i = n_{if}$.

For a simple case having no change in the number of moles with the reaction (or liquid-phase system), we can write

$$n_i = q C_i$$

giving

$$q \frac{dC_i}{dV} = \sum_{j=1}^{N} \sigma_{ij} r'_j$$

with initial condition at $V = 0$, $C_i = C_{if}$.

For multiple-input, multiple-output, the differential equation will not change; we will have only the equivalent of a mixer at the feed conditions and a splitter at the exit. The simplest way to handle multiple-input, multiple-output situations for distributed systems is to use the modular approach discussed earlier and as shown in Figure 4.2.

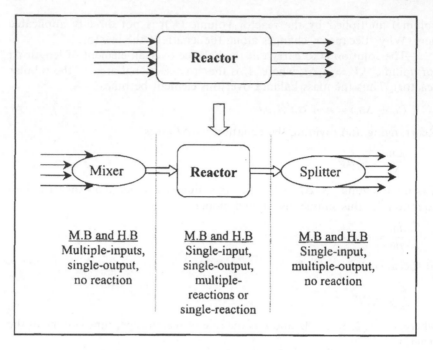

Figure 4.2   Modules and modular approach.

4.1.2 Nonisothermal Distributed Systems

Similarly, the heat balance for the distributed system is to be handled by carrying out the heat balance over an element (see Fig. 4.3).

For the nonadiabatic case, $Q' \neq 0$; Q' is the heat flux per unit length $(J/\min\,m)$.

A heat balance over the Δl element gives

$$\sum_{i=1}^{M} n_i H_i + Q'\Delta l = \sum_{i=1}^{M} n_i H_i + \sum_{i=1}^{M} \Delta(n_i H_i)$$

which, on rearrangement, gives

$$\sum_{i=1}^{M} \frac{\Delta(n_i H_i)}{\Delta l} = Q'$$

Taking the limiting case as $\Delta l \to 0$, we get

$$\sum_{i=1}^{M} \frac{d(n_i H_i)}{dl} = Q'$$

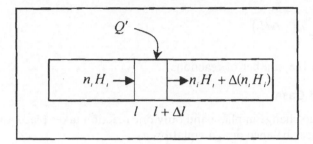

Figure 4.3   The heat balance setup for a tubular reactor.

On differentiating the left-hand side of the above equation, we get,

$$\sum_{i=1}^{M}\left(n_i \frac{dH_i}{dl} + H_i \frac{dn_i}{dl}\right) = Q'$$
(4.1)

From mass balance for the distributed system with a single reaction, we have

$$\frac{dn_i}{dl} = A_t \sigma_i r'$$
(4.2)

Substituting Eq. (4.2) into Eq. (4.1), we get

$$\sum_{i=1}^{M}\left(n_i \frac{dH_i}{dl} + H_i A_t \sigma_i r'\right) = Q'$$

Rearrangement gives

$$\sum_{i=1}^{M}\left(n_i \frac{dH_i}{dl}\right) + \sum_{i=1}^{M}\left(H_i A_t \sigma_i r'\right) = Q'$$

which can be written as,

$$\sum_{i=1}^{M}\left(n_i \frac{dH_i}{dl}\right) + A_t r' \underbrace{\sum_{i=1}^{M}(H_i \sigma_i)}_{\Delta H} = Q'$$

Defining ΔH as the heat of reaction (not at reference conditions), the above equation becomes

$$\sum_{i=1}^{M}\left(n_i \frac{dH_i}{dl}\right) = A_t r'(-\Delta H) + Q'$$

This derivation can be easily extended to multiple reactions by replacing the first term of the right-hand side (RHS) by

$$A_t \sum_{j=1}^{N} r'_j(-\Delta H_j)$$

where N is the number of reactions.

Simplified Case

If there is no change in phase and only one reaction takes place, then we can write the heat balance design equation as

$$\sum_{i=1}^{M} \left(n_i \frac{d(C_{pi}T)}{dl} \right) = A_t r'(-\Delta H) + Q'$$

If C_{pi} is taken as an average constant value for each i (not varying with temperature), then we get

$$\sum_{i=1}^{M} \left(n_i C_{pi} \frac{dT}{dl} \right) = A_t r'(-\Delta H) + Q'$$

Taking an average constant C'_{pmix} gives

$$\sum_{i=1}^{M} \left(n_i C'_{pmix} \frac{dT}{dl} \right) = A_t r'(-\Delta H) + Q'$$

which can be rearranged as

$$\underbrace{\left(\sum_{i=1}^{M} n_i \right)}_{n_t} C'_{pmix} \frac{dT}{dl} = A_t r'(-\Delta H) + Q'$$

The above equation gives

$$n_t C'_{pmix} \frac{dT}{dl} = A_t r'(-\Delta H) + Q'$$

where, n_t is the total molar flow rate and C'_{pmix} is the molar specific heat.

For constant volumetric flow rate (q) and specific heat per unit mass (C_p), this equation can be written in the more "popular" but "simplified" form

$$q\rho_{mix} C_p \frac{dT}{dl} = A_t r'(-\Delta H) + Q'$$

with initial condition at $l = 0, T = T_f$.

4.1.3 Batch Systems (Distributed in Time)

Batch reactors can be considered unsteady-state lumped systems with no input and no output, or it can be considered a distributed system in time (not in space) with the initial conditions as the feed conditions and the final conditions as the exit conditions.

In either case, the batch reactor design equations can be developed as follows. With the regard to the heat balance (see Fig. 4.4), we get

$$\sum_{i=1}^{M} n_i H_i + Q\Delta t = \sum_{i=1}^{M} n_i H_i + \sum_{i=1}^{M} \Delta(n_i H_i)$$

After the same kind of manipulation we did for the tubular reactor, we get

$$\sum_{i=1}^{M} \left(n_i \frac{dH_i}{dt} \right) = Vr'(-\Delta H) + Q$$

For multiple reactions,

$$\sum_{i=1}^{M} \left(n_i \frac{dH_i}{dt} \right) = \sum_{j=1}^{N} Vr_j'(-\Delta H)_j + Q$$

where M is the number of components and N is the number of reactions. We can simplify the above equation to (as we did earlier for the tubular reactor)

$$V\rho_{mix} C_p \frac{dT}{dt} = \sum_{j=1}^{M} Vr_j'(-\Delta H)_j + Q$$

where, C_p is an average constant specific heat per unit mass of the mixture.

The initial condition is at $t = 0$, $T = T_f$.

The mass and heat balance design (model) equations must be solved simultaneously (see Fig. 4.5, as a reminder of how we derived the mass

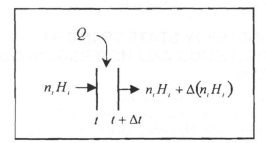

Figure 4.4 Heat balance for a batch reactor.

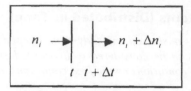

Figure 4.5 Mass balance for a batch reactor.

balance design equation for the batch reactor). The mass balance design (model) equation is easily obtained (for a single reaction):

$$n_i + \Delta n_i = n_i + \sigma_i V r' \Delta t$$

which, after some simple manipulations as discussed earlier, also gives

$$\frac{dn_i}{dt} = \sigma_i V r'$$

For multiple reactions, the equation becomes

$$\frac{dn_i}{dt} = V \sum_{j=1}^{N} \left(\sigma_{ij} r_j' \right)$$

The initial condition at $t = 0$ is given by $n_i = n_{if}$

For the special case of no change in the number of moles accompanying the reaction (or liquid phase) giving constant V, we can write

$$n_i = V C_i$$

Thus, the mass balance design equation becomes

$$\frac{dC_i}{dt} = \sum_{j=1}^{N} \left(\sigma_{ij} r_j' \right)$$

The initial condition at $t = 0$ is given by $C_i = C_{if}$.

4.2 THE UNSTEADY-STATE TERMS IN HOMOGENEOUS AND HETEROGENEOUS SYSTEMS

The heterogeneous system here is described by a pseudohomogeneous model. For complete heterogeneous systems, see Chapter 6.

The unsteady-state (dynamic) models differ from the steady-state model through the addition of the dynamic (accumulation/depletion) term. This term is always added to the side of the equation having the output stream. This term has many names, dynamic or unsteady state, or accumulation/depletion, or, sometimes, capacitance.

4.2.1 Lumped Systems

Isothermal Homogeneous Lumped Systems

We will start with the simplest isothermal lumped homogeneous case, a lumped system, as shown in Figure 4.6. The steady-state equation (for a single reaction) is given by

$$n_i = n_{if} + V\sigma_i r'$$

For the unsteady state, all that we will do is add the unsteady-state (dynamic) term to the side of equation having the output stream. In this case, the equation becomes

$$\underbrace{\frac{dn_i'}{dt}}_{\text{dynamic term}} + n_i = n_{if} + V\sigma_i r' \tag{4.3}$$

For the homogeneous system, we can write n_i' (the molar content of the tank) as

$$n_i' = VC_i$$

and if V is constant, then

$$\frac{dn_i'}{dt} = V\frac{dC_i}{dt}$$

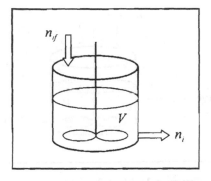

Figure 4.6 Mass flow diagram.

Now, Eq. (4.3) becomes

$$V\frac{dC_i}{dt} + n_i = n_{if} + V\sigma_i r'$$

Isothermal Heterogeneous Lumped Systems

Suppose that there are catalyst particles in the reactor (see Fig. 4.7). The total volume of the catalyst is $V(1 - \varepsilon)$ and the gas or liquid volume is $V\varepsilon$. The concentration of the component in the liquid or gas is C_i (in mol/L), whereas on the catalyst surface, the concentration is C_{is} in (mol/g (or cm^3 or cm^2) of catalyst). For illustration, we will use C_{is} with units of $gmol/cm^2$(surface) of catalyst). Therefore,

$$n_i' = V\varepsilon C_i + V(1 - \varepsilon)S_v C_{is}$$

where S_v is the specific surface area of the catalyst (in cm^2/cm^3). For constant V and ε,

$$\frac{dn_i'}{dt} = V\varepsilon\frac{dC_i}{dt} + V(1 - \varepsilon)S_v\frac{dC_{is}}{dt} \tag{4.4}$$

We can relate C_i and C_{is} through a simple linear adsorption isotherm, as follows:

$$C_{is} = K_i C_i$$

If we assume K_i to be constant, on differentiation we get

$$\frac{dC_{is}}{dt} = K_i\frac{dC_i}{dt} \tag{4.5}$$

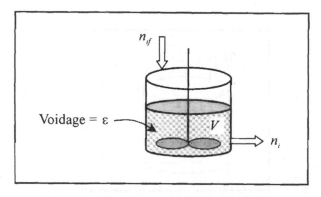

Figure 4.7 Mass flow diagram.

Using Eq. (4.5) in Eq. (4.4), we get

$$\frac{dn'_i}{dt} = \left(\underbrace{V\varepsilon}_{a_1} + \underbrace{V(1 - \varepsilon)S_vK_i}_{a_2} \right) \frac{dC_i}{dt}$$

Usually, a_2 is very large for fine or porous particles.
   In most practical cases,

$$a_2 >>> a_1$$

(a_2 is thousands of times greater than a_1), so a_1 can be neglected as compared to a_2:

$$\left(\underbrace{a_1}_{\text{neglected}} + a_2 \right) \frac{dC_i}{dt} + n_i = n_{if} + V\sigma_i r'$$

and the suitable dynamic (unsteady state) model equation for this heterogeneous system (described by pseudohomogeneous model, because mass transfer resistances are negligible) becomes

$$a_2 \frac{dC_i}{dt} + n_i = n_{if} + V\sigma_i r'$$

4.2.2 Distributed Systems

The same principles apply to distributed systems. However, all the balances are over differential elements, giving rise to partial differential equation as shown in Figure 4.8.

Homogeneous Isothermal Systems

The dynamic term is carried out over the element and is added to the term of the exit stream on the left-hand side of the equation, giving

$$n_i + \Delta n_i + A_t \Delta l \frac{\partial C_i}{\partial t} = n_i + \sigma_i A_t \Delta l r'$$

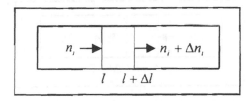

Figure 4.8 Mass flow diagram for a distributed system.

which can be put into the following partial differential equation form:

$$\frac{\partial n_i}{\partial l} + A_t \frac{\partial C_i}{\partial t} = \sigma_i A_t r'$$

For constant volumetric flow rate (q), the above equation can be written as

$$q \frac{\partial C_i}{\partial l} + A_t \frac{\partial C_i}{\partial t} = \sigma_i A_t r'$$

Dividing both sides by the cross-sectional area of the tubular reactor (A_t) and recognizing that $q/A_t = v$, which is velocity of the flow, we get

$$v \frac{\partial C_i}{\partial l} + \frac{\partial C_i}{\partial t} = \sigma_i r' \tag{4.6}$$

The initial conditions at $l = 0$ and $t = 0$ are, $C_i = C_{if}$, and $C_i = C_i(l)|_{t=0}$, respectively. Obviously for multiple reactions, Eq. (4.6) becomes

$$v \frac{\partial C_i}{\partial l} + \frac{\partial C_i}{\partial t} = \sum_{j=1}^{N} \left(\sigma_{ij} r'_j \right)$$

Heterogeneous Isothermal Systems

This is a case with negligible mass transfer resistances, as described by the pseudohomogeneous model. For a full heterogeneous system, see Chapter 6. This situation is a bit more complicated compared to the lumped system. We will consider a two-phase system with no mass transfer resistance between the phases and the voidage is equal to ε (see Fig. 4.9). The mass balance design equation over the element Δl is:

$$n_i + \Delta n_i + A_t \Delta l \varepsilon \frac{\partial C_i}{\partial t} + A_t \Delta l (1 - \varepsilon) S_v \frac{\partial C_{is}}{\partial t} = n_i + \sigma_i A_t \Delta l r'$$

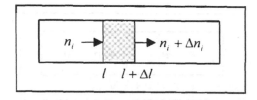

Figure 4.9 Mass flow diagram for a two-phase distributed system.

which can be put into the following partial differential equation form:

$$\frac{\partial n_i}{\partial l} + A_t \varepsilon \frac{\partial C_i}{\partial t} + A_t(1-\varepsilon)S_v \frac{\partial C_{is}}{\partial t} = \sigma_i A_t r' \tag{4.7}$$

Here, S_v is the solid specific surface area (per unit solid volume).

We can relate C_i and C_{is} through a simple linear adsorption isotherm:

$$C_{is} = K_i C_i$$

If we assume K_i to be constant, on differentiation we get

$$\frac{dC_{is}}{dt} = K_i \frac{dC_i}{dt} \tag{4.8}$$

Substituting Eq. (4.8) into Eq. (4.7), we get

$$\frac{\partial n_i}{\partial l} + A_t \varepsilon \frac{\partial C_i}{\partial t} + A_t(1-\varepsilon)S_v K_i \frac{\partial C_i}{\partial t} = \sigma_i A_t r'$$

For a constant flow rate q, we get

$$q \frac{\partial C_i}{\partial l} + (A_t \varepsilon + A_t(1-\varepsilon)S_v K_i) \frac{\partial C_i}{\partial t} = \sigma_i A_t r'$$

On dividing the equation by A_t, we obtain

$$v \frac{\partial C_i}{\partial l} + \left(\underbrace{\varepsilon}_{a_1} + \underbrace{(1-\varepsilon)S_v K_i}_{a_2} \right) \frac{\partial C_i}{\partial t} = \sigma_i r'$$

As discussed earlier, $a_2 >>> a_1$. So we can write

$$\left(\underbrace{a_1}_{\text{negligible}} + a_2 \right) \frac{\partial C_i}{\partial t} + v \frac{\partial C_i}{\partial l} = \sigma_i r' \tag{4.9}$$

With the initial conditions at $l = 0$ and $t = 0$, $C_i = C_{il}$ and $C_i = C_i(l)|_{t=0}$, respectively. Needless to mention, for multiple reactions equation (4.9) will become

$$\left(\underbrace{a_1}_{\text{negligible}} + a_2 \right) \frac{\partial C_i}{\partial t} + v \frac{\partial C_i}{\partial l} = \sum_{j=1}^{N} (\sigma_{ij} r_j')$$

The reader can now handle any dynamic (capacitance) case through a correct physical understanding, together with rational mathematical expression.

4.2.3 Nonisothermal Systems

Lumped Homogeneous Systems

For the steady state, the heat balance design equation for a single reaction is

$$q\rho C_p(T - T_f) = Vr'(-\Delta H)$$

For the unsteady state, after introducing the earlier assumption of no change in phase, average constant specific heats, constant volume and volumetric flow rate, and then adding the dynamic term with the output on the left-hand side of the equation, we get

$$V\rho C_p \frac{dT}{dt} + q\rho C_p(T - T_f) = Vr'(-\Delta H)$$

where, before the approximation, the first term on the left-hand side (the dynamic term) is $d\sum n_i H_i / dt$ and the second term is $\sum n_i(H_i - H_{ir}) - \sum n_{if}(H_{if} - H_{ir})$.

Lumped Heterogeneous Systems

Described by a pseudohomogeneous ($T_S = T$) system and solid is stationary inside the system with no solid input or output:

$$\left[V\varepsilon\rho C_p + V(1 - \varepsilon)\rho_s C_{ps}\right]\frac{dT}{dt} + q\rho C_p(T - T_f) = Vr'(-\Delta H)$$

For gas–solid systems,

$$\rho C_p << \rho_s C_{ps}$$

Distributed Homogeneous Systems

The steady state is given by

$$q\rho C_p \frac{dT}{dl} = A_t r'(-\Delta H)$$

The unsteady state is given by

$$A_t \rho C_p \frac{\partial T}{\partial t} + q\rho C_p \frac{\partial T}{\partial l} = A_t r'(-\Delta H)$$

at $l = 0, T = T_f$, and at $t = 0, T = T(l)\big|_{t=0}$.

Distributed Heterogeneous Systems

The distributed heterogeneous system (described by pseudohomogeneous $T = T_S$) is given by (with no input or output solids)

$$\left(A_t \varepsilon \rho C_p + A_t (1 - \varepsilon) \rho_s C_{ps}\right) \frac{\partial T}{\partial t} + q \rho C_p \frac{\partial T}{\partial l} = A_t r'(-\Delta H)$$

at $l = 0$, $T = T_f$, and at $t = 0$, $T = T(l)\big|_{t=0}$.

As an exercise the reader can develop equations for the case of continuous solid flow in and out of the reactor.

4.3   THE AXIAL DISPERSION MODEL

The assumption of plug flow is not always correct. The plug flow assumes that the convective flow (flow by velocity $q/A_t = v$, caused by a compressor or pump) is dominating over any other transport mode. In fact, this is not always correct, and it is sometimes important to include the dispersion of mass and heat driven by concentration and temperature gradients. However, the plug flow assumption is valid for most industrial units because of the high Peclet number. We will discuss this model in some detail, not only because of its importance but also because the techniques used to handle these two-point boundary-value differential equations are similar to that used for other diffusion–reaction problems (e.g., catalyst pellets) as well as countercurrent processes and processes with recycle. The analytical analysis as well as the numerical techniques for these systems are very similar to this axial dispersion model for tubular reactors.

4.3.1   Formulation and Solution Strategy for the Axial Dispersion Model

The axial dispersion of mass is described by Fick's law:

$$N_i = -D_i \frac{dC_i}{dl}$$

where, N_i is the mass flux of component i (in mol/cm^2 s), D_i is the diffusivity in component i (in cm^2/s), C_i is the concentration of component i (in mol/cm^3), and l is the length (in cm).

The axial dispersion of heat is described by Fourier's law:

$$\bar{q} = -\lambda \frac{dT}{dl}$$

where $\bar{q}$ is the heat flux (in J/cm^2 s), λ is the thermal conductivity (in J/cm s K), and T is the temperature (in K).

We will demonstrate the introduction of the axial dispersion of mass and heat for a single reaction in the tubular reactor operating at steady state with the generalized rate of reaction per unit volume (r').

In Figure 4.10, the mass balance for the convective and diffusion flows are shown. It should be noted that n_i is the convective flow, N_i is the diffu-

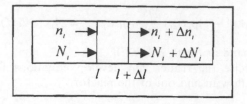

Figure 4.10 Convective and diffusion flows.

sion flow, and A is the cross-sectional area of the reactor tube. The steady-state mass balance with axial dispersion gives

$$n_i + \Delta n_i + A_t(N_i + \Delta N_i) = n_i + A_t N_i + \sigma_i A_t \Delta l r'$$

Dividing the equation by Δl and taking the limit $\Delta l \to 0$, we get

$$\frac{dn_i}{dl} + A_t \frac{dN_i}{dl} = \sigma_i A_t r' \tag{4.10}$$

This equation can be put in a dimensionless form after introducing simple assumptions (constant volumetric flow rate q). With this assumption we can write

$$q\frac{dC_i}{dl} + A_t \frac{dN_i}{dl} = \sigma_i A_t r'$$

Now, we use Fick's law,

$$N_i = -D_i \frac{dC_i}{dl}$$

Assuming D_i to be constant, we get

$$\frac{dN_i}{dl} = -D_i \frac{d^2 C_i}{dl^2}$$

and the mass balance design equation becomes

$$q\frac{dC_i}{dl} - A_t D_i \frac{d^2 C_i}{dl^2} = \sigma_i A_t r'$$

Dividing by A_t gives

$$v\frac{dC_i}{dl} - D_i \frac{d^2 C_i}{dl^2} = \sigma_i r'$$

where, v is the velocity of flow ($= q/A_t$).
Defining a dimensionless length

$$\omega = \frac{l}{L}$$

where L is the total length of the tubular reactor, we get

$$\frac{v}{L}\frac{dC_i}{d\omega} - \frac{D_i}{L^2}\frac{d^2C_i}{d\omega^2} = \sigma_i r'$$

On rearranging, we get,

$$\frac{1}{Pe_M}\frac{d^2C_i}{d\omega^2} - \frac{dC_i}{d\omega} + \sigma_i\frac{L}{v}r' = 0$$

where Pe_M is the Peclet number for mass ($= Lv/D_i$).

Applying the same procedure for heat balance with axial dispersion of heat, we get

$$\sum n_i H_i + A_t\bar{q} + Q'\Delta l = \sum [n_i H_i + \Delta(n_i H_i)] + A_t(\bar{q} + \Delta\bar{q})$$

The above equation can be rearranged to give

$$\sum\frac{d(n_i H_i)}{dl} + A_t\frac{d\bar{q}}{dl} = Q'$$

Differentiating the product $n_i H_i$ in the first term gives

$$\sum n_i\frac{dH_i}{dl} + \sum H_i\frac{dn_i}{dl} + A_t\frac{d\bar{q}}{dl} = Q' \qquad (4.11)$$

Equation (4.10) can be rewritten as

$$\frac{dn_i}{dl} = \sigma_i A_t r' - A_t\frac{dN_i}{dl}$$

Substitute this equation in the heat balance design equation (4.11), we obtain

$$\sum n_i\frac{dH_i}{dl} + \sum H_i\left(\sigma_i A_t r' - A_t\frac{dN_i}{dl}\right) + A_t\frac{d\bar{q}}{dl} = Q'$$

which can be rearranged to the following form:

$$\sum n_i\frac{dH_i}{dl} + A_t r'\sum\sigma_i H_i - A_t\sum H_i\left(\frac{dN_i}{dl}\right) - A_t\lambda\frac{d^2T}{dl^2} = Q'$$

After some approximations (for one-phase systems), as was used in previous chapters, we can approximate the above equation to the following form:

$$n_t C'_{p_{mix}}\frac{dT}{dl} + A_t r'(\Delta H) - A_t\sum H_i\frac{dN_i}{dl} - A_t\lambda\frac{d^2T}{dl^2} = Q'$$

where $C'_{p_{mix}}$ is the average molar specific heat of the mixture and n_i is the average total molar flow rate. This equation can also be written in the following form:

$$A_t \lambda \frac{d^2 T}{dl^2} - q\rho C_{p_{mix}} \frac{dT}{dl} + A_t r'(-\Delta H) + A_t \sum_i H_i \frac{dN_i}{dl} - = -Q' \qquad (4.12)$$

where $C_{p_{mix}}$ is the average mass specific heat for the mixture, q is the average volumetric flow rate, and ρ is the average density.

A Problematic Term Usually Neglected in All Books Without Ever Mentioning Anything About It

In Eq. (4.12), the term $A_t \sum H_i dN_i/dl)$ is the most problematic. It represents the enthalpy carried with mass axial dispersion. We make the following approximations in order to handle this term (without blindly neglecting it):

The First and Simplest Consider that the term $A_t \sum H_i dN_i/dl$, accounts for the heat transferred with the axial dispersion of mass to be accounted for through a small empirical correction in $\lambda \to \lambda_e$.

Second and Less Simple The term $A_t \sum H_i dN_i/dl$ can be written as follows:

$$-A_t \sum H_i D_i \frac{d^2 C_i}{dl^2}$$

which gives approximately

$$-A_t D_{av} \sum_i H_i \frac{d}{dl}\left(\frac{dC_i}{dl}\right) \qquad (4.13)$$

From the mass balance equation for plug flow (an assumption),

$$q\frac{dC_i}{dl} = \sigma_i A_t r'$$

which can be rewritten as

$$\frac{dC_i}{dl} = \left(\frac{A_t}{q}\right)\sigma_i r' \qquad (4.14)$$

Using Eq. (4.14) in Eq. (4.13) gives

$$-A_t D_{av} \sum_i H_i \frac{d}{dl}\left[\left(\frac{A_t}{q}\right)\sigma_i r'\right]$$

which simplifies to give

$$-A_t D_{av} \sum_i H_i \left(\frac{A_t}{q}\right)\sigma_i \frac{dr'}{dl}$$

It reduces to

$$-A_t D_{av}\left(\frac{A_t}{q}\right)\frac{dr'}{dl}\sum \sigma_i H_i$$

Replacing $\sum \sigma_i H_i$ with ΔH gives

$$-\frac{A_t^2 D_{av}}{q}\frac{dr'}{dl}(\Delta H) \tag{4.15}$$

From Eq. (4.15), we reach at the following expression for this problematic term:

$$A_t \sum H_i \frac{dN_i}{dl} = \frac{A_t^2 D_{av}}{q}(-\Delta H)\frac{dr'}{dl}$$

Using the above expression in the heat balance equation gives

$$A_t \lambda \frac{d^2 T}{dl^2} - q\rho C'_{p_{mix}}\frac{dT}{dl} + A_t r'(-\Delta H) + \frac{A_t^2 D_{av}}{q}(-\Delta H)\frac{dr'}{dl} = -Q'$$

which can be rearranged as

$$A_t \lambda \frac{d^2 T}{dl^2} - q\rho C'_{p_{mix}}\frac{dT}{dl} + A_t(-\Delta H)\left[r' + \underbrace{\left(\frac{A_t}{q}\right)D_{av}\frac{dr'}{dl}}_{1/v}\right] = -Q' \tag{4.16}$$

Dividing the equation by A_t gives

$$\lambda \frac{d^2 T}{dl^2} - v\rho C'_{p_{mix}}\frac{dT}{dl} + (-\Delta H)\left(r' + \frac{D_{av}}{v}\frac{dr'}{dl}\right) = -Q'$$

For most systems,

$$r' >>> \frac{D_{av}}{v}\frac{dr'}{dl}$$

So for this case, we can rewrite the above equation as

$$\lambda \frac{d^2 T}{dl^2} - v\rho C''_{p_{mix}}\frac{dT}{dl} + r'(-\Delta H) = -Q' \tag{4.17}$$

Note: For cases where r' is not much larger than $(D_{av}/v)(dr'/dl)$, this term can not be neglected and thus must be included in the equation.

Using the dimensionless form $\omega = l/L$ in Eq. (4.17), we get

$$\frac{\lambda}{L^2}\frac{d^2T}{d(l/L)^2} - \frac{\upsilon\rho C'_{p_{mix}}}{L}\frac{dT}{d(l/L)} + r'(-\Delta H) = -Q'$$

Multiplying both sides by $L/\upsilon\rho C'_{p_{mix}}$ gives

$$\frac{\lambda}{L^2}\frac{L}{\upsilon\rho C'_{p_{mix}}}\frac{d^2T}{d\omega^2} - \frac{\upsilon\rho C'_{p_{mix}}}{L}\frac{L}{\upsilon\rho C'_{p_{mix}}}\frac{dT}{d\omega} + r'(-\Delta H)\frac{L}{\upsilon\rho C'_{p_{mix}}} = -Q'\frac{L}{\upsilon\rho C'_{p_{mix}}}$$

which can be reorganized into the following form:

$$\frac{\lambda}{L\upsilon\rho C'_{p_{mix}}}\frac{d^2T}{d\omega^2} - \frac{dT}{d\omega} + r'(-\Delta H)\frac{L}{\upsilon\rho C'_{p_{mix}}} = -Q'\frac{L}{\upsilon\rho C'_{p_{mix}}}$$

Let

$$\frac{\lambda}{L\upsilon\rho C'_{p_{mix}}} = \frac{1}{Pe_H}$$

where Pe_H is the dimensionless Peclet number for heat transfer, and we get

$$\frac{1}{Pe_H}\frac{d^2T}{d\omega^2} - \frac{dT}{d\omega} + \frac{(-\Delta H)L}{\rho C'_{p_{mix}}}\frac{1}{\upsilon}r' = -\overline{Q}$$

where

$$\overline{Q} = Q'\frac{L}{\upsilon\rho C'_{p_{mix}}}$$

Now, let us consider that the rate of reaction is for a simple first-order irreversible reaction is

$$r' = k_0 e^{-(E/RT)}C_A$$

and let us define dimensionless temperature and concentration as follows:

$$y = \frac{T}{T_f} \quad \text{(dimensionless temperature)}$$

and

$$x_A = \frac{C_A}{C_{Af}} \quad \text{(dimensionless concentration)}$$

Using the rate of reaction and the dimensionless temperature and concentration, we get

$$\frac{1}{Pe_H}\frac{d^2y}{d\omega^2} - \frac{dy}{d\omega} + \underbrace{\frac{(-\Delta H)C_{Af}}{\rho C'_{p_{mix}}T_f}}_{\beta}D_a e^{-(\gamma/y)}x_A = -\frac{\overline{Q}}{T_f}$$

which can be simplified as

$$\frac{1}{Pe_H}\frac{d^2y}{d\omega^2} - \frac{dy}{d\omega} + \beta D_a e^{-(\gamma/y)} x_A = -\hat{Q} \tag{4.18}$$

Here,

$$\beta = \frac{(-\Delta H)C_{Af}}{\rho C'_{p_{mix}} T_f} \quad \text{(thermicity factor)}$$

$$D_a = \frac{k_0 L}{v} \quad \text{(Damkohler number)}$$

and

$$\hat{Q} = \frac{\overline{Q}}{T_f}$$

Similarly, the mass balance design equation for this case is

$$\frac{1}{Pe_M}\frac{d^2x_A}{d\omega^2} - \frac{dx_A}{d\omega} - D_a e^{-(\gamma/y)} x_A = 0 \tag{4.19}$$

Note that both mass and heat balance design equations (4.18) and (4.19) are two-point boundary-value differential equations. Thus, each one of them requires two boundary conditions. These boundary conditions are derived as shown in this solved example (Example 4.1). In Example 4.1, the boundary conditions are developed for the mass balance design equation. For the nonisothermal case, the boundary conditions for the heat balance design equations are left as an exercise for the reader.

Solved Example 4.1   Axial Dispersion Model

In this solved example, we present the development of the isothermal model and its boundary conditions for a case where the equation is linear and can be solved analytically. Also presented is another case, where the model is nonlinear and we describe its numerical solution using Fox's iterative method.

For a steady-state, isothermal, homogeneous tubular reactor consider the simple reaction

$$A \rightarrow B$$

taking place in a tubular reactor; axial dispersion is not negligible (Peclet number Pe = 15.0). Answer the following design questions:

   A.   If the reaction is first order, what is the value of D_a (Damkohler number) to achieve a conversion of 0.75 at the exit of the reactor?

B. If the reaction is second order and the numerical value of D_a (Damkohler number) is the same as in part 1 (although its definition is slightly different), find the exit conversion using the Fox's iterative method (explain your formulation of adjoint equations for the iterative solution of the nonlinear two-point boundary-value differential equation).

Solution (Fig. 4.11)

The mass balance design equation is

$$qC_A + A_t(N_A) = q(C_A + \Delta C_A) + A_t(N_A + \Delta N_A) + A_t\Delta lkC_A^n$$

where n is the order of reaction. Canceling similar terms from both sides of the equation gives

$$0 = q\Delta C_A + A_t\Delta N_A + A_t\Delta lkC_A^n$$

Dividing the equation by Δl and taking the limit as $\Delta l \to 0$ gives

$$0 = q\frac{dC_A}{dl} + A_t\frac{dN_A}{dl} + A_tkC_A^n \tag{4.20}$$

From Fick's law for diffusion we have

$$N_A = -D_A\frac{dC_A}{dl}$$

which, upon differentiation, gives

$$\frac{dN_A}{dl} = -D_A\frac{d^2C_A}{dl^2} \tag{4.21}$$

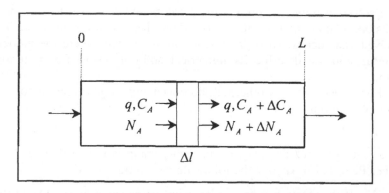

Figure 4.11 Axial dispersion mass flow.

From Eqs. (4.20) and (4.21), we get

$$0 = q\frac{dC_A}{dl} - A_t D_A \frac{d^2 C_A}{dl^2} + A_t k C_A^n$$

Divide both sides by A_t and note that v, the velocity of flow, equals q/A_t, we obtain

$$0 = v\frac{dC_A}{dl} - D_A \frac{d^2 C_A}{dl^2} + k C_A^n$$

We reorganize the equation to get

$$D_A \frac{d^2 C_A}{dl^2} - v\frac{dC_A}{dl} - k C_A^n = 0$$

We define the following dimensionless variables:

$$\omega = \frac{l}{L} \quad \text{(dimensionless length, where } L \text{ is the total length of the reactor tube)}$$

$$x = \frac{C_A}{C_{Af}} \quad \text{(dimensionless concentration, where } C_{Af} \text{ is the feed concentration of } A\text{)}$$

Using these dimensionless variables, we get

$$\frac{D_A}{L^2} \frac{d^2 (C_A/C_{Af})}{d(l/L)^2} - \frac{v}{L} \frac{d(C_A/C_{Af})}{d(l/L)} - k C_{Af}^{n-1} \frac{C_A^n}{C_{Af}^n} = 0$$

Multiplying the equation with L and dividing it by v we obtain

$$\frac{D_A}{Lv} \frac{d^2 x}{d\omega^2} - \frac{dx}{d\omega} - \left(\frac{k L C_{Af}^{n-1}}{v}\right) x^n = 0$$

Define $Pe = Lv/D_A$ and $D_a = k L C_{Af}^{n-1}/v$ to get

$$\frac{1}{Pe} \frac{d^2 x}{d\omega^2} - \frac{dx}{d\omega} - D_a x^n = 0 \tag{4.22}$$

where

for the first-order reaction

$$D_a = \frac{kL}{v}$$

for the second-order reaction

$$D_a = \frac{k L C_{Af}}{v}$$

Boundary Conditions

At the exit,

$$\omega = 1, \quad \frac{dx}{d\omega} = 0 \tag{4.23}$$

At the entrance,

$$\omega = 0, \quad \frac{1}{\text{Pe}} \frac{dx}{d\omega} = x - 1 \tag{4.24}$$

Boundary condition (4.24) can be obtained as shown in Figure 4.12.
A mass balance at the entrance at $l = 0$ gives

$$qC_{Af} = qC_A + A_t N_A$$

Dividing by A_t and using Fick's law, we get

$$vC_{Af} = vC_A - D_A \frac{dC_A}{dl}$$

Rearranging in dimensionless form gives

$$1 = x - \frac{D_A}{Lv} \frac{dx}{d\omega}$$

which can be written in its more popular form

$$\frac{1}{\text{Pe}} \frac{dx}{d\omega} = x - 1$$

Thus, the model equations are as follows

A. For the First-Order Reaction

$$\frac{1}{\text{Pe}} \frac{d^2 x}{d\omega^2} - \frac{dx}{d\omega} - D_a x = 0 \tag{4.22}$$

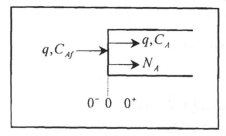

Figure 4.12   Mass flow at the entrance.

The boundary conditions are as follows:

At $\omega = 1$,

$$\frac{dx}{d\omega} = 0 \tag{4.23}$$

At $\omega = 0$,

$$\frac{1}{Pe} \frac{dx}{d\omega} = x - 1 \tag{4.24}$$

The present problem is a design problem because the required conversion is given and it is required to find the value of D_a. However, for educational purposes, before we present the solution of this design problem, we will first present the simulation problem when D_a is given and the conversion is unknown. For illustration, we will take $D_a = 1.4$. The solution steps for this simulation problem is as follows:

1. Solution of the second-order differential equation (4.22)
    Equation (4.22) can be rewritten as:

    $$\frac{d^2x}{d\omega^2} - (Pe) \frac{dx}{d\omega} - (PeD_a)x = 0$$

    This is a linear second-order ordinary differential equation which can be solved using the operator method (see Appendices B and C).

    Thus, the characteristic equation is

    $$\lambda^2 - Pe\lambda - (PeD_a) = 0$$

    The roots of this characteristic equation is obtained as:

    $$\lambda_{1,2} = \frac{Pe \pm \sqrt{Pe^2 + 4PeD_a}}{2}$$

    Thus, for $Pe = 15.0$ and $D_a = 1.4$, the characteristic roots (eigen-values) λ_1 and λ_2 are given by

    $$\lambda_{1,2} = \frac{15 \pm \sqrt{15^2 + 4(15)(1.4)}}{2}$$

    which simplifies to

    $$\lambda_{1,2} = \frac{15 \pm 17.58}{2}$$

    We get $\lambda_1 = 16.29$ and $\lambda_2 = -1.29$.

The solution of this second-order differential equation has the following general form:

$$x = C_1 e^{\lambda_1 \omega} + C_2 e^{\lambda_2 \omega}$$

For the present specific case, the dimensionless concentration x is given by

$$x = C_1 e^{16.29\omega} + C_2 e^{-1.29\omega}$$

The constants C_1 and C_2 can be calculated using the boundary conditions.

The differential of x is given by

$$\frac{dx}{d\omega} = C_1 \lambda_1 e^{\lambda_1 \omega} + C_2 \lambda_2 e^{\lambda_2 \omega}$$

First, the boundary condition at $\omega = 1.0$ is

$$\frac{dx}{d\omega} = C_1 \lambda_1 e^{\lambda_1} + C_2 \lambda_2 e^{\lambda_2} = 0$$

It gives

$$C_1 (16.29) e^{16.29} = -C_2(-1.29) e^{-1.29}$$

Further simplification yields

$$C_2 = (5.4488 \times 10^8) C_1$$

The second boundary condition at $\omega = 0$ is

$$\frac{1}{Pe} \frac{dx}{d\omega} = x - 1. \text{ Therefore,}$$

$$\frac{1}{Pe}[(C_1 \lambda_1 + C_2 \lambda_2) = (C_1 + C_2) - 1$$

which gives

$$\frac{1}{15}[(16.29)C_1 + (-1.29)(5.448 \times 10^8 C_1)]$$
$$= [C_1 + (5.448 \times 10^8 C_1)] - 1$$

Thus we can obtain C_1 as:

$$C_1 = 16.9 \times 10^{-10}$$

and because

$$C_2 = (5.4488 \times 10^8) C_1$$

C_2 is obtained as

$$C_2 = 0.9208$$

Thus, the solution is

$$x = (16.9 \times 10^{-10})e^{16.29\omega} + (0.9208)e^{-1.29\omega}$$

At the exit, $\omega = 1.0$,

$$\underbrace{x(\omega = 1.0)}_{\text{at exit}} = (16.9 \times 10^{-10})e^{16.29} + (0.9208)e^{-1.29}$$

giving

$$\underbrace{x(\omega = 1.0)}_{\text{at exit}} = 0.27248$$

As $x = C_A/C_{Af}$, the conversion is

$$C_{Af} - \frac{C_A}{C_{Af}} = 1 - x$$

so the value of conversion at the exit for $D_a = 1.4$ and Pe $= 15.0$ is

Conversion $= 1 - 0.27348 = 0.7265$

Note that we were just lucky that we took the value of $D_a = 1.4$, which gave us the conversion close to 0.75 in the design problem. This is just luck (do you know why?).

2. The design problem
   D_a is unknown, but $x(\omega = 1.0)$ at the exit is 0.25 (because the given conversion is 0.75). In this case, the same procedure is followed, but it is slightly more difficult and lengthier. Why?
   All the steps are the same as above until we reach the calculation of λ_1 and λ_2 with D_a unknown; thus, we have

$$\lambda_{1,2} = \frac{15 \pm \sqrt{15^2 + 4(15)D_a}}{2}$$

giving

$$\lambda_{1,2} = 7.5 \pm \sqrt{56.25 + 15D_a}$$

Thus, λ_1 and λ_2 are given by

$$\lambda_1 = 7.5 + \sqrt{56.25 + 15D_a} \quad \text{and} \quad \lambda_2 = 7.5 - \sqrt{56.25 + 15D_a}$$

Now, we have

$$x = C_1 e^{\lambda_1 \omega} + C_2 e^{\lambda_2 \omega}$$

and

$$\frac{dx}{d\omega} = C_1 \lambda_1 e^{\lambda_1 \omega} + C_2 \lambda_2 e^{\lambda_2 \omega}$$

The first boundary condition (at $\omega = 1.0$) is

$$\frac{dx}{d\omega} = C_1 \lambda_1 e^{\lambda_1} + C_2 \lambda_2 e^{\lambda_2} = 0$$

Therefore,

$$C_2 = -\frac{C_1 \lambda_1 e^{\lambda_1}}{\lambda_2 e^{\lambda_2}} \tag{4.25}$$

The other boundary condition is at $(\omega = 0)$

$$\frac{1}{Pe}\frac{dx}{d\omega} = x - 1$$

which can be written as

$$\frac{1}{15}(C_1 \lambda_1 + C_2 \lambda_2) = (C_1 + C_2) - 1 \tag{4.26}$$

Substituting Eq. (4.25) in Eq. (4.26), we get

$$\frac{1}{15}\left[C_1 \lambda_1 - \left(\frac{C_1 \lambda_1 e^{\lambda_1}}{\lambda_2 e^{\lambda_2}}\right)\lambda_2 \right] = \left(C_1 - \frac{C_1 \lambda_1 e^{\lambda_1}}{\lambda_2 e^{\lambda_2}}\right) - 1$$

which gives

$$C_1\left[\frac{\lambda_1}{15} - \left(\frac{\lambda_1}{15}\right)e^{\lambda_1 - \lambda_2}\right] = C_1\left[1 - \frac{\lambda_1}{\lambda_2}e^{\lambda_1 - \lambda_2}\right] - 1$$

Further rearrangement gives

$$C_1\left[\left(\frac{\lambda_1}{15} - 1\right) + \left(\frac{\lambda_1}{\lambda_2}e^{\lambda_1 - \lambda_2}\right)\left(1 - \frac{\lambda_2}{15}\right)\right] = -1$$

So we can write

$$C_1 = -\left[\left(\frac{\lambda_1}{15} - 1\right) + \left(\frac{\lambda_1}{\lambda_2}e^{\lambda_1 - \lambda_2}\right)\left(1 - \frac{\lambda_2}{15}\right)\right]^{-1}$$

Because

$$\lambda_1 - \lambda_2 = \left(7.5 + \sqrt{56.25 + 15D_a}\right) - \left(7.5 - \sqrt{56.25 + 15D_a}\right)$$

we have

$$\lambda_1 - \lambda_2 = 2\sqrt{56.25 + 15D_a} = \sqrt{225 + 60D_a}$$

Thus, C_1 is given as

$$
C_1 = -\left[\left(\frac{7.5 + \sqrt{56.25 + 15D_a}}{15} - 1\right) \right.
$$

$$
+ \left(\frac{7.5 + \sqrt{56.25 + 15D_a}}{7.5 - \sqrt{56.25 + 15D_a}\, e^{(225+60D_a)^{\frac{1}{2}}}}\right) \tag{4.27}
$$

$$
\left. \times \left(1 - \frac{7.5 - \sqrt{56.25 + 15D_a}}{15}\right)\right]^{-1}
$$

We can write Eq. (4.27) as

$$C_1 = F_1(D_a)$$

and from Eq. (4.25), we have

$$C_2 = -C_1 \frac{\lambda_1}{\lambda_2} e^{\lambda_1 - \lambda_2}$$

Therefore,

$$C_2 = -F_1(D_a) \frac{7.5 + \sqrt{56.25 + 15D_a}}{7.5 - \sqrt{56.25 + 15D_a}} e^{\sqrt{225+60D_a}}$$

which can be written as

$$C_2 = -F_2(D_a)$$

we return to the dimensionless concentration equation

$$x = C_1 e^{\lambda_1 \omega} + C_2 e^{\lambda_2 \omega}$$

which at exit ($\omega = 1.0$) can be written as

$$\underbrace{x(\omega = 1.0)}_{\text{at exit}} = C_1 e^{\lambda_1} + C_2 e^{\lambda_2}$$

Giving

$$\underbrace{x(\omega = 1.0)}_{\text{at exit}} = F_1(D_a) \underbrace{e^{7.5 + (56.25 + 15D_a)^{\frac{1}{2}}}}_{f_1(D_a)} + F_2(D_a) \underbrace{e^{7.5 - (56.25 + 15D_a)^{\frac{1}{2}}}}_{f_2(D_a)}$$

Now, for 75% conversion, the equation becomes

$$0.25 = F_1(D_a)f_1(D_a) + F_2(D_a)f_2(D_a)$$

which can be written as

$$\underbrace{F_1(D_a)f_1(D_a) + F_2(D_a)f_2(D_a) - 0.25}_{F(D_a)} = 0$$

Finally, we have this single equation in terms of D_a:

$$F(D_a) = 0$$

This last equation is a nonlinear algebraic equation in D_a; it can be solved using the Newton–Raphson method [if the differentiation is very lengthy and cumbersome, which is the case here, then in the Newton–Raphson method, you can use the modified Newton–Raphson by approximating $(\partial F/\partial D_a)^n$ by $(F^{n-1} - F^n)/(D_a^{n-1} - D_a^n)$ or, easier and sure to converge, use the bisectional method, which is straightforward and sure to converge and is applicable to this case, which is a single equation F in single variable D_a]. For Newton–Raphson, modified Newton–Raphson, and Bisectional methods, see Appendix B.

Hint: If all these very elementary techniques are too painful for the reader, then just plot $F(D_a)$ versus D_a for different values of D_a and find the value of D_a for which $F(D_a) = 0$.

B. For the Second-Order Reaction and Fox's Iterative Method for Nonlinear Two-Point Boundary-Value Differential Equation

For the second-order reaction, the equation is

$$\frac{1}{\text{Pe}} \frac{d^2x}{d\omega^2} - \frac{dx}{d\omega} - D_a x^2 = 0$$

with

$$D_a = \frac{kLC_{Af}}{v}$$

where D_a is the value obtained from the previous part (it will be close to or greater than 1.4) and the Peclet number is

$$\text{Pe} = 15.0$$

The boundary conditions are as follows:

At $\omega = 1$

$$\frac{dx}{d\omega} = 0$$

At $\omega = 0$

$$\frac{1}{Pe}\frac{dx}{d\omega} = x - 1$$

For the numerical solution of this nonlinear equation, the first step is to put the second-order differential equation in the form of two first-order differential equations. Let

$$x = x_1$$

so

$$\frac{dx}{d\omega} = \frac{dx_1}{d\omega}$$

and let

$$\frac{dx_1}{d\omega} = x_2$$

so

$$\frac{d^2x}{d\omega^2} = \frac{d^2x_1}{d\omega^2} = \frac{dx_2}{d\omega}$$

The second-order differential equation can now be rewritten as the set of the following two first-order differential equations:

$$\frac{1}{Pe}\frac{dx_2}{d\omega} - x_2 - D_a x_1^2 = 0$$

and

$$\frac{dx_1}{d\omega} = x_2$$

Thus, the equivalent two first-order differential equations are

$$\frac{dx_1}{d\omega} = x_2$$

$$\frac{dx_2}{d\omega} = Pex_2 + PeD_a x_1^2$$

with the boundary conditions:

At $\omega = 1$

$$x_2(1) = 0$$

At $\omega = 0$

$$x_2(0) = Pe[x_1(0) - 1].$$

If we want to start the integration using the marching technique (Euler, Runge–Kutta, or using any subroutine from Polymath, Matlab, or IMSL libraries for the solution of initial-value differential equations, see Appendix B), then we need to assume $x_1(0)$ or $x_2(0)$. Suppose we choose to assume $x_1(0) = x_1^n(0)$, where n refers to the nth iteration (for the initial guess, it will be $n = 1$). Then, $x_2^n(0)$ can be easily and directly computed from the relation

$$x_2^n(0) = \text{Pe}[x_1^n(0) - 1)]$$

Now, knowing $x_1^n(0)$ and $x_2^n(0)$, we can integrate forward using any of the subroutines until we reach $\omega = 1.0$, but we will find that $x_2^n(1) \neq 0$. So, our initial guess or choice is not correct and we have to choose another value, $x_1^{n+1}(0)$, and then try again and so on until we find a guess $x_1^{n+1}(0)$, which, after integration until $\omega = 1.0$, will give us $x_2^{n+1}(1) = 0$.

How can we accelerate this process? We can use Fox's method (which is a modified Newton–Raphson method for two-point boundary-value differential equations).

For a certain iteration $x_1^n(0)$, the next iteration $x_1^{n+1}(0)$ can be obtained using the following process.

Our objective is to have

$$x_2(1) = f(x_1(0)) = 0$$

However, for any guess of $x_1^n(0)$, we will not have this satisfied. Actually, we will have $f^n(x_1^n(0)) \neq 0$. Then we expand the function f, which is a function of $x(0)$ using Taylor series expansion, as follows:

$$f^{n+1}\left(x_1^{n+1}(0)\right) = f^n(x_1^n(0)) + \left(\frac{\partial f(x_1^n(0))}{\partial x_1(0)}\right)^n \left(x_1^{n+1}(0) - x_1^n(0)\right) + \cdots$$

Note: n and $n + 1$ are numbers of the iteration (not exponentials!)

Our objective is that

$$f^{n+1}\left(x_1^{n+1}(0)\right) = 0$$

and $f^n(x_1^n(0))$ is calculated from the integration at $\omega = 1.0$ as $x_2^n(1)$. Therefore, from the Taylor series expansion, we get

$$0 = x_2^n(1) + \left(\frac{\partial x_2(1)}{\partial x_1(0)}\right)^n [x_1^{n+1}(0) - x_1^n(0))] + \text{Neglect higher-order terms}$$

Rearrangement gives

$$x_1^{n+1}(0) = x_1^n(0) - \frac{x_2^n(1)}{(\partial x_2(1)/\partial x_1(0))^n}$$

In the above relation, if we calculate $(\partial x_2(1)/\partial x_1(0))^n$, then the iteration can go on. How do we calculate this differential $(\partial x_2(1)/\partial x_1(0))^n$?

We will do that through the following steps:

Define a variable (called adjoint variable)

$$y_1(\omega) = \left(\frac{\partial x_2(\omega)}{\partial x_1(0)} \right)$$

which leads to

$$\left(\frac{\partial x_2(1)}{\partial x_1(0)} \right) = y_1(1.0)$$

we also define

$$y_2(\omega) = \left(\frac{\partial x_1(\omega)}{\partial x_1(0)} \right)$$

How can we compute $y_1(\omega)$ so that we can get $x_1^{n+1}(0)$? This can be achieved by formulating adjoint equations for $y_1(\omega)$ that are solved simultaneously with the x_1 and x_2 differential equations. However, the adjoint equations must be initial-value differential equations. The method of doing this follows:

Because

$$\frac{dx_1(\omega)}{d\omega} = x_2(\omega)$$

if we differentiate both sides of the above equation partially with respect to $x_1(0)$, we get

$$\frac{d\left[\partial x_1(\omega) / \partial x_1(0) \right]}{d\omega} = \frac{\partial x_2(\omega)}{\partial x_1(0)}$$

By the definition of $y_1(\omega)$ and $y_2(\omega)$ given above, we get

$$\frac{dy_2(\omega)}{d\omega} = y_1(\omega)$$

which is the first adjoint equation. To get the second adjoint equation, we look at the second differential equation of the state variable x_2. Because

$$\frac{dx_2(\omega)}{d\omega} = Pe x_2(\omega) + Pe D_a x_1^2(\omega)$$

differentiating the above equation with respect to $x_1(0)$ gives

$$\frac{d\left[\partial x_2(\omega) / \partial x_1(0) \right]}{d\omega} = Pe \frac{\partial x_2(\omega)}{\partial x_1(0)} + Pe D_a \frac{\partial x_1^2(\omega)}{\partial x_1(0)}$$

which can be written as

$$\frac{dy_1(\omega)}{d\omega} = Pe(y_1(\omega)) + PeD_a \frac{\partial x_1^2(\omega)}{\partial x_1(\omega)} \frac{\partial x_1(\omega)}{\partial x_1(0)}$$

giving

$$\frac{dy_1(\omega)}{d\omega} = Pey_1(\omega) + 2PeD_a x_1(\omega) y_2(\omega)$$

which is the second adjoint equation.

The initial conditions at $\omega = 0$ for the adjoint variables are

$$y_1(0) = \frac{\partial x_2(0)}{\partial x_1(0)}$$

From the boundary condition at $\omega = 0$, we have

$$x_2(0) = Pe[x_1(0) - 1]$$

Differentiation gives

$$\frac{\partial x_2(0)}{\partial x_1(0)} = Pe = y_1(0)$$

and

$$y_2(0) = \frac{\partial x_1(0)}{\partial x_1(0)} = 1$$

Now the problem is ready for the iterative solution.

The system equations and the boundary conditions are

$$\frac{dx_1}{d\omega} = x_2$$

and

$$\frac{dx_2}{d\omega} = Pex_2 + PeD_a x_1^2$$

With the following boundary conditions:

At $\omega = 1$

$$x_2(1) = 0$$

At $\omega = 0$

$$x_2(0) = Pe(x_1(0) - 1)$$

The adjoint equations and their initial conditions are

$$\frac{dy_2(\omega)}{d\omega} = y_1(\omega)$$

and

$$\frac{dy_1(\omega)}{d\omega} = \text{Pe}\, y_1(\omega) + 2\text{Pe} D_a x_1(\omega) y_2(\omega)$$

with the initial conditions at $\omega = 0$ of

$$y_1(0) = \text{Pe} \quad \text{and} \quad y_2(0) = 1.0$$

The iterative formula is

$$x_1^{n+1}(0) = x_1^n(0) - \frac{x_2^n(1)}{y_1^n(1)}$$

Important Note: The iterative solution for this two-point boundary-value differential equation utilizing the adjoint equations is shown in Figure 4.13. Notice that this procedure may be unstable numerically and we may need to reverse the direction of the integration from the exit of the reactor toward the inlet as explained in the following subsection.

4.3.2 Solution of the Two-Point Boundary-Value Differential Equations and Numerical Instability Problems

Consider the axial dispersion model

$$\frac{1}{\text{Pe}} \frac{d^2x}{d\omega^2} - \frac{dx}{d\omega} - D_a x^2 = 0 \qquad (4.28)$$

with the following boundary conditions:

At $\omega = 1$

$$\frac{dx}{d\omega} = 0$$

At $\omega = 0$

$$\frac{1}{\text{Pe}} \frac{dx}{d\omega} = x - 1$$

The iterative procedure as explained earlier is to put the equation in the form of two first-order differential equations:

$$\frac{dx_1}{d\omega} = x_2$$

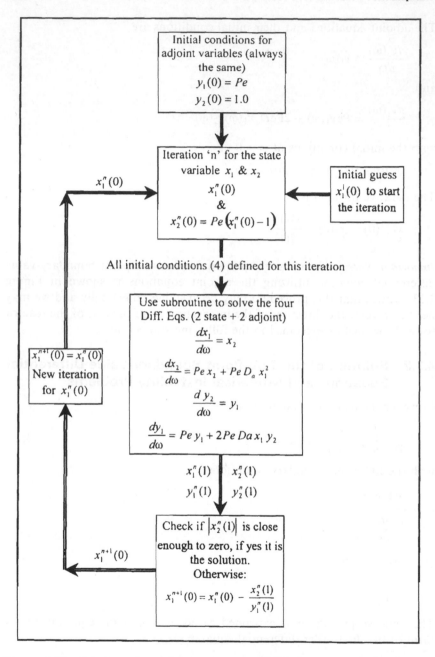

Figure 4.13 Solution scheme for the two-point boundary-value differential equations of the isothermal nonlinear axial dispersion model.

$$\frac{dx_2}{d\omega} = \mathrm{Pe}x_2 + \mathrm{Pe}D_a x_1^2$$

with the following boundary conditions:

At $\omega = 1$

$x_2(1) = 0$

At $\omega = 0$

$x_2(0) = \mathrm{Pe}(x_1(0) - 1)$ \hfill (4.29)

If we employ forward integration and iteration, then following two problems may arise:

1. You guess $x_1(0)$ with its errors; then, $x_2(0)$ is defined automatically from the boundary condition (4.29). For a large value of $\mathrm{Pe}(> 5.0)$ the error in $x_1(0)$ propagates to $x_2(0)$ amplified by Pe. Thus, the $x_1(0)$ error is clearly amplified.
2. Consider the linear version of this equation and inspect the characteristic roots (eigenvalues) of the differential equation. The characteristic equation for this linear case is

$$\lambda^2 - \mathrm{Pe}\lambda - D_a\mathrm{Pe} = 0$$

giving the characteristic roots (eigenvalues)

$$\lambda_{1,2} = \frac{\mathrm{Pe} \pm \sqrt{\mathrm{Pe}^2 + 4D_a\mathrm{Pe}}}{2}$$

This leads to one negative eigenvalue and one positive eigenvalue:

$$\lambda_1 = \frac{\mathrm{Pe} + \sqrt{\mathrm{Pe}^2 + 4D_a\mathrm{Pe}}}{2} \quad \text{which is positive}$$

$$\lambda_2 = \frac{\mathrm{Pe} - \sqrt{\mathrm{Pe}^2 + 4D_a\mathrm{Pe}}}{2} \quad \text{which is negative}$$

This is illustrated as follows:

Case 1

For $\mathrm{Pe} = 15.0$ and $D_a = 1.5$, the eigenvalues are

$$\lambda_1 = \frac{15 + \sqrt{15^2 + 4(1.5)(15)}}{2} = 16.37$$

$$\lambda_2 = \frac{15 - \sqrt{15^2 + 4(1.5)(15)}}{2} = -1.374$$

The solution of the differential equation is

$$x_1 = C_1 e^{16.37\omega} + C_2 e^{-1.374\omega}$$

Note that $|\lambda_1| >> |\lambda_2|$; $e^{16.37\omega}$ indicates strong instability in the ω direction and $e^{-1.374\omega}$ indicates stabilization in the ω direction. For such a case, it is very likely to have numerical instability.

Case 2

For $Pe = 0.1$ and $D_a = 1.5$, the eigenvalues are

$$\underbrace{\lambda_1 = 0.88}_{\text{Positive}} \quad \text{and} \quad \underbrace{\lambda_2 = -0.681}_{\text{Negative}}$$

Note that $|\lambda_1|$ and $|\lambda_2|$ are very comparable; therefore, numerical instability is less likely. Thus, it is clear that the integration in the positive direction from $\omega = 0$ to $\omega = 1.0$ is not very wise. The following backward integration is much more stable and avoids the complexities due to the above two reasons.

Backward Integration. Consider again the differential equations

$$\frac{dx_1}{d\omega} = x_2$$

$$\frac{dx_2}{d\omega} = Pex_2 + PeD_a x_1^2$$

with the following boundary conditions:

At $\omega = 1$

$$x_2(1) = 0$$

At $\omega = 0$,

$$x_2(0) = Pe(x_1(0) - 1)$$

We use the following simple transformation from ω to ω' in order to achieve backward integration:

$$\omega' = 1 - \omega$$

Thus, at $\omega = 1.0, \omega' = 0$, and at $\omega = 0, \omega' = 1$. The differential equations become

$$-\frac{dx_1}{d\omega'} = x_2$$

$$-\frac{dx_2}{d\omega'} = Pex_2 + PeD_a x_1^2$$

Now, the set of differential equations by its boundary conditions can be written as

$$\frac{dx_1}{d\omega'} = -x_2 \tag{4.30}$$

$$\frac{dx_2}{d\omega'} = -Pex_2 - PeD_a x_1^2 \tag{4.31}$$

with the following boundary conditions

At $\omega' = 0$

$$x_2(0) = 0 \tag{4.32}$$

At $\omega' = 1$

$$x_2(1) = Pe[x_1(1) - 1] \tag{4.33}$$

Now, the iteration proceeds by assuming that $x_1(0)$ (x_1 at $\omega' = 0$, which is at the exit) and the function $F(x_1(0))$ should be zero at $\omega' = 1$, which means that

$$\underbrace{x_2(1) - Pe(x_1(1) - 1)}_{F(x_1(0))} = 0 \tag{4.34}$$

Thus, the Taylor series expansion of $F(x_1(0))$ can be written as

$$\underbrace{F^{n+1}(x_1(0))}_{\text{equal to zero}} = F^n(x_1(0)) + \left(\frac{\partial F}{\partial(x_1(0))}\right)^n (x_1^{n+1}(0) - x_1^n(0))$$

From Eq. (4.34), differentiating with respect to $x_1(0)$ gives

$$\left(\frac{\partial F}{\partial(x_1(0))}\right)^n = \left(\frac{\partial x_2(1)}{\partial x_1(0)}\right) - Pe\left(\frac{\partial x_1(1)}{\partial x_1(0)}\right)$$

We define the following adjoint variables:

$$y_1(\omega') = \left(\frac{\partial x_1(\omega')}{\partial x_1(0)}\right)$$

$$y_2(\omega') = \left(\frac{\partial x_2(\omega')}{\partial x_1(0)}\right)$$

Differentiate the differential equation (4.30) with respect to $x_1(0)$ to get

$$\frac{d[\partial x_1(\omega')/\partial x_1(0)]}{d\omega'} = -\frac{\partial x_2(\omega')}{\partial x_1(0)}$$

which can be written in terms of the adjoint variables as follows:

$$\frac{dy_1(\omega')}{d\omega'} = -y_2(\omega')$$

Similarly, differentiating the differential equation (4.31) gives

$$\frac{d[\partial x_2(\omega')/\partial x_1(0)]}{d\omega'} = -Pe\frac{\partial x_2(\omega')}{\partial x_1(0)} - PeD_a\frac{\partial x_1^2(\omega')}{\partial x_1(0)}$$

which gives

$$\frac{dy_2(\omega')}{d\omega'} = -Pe[y_2(\omega')] - PeD_a\frac{\partial x_1^2(\omega')}{\partial x_1(\omega')}\frac{\partial x_1(\omega')}{\partial x_1(0)}$$

giving

$$\frac{dy_2(\omega')}{d\omega'} = -Pe[y_2(\omega')] - PeD_a(2x_1(\omega'))\frac{\partial x_1(\omega')}{\partial x_1(0)}$$

Further simplification gives

$$\frac{dy_2(\omega')}{d\omega'} = -Pe(y_2(\omega')) - PeD_a(2x_1(\omega'))y_1(\omega')$$

The final form is

$$\frac{dy_2(\omega')}{d\omega'} = -Pe[y_2(\omega')] - 2PeD_ax_1(\omega')y_1(\omega')$$

The two adjoint equations are thus given as

$$\frac{dy_1(\omega')}{d\omega'} = -y_2(\omega') \tag{4.35}$$

$$\frac{dy_2(\omega')}{d\omega'} = -Pe(y_2(\omega')) - 2PeD_ax_1(\omega')y_1(\omega') \tag{4.36}$$

The boundary conditions at $\omega' = 0$ for these adjoint equations are

$$y_1(0) = \frac{\partial x_1(0)}{\partial x_1(0)} = 1 \tag{4.37}$$

$$y_2(0) = \frac{\partial x_2(0)}{\partial x_1(0)} = 0 \tag{4.38}$$

The reader can use model equations (4.30)–(4.33) and these adjoint equations (4.35)–(4.38) to establish the iteration procedure as discussed earlier.

Solved Example 4.2

Two CSTRs in series are operated isothermally. The reaction is first order and the two tanks have equal volumes.

    (a) Formulate a steady-state mathematical model for this process.

    (b) Choose volumes of the two reactors and temperatures (isothermal and equal for both CSTRs) so that the conversion from the second CSTR is 0.78. The feed concentration is given as 0.45 mol/m^3, and the first-order reaction rate constant k is given by $k = k_0 e^{-E/RT}$, where $k_0 = 2.9 \times 10^7$ h^{-1} and $E/R = 5300$ K.

    (c) Formulate a steady-state mathematical model for the process when the second reactor is a distributed system (tubular, plug flow unit).

    (d) Find the process exit concentration for part (c), with all parameters being the same as part (b).

    (e) Derive an unsteady-state model for case (a) (two CSTRs).

Solution (see Fig. 4.14)

Consider the reaction as

$$A \rightarrow B$$

with the generalized rate of reaction as

$$r'_j = k_0 e^{-E/RT_j} C_{A_j}$$

where $j = 1$ and 2 (the two reactors).

Part (a)

The steady-state mathematical model is

$$n_{A_1} = n_{A_f} + \sigma_A V_1 r'_1 \tag{4.39}$$

$$n_{A_2} = n_{A_1} + \sigma_A V_2 r'_2 \tag{4.40}$$

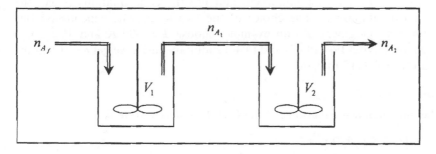

Figure 4.14 Mass flow diagram for CSTRs in series.

Part (b)

Given

$$V_1 = V_2 = V \quad \text{and} \quad T_1 = T_2 = T$$

and

$$r_1' = k_0 e^{-E/RT_1} C_{A_1} \quad \text{and} \quad r_2' = k_0 e^{-E/RT_2} C_{A_2}$$

From Eqs (4.39) and (4.40), we get

$$q C_{A_1} = q C_{A_f} - V k_0 e^{-E/RT} C_{A_1} \tag{4.41}$$

$$q C_{A_2} = q C_{A_1} - V k_0 e^{-E/RT} C_{A_2} \tag{4.42}$$

Given $k_0 = 2.9 \times 10^7 \text{ h}^{-1}$ and $E/R = 5300 \text{K}$, the overall conversion is

$$x_A = \frac{C_{A_f} - C_{A_2}}{C_{A_f}}$$

On substituting the given values, we get

$$0.78 = \frac{0.45 - C_{A_2}}{0.45}$$

Thus,

$$C_{A_2} = 0.099 \text{ mol/m}^3$$

Equations (4.41) and (4.42) can be rewritten respectively as

$$q C_{A_1} = q (0.45) - V (2.9 \times 10^7) e^{-5300/T} C_{A_1} \tag{4.43}$$

$$q (0.099) = q C_{A_1} - V (2.9 \times 10^7) e^{-5300/T} (0.099) \tag{4.44}$$

Note that these are two equations and the number of unknowns is four (q, C_{A_1}, V, T); thus, the degrees of freedom are 2. It means that we can choose two of the four unknowns and use the two equations to get the other two unknowns. The choice of the two unknowns depends on our engineering judgment. As an example, choose $T = 300$ K and $V = 2$ m^3. Then, on solving Eqs. (4.43) and (4.44) simultaneously, we get $q = 1.093$ m^3 and $C_{A_1} = 0.2115$ mol/m^3.

Part (c) (see Fig. 4.15)

The mass balance equation for the CSTR at steady state is,

$$n_{A_1} = n_{A_f} + \sigma_A V_1 r'$$

and for the plug flow reactor (PFR) it is

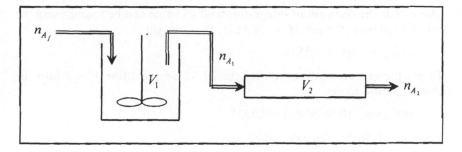

Figure 4.15   Mass flow diagram for CSTR followed by a plug flow reactor.

$$\frac{dn_A}{dV} = \sigma_A r'$$

The above two equations can be rewritten as

$$q C_{A_1} = q C_{A_f} - V k_0 e^{-E/RT} C_{A_1} \tag{4.45}$$

$$q \frac{dC_A}{dV} = -k_0 e^{-E/RT} C_A \tag{4.46}$$

with the initial condition at $V = 0$

$$C_A = C_{A_1}$$

[note that the value of C_{A_1} here is the same as C_{A_1} in part (b)].

Part (d)

C_A is the same as C_{A_1} in part (b). Now, we have to simply integrate Eq. (4.46) with the initial conditions to get the value of concentration at the exit of the PFR (volume of the PFR is equal to 2 m^3):

$$\frac{dC_A}{dV} = - \underbrace{k_0 e^{-E/RT}/q}_{\alpha = \text{constant}} C_A$$

with initial condition at $V = 0$

$$C_A|_{t=0} = 0.2115$$

It becomes

$$\frac{dC_A}{dV} = -\alpha C_A$$

On integration, we get

$$\ln C_A = -\alpha V + M$$

where M is the constant of integration and its value can be found using the initial condition. We get $M = -1.5535$; thus, we get

$$\ln C_A = -\alpha V - 1.5535$$

To get the exit concentration after the PFR, we substitute $V = 2$ into the above relation to get

$$\ln C_{A_2} = -(0.56395)(2) - 1.5535$$

where $\alpha = 0.56395$. On calculation, we get

$$C_{A_2} = 0.068 \ \text{mol/m}^3$$

Part (e)

The unsteady-state model for the arrangement is shown in Figure 4.14. For the first CSTR,

$$\underbrace{\frac{dn_{A_1}}{dt}}_{\substack{\text{Unsteady-}\\\text{state term}}} + n_{A_1} = n_{A_f} + \sigma_A V_1 r_1'$$

which can be rewritten as

$$V \frac{dC_{A_1}}{dt} + q\, C_{A_1} = q\, C_{A_f} - V k_0 e^{-E/RT} C_{A_1} \qquad (4.47)$$

Similarly for the second CSTR,

$$V \frac{dC_{A_2}}{dt} + q\, C_{A_2} = q\, C_{A_1} - V k_0 e^{-E/RT} C_{A_2} \qquad (4.48)$$

The initial conditions at $t = 0$ are

$$C_{A_1} = C_{A_1}(0) \quad \text{and} \quad C_{A_2} = C_{A_1}(0).$$

PROBLEMS

Problem 4.1

The hydrogenolysis of low-molecular-weight paraffins takes place in a tubular flow reactor. The kinetics of the propane reaction may be assumed to be first order in propane in the regime of interest. From the data given in Table P4.1, determine the reaction rate constants at the indicated temperatures and the activation energy of the reaction.

Feed ratio $H_2/C_3H_8 = 2.0$ in all cases
Reactor pressure = 7.0 MPa in all cases

Table P4.1 Experimental Data

Temperature, $T(°C)$	Space time (s)	Fraction propane converted
538	0	0
	42	0.018
	98	0.037
	171	0.110
593	40	0.260
	81	0.427
	147	0.635

For this problem, the stoichiometry of the main reaction may be considered to be of the form

$$H_2 + C_3H_8 \rightarrow CH_4 + C_2H_6$$

Problem 4.2

Consider the following homogeneous gas-phase reaction:

$$A + B \rightarrow C + D$$

The reaction is essentially "irreversible" and the rate of production of component A in a constant-volume batch reactor is given by

$$R_A = -kC_A C_B$$

At the temperature of interest, $k = 100$ m^3/mol s. Compounds A and B are available in the off-gas stream from an absorption column at concentrations of 21 mol/m^3 each. 15 m^3/s of this fluid is to be processed in a long iso-thermal tubular reactor. If the reactor is assumed to approximate a plug flow reactor, what volume of pipe is required to obtain 78% conversion of species A?

Problem 4.3

Many of the techniques used by chemical engineers are also helpful in the food processing industry. For example, consider the problem of sterilizing food after it has been placed in cylindrical cans. Normally, it is assumed that all harmful bacteria will be killed if the food temperature is raised to some value T_1. The heating process is accomplished by placing the can in a sterilization bath that is maintained at a high temperature T_0. Develop an equation for the sterilization time. Also, by selecting various values for the system parameters, see if you can determine whether or not it is necessary to consider the resistance to heat transfer in a stagnant film surrounding the can.

Problem 4.4

It would be possible to use a double-pipe heat exchanger as a reactor. For a single, irreversible, exothermic reaction, this unit would have the great advantage that the heat generated by the reaction could be used to raise the temperature of the reacting material, which would eliminate the need for any heating fluid. Derive a dynamic model for the system.

Problem 4.5

A simple first-order reaction

$$A \to B$$

takes place in a homogeneous tubular reactor. The reactor is isothermal and axial dispersion is important, with a Peclet number Pe = 20.0.

(a) Find the value of the Damkohler number (D_a) so that the exit conversion is 0.8. Note that the equation and its two-point boundary-value conditions are linear and therefore can be solved analytically (to get the characteristic equation, eigenvalues, etc.). You should be careful to note that this is a design equation, not a simulation one. Therefore, although the equations are linear, you will need some nonlinear iterations to obtain the solution at the end.

(b) Construct the numerical solution procedure using Fox's method when the reaction is second order for the same numerical value of the Damkohler number you obtained in part (a).

(c) Repeat the same you did in part (a) but for the Peclet number going once to zero and once to infinity. Compare and discuss both results.

Problem 4.6

Under appropriate conditions, compound A decomposes as follows:

$$A \to B \to C$$

The desired product is B and the rate constant for the first and second reactions are equal to the following values:

$$k_1 = 0.1 \text{ min}^{-1}$$
$$k_2 = 0.05 \text{ min}^{-1}$$

The feed flow rate is 1000 L/h, and the feed concentration of the reactant A is equal to $C_{Af} = 1.4$ mol/L, whereas $C_{Bf} = C_{Cf} = 0$.

(a) What size of PFR will maximize the yield of B, and what is the concentration of B in the effluent stream from this optimum reactor? What will be the effect of axial dispersion if the Peclet number is equal to 10.0, and what is the effect when the Peclet number is equal to 500.0?

(b) What size of CSTR will maximize the yield of B, and what is the concentration of B in the effluent stream from this optimized CSTR?

(c) Compare and comment on the results of previous two parts.

(d) What will be the effect if the reactors are nonisothermal, adiabatic, and both the reactions are exothermic.

Problem 4.7

A consecutive reaction takes place in different reactor configurations. Both reactions are first order. The rate constant for the first reaction is equal to 5 min^{-1}, and for the second reaction, it is equal to 2 min^{-1}. The feed is pure reactant. Calculate the following:

(a) For a batch reactor, the optimum operating time and maximum yield of the intermediate desired product

(b) For a CSTR, the optimum residence time and the maximum yield of the intermediate desired product

(c) For a tubular reactor, the optimum residence time and maximum yield of the intermediate desired product. What is the effect of axial dispersion when the Peclet number is equal to 0.1 and when it is equal to 1000.0?

(d) Compare the results obtained in previous parts and comment.

(e) Redo parts (a)–(d) if both the reactions follow second-order kinetics.

Problem 4.8

One approximate way to obtain the dynamics of a distributed system is to lump them into a number of perfectly mixed sections. Prove that a series of N mixed tanks is equivalent to a distributed system as N goes to infinity.

Problem 4.9

In an isothermal batch reactor, 70% of a liquid reactant is converted in 13 min. What space time and space velocity are needed to effect this conversion in a plug flow reactor?

Problem 4.10

Assuming a stoichiometry $A \rightarrow R$ for a first-order gas reaction, we calculate the size of plug flow reactor needed to achieve a required conversion of 99% of pure A feed to be 32 liters. In fact, however, the reaction stoichiometry is $A \rightarrow 3R$. With this corrected stoichiometry, what is the required reactor volume?

Problem 4.11

Consider the packed-bed tubular reactor whose schematic diagram is shown in Figure P4.11. It is a hollow cylindrical tube of uniform cross-sectional area A, packed with solid catalyst pellets, in which the exothermic reaction $A \rightarrow B$ is taking place. The packing is such that the ratio of void space to the total reactor volume—the void fraction—is known; let its value be represented by ε. The reactant flows in at one end at constant velocity v, and the reaction takes place within the reactor. Obtain a theoretical model that will represent the variation in the reactant concentration C and reactor temperature T as a function of time and spatial position z. Consider that the temperature on the surface of the catalyst pellets T_S is different from the temperature of the reacting fluid and that its variation with time and position is also to be modeled.

Use the following assumptions:

1. The reacting fluid flows through the reactor with a flat velocity profile (implying that there are no temperature, concentration, or velocity variations in the radial direction).

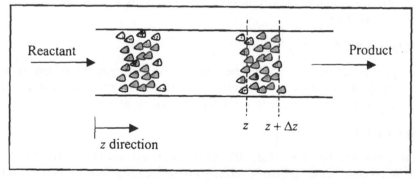

Figure P4.11   Schematic diagram of problem 4.11.

2. Transfer of material and energy in the axial (i.e., z) direction is by convective forces only; there is no diffusion in this direction.
3. The rate of reaction is first order with respect to the reactant concentration, with a rate constant k. It does not depend explicitly on position along the reactor; it only depends on reactant concentration and temperature. The heat of reaction may be taken as $-\Delta H$.
4. The heat transfer from the solid catalyst to the fluid is assumed to follow Newton's law of cooling; the area over which this heat transfer takes place is assumed to be a known constant A_S.
5. The fluid density ρ and specific heat capacity C_p are constant; the solid catalyst particles are all assumed to be identical, with identical density ρ_S and identical specific heat capacity C_{pS}.
6. The catalyst packing is assumed to be uniform, so that across any cross section of the reactor, the number and arrangement of particles are identical. (This allows the use of arbitrarily located microscopic element for developing the model).
7. There are no heat losses to the atmosphere.

If you need to make any additional assumptions, state them clearly.

Problem 4.12

A plug flow reactor is to be designed to produce the product D from A according to the following reaction:

$$A \rightarrow D, \qquad \text{rate } R_D = 60C_A \text{ mol/L s}$$

In the operating condition of this reactor, the following undesired reaction also takes place:

$$A \rightarrow U, \qquad \text{rate } R_U = \frac{0.003C_A}{1 + 10^5C_A} \text{ mol/L s}$$

The undesired product U is a pollutant and it costs \$10/mol U to dispose of it, whereas the desired product D has a value of \$35/mole D. What size reactor should be chosen in order to obtain an effluent stream at its maximum value?

Pure reactant A with volumetric flow rate of 15 L/s and molar flow rate of 0.1 mol/s enters the reactor. Value of pure A is \$5/mole A.

Problem 4.13

The vapor-phase cracking of acetone, described by the endothermic reaction

$$CH_3COCH_3 \rightarrow CH_2CO + CH_4$$

takes place in a jacketed tubular reactor. Pure acetone enters the reactor at a temperature of $T_0 = 1030$ K and a pressure of $P_0 = 160$ kPa, and the temperature of the external gas in the heat exchanger is constant at $T_e = 1200$ K. Other data are as follows:

Volumetric flow rate:	$q = 0.003 \text{ m}^3/\text{s}$
Volume of the reactor:	$V_R = 1.0 \text{ m}^3$
Overall heat transfer coefficient:	$U = 110 \text{ W/m}^2\text{K}$
Total heat transfer area:	$A = 160 \text{ m}^2/\text{m}^3\text{reactor}$
Reaction constant:	$k = 3.56 \ e^{[34200(1/1030-1/T)]} \text{ s}^{-1}$
Heat of reaction:	$\Delta H_R = 80700 + 6.7(T - 298)$
	$\quad -5.7 \times 10^{-3}(T^2 - 298^2)$
	$\quad -1.27 \times 10^{-6}(T^3 - 298^3)\text{J/mol}$
Heat capacity of acetone:	$C_{P_A} = 26.65 + 0.182T - 45.82$
	$\quad \times 10^{-6}T^2 \text{ J/mol K}$
Heat capacity of ketene:	$C_{P_K} = 20.05 + 0.095T - 31.01$
	$\quad \times 10^{-6}T^2\text{J/mol K}$
Heat capacity of methane:	$C_{P_M} = 13.59 + 0.076T - 18.82$
	$\quad \times 10^{-6}T^2\text{J/mol K}$

Determine the temperature profile of the gas along the length of the reactor. Assume constant pressure throughout the reactor.

Problem 4.14

Formulate steady-state and dynamic models for the tubular heat exchanger shown in Figure P4.14 (*Hint*: We will get a partial differential equation for the dynamic model).

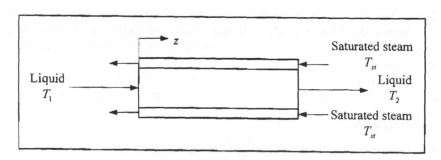

Figure P4.14   Tubular heat exchanger.

The following nomenclature is to be used:

Q = amount of heat transferred from the steam to the liquid per unit time and per unit of heat transfer area
A = cross-sectional area of the inner tube
v = average velocity of the liquid (assumed to be constant)
D = external diameter of the inner tube

List all of the assumptions made to derive the model and suggest an algorithm to solve the problem numerically.

5

Process Dynamics and Control

5.1 VARIOUS FORMS OF PROCESS DYNAMIC MODELS

The more logical manner to formulate dynamic mathematical models is to formulate it in terms of real time. However, sometimes it is convenient to express these dynamic models in a domain other than the real physical time domain. These situation arise either due to ease of solution and equation manipulation in another domain (the Laplace domain) or due to online measurement and control, as is shown in this chapter.

Therefore, the different dynamic models can be classified on a different basis than the previous classification.

A useful and interesting classification is the following:

1. State-space models

   The main characteristics of these physical/rigorous models are as follows:

   - The state variables occur explicitly along with the input and output variables.
   - The models are physically formulated from the process first principles.

2. Input–output models

   - They strictly relate only the input and output variables.

- They can occur in the Laplace or z-transform domain, or in the frequency domain as well as in the time domain.
- In the Laplace or z-transform domain, these input–output models usually occur in what is known as the "transform domain transfer function" form; in the frequency domain, they occur in the "frequency-response (or complex variable)" form; in the time domain, they occur in the "impulse-response (or convolution)" form.
- Input–output models can be obtained from the transformations of the state-space models, but they can also be obtained directly from input–output data correlations.

The above two main classifications can be further detailed into the following:

1. The state-space (differential equations or difference equations) form
2. The transform-domain (Laplace or z-transform) form
3. The frequency-response (complex variable) form
4. The impulse-response (convolution) form

The four forms are related to each other as shown in the following sections.

5.2 FORMULATION OF PROCESS DYNAMIC MODELS

5.2.1 The General Conservation Principles

In the previous chapters, we presented the general material and heat balance design equations in full detail and have shown the assumptions that reduce these equations into a simple form. Here, we will use the simplified form directly.

We can express the balance equation as follows:

"What remains accumulated within the boundaries of a system is the difference between
- What was added (input)
- What was taken out (output)
- What was generated by internal production, or disappeared by internal consumption"

In other, shorter terms, it is

Accumulation = Input − Output + Internal production (or − consumption)

We can also express this in terms of rate:

"The rate of accumulation of a conserved quantity q within the boundaries of a system is the difference between the rate at which this quantity is being added to the system and the rate at which it is being taken out (removed) plus the rate of internal production, or minus the rate of consumption."

$$\text{Rate of accumulation of } q = \begin{cases} \text{Rate of input of } q - \text{Rate of} \\ \text{output of } q + \text{Rate of} \\ \text{production of } q \end{cases}$$

5.2.2 Conservation of Mass, Momentum, and Energy

Such balances could be made over the entire system, to give "overall" or "macroscopic" balances, or they could be applied to portions of the system of differential size, giving "differential" or "microscopic" balances.

Mass Balance

We can write the conservation of mass equation as

$$\text{Rate of accumulation of mass} = \underbrace{\begin{matrix} \text{Rate input of mass} - \text{Rate output} \\ \text{of mass} + \text{Rate generation of} \\ \text{mass} - \text{Rate depletion of mass} \end{matrix}}_{\text{net rate of generation of mass}}$$

If there are n components, one can formulate a mass balance for each component (n equations) plus one overall mass balance equation, but from the $n + 1$ equations, only n equations are independent.

Note: The total mass balance equations will not have any generation/depletion terms, these will always be zero (as long as we do not have a nuclear reaction changing mass into energy). Of course, the same does not apply to the molar balance equations when there are chemical reactions associated with a change in the number of moles, as shown earlier.

Momentum Balance

Momentum balance is based on Newton's second law equation:

$$\text{Rate of accumulation or momentum} = \underbrace{\begin{matrix} \text{Rate input of momentum} - \text{Rate output} \\ \text{of momentum} + \text{Rate of forces acting on} \\ \text{the volume element} \end{matrix}}_{\text{net rate of generation of momentum}}$$

"Generation" of momentum must be due to forces acting on the volume element (over which the balance is carried out), and this is "done" at a rate equal to the total sum of these forces.

Energy Balance

Energy balance is based on the first law of thermodynamics:

$$\text{Rate of accumulation of energy} = \underbrace{\begin{array}{l}\text{Rate input of energy} - \text{Rate} \\ \text{output of energy} + \text{Rate of} \\ \text{generation of energy} - \text{Rate of} \\ \text{expenditure of energy}\end{array}}_{\text{net rate of generation of energy}}$$

5.2.3 Constitutive Equations

The next step after the conservation equations is the introduction of explicit expressions for the rates that appear in the balance equations. The following are the most widely used:

1. Equations of the properties of matter. Basic definitions of mass, momentum, and energy in terms of physical properties such as ρ, C_P, T, and so forth.

2. Transport rate equations

    *   Newton's law of viscosity (for momentum transfer)
    *   Fourier's heat conduction law (for heat transfer)
    *   Fick's law of diffusion (for mass transfer)

3. Chemical kinetic rate equations

    *   Law of mass action
    *   Arrhenius expression for temperature dependence of reaction rate constants

4. Thermodynamic relations

    *   Equations of state (e.g., ideal gas law, van der Waal's equation)
    *   Equations of chemical and phase equilibria

Example of Mathematical Model

Consider the heating (with no chemical reaction) tank shown in Figure 5.1. The unsteady mass balance equation is given by

$$n_i + \frac{dn_i^*}{dt} = n_{if}$$

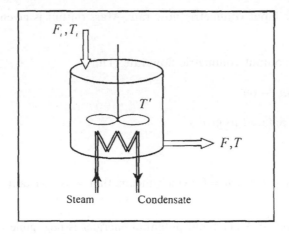

Figure 5.1 Continuous heating tank.

with no chemical reaction. The general heat balance equation is given by

$$\sum n_{if}(H_{if} - H_{ir}) + Q = \sum n_i(H_i - H_{ir}) + \frac{d(\sum H_i^*)}{dt}$$

We can introduce the following assumptions:

1. The tank is well mixed $(T' = T)$.
2. ρ (fluid density), C_p (fluid specific heat), and λ (heat of vaporization of the steam in the heating coil) are constant.
3. All of the heat of condensation of steam is given to the liquid (no heat accumulation in coils, stirrers, walls of tank, etc.).
4. There are negligible heat losses to the atmosphere.
5. There is no change of phase inside the tank.

Mass balance

Accumulation of mass is given by

$$\frac{d}{dt}(\rho V)$$

Which can be simplified as

$$\rho \frac{dV}{dt} \quad \text{(for constant density)}$$

Mass input is given by

$$F_i = \rho q_i$$

where q_i is the input volumetric flow rate. Mass output is given by

$$F = \rho q$$

where q is the output volumetric flow rate. Thus,

$$\rho \frac{dV}{dt} = \rho q_i - \rho q$$

which can be reduced to give

$$\frac{dV}{dt} = q_i - q$$

For $q_i = q$, we get $dV/dt = 0$, which means that V is constant.

Energy balance

 • Change in kinetic and potential energies is negligible.
 • T^* is the reference temperature.

The rate of accumulation of energy is given by

$$\rho C_p \frac{d}{dt}[V(T - T^*)]$$

The rate of heat input (feed) is equal to

$$\rho q_i C_p(T_i - T^*)$$

The rate of heat output (outlet stream) is given by

$$\rho q C_p(T - T^*)$$

The rate of heat input (heating coil) is equal to

$$\lambda \rho_s q_s$$

where q_s is the steam volumetric flow rate in the steam heating coil and ρ_s is the steam density and λ is the latent heat of steam. Thus, the overall energy balance will be given by

$$\rho C_p \frac{d}{dt}[V(T - T^*)] = \rho q_i C_p(T_i - T^*) + \lambda \rho_s q_s - \rho q C_p(T - T^*)$$

For $q_i = q$, and $V = $ constant, the above equation becomes

$$\rho V C_p \frac{dT}{dt} = \rho q C_p(T_i - T) + \lambda \rho_s q_s$$

This equation can be put in the following form:

$$\frac{dT}{dt} = \frac{\rho q C_p}{\rho V C_p}(T_i - T) + \frac{\lambda \rho_s}{\rho V C_p} q_s$$

Now, we define the following normalized parameters:

$$\theta = \frac{V}{q} = \text{Residence time} \quad \text{and} \quad \beta = \frac{\lambda \rho_s}{\rho V C_p}$$

Thus, the equation can be put in the following form:

$$\frac{dT}{dt} = -\frac{1}{\theta}T + \beta q_s + \frac{1}{\theta}T_i \tag{5.1}$$

The steady-state relation is given by

$$0 = -\frac{1}{\theta}T_S + \beta q_{sS} + \frac{1}{\theta}T_{iS}$$

Exercise

1. Solve the above equation for certain values of T_{iS} and q_{sS} to obtain T_S; the subscript S denotes steady state.
2. Calculate the effect of the change in β, θ, T_{iS}, and q_{sS} on the steady-state temperature T_S.

Deviation Variables

Equation (5.1) can be written in terms of the deviation variables (deviation from the steady state equations) in the following form

$$y = T - T_S$$
$$u = q_s - q_{sS}$$
$$d = T_i - T_{iS}$$

Thus,

$$\frac{dy}{dt} = -\frac{1}{\theta}y + \beta u + \frac{1}{\theta}d \tag{5.2}$$

If the process is initially at steady state, then $y(0) = 0$.

5.2.4 The Laplace Transform Domain Models

Laplace transformation is a simple mathematical technique that transforms differential equations from the time domain (t) to the Laplace domain (s). Some detail about Laplace transformation is given later in this chapter. One of the main characteristics of Laplace transformation is to transform derivatives in the time domain (t) to algebraic form in the Laplace domain [i.e., $dy(t)/dt \Rightarrow s\,y(s)$]. Using this property, we find that the Laplace transformation of Eq. (5.2) gives

$$sy(s) = -\frac{1}{\theta}y(s) + \beta u(s) + \frac{1}{\theta}d(s)$$

Rearranging to put the output $y(s)$ on one side of the equation gives

$$\left(s + \frac{1}{\theta}\right)y(s) = \beta u(s) + \frac{1}{\theta}d(s)$$

Again, on rearranging, we obtain the following input–output relation:

$$y(s) = \left(\frac{\beta\theta}{1 + \theta s}\right)u(s) + \left(\frac{1}{1 + \theta s}\right)d(s)$$

Now, we introduce the following definition of what we call "transfer functions" [$g_u(s)$ and $g_d(s)$):

$$\left(\frac{\beta\theta}{1 + \theta s}\right) = g_u(s)$$

and

$$\left(\frac{1}{1 + \theta s}\right) = g_d(s)$$

Finally, we get,

$$\boxed{y(s) = g_u(s)u(s) + g_d(s)d(s)}$$

This is called the "transfer-function model," where the output $y(s)$ is related to the inputs $u(s)$ and $d(s)$ through a relation involving the transfer functions $g_u(s)$ and $g_d(s)$. This is the basis of block diagram algebra, which will be discussed in a later section.

5.2.5   The Frequency-Response Models

This is still a transfer-function model, but in the frequency domain, where we introduce

$$s = j\omega$$

where ω is the frequency and $j = \sqrt{-1}$. If we put $j\omega$ instead of s in $g_u(s)$, the Laplace transform becomes a Fourier transform $g_u(j\omega)$ and we can convert the transform domain model to a frequency-response model as shown in Table 5.1 for both $g_u(s)$ and $g_d(s)$. Therefore, the frequency-domain model becomes

$$y(j\omega) = g_u(j\omega)u(j\omega) + g_d(j\omega)d(j\omega)$$

5.2.6   Discrete Time Models

The linear discrete single-input, single-output (SISO) processes are described by $x(k)$, $u(k)$, $d(k)$, and $y(k)$, which are the state, control, disturbance, and

Table 5.1 The Transfer Function in the Frequency Domain

The $g_u(s)$ transfer function	The $g_d(s)$ transfer function
Simply substitute s by $j\omega$ to get	Simply substitute s by $j\omega$ to get
$$g_u((\omega)) = \frac{\beta\theta}{j\theta\omega + 1}$$	$$g_d((\omega)) = \frac{1}{j\theta\omega + 1}$$
It can be arranged in the form	It can be arranged in the form
$$g_u((\omega)) = \frac{\beta\theta(1 - j\theta\omega)}{(j\theta\omega + 1)(1 - j\theta\omega)}$$	$$g_d((\omega)) = \frac{(1 - j\theta\omega)}{(j\theta\omega + 1)(1 - j\theta\omega)}$$
which gives	which can be rewritten as
$$g_u((\omega)) = \frac{\beta\theta(1 - j\theta\omega)}{1 - j\theta\omega + j\theta\omega - j^2\theta^2\omega^2}$$	$$g_d((\omega)) = \frac{1 - j\theta\omega}{1 + \theta^2\omega^2}$$
which can be rewritten as	and, finally, can be divided into the real part (Re) and the Imaginary part (Im) to give
$$g_u((\omega)) = \frac{\beta\theta - j\theta^2\omega\beta}{1 + \theta^2\omega^2}$$	$$\text{Re}[g_d((\omega))] = \frac{1}{1 + (\theta\omega)^2}$$
and, finally, can be divided into the real part (Re) and the Imaginary part (Im) to give	$$\text{Im}[g_d((\omega))] = \frac{-\omega\theta}{1 + (\theta\omega)^2}$$
$$\text{Re}[g_u(\omega)] = \frac{\beta\theta}{1 + (\theta\omega)^2}$$	
$$\text{Im}[g_u(\omega)] = \frac{-\beta\omega\theta^2}{1 + (\theta\omega)^2}$$	

output variables, respectively, at the discrete time instant $t_k = k\Delta t$, where Δt is the sampling interval. Thus, the model is given by

$$x(k + 1) = ax(k) + bu(k) + \gamma d(k)$$

and

$$y(k) = cx(k)$$

The multiple-input, multiple-output (MIMO) version is

$$\underline{X}(k + 1) = \underline{A}\ \underline{X}(k) + \underline{B}\ \underline{u}(k) + \underline{\Gamma}\ \underline{d}(k)$$

and

$$\underline{Y}(k) = \underline{C}\ \underline{X}(k)$$

The nonlinear discrete model is

$$\underline{X}(k) = \underline{f}(\underline{X}(k), \underline{u}(k), \underline{d}(k))$$

and

$$\underline{Y}(k) = \underline{h}(\underline{X}(k))$$

The state-space models are used most frequently in dynamic analysis because of the following:

- They give real-time behavior, suitable for computer simulation of process behavior.
- They are used almost exclusively for the analysis of nonlinear systems behavior.
- Another important use of state-space models is in representing processes in which some states are not measured.

All the other model forms are based on input/output relationships so that any aspect of the process not manifested as either input or measured output is not likely to be represented.

5.2.7 SISO and MIMO State-Space Models

The state-space models (in terms of real time) can be classified as follows:

Linear Lumped SISO Processes

One input variable $u(t)$, one disturbance $d(t)$, and one state variable $x(t)$ are modeled by

$$\frac{dx(t)}{dt} = ax(t) + bu(t) + \gamma d(t)$$

Note: The output is not necessarily $x(t)$ [it can be $y(t)$], but is related to $x(t)$ through the linear relation

$$y(t) = cx(t)$$

Linear Lumped MIMO Processes

The MIMO processes can be two dimensional (the simplest MIMO case), as in the following simple example:

$$\frac{dx_1(t)}{dt} = a_{11}x_1(t) + a_{12}x_2(t) + b_{11}u_1(t) + b_{12}u_2(t) + \gamma_{11}d_1(t) + \gamma_{12}d_2(t)$$

$$\frac{dx_2(t)}{dt} = a_{21}x_1(t) + a_{22}x_2(t) + b_{21}u_1(t) + b_{22}u_2(t) + \gamma_{21}d_1(t) + \gamma_{22}d_2(t)$$

and the single output is,

$$y(t) = c_1 x_1(t) + c_2 x_2(t)$$

In the matrix form, this can be rewritten as

$$\frac{d}{dt}\begin{pmatrix} x_1 \\ x_2 \end{pmatrix} = \begin{pmatrix} a_{11} & a_{12} \\ a_{21} & a_{22} \end{pmatrix}\begin{pmatrix} x_1 \\ x_2 \end{pmatrix} + \begin{pmatrix} b_{11} & b_{12} \\ b_{21} & b_{22} \end{pmatrix}\begin{pmatrix} u_1 \\ u_2 \end{pmatrix} + \begin{pmatrix} \gamma_{11} & \gamma_{12} \\ \gamma_{21} & \gamma_{22} \end{pmatrix}\begin{pmatrix} d_1 \\ d_2 \end{pmatrix}$$

and

$$y(t) = \begin{pmatrix} c_1 & c_2 \end{pmatrix}\begin{pmatrix} x_1 \\ x_2 \end{pmatrix}$$

or in more general form for any n-dimensional system, we can write

$$\frac{d\underline{X}}{dt} = \underline{A}\,\underline{X} + \underline{B}\,\underline{u} + \underline{\Gamma}\,\underline{d}$$

and

$$\underline{Y}(t) = \underline{C}\,\underline{X}$$

where

$$\underline{X}_{(n\times 1)} = \begin{pmatrix} x_1 \\ x_2 \\ \vdots \\ x_n \end{pmatrix}, \qquad \underline{A}_{(n\times n)} = \begin{pmatrix} a_{11} & a_{12} & \cdots & a_{1n} \\ a_{21} & a_{22} & \cdots & a_{2n} \\ \vdots & \vdots & \ddots & \vdots \\ a_{n1} & a_{n2} & \cdots & a_{nn} \end{pmatrix}$$

$$\underline{B}_{(n\times n)} = \begin{pmatrix} b_{11} & b_{12} & \cdots & b_{1n} \\ b_{21} & b_{22} & \cdots & b_{2n} \\ \vdots & \vdots & \ddots & \vdots \\ b_{n1} & b_{n2} & \cdots & b_{nn} \end{pmatrix}, \qquad \underline{\Gamma}_{(n\times n)} = \begin{pmatrix} \gamma_{11} & \gamma_{12} & \cdots & \gamma_{1n} \\ \gamma_{21} & \gamma_{22} & \cdots & \gamma_{2n} \\ \vdots & \vdots & \ddots & \vdots \\ \gamma_{n1} & \gamma_{n2} & \cdots & \gamma_{nn} \end{pmatrix}$$

$$\underline{C}_{(m\times n)} = \begin{pmatrix} c_{11} & c_{12} & \cdots & c_{1n} \\ c_{21} & c_{22} & \cdots & c_{2n} \\ \vdots & \vdots & \ddots & \vdots \\ c_{m1} & c_{m2} & \cdots & c_{mn} \end{pmatrix}, \qquad \underline{u}_{(n\times 1)} = \begin{pmatrix} u_1 \\ u_2 \\ \vdots \\ u_n \end{pmatrix}$$

$$d_{(n\times 1)} = \begin{pmatrix} d_1 \\ d_2 \\ \vdots \\ d_n \end{pmatrix}, \qquad Y_{(m\times 1)} = \begin{pmatrix} y_1 \\ y_2 \\ \vdots \\ y_m \end{pmatrix}$$

Nonlinear Lumped Systems

$$\frac{dX}{dt} = f(X, u, d)$$

where $f(X, u, d)$ are nonlinear functions, and the output vector is given by

$$Y(t) = h(X(t))$$

Distributed Systems

These are systems where the state variables vary with one or more of the state coordinates.

One-dimensional model

$$\frac{\partial T}{\partial t} + v\frac{\partial T}{\partial z} = \frac{UA_S}{\rho A C_P}(T_S - T)$$

This partial differential equation describes one side of the steam-heated shell and tube heat exchanger.

High-dimensional models

$$\frac{\partial X}{\partial t} = f(X, u, \nabla X, \nabla^2 X, \ldots)$$

∇ is the Laplacian operator representing the vector of partial derivative operators in the spatial directions.

5.2.8   SISO and MIMO Transform Domain Models

SISO

The SISO linear model will be represented after applying Laplace transformation as

$$y(s) = g_u(s)u(s) + g_d(s)d(s)$$

where $y(s)$ is the process output, $g_u(s)$ is the process transfer function (in the transform domain), $u(s)$ is the process input, $g_d(s)$ is the disturbance transfer function (in the transfer domain), and $d(s)$ is the process disturbance.

MIMO

The MIMO linear model will be represented after applying the Laplace transformation as

$$\underline{Y}(s) = \underline{G}_u(s)\underline{u}(s) + \underline{G}_d(s)\underline{d}(s)$$

where

$$\underline{Y}(s)_{(n\times 1)} = \begin{pmatrix} y_1 \\ y_2 \\ \vdots \\ y_n \end{pmatrix} \qquad \underline{G}_u(s)_{(n\times m)} = \begin{pmatrix} g_{u_{11}} & g_{u_{12}} & \cdots & g_{u_{1m}} \\ g_{u_{21}} & g_{u_{22}} & \cdots & g_{u_{2n}} \\ \vdots & \vdots & \vdots & \vdots \\ g_{u_{n1}} & g_{u_{n2}} & \cdots & g_{u_{nm}} \end{pmatrix}$$

$$\underline{G}_d(s)_{(n\times l)} = \begin{pmatrix} g_{d_{11}} & g_{d_{12}} & \cdots & g_{d_{1l}} \\ g_{d_{21}} & g_{d_{22}} & \cdots & g_{d_{2l}} \\ \vdots & \vdots & \vdots & \vdots \\ g_{d_{n1}} & g_{d_{n2}} & \cdots & g_{d_{nl}} \end{pmatrix}$$

$$\underline{u}(s)_{(m\times 1)} = \begin{pmatrix} u_1 \\ u_2 \\ \vdots \\ u_m \end{pmatrix}, \qquad \underline{d}(s)_{(l\times 1)} = \begin{pmatrix} d_1 \\ d_2 \\ \vdots \\ d_l \end{pmatrix}$$

Here, $\underline{G}_u(s)$ and $\underline{G}_d(s)$ are the matrix transfer functions.

5.2.9   SISO and MIMO Frequency-Response Models

For continuous time systems, we have the following.

SISO

$$y(j\omega) = g_u(j\omega)u(j\omega) + g_d(j\omega)d(j\omega)$$

where $g_u(j\omega)$ and $g_d(j\omega)$ are frequency-response transfer functions. Also

$$g(j\omega) = \text{Re}[g(\omega)] + j\text{Im}[g(\omega)]$$

MIMO

$$\underline{Y}(j\omega) = \underline{G}_u(j\omega)\underline{u}(j\omega) + \underline{G}_d(j\omega)\underline{d}(j\omega)$$

5.2.10 SISO and MIMO Discrete Time Models

SISO

It relates the z-transform of the sampled point signal to that of the output signal by the following relation:

$$\hat{y}(z) = \hat{g}_u(z)\hat{u}(z) + \hat{g}_d(z)\hat{d}(z)$$

where $\hat{g}_u(z)$ and $\hat{g}_d(z)$ are the z-transform functions of the discrete time system.

MIMO

$$\underline{\hat{Y}}(z) = \underline{\hat{G}}_u(z)\underline{\hat{u}}(z) + \underline{\hat{G}}_d(z)\underline{\hat{d}}(z)$$

where $\underline{\hat{G}}_u(z)$ and $\underline{\hat{G}}_d(z)$ are the transfer-function matrices. Transform-domain transfer functions are used extensively in process dynamics and control.

Figure 5.2 shows the interrelationship between the discussed process model forms in a nice, concise manner. Uses of the three different types of models are as follows

1. State space: Real-time simulation of process behavior, nonlinear dynamic analysis.
2. Transform domain: Linear dynamic analysis involving well-characterized input functions; control system design.
3. Frequency response: Linear nonparametric models for processes of arbitrary mathematical structure; control system design.

Continuous and Discrete Time Models

The conversion of continuous-time models to discrete-time form is achieved by discretizing the continuous process model. This will be discussed in some detail later.

5.3 STATE-SPACE AND TRANSFER DOMAIN MODELS

These are the most important types of model. Most process models formulated from first principles occur in the state-space form, but the preferred form for dynamic analysis and controller design is the transfer domain form. However, after dynamic analysis and controller design, the task of control system simulation and real-time controller implementation are carried out in the time domain, requiring the state-space form again. The interrelation

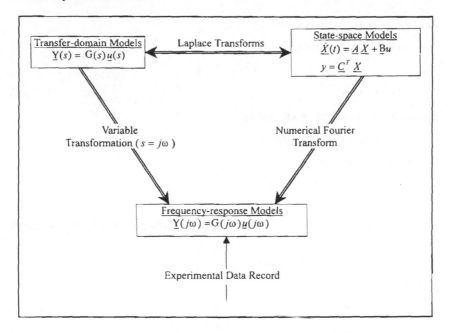

Figure 5.2 Interrelationship between process model forms.

between state-space and transfer domain models is shown in Figure 5.3. The SISO process is given by

$$\frac{dx(t)}{dt} = ax(t) + bu(t) + \gamma d(t)$$

and

$$y(t) = cx(t)$$

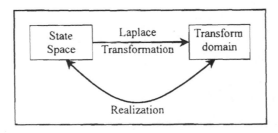

Figure 5.3 Interrelation between state-space and transform domain models.

Laplace transformation gives

$$sx(s) = ax(s) + bu(s) + \gamma d(s) \tag{5.3}$$

and

$$y(s) = cx(s) \tag{5.4}$$

Equation (5.3) can be rewritten as

$$(s - a)x(s) = bu(s) + \gamma d(s)$$

On rearrangement, it gives

$$x(s) = \frac{b}{s - a}u(s) + \gamma\frac{d(s)}{s - a}$$

Consider the following more general input–output Laplace transform model:

$$y(s) = g_u(s)u(s) + g_d(s)d(s)$$

where

$$g_u(s) = \frac{K_u}{\tau_u s + 1} \quad \text{and} \quad g_d(s) = \frac{K_d}{\tau_d s + 1}$$

The solution (realization) is given by

$$y(s) = \left(\frac{K_u}{\tau_u s + 1}\right)u(s) + \left(\frac{K_d}{\tau_d s + 1}\right)d(s) \tag{5.5}$$

We can write $y(s)$ in terms of $x_1(s)$ and $x_2(s)$ in a simple form as follows:

$$y(s) = x_1(s) + x_2(s) \tag{5.6}$$

Note:

$$y(s) = \underline{C}^T x(s)$$

where

$$\underline{C}^T = \begin{pmatrix} 1 \\ 1 \end{pmatrix}$$

Now, $x_1(s)$ is given by

$$x_1(s) = \left(\frac{K_u}{\tau_u s + 1}\right)u(s)$$

The above relation, on rearrangement, gives

$$\tau_u s x_1(s) + x_1(s) = K_u u(s)$$

$x_2(s)$ is given by

$$x_2(s) = \left(\frac{K_d}{\tau_d s + 1}\right) d(s)$$

which, on rearrangement, gives

$$\tau_d s x_2(s) + x_2(s) = K_d\, d(s)$$

Thus, realization in the time domain gives

$$\tau_u \frac{dx_1}{dt} + x_1 = K_u u(t)$$

and

$$\tau_d \frac{dx_2}{dt} + x_2 = K_d\, d(t)$$

together with

$$y(t) = x_1(t) + x_2(t)$$

at $t = 0$, $x_1(0) = 0$, and $x_2(0) = 0$. This is one possible realization, and it is not unique.

Realization is one of the more popular names for this process of transferring from Laplace domain to the time domain. It actually uses inverse transformation from the Laplace domain to the time domain. An even simpler example will make this process very clear. Consider the following very simple input–output relation in the Laplace domain:

$$y(s) = g(s)u(s)$$

where

$$g(s) = \frac{K}{\tau s + 1}$$

Thus, we can write

$$y(s) = \left(\frac{K}{\tau s + 1}\right) u(s)$$

which can be written as

$$(\tau s + 1)y(s) = K u(s)$$

On rearrangement, we get

$$(\tau s)y(s) + y(s) = K u(s)$$

Inverse Laplace transformation of the above relation gives

$$\tau \frac{dy}{dt} + y(t) = Ku(t)$$

The **nonuniqueness** (it is a linear nonuniqueness in contrast to nonlinear nonuniqueness discussed in Chapter 7) can be shown in a more general form:

$$y(s) = g_u(s)u(s) + g_d(s)d(s)$$

where

$$g_u(s) = \frac{K_u}{\tau_u s + 1} \quad \text{and} \quad g_d(s) = \frac{K_d}{\tau_d s + 1}$$

Now, we can write

$$y(s) = c_1 x_1(s) + c_2 x_2(s) \tag{5.7}$$

and

$$x_1(s) = \frac{a_1 u(s)}{\tau_u s + 1} \quad \text{and} \quad x_2(s) = \frac{a_2 d(s)}{\tau_d s + 1}$$

On substituting the values of $x_1(s)$ and $x_2(s)$ in Eq. (5.7), we get

$$y(s) = \left(\frac{a_1 c_1}{\tau_u s + 1} \right) u(s) + \left(\frac{a_2 c_2}{\tau_d s + 1} \right) d(s)$$

Now,

$$\frac{a_1 c_1}{\tau_u s + 1} = \frac{K_u}{\tau_u s + 1} \quad \text{and} \quad \frac{a_2 c_2}{\tau_d s + 1} = \frac{K_d}{\tau_d s + 1}$$

which gives

$$K_u = a_1 c_1 \quad \text{and} \quad K_d = a_2 c_2$$

This shows that the degree of freedom is equal to 2.

Thus, there is an infinite combination of $a_1 c_1$ and $a_2 c_2$ that satisfy the above equations. This is the linear nonuniqueness.

A special case: For the input–output Laplace transform model,

$$y(s) = \left(\frac{K_u}{\tau_u s + 1} \right) u(s) + \left(\frac{K_d}{\tau_d s + 1} \right) d(s)$$

If we consider the special case of $\tau_u = \tau_d = \tau$, we get

$$y(s) = \frac{1}{\tau s + 1} [K_u u(s) + K_d d(s)]$$

On rearrangement, we get

$$(\tau s + 1)y(s) = K_u u(s) + K_d d(s)$$

and inverse transformation to the time domain gives

$$\tau \frac{dy(t)}{dt} + y(t) = K_u u(t) + K_d d(t)$$

We can set $y(t) = x(t)$ and write

$$\tau \frac{dx(t)}{dt} + x(t) = K_u u(t) + K_d d(t)$$

and

$$y(t) = x(t)$$

5.4 INTRODUCTORY PROCESS CONTROL CONCEPTS

In this part of the chapter, we present in a very condensed/simple manner the basic process control concepts.

5.4.1 Definitions

The process control system is the entity that is charged with the responsibility for the following important tasks: monitoring outputs, making decisions about how best to manipulate inputs so as to obtain desired output behavior, and effectively implement such decisions on the process. Thus, it demands that three different tasks be done:

1. Monitoring process output variables by measurements
2. Making rational decisions regarding what corrective action is needed on the basis of the information about the current (output variables) and the desired (set point) state of the process
3. Effectively implementing these decisions on the process

It can have different levels of sophistication as follows:

- The simplest are, of course, the manual control systems.
- More advanced are the automatic control systems.
- The most advanced are the digital control systems.

The control hardware elements are as follows:

1. Sensors. Sensors are also called measuring devices or primary elements. Examples are the following:

    - Thermocouples
    - Differential pressure cells (for liquid level measurements)
    - Gas/liquid chromatographs (for analysis of gas samples)

2. Controllers. The controller is the decision-maker or the heart of the control system. It requires some form of intelligence. Controllers can be of the following types:

- Pneumatic (now almost obsolete)
- Electronic (more modern)
- Digital (for more complex operations, they are some kind of a special-purpose small digital computer)

Currently, the first two types of controller (pneumatic and electronic) are limited to simple operations.

3. Transmitters. Examples are pneumatic (air pressure), electrical signals, and digital signals.

4. Final control elements

- They implement the control command issued by the controller.
- They are mostly "control valves" [usually pneumatic (i.e., air driven)]
- Other types include variable-speed fans, pumps, compressors, conveyors, relay switches, and so forth.

5. Other hardware elements

- Transducers: to convert an electric signal from an electronic controller to a pneumatic signal needed for a control valve.
- A/D and D/A converters (for computer devices): A/D is analog to digital and D/A is for digital to analog. They are needed simply because the control system operates on analog signals (electric voltage or pneumatic pressures), whereas the computer operates digitally (giving out and receiving only binary numbers).

Figure 5.4 shows a typical sequential setup for the control system components.

Some additional process control terminologies are as follows:

Set point: The value at which the output is to be maintained, fed to the controller to compare with the measurement, and make decisions.

Regulatory control: This is used when it is required to keep the process at a specific fixed set point.

Servo control: This is used when it is required to make the output track a changing set point.

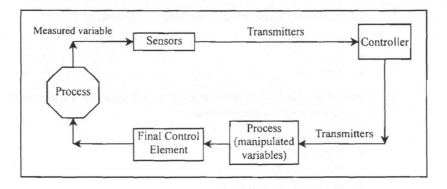

Figure 5.4 Logical representation of the control system components.

5.4.2 Introductory Concepts of Process Control

We will illustrate the basic concepts of process control using a typical generic chemical process as shown in Figure 5.5. A typical chemical process can be a petrochemical plant, a refinery (or a section of it), a biochemical plant, and so forth. Let us consider it to be a part of a refinery (the fractionating part). It will typically consist of the following:

1. Processing units. Typically these are the following:

   - The storage tanks
   - The furnaces
   - The fractionation towers

   and their auxiliary equipments.

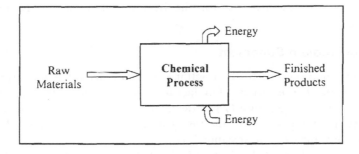

Figure 5.5 A typical chemical process.

2. Raw materials. These usually include the following:
   - Crude oil
   - Air
   - Fuel gas

3. Finished products. A fractionating unit will produce a wide range of products such as the following:
   - Naphtha
   - Light gas oil
   - Heavy gas oil
   - High-boiling residues

4. Heat input. It is mainly the furnace generated heat.

5. Heat output. It is usually the heat removed from the condenser.

Basic Principles Guiding the Operation of Processing Units

1. It is desirable to operate the processing unit safely (including environmental safety).
2. Specific production rates must be maintained.
3. Product quality specifications must be maintained.

Process dynamics and control is that aspect of chemical engineering which is concerned with the analysis, design, and implementation of control systems that facilitate the achievement of specified objectives of process safety, production rate, and product quality.

Here, we examine a typical industrial process control problem. Consider the furnace shown in Figure 5.6, where crude oil is preheated to be fed to the fractionating column. The crude oil feed flow rate fluctuates considerably, however we need to supply it to the fractionator at a constant temperature T^*. Consider T_m to be the highest safe temperature for the type of metals used for the heating tubes and T_t is the tube temperature.

Control Problem Statement

"Deliver crude oil feed to the fractionator at a constant temperature (T^*) and flow rate (F_0) regardless of all factors potentially capable of causing the furnace outlet temperature (T) to deviate from this desired value, making sure that the temperature of the tube surfaces within the furnace does not exceed the value (T_m) at any time."

In short, keep $T = T^*$, $F = F_0$, and $T_t < T_m$ (where T_t is the tube temperature).

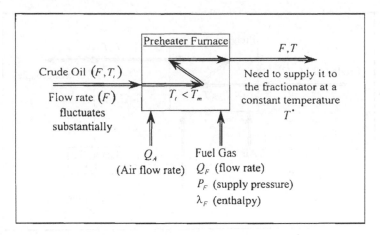

Figure 5.6 Preheater furnace for crude oil.

A dialogue between the plant engineer (PE) and the control engineer (CE) regarding this problem involves the following phases:

Phase 1. Keep the outlet temperature T at the desired value T^*. This is to be achieved through a controller (e.g., PID = proportional integral derivative, as discussed in the next part of this chapter) which measures the output temperature using a thermocouple and compares it with the desired temperature (set point, T^*) and then the deviation is used to control the furnace heating fuel gas. If $T > T^*$, then we decrease the fuel gas flow rate, whereas if $T < T^*$, then we increase the fuel gas flow rate, and, finally, if $T = T^*$ then we keep the fuel gas flow rate constant. This is what we call a feedback control system. Its main advantage is that it can compensate for the deviation of T from T^* regardless of the cause of this deviation. Its disadvantage is that it takes action only after the disturbance(s) has (have) caused measurable deviation of T from T^*. This feedback control system is shown in Figure 5.7.

Phase 2. Suppose that we cannot tolerate the time lag associated with the effect of the input disturbance(s) appearing as deviation in T. In this case, we can use what we call the feed-forward control system. This is based on measuring the input flow rate, and as soon as this flow rate deviates from the design value F_0, we take control action to manipulate the fuel gas flow rate before the effect of disturbance in F appears as a deviation in T. The advantage of feed-forward control is that the control ac-

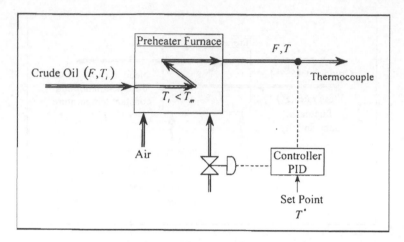

Figure 5.7   Feedback control system.

tion is performed before the disturbance starts to cause the deviation in the output variables. However, the main disadvantages are as follows:

1. There has to be some relation to relate feed flow rate to fuel gas feed rate.
2. The control system will not respond to input disturbances other than F.

The feed-forward control arrangement for this process is shown in Figure 5.8.

Phase 3. The two types of feedback and feed-forward can be combined as shown in Figure 5.9 (the feed-forward controller sets the setpoint for the feedback controller) or the cascade system as shown in Figure 5.10.

5.4.3   Variables of a Process

It is important to distinguish the different types of variables of a process. There are many classifications for these variables.

Process variables. Process variables include variables such as temperature, flow rate in/out, pressure, concentration, and so forth. These can be divided into two types:

1. Input variables: These are capable of influencing the process conditions.

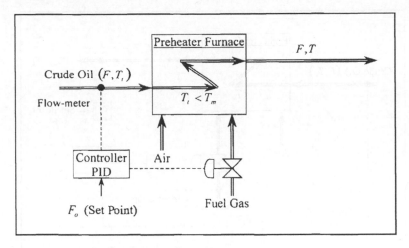

Figure 5.8 Feed-forward control system.

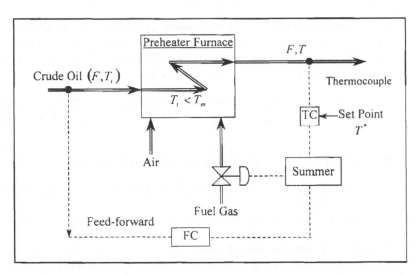

Figure 5.9 The feed-forward controller sets the set point for the feedback controller.

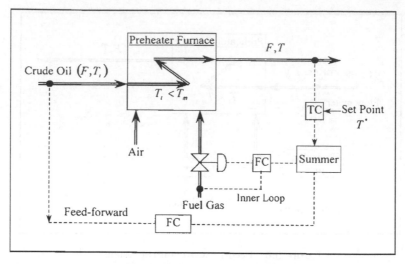

Figure 5.10 Cascade control.

2. Output variables: These provide information about the process conditions.

State variables versus output variables

State variables. They are the minimum set of variables essential for completely describing the internal state (or condition) of a process. In other words, they are the variables which describe the state of the system (e.g., temperature, concentration, pH, etc.)

Output variables. They are some measurements, either of a single state variable or a combination of state variables at the outlet of the process.

Input variables. They can be divided into two main types:

- Manipulated or control variables (we can have control over them)
- Disturbance variables (we do not have control over them)

Measured and unmeasured variables. Output variables which are characterized by the fact that we need to take samples for analysis are not called measured variables because they are not measured on-line.

Figure 5.11 shows a typical process, where the input is divided into two types: manipulated variables and input disturbances. Input disturbances can be divided further, into measured and unmeasured disturbances. The output variables to be controlled are also divided into measured and unmeasured variables.

Examples

Stirred heating tank process. The input variables are as shown in Figure 5.12 include the feed flow rate F_i and feed temperature T_i, whereas the output variables are F and T. The steam flow rate Q is the input manipulated variable.

A furnace. The input disturbances for this furnace are the input flow rate F_i, the input temperature T_i, the fuel supply pressure P_F, and the feed heat content λ_F. The manipulated variables are airflow rate Q_A and fuel flow rate Q_F, whereas F and T are the output variables. (see Fig. 5.13).

5.4.4 Control Systems and Their Possible Configurations

In this subsection, the reader is introduced to a number of the more typical control system configurations (briefly discussed earlier for a specific example).

1.   Feedback control. For this configuration, the control action is taken after the effect; that is, for any disturbance entering the process, no control action is taken except after the effect of the disturbance appears in the output (see Fig. 5.14).

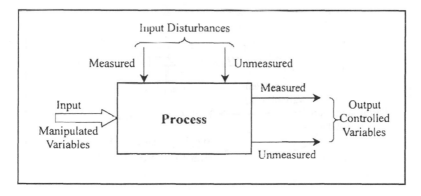

Figure 5.11 Types of input–output variables.

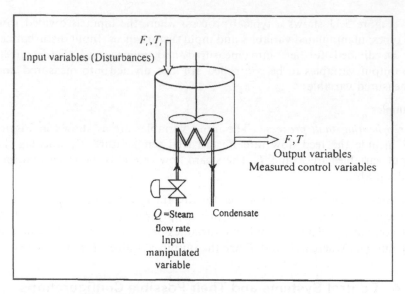

Figure 5.12 Variables of a stirred heating tank.

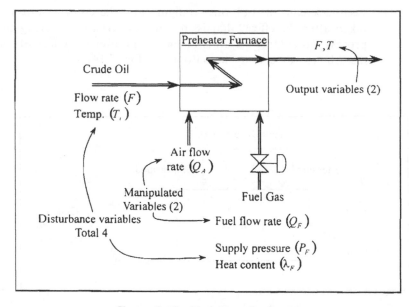

Figure 5.13 Variables of a furnace.

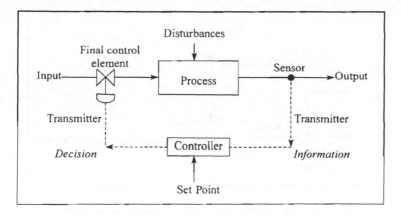

Figure 5.14 Typical feedback control system.

2. Feed-forward control. For this configuration action is taken before the effect (i.e., before the process is affected). The main advantage is that the decision of the controller is taken using a model for the process. The model relates the input disturbance and output variables in order to calculate the input changes that compensates for the disturbances to keep the output at its desired value.

The disadvantages are as follows:

- The negative effects associated with inaccuracy of models used.
- The choice of the disturbances to be measured and the possible effect of the unmeasured disturbances. It does not act to compensate for any unmeasured disturbances.
- The controller has no information about the conditions existing at the process output.

A typical feed-forward control configuration is shown in Figure 5.15.

3. Open-loop control. Open-loop control is where the set point to the controller is pre-programmed to follow a certain path (servo control) as shown in Figure 5.16.

Other control configurations, such as cascade control have been shown earlier in the book.

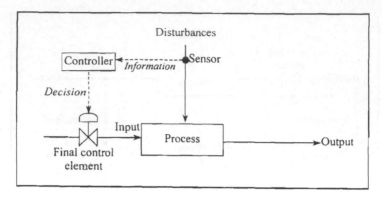

Figure 5.15 Typical feed-forward control arrangement.

5.4.5 Overview of Control Systems Design

Here, we will present a preliminary introduction to the design of control systems using a very simple example.

General Principles Involved in Designing a Control System

Step 1. Assess the process and define the control objectives to answer the following questions:

1. Why is there a need for control?

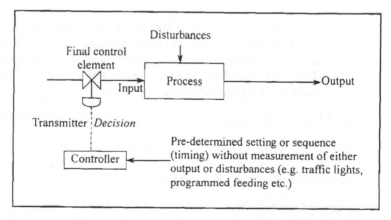

Figure 5.16 Open-loop control arrangement.

2. Can the problem be solved only by control, or there is another alternative (such as redesigning part of the process)?
3. What is the control of the process expected to achieve?

Step 2. Selection of the process variables (output and manipulated variables as well as disturbances in the case of feed-forward control) in order to answer the following questions:

1. What is the crucial output variables to be measured (for feedback control)?
2. What are the most serious disturbances, and which ones are measurable (for the feed-forward control)?
3. Which inputs to be chosen as the input manipulated variables (for both feed-forward and feed-backward controls).

Step 3. Selection of the control structure, whether open-loop, feed-back, feed-forward, or a combination of feed-backward and feed-forward control is required.

Step 4. Design of the controller(s) may have different degrees of complexity, but it basically involves the formulation of a control law which utilizes as much information as possible from the process. This control law is used to produce a control decision to be used to adjust the manipulated variables.

Process control engineers must have a good understanding of the process and its dynamics. They must also understand the steady-state characteristics of the process.

Example

We will follow the above four steps and apply them to a very simple example. This example is the filling/emptying of a liquid tank with valves at both the inlet and outlet, as shown in Figure 5.17.

For the process model without control, it is very easy to write the differential equations (the mathematical model) describing the dynamic behavior of the tank. It will be

$$A_C \frac{dh}{dt} = F_i - F \tag{5.8}$$

with the initial conditions at $t = 0$, $h = h_0$. F and F_i should be certain constant values or functions of h. For example, $F = C\sqrt{h}$, where C is called the valve coefficient, and $F_i = B$, where B is some constant value. Thus, Eq. (5.8) can be rewritten

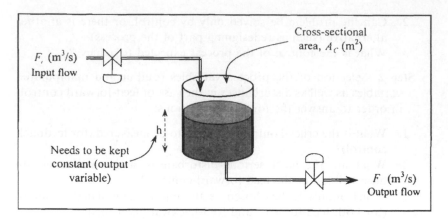

Figure 5.17 Continuously filling and emptying tank.

$$A_C \frac{dh}{dt} = B - C\sqrt{h} \qquad (5.9)$$

Let us consider the following steady-state situation: At steady state, $dh/dt = 0$ (there is no height change with time). So,

$$B = C\sqrt{h_{SS}}$$

which gives

$$\sqrt{h_{SS}} = \frac{B}{C}$$

Therefore, the steady-state height h_{SS} is given by

$$h_{SS} = \left(\frac{B}{C}\right)^2$$

The dynamic behavior is described by Eq. (5.9).

Equation (5.9) is a nonlinear differential equation. It can be solved numerically by using any of the well-known subroutine packages such as Polymath, MathCAD, IMSL libraries, and so forth. However, can we use some transformations to solve this equation analytically? Yes, it can be done as follows. Let

$$B - C\sqrt{h} = y$$

Thus, we can write

$$\sqrt{h} = \frac{B - y}{C}$$

On differentiation of y with respect to h, we get

$$\frac{dy}{dh} = -c'\left(\frac{1}{2}h^{-1/2}\right) \tag{5.10}$$

We can also write

$$\frac{dy}{dt} = \frac{dy}{dh}\frac{dh}{dt}$$

Therefore,

$$\frac{dy}{dt} = \left(-\frac{C}{2}h^{-1/2}\right)\frac{dh}{dt} \tag{5.11}$$

Rearranging gives,

$$\frac{dh}{dt} = \left(-\frac{2\sqrt{h}}{C}\right)\frac{dy}{dt}$$

Substituting the value of $\sqrt{h}$ gives,

$$\frac{dh}{dt} = -\frac{2}{C^2}(B-y)\frac{dy}{dt} \tag{5.12}$$

Substitution from Eq. (5.12) into Eq. (5.9), gives

$$A_C\left(-\frac{2}{C^2}(B-y)\right)\frac{dy}{dt} = B - C\left(\frac{(B-y)}{C}\right)$$

Rearranging gives

$$\underbrace{\frac{2A_C}{C^2}}_{\alpha}(y-B)\frac{dy}{dt} = y$$

which can be rewritten as

$$\frac{y-B}{y}dy = \frac{1}{\alpha}dt \tag{5.13}$$

where

$$\alpha = \frac{2A_C}{C^2} \quad \text{(a constant value)}$$

Rearrangement of Eq. (5.13) gives

$$\left(1 - \frac{B}{y}\right)dy = \frac{1}{\alpha}dt$$

On integration, we get

$$y - B\ln y = \frac{1}{\alpha}t + C_1$$

Boundary condition is at $t = 0$, $y = y_0$, where $y_0 = B - C\sqrt{h_0}$. So we get the value of constant of integration as

$$C_1 = y_0 - B\ln y_0$$

Using this value of the constant of integration, we obtain the solution

$$y - B\ln y = \frac{1}{\alpha}t + y_0 - B\ln y_0$$

which can be rearranged in the following form:

$$(y - y_0) - B\ln\left(\frac{y}{y_0}\right) = \frac{1}{\alpha}t \tag{5.14}$$

We can easily find the change of y with t using this simple equation. Obviously, we can transfer y back to h very easily. This completes the example.

Let us now look at the effect of different control-loop configurations on the model equations and the behavior of the system.

First configuration

The *first configuration* we look at is the simple feedback-level control using the tank level h as the measured variable and the output flow rate F as the control (or manipulated) variable as shown in Figure 5.18, where F_i is the input variable (disturbance), F is the output variable (control or manipulated variable), and h is the output variable (liquid-level measurement). The model equation is given by

$$A_C \frac{dh}{dt} = F_i - F \tag{5.15}$$

At steady state, we can write

$$F_{iSS} - F_{SS} = 0 \tag{5.16}$$

We can write Eq. (5.15) at steady state as

$$A_C \frac{dh_{SS}}{dt} = F_{iSS} - F_{SS} = 0 \tag{5.17}$$

To convert the variables into deviation variables, subtract Eq. (5.17) from Eq. (5.15) to get

$$A_C \frac{d(h - h_{SS})}{dt} = (F_i - F_{iSS}) - (F - F_{SS}) \tag{5.18}$$

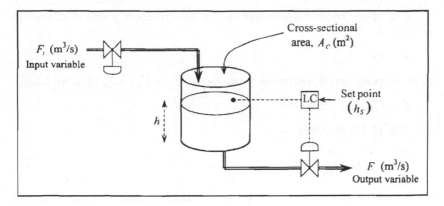

Figure 5.18   Feedback-level control.

Define the deviation variables as

$$y = (h - h_{SS})$$

$$d = (F_i - F_{iSS})$$

$$u = (F - F_{SS})$$

Thus Eq. (5.18) can be rewritten in terms of deviation variables as

$$A_C \frac{dy}{dt} = d - u$$

The above equation can be rearranged to give

$$\frac{dy}{dt} = \frac{1}{A_C} d - \frac{1}{A_C} u \qquad\qquad (5.19)$$

The simple control law is a proportional control on F; that is

$$F = F_{SS} + K(h - h_{SS}) \qquad\qquad (5.20)$$

where K is the proportional controller gain. Equation (5.20) can be re-arranged as

$$F - F_{SS} = K(h - h_{SS})$$

Note: Notice that this linear formulation is restricted by the fact that when $K = 0$, then $F = F_{SS}$, regardless of the height inside the tank, meaning that h increases to infinity for any positive disturbance.

The above control law equation can be written in terms of deviation variables as

$$u = Ky$$

Let us suppose that F_i increases from F_{iss} to $F_i = F_{iss} + \xi$, then we have

$$d = F_{iss} + \xi - F_{iss} = \xi$$

Thus, Eq. (5.19) becomes

$$\frac{dy}{dt} = \frac{1}{A_C}\xi - \frac{1}{A_C}(Ky) \tag{5.21}$$

with the initial condition at $t = 0$, $y = 0$ (which means that at $t = 0$, the system is at its steady state). If ξ is a constant, then we can write

$$\frac{\xi}{A_C} = a_1 \quad \text{and} \quad \frac{K}{A_C} = a_2$$

Thus, Eq. (5.21) becomes

$$\frac{dy}{dt} = a_1 - a_2 y \tag{5.22}$$

which can be rewritten as

$$\frac{dy}{a_1 - a_2 y} = dt$$

On integrating the above equation, we get

$$-\frac{1}{a_2}\ln(a_1 - a_2 y) = t + C_1$$

We obtain the value of constant C_1 from the initial condition at $t = 0$, $y = 0$, Therefore the value of integration constant is equal to

$$C_1 = -\frac{1}{a_2}\ln(a_1)$$

and we get

$$-\frac{1}{a_2}\ln(a_1 - a_2 y) = t - \frac{1}{a_2}\ln(a_1)$$

On rearrangement, we get

$$\frac{1}{a_2}[\ln(a_1) - \ln(a_1 - a_2 y)] = t$$

Further rearrangement gives

$$\ln\left(\frac{a_1}{a_1 - a_2 y}\right) = a_2 t$$

Alternatively, we can rearrange it as

$$\ln\left(\frac{a_1 - a_2 y}{a_1}\right) = -a_2 t$$

After some simple manipulations, we get

$$y = \frac{a_1}{a_2}\left(1 - e^{-a_2 t}\right) \tag{5.23}$$

With Eq. (5.23), we can compute the value of y for any value of t.
Substituting the values of a_1 and a_2 in Eq. (5.23) gives

$$y(t) = \frac{\xi}{A_C} \frac{A_C}{K}\left(1 - e^{-(K/A_C)t}\right)$$

On rearrangement, we get

$$y(t) = \frac{\xi}{K}\left(1 - e^{-(K/A_C)t}\right) \tag{5.24}$$

Note: We cannot put $K = 0$ in this solution, as it will give us the 0/0 form, which is indeterminate. Thus, if $K = 0$ ($a_2 = 0$), then there is no control. For this case, Eq. (5.22) reduces to

$$\frac{dy}{dt} = a_1$$

which actually is

$$\frac{dy}{dt} = \frac{\xi}{A_C}$$

The above equation can be integrated to give

$$y = \frac{\xi}{A_C} t + C_1$$

with initial condition at $t = 0$, $y = 0$. The constant of integration $C_1 = 0$ for this initial condition. So we get the solution as

$$y = \frac{\xi}{A_C} t \quad \text{(for no control)}$$

Figure 5.19 shows the behavior of system with and without the control. Note that the "no control" (by putting $K = 0$) means that the valve is

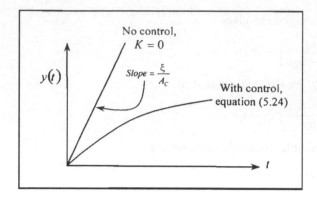

Figure 5.19   Behavior of system with and without control.

not affected by the change in height of liquid in the tank as explained earlier. Of course, this is not the real situation when there is no control. When there is no control, the flow output will increase (proportional to the square root of the height of liquid in the tank) until the system reaches a new steady state.

Second configuration

The *second configuration* we look at is the simple feed-forward-level control with the input flow rate F_i as the measured variable and the output flow rate F as the control variable, as shown in Figure 5.20. If inflow is equal to outflow, then whatever happens to F_i, F follows and, therefore, h remains constant. Thus, the control law is

$$F = F_i$$

Thus,

$$A_C \frac{dh}{dt} = 0$$

where h = constant, depending on the accuracy of measurement, regulators, and different time lags.

Third configuration

The third configuration (Fig. 5.21) is feedback control strategy, but it is different from the first feedback configuration. The inflow F_i is used as the control variable rather than the outlet flow, but still measuring the level (h), which is an output variable. This is the reason it is still a feedback strategy.

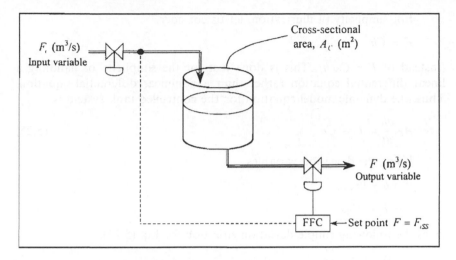

Figure 5.20 Feed-forward control.

Here, F_i lost its status as a disturbance variable and became a control (or manipulated) variable, whereas F was solely dependent upon h (which, in turn, depends on the C value of the valve coefficient) and lost its identity as a control (or manipulated) variable. Thus, the output variable h is the measured variable and the input variable F_i is the control (or manipulated) variable.

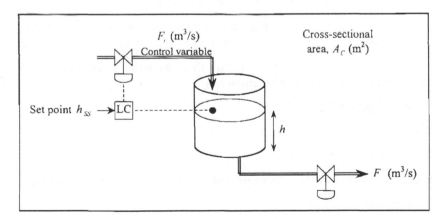

Figure 5.21 Special type of feedback configuration.

For simplicity of illustration, let us consider

$$F = Ch$$

(instead of $F = C\sqrt{h}$). This is done just for the simplicity of handling a linear differential equation rather than a nonlinear differential equation. Thus, the dynamic model equation for the controlled tank system is

$$A_C \frac{dh}{dt} = F_i - Ch \tag{5.25}$$

Using the deviation variables

$$y = h - h_{SS}$$

$$u = F_i - F_{iSS}$$

gives the following simple deviation equation for Eq. (5.25):

$$A_C \frac{dy}{dt} = u - Cy$$

The above differential equation can be rearranged as

$$\frac{dy}{dt} = \left(-\frac{C}{A_C}\right)y + \frac{1}{A_C}u \tag{5.26}$$

If this system is originally at steady state, then the initial condition will be at $t = 0$, $h = h_{SS}$ or, in terms of deviation variable, $y = 0$. Now, if we use proportional control, the control law equation will be

$$F_i = F_{iSS} + K(h - h_{SS})$$

Consider the following physical implications for $K > 0$:

- If $h > h_{SS}$, then $F_i > F_{iSS}$ and, therefore, h continues to increase.
- If $h < h_{SS}$, then $F_i < F_{iSS}$ and, therefore, h continues to decrease.

This is obviously the opposite of control; thus, we either must have $K < 0$, or a much better choice is the control law

$$F_i = F_{iSS} + K(h_{SS} - h)$$

Then,

$$F_i - F_{iSS} = -K(h - h_{SS})$$

which can be rewritten in a more simplified form as deviation variables:

$$u = -Ky$$

Differential Eq. (5.26) becomes

$$\frac{dy}{dt} = \left(-\frac{C}{A_C}\right)y + \frac{1}{A_C}(-Ky) \qquad (5.27)$$

which can be rearranged as

$$\frac{dy}{dt} = -\underbrace{\left(\frac{C+K}{A_C}\right)}_{\alpha}y$$

and can be thus written in the very simple form

$$\frac{dy}{dt} = -\alpha y$$

With the initial condition at $t = 0$, $y = 0$. On integration of the differential equation, we obtain

$$\ln y = -\alpha t + \ln B$$

where B is the constant of integration. On rearrangement we get,

$$\ln\left(\frac{y}{B}\right) = -\alpha t$$

which can be written as

$$y = Be^{-\alpha t}$$

To get the value of integration constant B, the initial condition is used to get

$$B = 0$$

Thus, the solution of the differential equation is thus

$$y = 0$$

This means that the system is always at the steady state (except for the effect of delays in the control loop).

Because of this integration constant problem (it is not really a problem, it is actually correct), we go back to the original Eq. (5.25):

$$A_C \cdot \frac{dh}{dt} = F_i - Ch$$

At steady state, the equation is

$$A_C \frac{dh_{SS}}{dt} = F_{iSS} - Ch_{SS} = 0$$

This gives the steady-state height as

$$h_{SS} = \frac{F_{iSS}}{C}$$

Introduction of the proportional controller gives

$$F_i = F_{iSS} + K(h_{SS} - h) \quad \text{or} \quad F_i = F_{iSS} - K(h - h_{SS})$$

Thus, we get

$$A_C \frac{dh}{dt} = F_{iSS} - K(h - h_{SS}) - Ch \tag{5.28}$$

At steady state,

$$A_C \frac{dh_{SS}}{dt} = F_{iSS} - K(h_{SS} - h_{SS}) - Ch_{SS} = 0$$

From the dynamics point of view for this special feedback control loop, we have

$$\frac{dh}{dt} = \frac{F_{iSS}}{A_C} - \frac{1}{A_C}(Kh - Kh_{SS} + Ch)$$

which can be rearranged as

$$\frac{dh}{dt} = \underbrace{\left(\frac{F_{iSS} + Kh_{SS}}{A_C}\right)}_{a_1} - \underbrace{\frac{(K + C)}{A_C}}_{a_2} h$$

The above equation can be written in a simplified form as

$$\frac{dh}{dt} = a_1 - a_2 h \tag{5.29}$$

where

$$a_1 = \left(\frac{F_{iSS} + Kh_{SS}}{A_C}\right)$$

$$a_2 = \frac{K + C}{A_C}$$

with the initial condition at $t = 0, h = h_{SS}$. We can write the differential Eq. (5.29) as

$$\frac{dh}{a_1 - a_2 h} = dt$$

Integration gives

$$-\frac{1}{a_2}\ln(a_1 - a_2 h) = t + C_1 \tag{5.30}$$

where C_1 is the constant of integration and the initial condition is at $t = 0, h = h_{SS}$. Thus, we obtain the integration constant C_1 as

$$-\frac{1}{a_2}\ln(a_1 - a_2 h_{SS}) = C_1$$

On substituting the value of C_1, the solution of differential equation becomes

$$-\frac{1}{a_2}\ln(a_1 - a_2 h) = t - \frac{1}{a_2}\ln(a_1 - a_2 h_{SS})$$

which can be rearranged as

$$\frac{1}{a_2}[\ln(a_1 - a_2 h_{SS}) - \ln(a_1 - a_2 h)] = t$$

Further rearrangement yields

$$\ln\left(\frac{a_1 - a_2 h}{a_1 - a_2 h_{SS}}\right) = -a_2 t$$

After some manipulations, we get

$$h = \frac{a_1}{a_2} - \left(\frac{a_1 - a_2 h_{SS}}{a_2}\right)e^{-a_2 t} \tag{5.31}$$

Now, we calculate the following and substitute these values in Eq. (5.31):

$$\frac{a_1}{a_2} = \frac{F_{iSS} + K h_{SS}}{A_C}\frac{A_C}{K + C} = \frac{F_{iSS} + K h_{SS}}{K + C}$$

and

$$\frac{a_1 - a_2 h_{SS}}{a_2} = \frac{F_{iSS} + K h_{SS}}{K + C} - h_{SS} = F_{iSS} - C h_{SS} = 0$$

Thus, the solution (5.31) can be written in following form:

$$h = \frac{F_{iSS} + K h_{SS}}{K + C} - \underbrace{(F_{iSS} - C h_{SS})}_{\text{equal to zero}}e^{-a_2 t}$$

which gives

$$h = \left(\frac{F_{iSS} + K h_{SS}}{K + C}\right)$$

and this can be rearranged to give

$$h = \frac{C h_{SS} + K h_{SS}}{K + C} = \left(\frac{K + C}{K + C}\right)h_{SS}$$

which basically gives

$$h = h_{SS}$$

Thus, with this feed-forward control strategy and with F linearly dependent on h together with the assumption that all loop delays are neglected, then h remains at h_{SS} all the time.

Some Concluding Remarks

1. Note that we used a simplifying assumption, $F = Ch$, and not $F = C\sqrt{h}$, just for the simplicity of using linear differential equation.

2. Modeling errors. Intrinsic (inherent) degree of errors in mathematical models are unavoidable and, of course, they affect the control policy.

3. Implementation problems. These include the following:

    *   Time delays
    *   Imperfect measurements
    *   Inaccurate transmission
    *   Control valve inertia (leading to inaccurate valve actuation)

4. Complicated process control structure. Consider the example shown in Figure 5.22. A number of elementary questions arise with regard to this relatively complicated control structure. For example, how did we choose that the hot stream is not the effective control (manipulated) variable for the control of the level (rather than the control of the temperature)? In more general

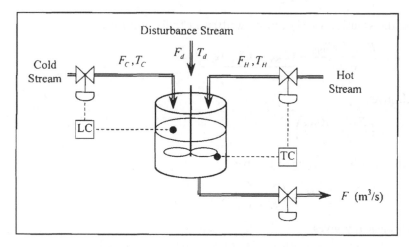

Figure 5.22 A relatively complicated control structure.

terms, which control variable should be manipulated (regulated/controlled) to control which output variable (T, level) to achieve maximum effectiveness? Assuming that the shown input/output pairing is the proper one, the question still remains, can the cold stream regulate the liquid level without upsetting the hot stream's task of regulating the liquid temperature, and vice versa? This mutual interference is called *interaction* and it is very essential for MIMO design.

5.5 PROCESS DYNAMICS AND MATHEMATICAL TOOLS

In this section, we discuss some basic ideas associated with process dynamics. Consider the process shown in Figure 5.23. A number of possible responses [output variable change with time $y(t)$] to a step change in input variable $u(t)$ depending on the characteristic of the system or the process are shown in Figure 5.24.

5.5.1 Tools of Dynamic Models

Process Model and Ideal Forcing Functions

The input–output relation of the process can be obtained by disturbing the input and recording the corresponding output. However this is expensive, time-consuming, and, most importantly, assumes that the unit does exist and is not in the design stage. To avoid these main three limitations, we should use mathematical models. This is a task which can be stated in a clearer form as follows: Given some form of mathematical representation of the process, investigate the process response to various input changes; that is, given a process model, find $y(t)$ in response to inputs $u(t)$ and $d(t)$. This task needs (1) a process model and (2) well-characterized input functions (forcing functions).

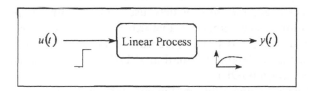

Figure 5.23 Input-output process (response to step input disturbance).

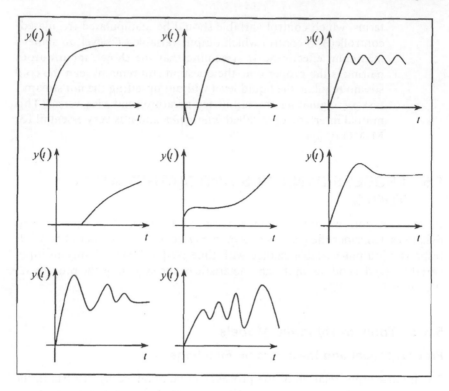

Figure 5.24 Some of the different forms of response.

Mathematical Tools

The mathematical tools available for this task include the following:

- Dynamic mathematical models: usually in the form of differential equations (ordinary and partial, linear, and nonlinear).
- Laplace transform: used in the development of transfer functions, which are the most widely used model form in process control studies. The Laplace transform converts an ordinary differential equation (ODE) to an algebraic equation and, likewise, converts a partial differential equation (PDE) into an ordinary differential equation (ODE).
- For a nonlinear differential equation, either numerical solutions or linearization in the neighborhood of a certain state is used.

- The z-transform may be used to convert a difference equation to an algebraic equation (it is used in digital control, which is the dominant form of control for last three to four decades).
- Multivariable systems utilize the short-hand methods of matrices.

Digital computers are used to obtain numerical solutions for systems which are described by equations that cannot be solved analytically.

5.6 THE LAPLACE TRANSFORMATION

As we mentioned earlier, some transformations are useful in handling process dynamics and control problems. The most popular for linear (or linearized) systems is the Laplace transformation. The basic definition of Laplace transformation is

$$\bar{f}(s) = \int_0^\infty e^{-st} f(t) \, dt = L\{f(t)\}$$

where s is a complex variable. Laplace transformation is a mapping from the t-domain (time domain) to the s-domain (Laplace domain). Later, $f(s)$ is used instead of $\bar{f}(s)$ for easier representation.

5.6.1 Some Typical Laplace Transforms

If $f(t) = 1$, then

$$\bar{f}(s) = L\{f(t)\} = L\{1\} = \int_0^\infty e^{-st}(1) \, dt = \left[-\frac{1}{s} e^{-st} \right]_0^\infty = -\frac{1}{s} \left[e^{-\infty} - e^{-0} \right]$$

$$= -\frac{1}{s} [0 - 1]$$

giving

$$L\{1\} = \frac{1}{s}$$

If $f(t) = e^{-at}$

then

$$\bar{f}(s) = L\{f(t)\} = L\{e^{-at}\} = \int_0^\infty e^{-st} e^{-at} \, dt = \int_0^\infty e^{-(s+a)t} \, dt = \left[\frac{-1}{s+a} e^{-(s+a)t} \right]_0^\infty$$

$$= \frac{-1}{s+a} \left[e^{-\infty} - e^{-0} \right] = \frac{-1}{s+a} [0 - 1]$$

finally giving

$$L\{e^{-at}\} = \frac{1}{s+a}$$

We usually use tables and some properties to get the Laplace transforms. A table of the most important Laplace transforms is given in Appendix D. In order to cover a wider range of functions using a limited table like the one in Appendix D, we must utilize some important properties of the Laplace transformation. The main properties of Laplace transform are as follows:

1. All functions necessary in the study of process dynamics and control have Laplace transforms.
2. $\bar{f}(s)$ has no information for $t < 0$, because the integral starts at $t = 0$. However, this fact practically represents no problem, because t is always time and we are interested in $t > t_0$, where t_0 is the initial time that can arbitrarily be set to zero (for $t < 0$, we can put $f(t) = 0$).
3. It is not possible for two different functions $f(t)$ and $g(t)$ to have the same Laplace transform. $f(t)$ and $\bar{f}(s)$ are called the transfer pair and this transfer pair is unique.
4. The Laplace transform operation is linear:

$$L\{C_1 f_1(t) + C_2 f_2(t)\} = C_1 L\{f_1(t)\} + C_2 L\{f_2(t)\}$$

5.6.2   The Inverse Laplace Transform

The inverse Laplace formula is

$$f(t) = L^{-1}\{\bar{f}(s)\} = \frac{1}{2\pi j} \oint_C e^{st} \bar{f}(s)\, ds$$

which is a complex contour integral over the path represented by C (called the Bromwich path). Tables of the inverse transform are also available (see Appendix D).

5.6.3   The Transform of Derivatives

Because

$$L\{f(t)\} = \int_0^\infty e^{-st} f(t)\, dt \tag{5.32}$$

we have

$$L\left\{\frac{df(t)}{dt}\right\} = \int_0^\infty e^{-st}\left(\frac{df(t)}{dt}\right)dt = s\bar{f}(s) - \underbrace{f(0)}_{\text{at } t=0}$$

Simple proof. Let

$$e^{-st}f(t) = y$$

Then,

$$\frac{dy}{dt} = -se^{-st}f(t) + e^{-st}\frac{df(t)}{dt}$$

Rearranging gives

$$\frac{df(t)}{dt} = \frac{1}{e^{-st}}\left(\frac{dy}{dt} + se^{-st}f(t)\right) \tag{5.33}$$

Substituting Eq. (5.33) into Eq. (5.32) gives

$$\int_0^\infty e^{-st}\left(\frac{df(t)}{dt}\right)dt = \int_0^\infty e^{-st}\frac{1}{e^{-st}}\left(\frac{dy}{dt} + se^{-st}f(t)\right)dt$$

$$= \int_0^\infty \left(\frac{dy}{dt}\right)dt + s\int_0^\infty e^{-st}f(t)\,dt$$

$$= \int_0^\infty dy + s\int_0^\infty e^{-st}f(t)\,dt$$

$$= y(\infty) - y(0) + s\bar{f}(s)$$

Because

$$y(\infty) = e^{-\infty}f(\infty) = 0$$
$$y(0) = e^{-0}f(0) = f(0)$$

we have

$$L\left\{\frac{df(t)}{dt}\right\} = s\bar{f}(s) - \underbrace{f(0)}_{\text{at } t=0}$$

Similarly, we can obtain the Laplace transformation for higher derivatives:

$$L\left\{\frac{d^2 f(t)}{dt^2}\right\} = s^2\bar{f}(s) - sf(0) - f'(0)$$

where

$$f^1(0) = \frac{df(t)}{dt}\bigg|_{t=0}$$

The general form is

$$L\left\{\frac{d^n f(t)}{dt^n}\right\} = s^n \overline{f}(s) - s^{n-1} f(0) - s^{n-2} f^1(0) - s^{n-3} f^2(0) - \cdots - f^{n-1}(0)$$

(5.34)

where

$$f^{n-1}(0) = \frac{d^{n-1} f(t)}{dt^{n-1}}\bigg|_{t=0}$$

Equation (5.34) can also be written in the form

$$L\left\{\frac{d^n f(t)}{dt^n}\right\} = s^n \overline{f}(s) - \sum_{k=0}^{n-1} s^k f^{(n-1-k)}(0)$$

where $f^i(0)$ is the ith derivative of $f(t)$ at $t = 0$ and $f^0(0)$ is the function $f(t)$ at $t = 0$.

If all the initial conditions of $f(t)$ and its derivatives are zero (disturbance variables), then

$$L\left\{\frac{d^n f(t)}{dt^n}\right\} = s^n \overline{f}(s) - \underbrace{\sum_{k=0}^{n-1} s^k f^{(n-1-k)}(0)}_{\text{each term equal to zero}}$$

And, therefore, the Laplace transformation of the derivatives is given by

$$L\left\{\frac{d^n f(t)}{dt^n}\right\} = s^n \overline{f}(s)$$

The transforms of integrals is

$$L\left\{\int_0^t f(t')dt'\right\} = \frac{1}{s}\,\overline{f}(s)$$

Proof: Reader is advised to see the proof in any elementary mathematics text.

5.6.4   Shift Properties of the Laplace Transform

If

$$L\{f(t)\} = \overline{f}(s)$$

then

$$L\{e^{at}f(t)\} = \int_0^\infty e^{at} e^{-st} f(t)\, dt$$

$$= \int_0^\infty e^{-(s-a)t} f(t)\, dt$$

$$= \bar{f}(s') \equiv \bar{f}(s-a)$$

where

$$s' = s - a$$

that is to say,

$$L\{e^{at}f(t)\} = \bar{f}(s-a)$$

Thus, the s variable in the transform has been shifted by a units.

A more important result is the *shift with regard to the inverse of the Laplace transform:*

$$L^{-1}\{\bar{f}(s-a)\} = e^{at} f(t)$$

Let us use a very simple example:

$$f(t) = 1$$

then

$$\bar{f}(s) = L\{1\} = \frac{1}{s}$$

Find

$$L^{-1}\left\{\frac{1}{s+a}\right\}$$

The solution is

$$L^{-1}\left\{\frac{1}{s-(-a)}\right\} = f(t) = e^{-at}$$

The shift in time is given by

$$L\{f(t-a)\} = e^{-as}\bar{f}(s)$$

The proof is as follows:

$$L\{f(t-a)\} = \int_0^\infty e^{-st} f(t-a)\, dt$$

Define

$$(t-a) = t'$$

which gives $dt = dt'$. For the limit $t = 0$, we get $t' = -a$; for $t = \infty$, we get $t' = \infty$. Thus, we can write,

$$L\{f(t-a)\} = \int_{-a}^{\infty} e^{-s(t'+a)} f(t') \, dt'$$

$$= \int_{-a}^{\infty} e^{-st'} e^{-as} f(t') \, dt'$$

$$= e^{-as} \left[\underbrace{\int_{-a}^{0} e^{-st'} f(t') \, dt'}_{\text{equal to zero}} + \underbrace{\int_{0}^{\infty} e^{-st'} f(t') \, dt'}_{\bar{f}(s)} \right]$$

It is always assumed that $f(t')$ for $t' < 0$ is equal to zero. Thus

$$L\{f(t-a)\} = e^{-as} \bar{f}(s)$$

Figure 5.25 shows the change in the behavior of a function due to the shift in time.

5.6.5 The Initial- and Final-Value Theorems

These theorems relate the limits of $t \to 0$ and $t \to \infty$ in the time domain to $s \to \infty$ and $s \to 0$, respectively, in the Laplace domain.

Initial-value theorem

$$\lim_{t \to 0} [f(t)] = \lim_{s \to \infty} \left[s \bar{f}(s) \right]$$

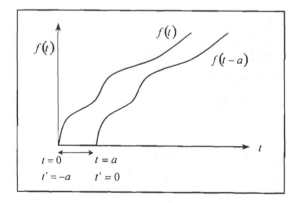

Figure 5.25   Behavior due to shift in time.

Final-value theorem

$$\lim_{t \to \infty} [f(t)] = \lim_{s \to 0} [s\overline{f}(s)]$$

5.6.6 Use of Laplace Transformation for the Solution of Differential Equations

The condensed information regarding the Laplace transformation in the previous sections is sufficient to utilize it in solving linear differential equations, as shown in the following subsection.

Consider the following differential equation:

$$\tau \frac{dy}{dt} + y = Ku(t) \tag{5.35}$$

For $u(t) = 1$, $y(0) = 0$, we get, by Laplace transformation,

$$\tau s y(s) + y(s) = \frac{K}{s}$$

On rearrangement, we get

$$(\tau s + 1)y(s) = \frac{K}{s}$$

which can be rewritten as

$$y(s) = \frac{K}{s(\tau s + 1)} \tag{5.36}$$

Using partial fractions, we can write the right-hand side of Eq. (5.36) in the form

$$\frac{K}{s(\tau s + 1)} = \frac{a}{s} + \frac{b}{\tau s + 1}$$

We can start to obtain the constants a and b by putting the equation in the form

$$\frac{K}{s(\tau s + 1)} = \frac{a(\tau s + 1) + bs}{s(\tau s + 1)}$$

which gives

$$K = (a\tau + b)s + a$$

On equating the coefficients, we get

$$a = K$$
$$b = -K\tau$$

Thus, we can write

$$y(s) = \frac{K}{s} - \frac{K\tau}{(\tau s + 1)}$$

In a more simplified form,

$$y(s) = \frac{K}{s} - \frac{K}{s - (-1/\tau)}$$

On taking the inverse Laplace transform of the two parts, we get

$$y(t) = K - Ke^{-t/\tau}$$

which can be rewritten as

$$y(t) = K(1 - e^{-t/\tau}) \qquad\qquad (5.37)$$

This is the solution of our differential equation.

The reader is advised to solve the same differential equation using the operator method (see Appendix C).

Solution of Higher-Order ODE by Laplace Transformation

Consider the second-order ODE for which y'' represents the second derivative of y and y' represents the first derivative of y:

$$y''(t) + 5y'(t) + 6y(t) = f(t) \qquad\qquad (5.38)$$

with

$$f(t) = 1, \quad y(0) = 1.0, \quad y'(0) = 0$$

The Laplace transformation of the differential Eq. (5.38) gives

$$[s^2\bar{y}(s) - sy(0) - y'(0)] + [5(s\bar{y}(s) - y(0))] + 6\bar{y}(s) = \frac{1}{s}$$

For $y(0) = 1.0$ and $y'(0) = 0$, this equation becomes,

$$s^2\bar{y}(s) - s - 0 + 5([s\bar{y}(s) - 1]) + 6\bar{y}(s) = \frac{1}{s}$$

Further simplification yields

$$s^2\bar{y}(s) + 5s\bar{y}(s) + 6\bar{y}(s) = \frac{1}{s} + (s + 5)$$

Rearrangement gives

$$(s^2 + 5s + 6)\bar{y}(s) = \frac{1}{s} + (s + 5)$$

From the above equation, we can obtain the expression for $\bar{y}(s)$ as

$$\bar{y}(s) = \frac{(1/s) + (s + 5)}{s^2 + 5s + 6}$$

which finally simplifies to,

$$\bar{y}(s) = \frac{1 + s^2 + 5s}{s(s^2 + 5s + 6)} \tag{5.39}$$

Using partial fractions and employing the fact that

$$s^2 + 5s + 6 = (s + 3)(s + 2)$$

we can write the right-hand side of Eq. (5.39) as

$$\frac{1 + s^2 + 5s}{s(s^2 + 5s + 6)} = \frac{A}{s} + \frac{B}{s + 3} + \frac{C}{s + 2} \tag{5.40}$$

Rearrangement gives

$$\frac{1 + s^2 + 5s}{s(s + 3)(s + 2)} = \frac{A(s + 3)(s + 2) + Bs(s + 2) + Cs(s + 3)}{s(s + 3)(s + 2)}$$

From the above equation, we get

$$1 + s^2 + 5s = A(s^2 + 5s + 6) + B(s^2 + 2s) + C(s^2 + 3s)$$

which can be written as

$$1 + s^2 + 5s = s^2(A + B + C) + s(5A + 2B + 3C) + 6A$$

Equating the coefficients give

$$6A = 1$$
$$A + B + C = 1$$
$$5A + 2B + 3C = 5$$

On solving the above three linear algebraic equations in terms of A, B, and C, we get

$$A = \frac{1}{6}, \quad B = -\frac{5}{3}, \quad C = \frac{5}{2}$$

Using the inverse Laplace transformation on Eq. (5.40), we get the solution of the differential equation as

$$y(t) = \frac{1}{6} - \frac{5}{3}e^{-3t} + \frac{5}{2}e^{-2t}$$

5.6.7 Main Process Control Applications of Laplace and Inverse Transformations

Figure 5.26 shows a flow diagram for the Laplace transformation and inverse transformation. It is clear that the main function of the Laplace transformation is to put the differential equation (in the time domain) into an algebraic form (in the s-domain). These s-domain algebraic equations can be easily manipulated as input–output relations.

5.7 CHARACTERISTICS OF IDEAL FORCING FUNCTIONS

The ideal forcing functions are the following:

1. Ideal step function (Fig. 5.27)

$$u(t) = \begin{cases} 0 & \text{for} \quad t < 0 \\ A & \text{for} \quad t > 0 \end{cases}$$

or

$$u(t) = AH(t)$$

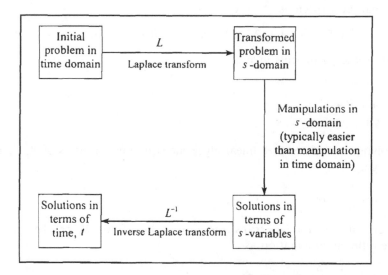

Figure 5.26 Laplace transformation and the inverse transformation.

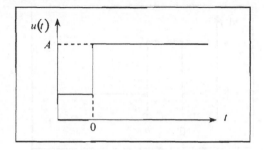

Figure 5.27   Ideal step function.

where $H(t)$ is the Heaviside function defined as

$$H(t) = \begin{cases} 0 & \text{for} \quad t < 0 \\ 1 & \text{for} \quad t > 0 \end{cases}$$

The Laplace transformation of the step function is

$$\bar{u}(s) = \frac{A}{s}$$

2. Dirac delta function (Fig. 5.30). It is the limit of the rectangular pulse function (Fig. 5.28). Let

$$A = \frac{1}{b}$$

Thus,

$$A\,b = \frac{1}{b}b = 1$$

So, as $b \to 0$, $A \to \infty$.

The Dirac delta function is a rectangular-pulse function of zero width and unit area:

$$L\{\delta(t)\} = \int_0^\infty e^{-st}\delta(t)\,dt$$

$$= \int_{-\varepsilon}^{\varepsilon} e^{-st}\delta(t)\,dt$$

$$= e^{-0}$$

Thus,

$$L\{\delta(t)\} = 1$$

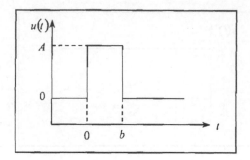

Figure 5.28   Ideal rectangular pulse function which becomes Dirac as $b \to \infty$.

For any function $f(t)$

$$\int_{t_0-\varepsilon}^{t_0+\varepsilon} \delta(t-t_0)f(t)\,dt = f(t_0)$$

Here, $\delta(t-t_0)$ is the general shifted delta function (zero everywhere except at $t = t_0$). For $t_0 = 0$, we get

$$\int_{-\varepsilon}^{\varepsilon} \delta(t)f(t)\,dt = f(0)$$

For $f(t) = A$

$$\int_{-\varepsilon}^{\varepsilon} \delta(t)A\,dt = A$$

3. The ideal rectangular-pulse function (Fig. 5.29)

$$u(t) = \begin{cases} 0 & \text{for} \quad t < 0 \\ A & \text{for} \quad 0 < t < b \\ 0 & \text{for} \quad t > b \end{cases}$$

$$H(t-b) = \begin{cases} 0 & \text{for} \quad t < b \\ 1 & \text{for} \quad t > b \end{cases}$$

So

$$u(t) = A[H(t) - H(t-b)]$$

Thus

$$\bar{u}(s) = A\left(\frac{1}{s} - e^{-bs}\frac{1}{s}\right) = \frac{A}{s}\left(1 - e^{-bs}\right)$$

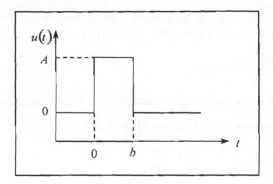

Figure 5.29 Ideal rectangular-pulse function.

4. Ideal impulse function (Dirac delta function) (Fig. 5.30)

$$u(t) = A\delta(t)$$

It is infinite at the point $t = 0$ and zero elsewhere,

$$\delta(t) = \begin{cases} \infty & \text{for } t = 0 \\ 0 & \text{elsewhere} \end{cases}$$

and

$$\int_{-\infty}^{\infty} \delta(t)dt = \int_{-\varepsilon}^{\varepsilon} \delta(t)dt = 1$$
$$\bar{u}(s) = L\{u(t)\} = A$$

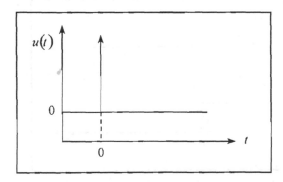

Figure 5.30 Ideal impulse function.

The Dirac delta function is the derivative of the unit step (or Heaviside function) (Fig. 5.31). $H(t)$ is the unit step function:

$$\frac{d[H(t)]}{dt} = \delta(t)$$

It is not possible to implement it exactly experimentally.

5. The ideal ramp function (Fig. 5.32)

$$u(t) = \begin{cases} 0 & \text{for } t < 0 \\ At & \text{for } t > 0 \end{cases}$$

The Laplace transformation is

$$\bar{u}(s) = L\{u(t)\} = \frac{A}{s^2}$$

6. The ideal sinusoidal function (Fig. 5.33)

$$u(t) = \begin{cases} 0 & \text{for } t < 0 \\ A \sin(\omega t) & \text{for } t > 0 \end{cases}$$

The Laplace transformation is

$$\bar{u}(s) = L\{u(t)\} = \frac{A\omega}{s^2 + \omega^2}$$

It is difficult to implement exactly experimentally. However, the response to a theoretical sinusoidal function is very useful in process dynamics and design of controllers.

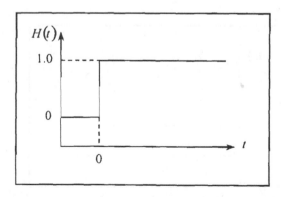

Figure 5.31 Unit step function.

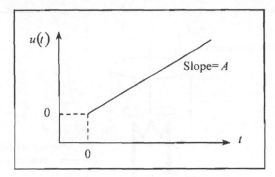

Figure 5.32 Ideal ramp function.

Realization of Ideal Forcing Functions

If the input of interest is the steam flow rate (Figure 5.34), then,

1. Step input. Open the steam valve a given percentage at $t = 0$ such that Q changes by A units.
2. Pulse input. Open the steam valve at $t = 0$, hold at the new value for a duration of b time units, and then return to the old value.
3. Impulse (impossible to realize perfectly). Open the steam valve (wide open) at $t = 0$ and instantaneously (or as soon as possible thereafter) return to the initial position.
4. Ramp input. Gradually open the steam valve such that Q increases linearly. Ramp ends when the steam valve is fully open.
5. Sinusoidal input. The only practical way to achieve this is to connect a sine-wave generator to the steam valve. Realizing high-frequency sinusoidal input may be limited by the valve dynamics.

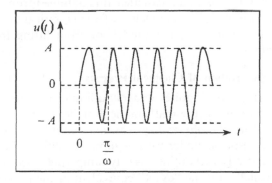

Figure 5.33 Ideal sinusoidal function.

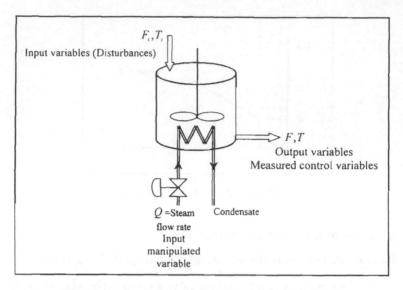

Figure 5.34   Variables of a stirred heating tank.

Important Highlights

Process Model

- The idea of utilizing a collection of mathematical equation as a "surrogate" for a physical process is at once ingenious as well as expedient. More importantly, however, this approach has now become indispensable to proper analysis and design of process control systems (and also design of equipment).
- Classical control theory concentrates on the transfer-function model in the s-domain (whether it is obtained from Laplace transformation of rigorous models, or from empirical fitting to the Bode diagram of the rigorous model, or from fitting to experimental responses).

Mathematical Description of Chemical Process

- Process control requires understanding of process dynamics.
- The main use of the mathematical model is as a convenient "surrogate" for the physical system, making it possible to investigate system response under various input conditions, both rapidly and inexpensively without necessarily tampering with the actual entity.
- It is usually not the "exact" equivalent of the process (but we can make it so, and we should use "optimum degree of sophistication").

- Note the title "lumped/distributed parameter systems (models)" is not correct because it is not the "parameter" that is distributed, it is the state variable in fact.

Process Characteristics and Process Models

- Dependent variables: all inputs, output, disturbances, and so forth.
- Independent variables: time and space
- Linear and nonlinear systems.
- Lumped (ODEs) and distributed (PDEs) systems
- Discrete-time system

Even though the tacit assumption is that the variables of a process do not ordinarily change in "jumps," that they normally behave as smooth continuous functions of time (and position), there are situations in which output variables are deliberately sampled and control action implemented only at discrete points at time. The process variables then appear to change in a piecewise constant fashion with respect to the now discretized time. Such processes are modeled by difference equations and are referred to as discrete-time systems.

5.8 BASIC PRINCIPLES OF BLOCK DIAGRAMS, CONTROL LOOPS, AND TYPES OF CLASSICAL CONTROL

Laplace transformation is the best way to handle classical linear control problems using P (proportional), I (integral), and D (derivative) controllers. The controller is activated by the error $e(t)$ (the difference between the measured variable and the set point) to give its output signal $y(t)$. The relation between $e(t)$ and $y(t)$ depends on the type of controller used. It is much easier to handle this relation in the Laplace domain. In the Laplace domain, $e(t)$ becomes $\bar{E}(s)$ and $y(t)$ becomes $\bar{Y}(s)$ and they are related to each other through the transfer function $G(s)$.

Types of Classical Control (Fig. 5.35)

Proportional Controller

For this type of control, the time domain relation between the input and output is

$$y(t) = K_C e(t)$$

In the Laplace domain, the relation becomes

$$\bar{Y}(s) = K_C \bar{E}(s)$$

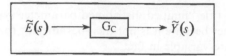

Figure 5.35 Input–output relation.

Thus, the transfer function G_C is given by $G_C = K_C$ for a proportional controller.

Integral Controller

For this type of controller, the input–output relation in the time domain is

$$y(t) = K_I \int_0^t e(t)\, dt$$

In the Laplace domain, it becomes

$$\tilde{Y}(s) = \frac{K_I}{s}\, \tilde{E}(s)$$

Thus, $G_C = K_I/s$ for an integral controller.

Note that the integral relation in the time domain has changed into an algebraic relation in the Laplace domain.

Derivative Controller

In the time domain, the input–output relation is

$$y(t) = K_D \frac{de(t)}{dt}$$

In the Laplace domain, it is

$$\tilde{Y}(s) = K_D s \tilde{E}(s)$$

Thus, $G_C = K_D s$ for a derivative controller.

Note that the differential relation in the time domain has changed into an algebraic relation in the Laplace domain.

Three-Mode Controller (PID: Proportional, Integral, and Derivative)

The input–output proportional-integral-derivative relation in the time domain is

$$y(t) = K_C e(t) + K_I \int_0^t e(t)\, dt + K_D \frac{de(t)}{dt}$$

In the Laplace domain, the above relation becomes an algebraic relation:

$$\tilde{Y}(s) = K_C \tilde{E}(s) + \frac{K_I}{s}\tilde{E}(s) + K_D s \tilde{E}(s)$$

which can be rearranged in the simpler form

$$\tilde{Y}(s) = \left(K_C + \frac{K_I}{s} + K_D s\right)\tilde{E}(s)$$

Thus,

$$G_C = \left(K_C + \frac{K_I}{s} + K_D s\right)$$

for a PID controller.

A Very Simple Example

Consider the tank shown in Figure 5.36. The unsteady-state mass balance equation is given by the differential equation

$$A\frac{dh}{dt} = q_i - q_0 \qquad (5.41)$$

where A is the cross-sectional area of the tank and h is the liquid height inside the tank.

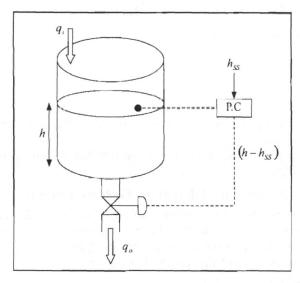

Figure 5.36 Filling tank with feedback proportional controller.

With the initial condition (at steady state) given by $t = 0$, $h = h_{SS}$, $q_i = q_{iSS}$, $q_0 = q_{0SS}$ which means that at $t = 0$, the system is at steady state.

We define deviation (or hat) variables as follows:

$$\hat{h} = h - h_{SS}$$

$$\hat{q}_i = q_i - q_{iSS}$$

$$\hat{q}_0 = q_0 - q_{0SS}$$

Thus, the differential Eq. (5.41) becomes

$$A\frac{d\hat{h}}{dt} = (\hat{q}_i + q_{iSS}) - (\hat{q}_0 + q_{0SS})$$

which can be rearranged as

$$A\frac{d\hat{h}}{dt} = \hat{q}_i - \hat{q}_0 + \underbrace{(q_{iSS} - q_{0SS})}_{\text{zero}}$$

Thus, the above equation simplifies to give

$$A\frac{d\hat{h}}{dt} = \hat{q}_i - \hat{q}_0 \tag{5.42}$$

with the initial condition at $t = 0$, $\hat{h} = 0$, $\hat{q}_i = 0$, and $\hat{q}_0 = 0$, which means that all deviation (hat) variables are equal to zero at initial time.

The Laplace transform of Eq. (5.42) gives

$$As\tilde{h}(s) = \tilde{q}_i(s) - \tilde{q}_0(s)$$

which can be written as

$$\tilde{h}(s) = \frac{1}{As}\tilde{q}_i(s) - \frac{1}{As}\tilde{q}_0(s) \tag{5.43}$$

This is the open-loop equation (without the effect of feedback control) in the Laplace domain.

We can put the equation in the form of the shown open-loop block diagram in Figure 5.37,

Closed Loop (Including the Effect of the Feedback Control)

Consider a proportional controller relating $\tilde{h}(s)$ to $\tilde{q}_0(s)$. In time scale, we consider a very simple proportional controller making the flow rate out of the tank proportional to the height in the tank (this is an approximation to explain the steps without the lengthy linearization steps):

$$q_0 = K_C h$$

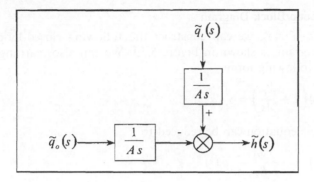

Figure 5.37 Block diagram for the open-loop system.

Note: We consider $q_0 \propto h$ rather than $q_0 \propto \sqrt{h}$ for simplicity of illustration. We will discuss the nonlinear case and the associated linearization process later.

Using deviation variables, we can write

$$q_0 - q_{0ss} = K_C h - K_C h_{ss}$$

In terms of deviation (hat) variables,

$$\hat{q}_0 = K_C \hat{h}$$

Taking the Laplace transform of the above equation gives

$$\tilde{q}_0(s) = K_C \tilde{h}(s).$$

Because $\hat{q}_0(0) = 0$ and $\hat{h}(0) = 0$ (5.44)

We can represent this in a block diagram as shown in Figure 5.38.
From equations (5.43) and (5.44), we get

$$\tilde{h}(s) = \frac{1}{As} \tilde{q}_i(s) - \frac{1}{As} K_C \tilde{h}(s) \tag{5.45}$$

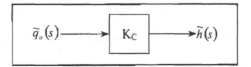

Figure 5.38 Block diagram for the closed-loop control.

Closed-Loop Block Diagram

From Eq. (5.45), we can construct the following closed-loop (feedback) block diagram as shown in Figure 5.39. We can also rearrange Eq. (5.45) into the following form:

$$\tilde{h}(s)\left(1 + \frac{K_C}{As}\right) = \frac{1}{As}\tilde{q}_i(s)$$

The above equation can be reduced to

$$\tilde{h}(s) = \frac{1/As}{1 + K_C/As}\tilde{q}_i(s)$$

which can be written as

$$\tilde{h}(s) = G_C(s)\tilde{q}_i(s) \tag{5.46}$$

where

$$G_C(s) = \frac{1/As}{1 + K_C/As}$$

Equation (5.46) can be drawn as shown in Figure 5.40.

A more general form for the closed loop (feedback) controlled process can be drawn as in Figure 5.41.

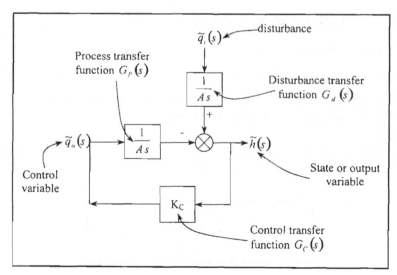

Figure 5.39 Alternate block diagram for the closed-loop control.

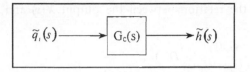

Figure 5.40   Block diagram for Eq. (5.46).

The transfer functions for the different parts can be written as follows:

Process, Open-Loop Equation

$$X(s) = G_d(s)D(s) - G_P(s)C(s) \tag{5.47}$$

Controller Equation

$$C(s) = G_C(s)X(s) \tag{5.48}$$

From relations (5.47) and (5.48), we obtain

$$X(s) = G_d(s)D(s) - G_P(s)G_c(s)X(s)$$

which can be rearranged in the form

$$X(s)[1 + G_P(s)G_C(s)] = G_d(s)D(s)$$

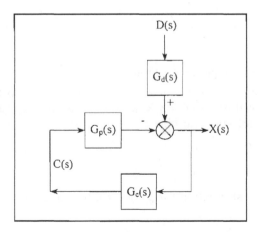

Figure 5.41   A more general block diagram.

Finally, we get the relation between the output $X(s)$ and the disturbance $D(s)$:

$$X(s) = \frac{G_d(s)}{1 + G_P(s)G_C(s)} D(s)$$

Of course, this relation can be written as

$$X(s) = G(s)D(s)$$

where $G(s)$ is the overall transfer function given by

$$G(s) = \frac{G_d(s)}{1 + G_P(s)G_C(s)}$$

For the Tank Problem

$$G_d(s) = \frac{1}{As}$$

and

$$G_P(s) = \frac{1}{As}$$

For the proportional controller,

$$G_C(s) = \overline{K}_C \ (= C_{SS} + K_C, \text{ where } C_{SS} \text{ is the valve coefficient and }$$
$$K_C \text{ is the proportional gain of the controller})$$

so we get

$$\underbrace{X(s)}_{\tilde{h}(s)} = \frac{1/As}{1 + \overline{K}_C(1/As)} \underbrace{D(s)}_{\tilde{q}_i(s)}$$

Response to a Step Input in the Feed

Figure 5.42 shows the step change in input flow rate q_i to the tank and its effect on the deviation variable $\hat{q}_i = q_i - q_{iss}$. For this step function, we have in term of the deviation variable $\hat{q}_i$:

$$\text{For} \quad t < 0, \quad \hat{q}_i = 0.$$

$$\text{For} \quad t \geq 0, \quad \hat{q}_i = \overline{A}$$

The Laplace transformation for $\tilde{q}_i(s) [\equiv D(s)]$ is given by

$$D(s) = \frac{\overline{A}}{s}$$

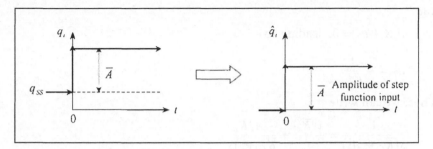

Figure 5.42 Step change in input flow rate and corresponding change in deviation variable (not to scale).

Therefore, $X(s)[\equiv \tilde{h}(s)]$ is given by

$$X(s) = \frac{1/As}{1 + \overline{K}_C(1/As)} \frac{\overline{A}}{s} = \frac{1}{As} \frac{1}{1 + \overline{K}_C(1/As)} \frac{\overline{A}}{s}$$

$$= \frac{\overline{A}}{A} \frac{1}{s} \frac{1}{s} \frac{1}{(As + \overline{K}_C)/As} = \frac{\overline{A}}{A} \frac{1}{s^2} \frac{As}{As + \overline{K}_C}$$

Thus, after some simple manipulations, $X(s)$ is given by

$$X(s) = \overline{A} \frac{1}{s(\overline{K}_C + As)} \tag{5.49}$$

Using partial fractions, we get

$$\frac{1}{s(\overline{K}_C + As)} = \frac{a_1}{s} + \frac{a_2}{\overline{K}_C + As}$$

$$= \frac{a_1(\overline{K}_C + As) + a_2 s}{s(\overline{K}_C + As)}$$

$$= \frac{a_1 \overline{K}_C + (Aa_1 + a_2)s}{s(\overline{K}_C + As)}$$

From the above relation, we get

$$a_1 \overline{K}_C = 1$$

giving

$$a_1 = \frac{1}{\overline{K}_C}$$

and

$$a_1 A + a_2 = 0, \text{ leading to}$$

$$a_2 = -\frac{A}{K_C}$$

Thus, we can write

$$\frac{1}{s(\overline{K}_C + As)} = \frac{1/\overline{K}_C}{s} + \frac{-A/\overline{K}_C}{\overline{K}_C + As}$$

Therefore, relation (5.49) becomes

$$X(s) = \left(\frac{1}{\overline{K}_C} \frac{1}{(s-0)} - \frac{1}{\overline{K}_C} \frac{1}{s - (-\overline{K}_C/A)} \right) \overline{A} \qquad (5.50)$$

Taking the inverse Laplace transformation of Eq. (5.50), we get

$$X(t) = \hat{h}(t) = \frac{\overline{A}}{\overline{K}_C} \left(e^{0t} - e^{-(\overline{K}_C/A)t} \right) \qquad (5.51)$$

Let us say

$$\frac{\overline{K}_C}{A} = \frac{1}{\tau}$$

We get

$$\hat{h}(t) = \frac{\overline{A}}{\overline{K}_C} \left(1 - e^{-t/\tau} \right) \qquad (5.52)$$

Does $\hat{h}(t)$ go to zero as $t \to \infty$ (i.e., $h \to h_{SS}$)?

The answer is "no"; there is an offset which decreases as $\overline{K}_C$ increases,

$$\hat{h}(\infty) = \frac{\overline{A}}{\overline{K}_C} \quad \text{(offset)}$$

Thus, as $\overline{K}_C$ increases, the offset decreases, as can be seen in Figure 5.43. Note that at steady state, the deviation variables (after the step increase in $\hat{q}_i$ and without controller) will be given by the simple relation

$$\hat{q}_i = \overline{A} = C_{SS}\left(h_{SS_{new}} - h_{SS} \right)$$

Thus, the new steady state due to the input increase (without controller) will be

$$h_{SS_{new}} - h_{SS} = \hat{h}_{SS_{new}} = \frac{\overline{A}}{C_{SS}}$$

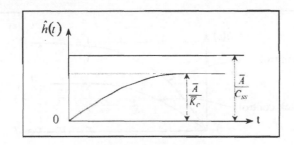

Figure 5.43 Effect on offset.

Now, when we have a proportional controller with a proportional gain $= K_C$, we get

$$q_0 = C_{SS}h + K_C(h - h_{SS})$$

Here K_C is the controller proportional gain. When $h = h_{SS}$ (at steady state), we have

$$q_{0SS} = C_{SS}h_{SS}$$

Using the above two equations, we get the deviation variable

$$\hat{q}_0 = (C_{SS} + K_C)\hat{h} = \overline{K}_c \cdot \hat{h}$$

Thus, as K_C increases, $\overline{K}_C$ increases over C_{SS}. The response becomes faster and the offset smaller, as seen in Figure 5.44.

If $K_C = 0$, then $\overline{K}_C = C_{SS}$. Note that in all of the above calculations, we have used the relation $q_0 = Ch$ for simplicity to illustrate the main ideas (in fact, we should have used $q_0 = C\sqrt{h}$). For $K_C > 0$, it goes toward $h(\infty) = \overline{A}/\overline{K}_C$ (offset). As $\overline{K}_C$ increases, the offset decreases. When $\overline{K}_C \gg 1.0$, then the system tends to $\hat{h} = 0$, which is the original steady state or the set point.

What if the Controller Is PI?

In this case, the controller transfer function is given by

$$G_C(s) = \left(\overline{K}_C + \frac{K_I}{s}\right)$$

Therefore,

$$\underbrace{X(s)}_{\hat{h}(s)} = \frac{\overbrace{1/As}^{G_d(s)}}{1 + \underbrace{(1/As)}_{G_P(s)}\underbrace{\left(\overline{K}_C + K_I/s\right)}_{G_C(s)}}\underbrace{D(s)}_{\hat{q}_i(s)} \qquad (5.53)$$

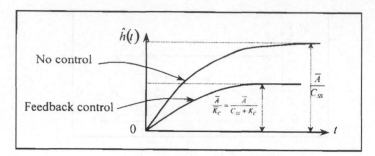

Figure 5.44 Effect of controller.

For a step input, $\tilde{q}_i(s) = \overline{A}/s$

Thus, from Eq. (5.53), we get

$$\tilde{h}(s) = \frac{1/As}{1 + (1/As)(\overline{K}_C + K_I/s)} \frac{\overline{A}}{s} \tag{5.54}$$

On simplification, Eq. (5.54) gives

$$\tilde{h}(s) = \frac{\overline{A}}{As^2} \left(\frac{As^2 + \overline{K}_C s + K_I}{As^2} \right)^{-1} = \frac{\overline{A}}{As^2 + \overline{K}_C s + K_I} \tag{5.55}$$

Using partial fractions for Eq. (5.55), we get

$$\tilde{h}(s) = \frac{\overline{A}}{(s - \lambda_1)(s - \lambda_2)} = \frac{a_1}{s - \lambda_1} + \frac{a_2}{s - \lambda_2} = \frac{a_1 s - a_1 \lambda_2 + a_2 s - a_2 \lambda_1}{(s - \lambda_1)(s - \lambda_2)}$$

On equating the coefficients, we get

$$a_1 + a_2 = 0$$

thus giving $a_1 = -a_2$, and

$$-a_1 \lambda_2 - a_2 \lambda_1 = \overline{A}$$

which gives $a_1 = \overline{A}/(\lambda_2 - \lambda_1)$. From the above two equations, we obtain

$$a_1 = \frac{\overline{A}}{\lambda_2 - \lambda_1} \quad \text{and} \quad a_2 = \frac{\overline{A}}{\lambda_1 - \lambda_2}$$

Now, to calculate λ_1 and λ_2,

$$\lambda_{1,2} = \frac{-\overline{K}_C \pm \sqrt{\overline{K}_C^2 - 4\overline{A}K_I}}{2\overline{A}}$$

So,

$$\lambda_1 - \lambda_2 = \frac{1}{2\overline{A}}\left(-\overline{K}_C + \sqrt{\overline{K}_C^2 - 4\overline{A}K_I} + \overline{K}_C + \sqrt{\overline{K}_C^2 - 4\overline{A}K_I}\right)$$

giving

$$\lambda_1 - \lambda_2 = \frac{1}{\overline{A}}\left(\sqrt{\overline{K}_C^2 - 4\overline{A}K_I}\right)$$

To have real values, choose $4\overline{A}K_I < \overline{K}_C^2$. Thus,

$$\lambda_1 - \lambda_2 = \frac{1}{\overline{A}}\left(\sqrt{\overline{K}_C^2 - 4\overline{A}K_I}\right) \quad \text{and} \quad \lambda_2 - \lambda_1 = -\frac{1}{\overline{A}}\left(\sqrt{\overline{K}_C^2 - 4\overline{A}K_I}\right)$$

On taking the inverse Laplace transformation of

$$\tilde{h}(s) = \frac{a_1}{(s - \lambda_1)} + \frac{a_2}{(s - \lambda_2)}$$

we get

$$\hat{h}(t) = a_1 e^{\lambda_1 t} + a_2 e^{\lambda_2 t}$$

Substitution of the values of a_1 and a_2 gives

$$\hat{h}(t) = \frac{\overline{A}}{\lambda_2 - \lambda_1} e^{\lambda_1 t} + \frac{\overline{A}}{\lambda_1 - \lambda_2} e^{\lambda_2 t}$$

which can be rearranged to give

$$\hat{h}(t) = \frac{\overline{A}}{\lambda_1 - \lambda_2}\left(e^{\lambda_2 t} - e^{\lambda_1 t}\right) \tag{5.56}$$

If both λ_1 and λ_2 are negative, then the closed-loop control system is stable, and as $t \to \infty$, the value of $\hat{h}(t) \to 0$, as can be observed from Eq. (5.56). What is the condition that λ_1 and λ_2 are negative? It is that $\overline{K}_C > 0$ and $4\overline{A}K_I > 0$ (and the condition of $4\overline{A}K_I < \overline{K}_C^2$ is also satisfied) which is, of course, always satisfied. Thus, it is clear that the integral control action always prevents the offset, since λ_1 and λ_2 are always negative.

Another Example: Feedback Control of a CSTR

Consider the first-order irreversible chemical reaction

$$A \to B$$

with rate of reaction equal to $r = kC_A$ taking place in the continuous-stirred tank reactor (CSTR) shown in Figure 5.45. The CSTR is assumed to be isothermal.

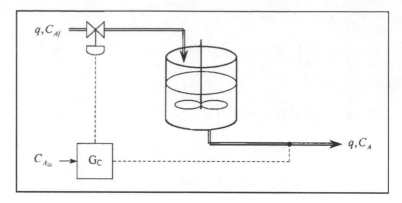

Figure 5.45   Continuous-stirred tank reactor.

Open Loop, Uncontrolled

We start by looking at the open-loop (uncontrolled) case. The first step is to develop the dynamic model. The dynamic mass balance equation is given by the relation

$$n_i + \frac{d\bar{n}_i}{dt} = n_{if} + V\sigma_i r$$

Considering a constant flow rate and constant volume, we get

$$n_i = qC_i, \quad n_{if} = qC_{if}, \quad \frac{d\bar{n}_i}{dt} = V\frac{dC_i}{dt}$$

Writing the balance for component A gives

$$qC_{Af} = VkC_A + qC_A + V\frac{dC_A}{dt}$$

The above equation can be rewritten as

$$\underbrace{\left(\frac{V}{q}\right)}_{\tau}^{\overbrace{}^{\substack{\text{residence}\\\text{time}}}} \frac{dC_A}{dt} = (C_{Af} - C_A) - \underbrace{\left(\frac{Vk}{q}\right)}_{\alpha} C_A$$

where

$$\tau = \frac{V}{q}$$

$$\alpha = \frac{Vk}{q}$$

In terms of normalized parameters, it can be written as

$$\tau\frac{dC_A}{dt} = (C_{Af} - C_A) - \alpha C_A \tag{5.57}$$

For the control problem the initial condition is at $t = 0$ and $C_A = C_{Ass}$. We obtain the steady state C_{Ass} from the steady-state equation

$$C_{Af} = (1 + \alpha)C_{Ass}$$

which can be rearranged to give

$$C_{Ass} = \frac{C_{Af}}{(1 + \alpha)} = \frac{C_{Af}}{(1 + Vk/q)} = \frac{qC_{Af}}{(q + Vk)}$$

Subtracting

$$V\frac{dC_A}{dt} = q(C_{Af} - C_A) - VkC_A$$

from

$$V\frac{dC_{Ass}}{dt} = q(C_{Afss} - C_{Ass}) - VkC_{Ass}$$

to get the unsteady-state equation in terms of deviation variables, we get the following equation in terms of deviation variables:

$$V\frac{d\hat{C}_A}{dt} = q(\hat{C}_{Af} - \hat{C}_A) - Vk\hat{C}_A \tag{5.58}$$

where

$$\hat{C}_A = C_A - C_{Ass}$$

$$\hat{C}_{Af} = C_{Af} - C_{Afss}$$

with initial condition at $t = 0$, $\hat{C}_A = 0$ and $\hat{C}_{Af} = 0$.
Laplace transformation of Eq. (5.58) gives

$$Vs\tilde{C}_A(s) = q\tilde{C}_{Af}(s) - (q + Vk)\tilde{C}_A(s)$$

The above equation can be rearranged as

$$\tilde{C}_A(s)(Vs + q + Vk) = q\tilde{C}_{Af}(s)$$

Finally, the relation between the output $\tilde{C}_A(s)$ and the input $\tilde{C}_{Af}(s)$ is given by

$$\tilde{C}_A(s) = \frac{q}{(Vs + q + Vk)} \tilde{C}_{Af}(s) \tag{5.59}$$

Equation (5.59) can be put in the form of a block diagram, as shown in Figure 5.46.

Now, if

$$\frac{q + Vk}{q} = \alpha'$$

we get

$$\tilde{C}_A(s) = \left(\underbrace{\frac{V}{q}}_{\tau} s + \underbrace{\frac{q + Vk}{q}}_{\alpha'} \right)^{-1} \tilde{C}_{Af}(s)$$

Thus, the above equation becomes

$$\tilde{C}_A(s) = \frac{1}{\tau s + \alpha'} \tilde{C}_{Af}(s) \tag{5.60}$$

Therefore, the input–output relation in the Laplace domain is given by

$$\tilde{C}_A(s) = \frac{1}{\alpha' + \tau s} \tilde{C}_{Af}(s) \tag{5.61}$$

Consider a step function in the feed concentration; thus

$$\tilde{C}_{Af}(s) = \frac{\overline{A}}{s}$$

where $\overline{A}$ is the size of the step. So, for this input, we get the following response (output) using Eq. (5.61):

$$\tilde{C}_A(s) = \frac{1}{\alpha' + \tau s} \frac{\overline{A}}{s}$$

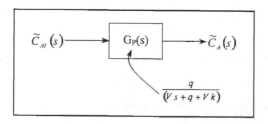

Figure 5.46 Block diagram for Eq. (5.59).

which, on rearrangement, gives

$$\tilde{C}_A(s) = \frac{\overline{A}}{\tau}\left[\frac{1}{(s + \alpha'/\tau)(s - 0)}\right]$$

Using partial fractions, we get

$$\frac{1}{\left(s + \underbrace{\alpha'/\tau}_{\gamma}\right)(s - 0)} = \frac{a_1}{s + \gamma} + \frac{a_2}{s} = \frac{a_1 s + a_2 s + a_2 \gamma}{(s + \gamma)s}$$

where

$$\gamma = \frac{\alpha'}{\tau}$$

On equating the coefficients, we get

$$a_2\gamma = 1$$

therefore,

$$a_2 = \frac{1}{\gamma}$$

and

$$a_1 + a_2 = 0$$

so it gives

$$a_1 = -\frac{1}{\gamma}$$

Thus, we get the response (still in the Laplace domain but rearranged for the ease of inversion)

$$\tilde{C}_A(s) = \frac{\overline{A}}{\tau}\left(\frac{-1/\gamma}{s - (-\gamma)} + \frac{1/\gamma}{s - 0}\right) \tag{5.62}$$

On taking the inverse Laplace transformation of Eq. (5.62), we get

$$\tilde{C}_A(t) = \frac{\overline{A}}{\alpha'}\left(1 - e^{-(\alpha'/\tau)t}\right) \tag{5.63}$$

Old Steady State (for $\hat{C}_{Af} = 0$) and New Steady State (for $\hat{C}_{Af} = \overline{A}$)

The old steady state $\left(\hat{C}_{A\text{ssold}}\right)$ is given for the following conditions:

$$\hat{C}_{Af} = 0 \quad \text{and} \quad \hat{C}_{Ass} = 0$$

The new steady state $\left(\hat{C}_{A_{SS}new}\right)$ (Fig. 5.47) can be obtained from the steady state equation

$$0 = q\left(\underbrace{\hat{C}_{Af} - \hat{C}_{A_{SS}new}}_{\overline{A}}\right) - Vk\hat{C}_{A_{SS}new}$$

which gives

$$(q + Vk)\hat{C}_{A_{SS}new} = \overline{A}q$$

thus giving the new steady state as

$$\hat{C}_{A_{SS}new} = \frac{\overline{A}q}{q + Vk}$$

Of course, the same can be obtained by setting $t \to \infty$ in Eq. (5.63).

What if we apply a feedback control?

- Deviations of C_A from $C_{A_{SS}}$ are used to manipulate q (assume that the flow rate response is constant, with no time delays, i.e., $q_{in} = q_{out} = q$ always and the volume V is also constant).
- Note that as the flow rate increases, C_A increases and vice versa. In other words, when we want to compensate for an increase in C_A, we should decrease q.

Thus, for proportional control,

$$q = q_{SS} + K\left(C_{A_{SS}} - C_A\right)$$

because when C_A is higher than $C_{A_{SS}}$, we want to decrease C_A, of course, and in order to decrease it, we should decrease q and vice versa. Thus, the dynamic equation with the feedback controller becomes,

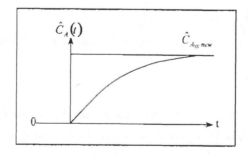

Figure 5.47   Attainment of new steady state.

$$\underbrace{\left[q_{ss} + K\left(C_{A_{ss}} - C_A\right)\right]}_{q_r} C_{Af} = VkC_A + V\frac{dC_A}{dt} + C_A \underbrace{\left[q_{ss} + K\left(C_{A_{ss}} - C_A\right)\right]}_{q_c}$$

(5.64)

Note the weak nonlinearity resulting from the control in the left-hand side as well as the last term of the equation on the right-hand side.

The Steady-State Equation

The open-loop steady state is obtained by putting $C_{A_{ss}} - C_A = 0$ in Eq. (5.64):

$$q_{ss}C_{Af_{ss}} = VkC_{A_{ss}} + V\frac{dC_{A_{ss}}}{dt} + q_{ss}C_{A_{ss}}$$

(5.65)

The closed-loop (controlled) steady-state equation is given by

$$q_{ss}C_{Af} + K\left(C_{A_{ss}} - C_A\right)C_{Af} = VkC_A + V\frac{dC_A}{dt} + q_{ss}C_A$$
$$+ K\left(C_{A_{ss}} - C_A\right)C_A$$

(5.66)

Subtract Eq. (5.65) from Eq. (5.66) and define the following deviation variables:

$$\hat{C}_A = C_A - C_{A_{ss}}$$

$$\hat{C}_{Af} = C_{Af} - C_{Af_{ss}}$$

Equation (5.66) in terms of deviation (or hat) variables becomes

$$q_{ss}\hat{C}_{Af} + K\underbrace{C_{Af}\hat{C}_A}_{a} = Vk\hat{C}_A + V\frac{d\hat{C}_A}{dt} + q_{ss}\hat{C}_A + K\underbrace{C_A\hat{C}_A}_{b}$$

with initial condition at $t = 0$, $\hat{C}_A = 0$ and $\hat{C}_{Af} = 0$. Note that we have two nonlinear terms (a and b). For linear analysis, we need to linearize them. For nonlinear analysis, we can solve numerically and we do not need to linearize them.

5.9 LINEARIZATION

For the term labeled a in the above equation,

$$C_{Af}\hat{C}_A = C_{Af}\left(C_{A_{ss}} - C_A\right) = \underbrace{C_{Af}C_{A_{ss}}}_{\text{linear in } C_{Af}} - \underbrace{C_{Af}C_A}_{\text{nonlinear}}$$

To linearize the nonlinear term (last term of the above relation), we use the Taylor's series expansion (and neglecting the higher-order terms), we get

$$C_{Af}C_A = C_{Af_{ss}}C_{A_{ss}} + \frac{\partial(C_{Af}C_A)}{\partial C_A}\bigg|_{ss}(C_{A_{ss}} - C_A) + \frac{\partial(C_{Af}C_A)}{\partial C_{Af}}\bigg|_{ss}(C_{Af_{ss}} - C_{Af})$$

Thus, the nonlinear term a is given in the following linearized form:

$$C_{Af}C_A = \underbrace{C_{Af_{ss}}C_{A_{ss}} + C_{Af_{ss}}\hat{C}_A + C_{A_{ss}}\hat{C}_{Af}}_{\text{linearized form of } C_{Af}C_A} \tag{5.67}$$

Therefore, using Eq. (5.67), we get the expression of $C_{Af}\hat{C}_A = a$ as follows,

$$C_{Af}\hat{C}_A = \underbrace{C_{Af}C_{A_{ss}}}_{\text{linear in } C_{Af}} - \underbrace{C_{Af}C_A}_{\text{non-linear}}$$

$$= \underbrace{C_{Af}C_{A_{ss}} - C_{Af_{ss}}C_{A_{ss}}}_{-\hat{C}_{Af}C_{A_{ss}}} - C_{Af_{ss}}\hat{C}_A - C_{A_{ss}}\hat{C}_{Af}$$

$$= -\hat{C}_{Af}C_{A_{ss}} - C_{Af_{ss}}\hat{C}_A - C_{A_{ss}}\hat{C}_{Af}$$

$$= -2\hat{C}_{Af}C_{A_{ss}} - C_{Af_{ss}}\hat{C}_A$$

Thus, the nonlinear term a has been linearized.

The reader is encouraged to do the same for the other nonlinear term (b) and analyze the resulting linear differential equation.

The reader should show the conditions for the stability of the system also show the effect of the proportional controller gain on stability, speed of response, and offset.

5.10   SECOND-ORDER SYSTEMS

Second-order models are usually used to empirically represent some processes (D.J. Cooper, Control Station for Windows—A Software for Process Control Analysis, Tuning and Training. *http://www.ControlStation.com*).

Consider a typical second-order system described by the following second-order differential equation:

$$a_2 \frac{d^2 y}{dt^2} + a_1 \frac{dy}{dt} + a_0 y = bf(t) \tag{5.68}$$

Here, y is the state variable, t is an independent variable (time), $f(t)$ is the forcing function (nonhomogeneous term), and a_0, a_1, a_2 and b are the parameters. Usually, the typical differential Eq. (5.68) for second-order systems is written in a different form:

$$\tau^2 \frac{d^2y}{dt^2} + 2\xi\tau \frac{dy}{dt} + y = K_P f(t) \tag{5.69}$$

where

$$\tau = \sqrt{\frac{a_2}{a_0}}$$

$$\xi = \frac{a_1}{2\sqrt{a_0 a_2}}$$

$$K_P = \frac{b}{a_0}$$

The physical significance of the above parameters are as follows: τ is the natural period of oscillation of the system [the autonomous system when $f(t) = 0$], ξ is damping factor, and K_P is static (steady state) gain of the system

If Eq. (5.69) is in terms of deviation variables (or the hat variables) and

$$\hat{y}(0) = 0, \quad \frac{d\hat{y}(0)}{dt} = 0, \quad \text{and} \quad \frac{d^2\hat{y}(0)}{dt^2} = 0$$

then the Laplace transformation of Eq. (5.69) gives

$$\tau^2 s^2 \tilde{y}(s) + 2\xi\tau s \tilde{y}(s) + \tilde{y}(s) = K_P \tilde{f}(s)$$

which can be arranged in the following input $\tilde{f}(s)$ and output $\tilde{y}(s)$ form:

$$\tilde{y}(s) = \underbrace{\left(\frac{K_P}{\tau^2 s^2 + 2\xi\tau s + 1}\right)}_{\text{transfer function} = G(S)} \tilde{f}(s) \tag{5.70}$$

The transfer function is given by

$$G(s) = \left(\frac{K_P}{\tau^2 s^2 + 2\xi\tau s + 1}\right) \tag{5.71}$$

Response to Unit Step Input

For a unit step input, $\tilde{f}(s)$ is given by

$$\tilde{f}(s) = \frac{1}{s}$$

So, using Eq. (5.70), we get the output $\tilde{y}(s)$ in the form

$$\tilde{y}(s) = \left(\frac{K_P}{\tau^2 s^2 + 2\xi\tau s + 1}\right) \frac{1}{s}$$

Using the partial fractions (or Heaviside theorem), we can obtain the inverse
Laplace transform.

Suppose the roots of $\tau^2 s^2 + 2\xi\tau s + 1$ are λ_1 and λ_2; we can write

$$\tilde{y}(s) = \frac{K_P}{(s-0)(s-\lambda_1)(s-\lambda_2)}$$

and

$$\lambda_1, \lambda_2 = -\frac{2\xi\tau \pm \sqrt{4\xi^2\tau^2 - 4\tau^2}}{2\tau^2}$$

which gives

$$\lambda_1 = -\frac{\xi}{\tau} + \frac{\sqrt{\xi^2 - 1}}{\tau} \quad \text{and} \quad \lambda_2 = -\frac{\xi}{\tau} - \frac{\sqrt{\xi^2 - 1}}{\tau}$$

It is clear that the character of λ_1 and λ_2 will depend on the value of ξ:

1. For $\xi > 1.0$, both λ_1 and λ_2 are real (the system is called over-
    damped).
2. For $\xi = 1.0$, λ_1 and λ_2 are equal to $-\xi/\tau$ (the system is called
    critically damped). *These are repeated roots, so the reader is
    advised to be careful while taking the inverse transform.*
3. For $\xi < 1.0$, λ_1 and λ_2 are complex in nature (the system is called
    underdamped or oscillatory).

5.10.1 Overdamped, Critically Damped, and Underdamped Responses

The reader is advised to do the complete analytical manipulation for the
three cases.

* $\xi > 1.0$, distinct real roots
* $\xi = 1.0$, repeated real roots
* $\xi < 1.0$, complex roots

The typical responses for the above-mentioned cases are shown in Figure
5.48.

5.10.2 Some Details Regarding the Underdamped Response

As an example for the reader, some analysis of the underdamped system is
given here (refer to Figure 5.49). Note the following:

* ω = radian frequency (radians per unit time)
* f = cyclic frequency (cycles per unit time)
* T = period of one cycle = $1/f$
* $\omega = 2\pi f = 2\pi/T$

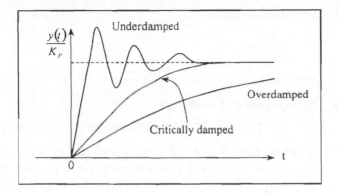

Figure 5.48 Typical overdamped, critically damped, and underdamped systems.

The main characteristics of the response are as follows:

1. **Overshoot**

$$\frac{A}{B} = \exp\left(-\frac{\pi\xi}{\sqrt{1-\xi^2}}\right)$$

2. **Decay ratio**

$$\frac{C}{A} = \exp\left(-\frac{2\pi\xi}{\sqrt{1-\xi^2}}\right) = (\text{overshoot})^2$$

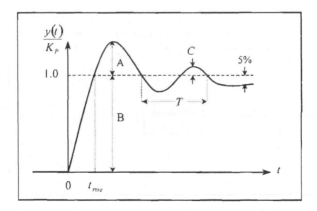

Figure 5.49 Typical underdamped system.

3. **Period of oscillation**

$$T = \frac{2\pi\tau}{\sqrt{1 - \xi^2}} = \frac{2\pi}{\omega},$$

where $\omega = \sqrt{1 - \xi^2}/\tau$

4. **Natural period of oscillation**

$$T_{natural} = T \quad \text{when } \xi = 0 \ (\text{ i.e., } T_{natural} = 2\pi\tau)$$

5. **Response time**: When the system approaches ±5% of its final value, it is considered that the system has reached practically the steady state and the time elapsed is the response time.

6. **Rise time**: It is the time when the response curve cuts the horizontal steady-state line for the first time.

5.11 COMPONENTS OF FEEDBACK CONTROL LOOPS

Figure 5.50 shows the typical components found in a control loop. It can be seen that there are four main components involved: process, measuring device, controller, and the final control element. The controller mechanism gets the value of the set point and directs the final control element to carry out the actions. Disturbances enter the system through the process, which affect the process variables.

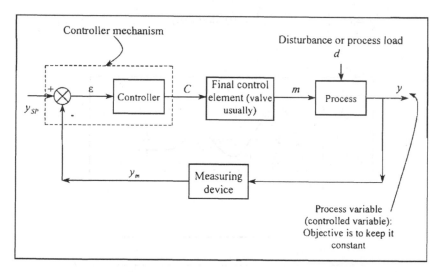

Figure 5.50 Typical components of a control loop.

Example 1: Pressure control

Figure 5.51 shows an example of a tank pressure control mechanism. It can be seen that the measured variable is the pressure inside the tank. The controller mechanism receives the pressure value and its set point and accordingly directs the control valve to either open or close to keep the pressure constant inside the tank.

Example 2: Temperature control

Figure 5.52 shows the control loop for a heat exchanger. The outlet temperature of the fluid to be heated is the measured variable and it is sent to the controller mechanism to be compared with the temperature set point. Based on the error, the controller takes action to either increase or decrease the steam flow rate into the heat exchanger.

5.12 BLOCK DIAGRAM ALGEBRA

5.12.1 Typical Feedback Control Loop and the Transfer Functions

In this subsection, we present the typical control loop (as shown in Fig. 5.53) with its transfer functions, inputs–outputs, and their relations. It is also a simple introduction to block diagram algebra.

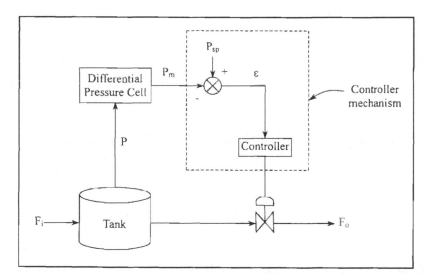

Figure 5.51 Typical tank pressure control loop.

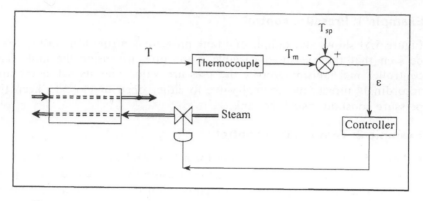

Figure 5.52 Typical temperature control loop for heat exchanger.

The process: The input variables [$\overline{m}(s)$ and $\overline{d}[s]$] enter the process and the transfer functions are related through the relation

$$\overline{y}(s) = G_P(s)\overline{m}(s) + G_d(s)\,\overline{d}(s)$$

The measuring device: The input, output, and transfer function of the measuring instrument are related through the relation

$$\overline{y}_m(s) = G_m(s)\overline{y}(s)$$

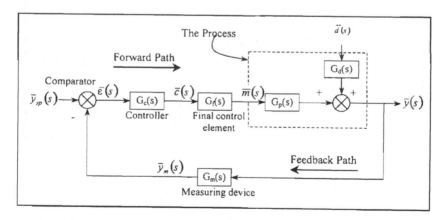

Figure 5.53 Typical control loop with transfer functions and input–output variables.

Controller mechanism: The controller is formed of the following two parts:

> *Comparator:* This part has the following relation producing the error between the measured variable and the set point:
>
> $$\bar{\varepsilon}(s) = \bar{y}_{sp}(s) - \bar{y}_m(s)$$
>
> *Controller:* This part produces the controller action $\bar{C}(s)$ from the error $\bar{\varepsilon}(s)$, where $G_c(s)$ is the transfer function of the controller depending on the mode of the controller, as shown earlier:
>
> $$\bar{C}(s) = G_c(s)\bar{\varepsilon}(s)$$

Final control element: The relation for the final control element is given by (usually the final control element is a valve)

$$\bar{m}(s) = G_f(s)\bar{C}(s)$$

5.12.2 Algebraic Manipulation of the Loop Transfer Functions

Backward substitution into the process equation gives

$$\bar{m}(s) = G_f(s)G_c(s)\bar{\varepsilon}(s)$$

Substitution of the $\bar{\varepsilon}(s)$ value gives

$$\bar{m}(s) = G_f(s)G_c(s)\{\bar{y}_{sp}(s) - \bar{y}_m(s)\}$$

Substitution of $\bar{y}_m(s)$ gives

$$\bar{m}(s) = G_f(s)G_c(s)\lfloor\bar{y}_{sp}(s) - G_m(s)\bar{y}(s)\rfloor$$

Substituting into the process equation gives the relation

$$\bar{y}(s) = G_P(s)\{G_f(s)G_c(s)[\bar{y}_{sp}(s) - G_m(s)\bar{y}(s)]\} + G_d(s)\,\bar{d}(s)$$

which is further expanded to give

$$\bar{y}(s) = [G_P(s)G_f(s)G_c(s)]\bar{y}_{sp}(s) - [G_P(s)G_f(s)G_c(s)G_m(s)]\bar{y}(s) + G_d(s)\,\bar{d}(s)$$

Rearrangement of the above equation gives

$$\bar{y}(s)[1 + G_P(s)G_f(s)G_c(s)G_m(s)] = \underbrace{[G_P(s)G_f(s)G_c(s)]}_{G(s)}\bar{y}_{sp}(s) + G_d(s)\,\bar{d}(s)$$

$$(5.72)$$

The forward path overall transfer function is equal to

$$G(s) = G_P(s)G_f(s)G_c(s)$$

Thus, Eq. (5.72) can be written as

$$\bar{y}(s)[1 + G(s)G_m(s)] = G(s)\bar{y}_{sp}(s) + G_d(s)\,\bar{d}(s)$$

which can be rearranged to give

$$\bar{y}(s) = \underbrace{\left(\frac{G(s)}{1 + G(s)G_m(s)}\right)}_{G_{sp}(s)}\bar{y}_{sp}(s) + \underbrace{\left(\frac{G_d(s)}{1 + G(s)G_m(s)}\right)}_{G_{load}(s)}\bar{d}(s)$$

Now, we define

$$G_{sp}(s) = \frac{G(s)}{1 + G(s)G_m(s)} \quad \text{and} \quad G_{load}(s) = \frac{G_d(s)}{1 + G(s)G_m(s)}$$

Thus, we get

$$\bar{y}(s) = G_{sp}(s)\bar{y}_{sp}(s) + G_{load}(s)\,\bar{d}(s) \tag{5.73}$$

Equation (5.73) can be illustrated as a block diagram, as shown in Figure 5.54.

Two types of closed-loop controls use the following:

1. The disturbance does not change, it is the set point which changes $[\bar{d}(s) = 0]$. This is called the "servo problem" and the process equation becomes

$$\bar{y}(s) = G_{sp}(s)\bar{y}_{sp}(s)$$

2. The set point is constant $[\bar{y}_{sp}(s) = 0]$, but the disturbance changes. This is called the "regulatory problem" and the process equation becomes

$$\bar{y}(s) = G_{load}(s)\,\bar{d}(s)$$

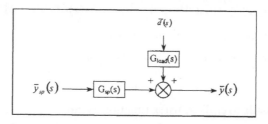

Figure 5.54   Block diagram for Eq. (5.73).

Example: Control of the Liquid Level in a Tank

In this example, the liquid level inside a tank is controlled using a DPC (differential pressure cell). The DPC sends out the measured variable to the comparator, where it is compared with the set point. Based on the error, the control action is transmitted to the final control element (a control valve) in order to either increase or decrease the outflow rate of liquid from the tank, as shown in Figure 5.55.

5.12.3 Block Diagram and Transfer Functions

These are related to the model equations shown earlier and are repeated again here (see Fig. 5.56).

The transfer functions in the process part of the loop (G_p and G_d): The dynamic equation can be simply written as

$$A\frac{dh}{dt} = q_i - q_0 \tag{5.74}$$

In terms of deviation variables

$$A\frac{d\hat{h}}{dt} = \hat{q}_i - \hat{q}_0 \tag{5.75}$$

with initial condition, at $t = 0$, $\hat{h} = 0$, $\hat{q}_i = 0$, and $\hat{q}_0 = 0$. Taking the Laplace transformation of Eq. (5.75) yields

$$As\,\tilde{h}(s) = \tilde{q}_i(s) - \tilde{q}_0(s)$$

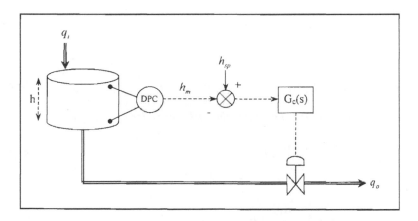

Figure 5.55 Liquid-level control in a tank.

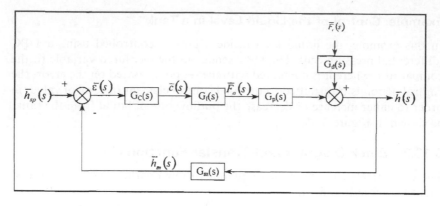

Figure 5.56 Block diagrams and transfer functions.

and rearrangement gives

$$\tilde{h}(s) = \frac{1}{As}\tilde{q}_i(s) - \frac{1}{As}\tilde{q}_0(s) \qquad (5.76)$$

The transfer function for the measuring instrument $[G_m(s)]$: The measuring device is a DPC. It will be described in most cases by a second-order system as follows:

$$\tau^2\frac{d^2h_m}{dt^2} + 2\xi\tau\frac{dh_m}{dt} + h_m = K_P\Delta P = K_P\alpha h \qquad (5.77)$$

After Laplace transformation, Eq. (5.77) can be put in the following form:

$$\frac{\tilde{h}_m(s)}{\tilde{h}(s)} = \frac{\alpha K_P}{\tau^2 s^2 + 2\xi\tau s + 1} = G_m(s) \qquad (5.78)$$

Comparator:

$$\tilde{\varepsilon}(s) = \tilde{h}_m(s) - \tilde{h}_{sp}(s) \qquad (5.79)$$

The transfer function for the controller (PI controller):

$$\frac{\tilde{C}(s)}{\tilde{\varepsilon}(s)} = K_C\left(1 + \frac{1}{\tau_I s}\right) = G_C(s) \qquad (5.80)$$

where

$$\frac{\tau_I}{K_C} = \frac{1}{K_I} \quad \text{gives} \quad \frac{K_C}{\tau_I} = K_I$$

Control valve (final control element) for a first-order system

$$\frac{\tilde{q}_0(s)}{\bar{C}(s)} = \frac{K_V}{1 + \tau_f s} = G_f(s) \tag{5.81}$$

Algebraic manipulations: Equation (5.76) can be rewritten as

$$\tilde{h}(s) = \underbrace{\frac{1}{As}}_{G_d(s)} \underbrace{\tilde{q}_i(s)}_{\bar{d}_i(s)} - \underbrace{\frac{1}{As}}_{G_p(s)} \tilde{q}_0(s)$$

Making use of relations (5.78)–(5.81), we get

$$\tilde{h}(s) = G_d(s)\tilde{d}_i(s) - G_p(s)G_f(s)\bar{C}(s)$$
$$= G_d(s)\tilde{d}_i(s) - G_p(s)G_f(s)G_c(s)\tilde{\varepsilon}(s)$$
$$= G_d(s)\tilde{d}_i(s) - G_p(s)G_f(s)G_c(s)[\tilde{h}_m(s) - \tilde{h}_{sp}(s)]$$
$$= G_d(s)\tilde{d}_i(s) - G_p(s)G_f(s)G_c(s)G_m(s)\tilde{h}(s) + G_p(s)G_f(s)G_c(s)\tilde{h}_{sp}(s)$$

Rearrangement gives

$$[1 + G_p(s)G_f(s)G_c(s)G_m(s)]\tilde{h}(s) = \underbrace{G_p(s)G_f(s)G_c(s)}_{G(s)} \tilde{h}_{sp}(s) + G_d(s)\tilde{d}_i(s)$$

Set

$$G(s) = G_p(s)G_f(s)G_c(s)$$

to get

$$[1 + G(s)G_m(s)]\tilde{h}(s) = G(s)\tilde{h}_{sp}(s) + G_d(s)\tilde{d}_i(s)$$

Rearranging gives

$$\tilde{h}(s) = \underbrace{\frac{G(s)}{1 + G(s)G_m(s)}}_{G_{sp}(s)} \tilde{h}_{sp}(s) + \underbrace{\frac{G_d(s)}{1 + G(s)G_m(s)}}_{G_{load}(s)} \tilde{d}_i(s) \tag{5.82}$$

In shorter notation

$$\tilde{h}(s) = G_{sp}(s)\tilde{h}_{sp}(s) + G_{load}(s)\tilde{d}_i(s) \tag{5.83}$$

The above relation can be drawn as shown in Figure 5.57.

Servo problem:

$$\tilde{d}_i(s) = 0$$

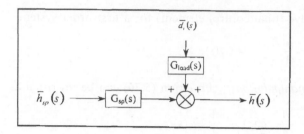

Figure 5.57 Block diagrams and transfer functions for Eq. (5.83).

Regulatory control:

$$\tilde{y}_{\text{sp}}(s) = 0$$

For the regulatory problem

$$\frac{\tilde{h}(s)}{\tilde{d}_i(s)} = \frac{G_d(s)}{1 + G(s)G_m(s)}$$

Substitution of the values of $G_d(s)$, $G(s)$, and $G_m(s)$ gives

$$\frac{\tilde{h}(s)}{\tilde{d}_i(s)} = \left(\frac{1}{As}\right)\left[1 + \frac{1}{As}K_C\left(1 + \frac{1}{\tau_I s}\right)\left(\frac{K_V}{1 + \tau_V s}\right)\left(\frac{\alpha K_p}{\tau_p^2 s^2 + 2\xi\tau_p s + 1}\right)\right]^{-1}$$

$$(5.84)$$

5.13 SOME TECHNIQUES FOR CHOOSING THE CONTROLLER SETTINGS

There are a wide range of techniques for finding the optimal settings of controllers. With the advent of digital control, most of these techniques are now obsolete; however, some of them are still used as guidelines. We give a very brief idea regarding these techniques in this section.

5.13.1 Choosing the Controller Settings

Many criteria can be used to find the optimal controller settings. Some of them are based on the time response and others on the frequency response.

We will give an example of the time response criteria. The different performance criteria that can be used include the following:

1. Keep the maximum deviation (error) as small as possible.
2. Achieve short settling times.

3. Minimize the integral of the errors until the process settles down to the desired set point.
4. And others depending on the process and its requirements.

How does one achieve that? It may be through the following:

1. Certain limit on overshoot.
2. Criterion for rise time and/or settling time.
3. Choosing an optimal value for the decay ratio.
4. Some characteristics for the frequency response.
5. Other techniques not covered here.

Sometimes, a combination of a number of different criteria is used.

Figure 5.58 shows the response of system to a unit step change in load for different controllers, namely proportional (P), integral (I) and derivative (D).

Example

We will illustrate one of the techniques using the tank problem. For the tank problem explained earlier, take $G_m = G_f = 1.0$ and look at the servo problem $[\bar{d}_i(s) = 0]$; we substitute the values of $G(s)$ and $G_m(s)$ to get

$$\bar{y}(s) = \frac{\tau_I s + 1}{\tau^2 s^2 + 2\xi\tau s + 1} \bar{y}_{sp}(s) \tag{5.85}$$

where

$$\tau = \sqrt{\frac{\tau_I \tau_p}{K_P K_C}} \quad \text{and} \quad \xi = \frac{1}{2}\sqrt{\frac{\tau_I}{\tau_p K_P K_C}} (1 + K_P K_C)$$

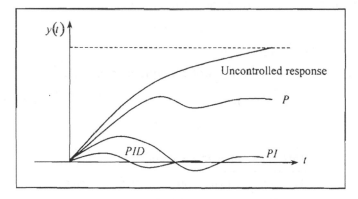

Figure 5.58 Response to unit step change in load.

A step response of $\bar{y}(s)$ for a unit step input in $\bar{y}_{sp}(s)$ gives that the decay ratio is equal to

$$\text{Decay ratio} = \frac{C}{A} = \exp\left(-\frac{2\pi\xi}{\sqrt{1-\xi^2}}\right)$$

On substituting the values, we get

$$\frac{C}{A} = \exp\left(\frac{-2\pi\frac{1}{2}\sqrt{\tau_I/\tau_p K_P K_C(1+K_P K_C)}}{\sqrt{1-\frac{1}{4}(\tau_I/\tau_p K_P K_C)(1+K_P K_C)^2}}\right) \tag{5.86}$$

One of the most widely used criterion is that

$$\frac{C}{A} = \frac{1}{4} \tag{5.87}$$

Using criterion (5.87) together with Eq. (5.86) and after some manipulations, we get

$$-2\pi\left(\sqrt{\frac{\tau_I}{4\tau_p K_P K_C - \tau(1+K_P K_C)^2}}\right)(1+K_P K_C) = \ln\left(\frac{1}{4}\right) \tag{5.88}$$

For a given process, generally τ_p and K_P are known (e.g., take $\tau_p = 10$ and $K_P = 0.1$). Now, we have an equation in τ_I and K_C, and for each value of K_C, we will get a corresponding value of τ_I, as shown in Table 5.2. The final choice depends upon good chemical engineering understanding of the process.

5.13.2 Criteria for Choosing the Controller Settings from the Time Response

Mainly time integral performance criteria are used:

$$\varepsilon(t) = y_{sp}(t) - y(t) \quad \text{or} \quad y(t) - y_{sp}(t)$$

The following are typical examples:

1. Minimize the integral of the square error (ISE)

$$ISE = \int_0^\infty \varepsilon^2(t)\,dt$$

Table 5.2 Different Values of τ_I and K_C

K_C	1.0	10.0	30.0	50.0	100.0
τ_I	0.127	0.12	0.022	0.007	0.001

2. Minimize the integral of the absolute value of the error (IEA)

$$IEA = \int_0^\infty |\varepsilon(t)| \, dt$$

3. Minimize the integral of the time-weighted absolute error (ITAE)

$$ITAE = \int_0^\infty t |\varepsilon(t)| \, dt$$

Which one of the above criterion has to be chosen? Again, it depends on the process characteristics.

5.13.3 Cohen and Coon Process Reaction Curve Method

This is one of popular techniques for choosing controller settings. From Figure 5.59,

$$\frac{\tilde{y}_m(s)}{\tilde{C}(s)} = \underbrace{G_f G_p G_m}_{G_{PRC}}$$

where

$$G_{PRC} = G_f G_p G_m$$

Thus

$$\tilde{y}_m(s) = G_{PRC}(s)\tilde{C}(s)$$

For a step input in $\tilde{C}(s) = A/s$ (where A is the amplitude of the step input), we get a response like the one shown in Figure 5.60.

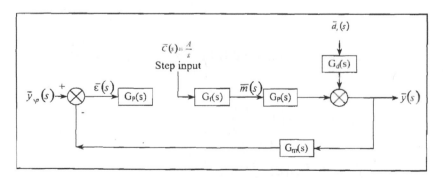

Figure 5.59 Block diagram.

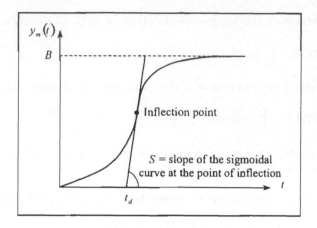

Figure 5.60   Response to a step input.

$G_{PRC}(s)$ can be approximated by

$$G_{PRC}(s) = \frac{\tilde{y}_m(s)}{\tilde{C}(s)} \cong \frac{Ke^{-t_d s}}{1 + \tau s}$$

where

$$K = \frac{B}{A} \quad \text{and} \quad \tau = \frac{B}{S}$$

Choosing the Controller Setting According to COHEN and COON

1. For proportional (P),

$$G_{PRC}(s) = K_C = \frac{1}{K} \frac{\tau}{t_d} \left(1 + \frac{t_d}{3\tau}\right)$$

2. For proportional-integral (PI),

    where,

$$G_{PRC}(s) = K_C \left(1 + \frac{1}{\tau_I s}\right)$$

$$K_C = \frac{1}{K} \frac{\tau}{t_d} \left(0.9 + \frac{t_d}{12\tau}\right)$$

And,

$$\tau_I = t_d \left(\frac{30 + 3t_d/\tau}{9 + 20t_d/\tau} \right)$$

3. For proportional–integral–differential (*PID*),

$$G_{PRC}(s) = K_C \left(1 + \frac{1}{\tau_I s} + \tau_D s \right)$$

and we get,

$$K_C = \frac{1}{K} \frac{\tau}{t_d} \left(\frac{4}{3} + \frac{t_d}{4\tau} \right)$$

where,

$$\tau_I = t_d \left(\frac{32 + 6t_d/\tau}{13 + 8t_d/\tau} \right) \quad \text{for } G_c(s) = K_C + \frac{K_I}{s} + K_D s$$

and

$$\tau_D = t_d \left(\frac{4}{11 + 2t_d/\tau} \right)$$

This empirical technique can also be used without development of a model, through obtaining the response curve experimentally.

SOLVED EXAMPLES

Solved Example 5.1

Figure 5.61 shows an underground tank used for storing gasoline for sale to the public. Recently, engineers believe that a leak has developed, which

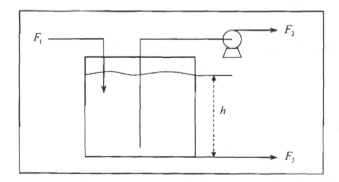

Figure 5.61 Schematic diagram for Solved Example 5.1.

threatens the environment. Your supervisor has assigned you the task of modeling the height in the tank as a function of the supply flow F_1, the sales flow F_2, and the unmeasured leakage F_3. It is assumed that this leakage is proportional to the height of liquid in the tank (i.e., $F_3 = \beta h$). The tank is a cylinder with constant cross-sectional area A. The density of gasoline is ρ and assumed to be constant. Find the transfer-function model for the level in the tank as a function of the supply flow and the sales flow. Which of these input variables would you describe as a control variable and which you describe as a disturbance variable?

Solution

- We have to get the relation between h, F_1, and F_2 [i.e., $h = f(F_1, F_2)$].
- Unmeasured leakage $F_3 = \beta h$; usually, it is $F_3 = \beta \sqrt{h}$. However, to keep equation linear we use this linear relation.
- Density and area of the cross section of the tank are constant.

We write the generalized mass balance as

$$n_i + \frac{d\bar{n}_i}{dt} = n_{if} + \underbrace{\text{Reaction}}_{\text{zero}}$$

Thus, we get

$$n_2 + n_3 + \frac{d\bar{n}}{dt} = n_1$$

Because the density and molecular weight of gasoline coming in and going out are same, we get

$$F_2 + F_3 + \frac{dV}{dt} = F_1$$

and we can write

$$V = Ah$$

where $A = $ constant. On differentiation, we get

$$\frac{dV}{dt} = A\frac{dh}{dt}$$

Also,

$$F_3 = \beta h$$

Thus, we get

$$A\frac{dh}{dt} = F_1 - F_2 - \beta h$$

with initial condition at $t = 0$, $h = h_0$. The disturbance variable will be F_1 and the control variable is h.

Note 1: You can have the control variable as F_2 also, because you can have control over the pumping out of gasoline. Also, if you have a measurement of h, you can use F_2 as the manipulated variable to keep h close to the value you want.

Note 2: If we fix the leakage, then $F_3 = 0$, and the model equation becomes

$$A \frac{dh}{dt} = F_1 - F_2$$

with initial condition, at $t = 0$, $h = h_0$.

The steady state of the tank is when the differential term is equal to zero; hence, we get

$$F_{1SS} - F_{2SS} - \beta h_{SS} = 0$$

Let us define the following deviation variables:

$$\hat{h} = h - h_{SS}$$
$$\hat{F}_1 = F_1 - F_{1SS}$$
$$\hat{F}_2 = F_2 - F_{2SS}$$

Thus, the unsteady-state equation can be written in terms of deviation variables as follows:

$$A \frac{d\hat{h}}{dt} = \hat{F}_1 - \hat{F}_2 - \beta \hat{h}$$

with initial condition at $t = 0$, $\hat{h} = 0$.

Taking the Laplace transforms of the above equation and rearrangement gives the required transfer-function model:

$$\tilde{h}(s) = \frac{\tilde{F}_1(s)}{As + \beta} - \frac{\tilde{F}_2(s)}{As + \beta}$$

Or we can write

$$\tilde{h}(s) = \frac{(1/\beta)\tilde{F}_1(s)}{(A/\beta)s + 1} - \frac{(1/\beta)\tilde{F}_2(s)}{(A/\beta)s + 1}$$

Solved Example 5.2

(a) Figure 5.62 shows a system for heating a continuous-flow water kettle using a hot plate. Assuming that the hot-plate temperature

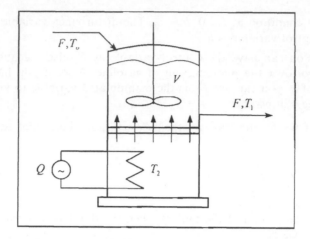

Figure 5.62   Schematic diagram for Solved Example 5.2.

can be changed instantaneously by adjusting the hot-plate rate of heat input and assuming uniform water kettle temperature, show that the following is a reasonable model for the process:

$$\rho V C_p \frac{dT_1}{dt} = c(T_2 - T_1) + F\rho C_p T_0 - F\rho C_p T_1$$

where ρ is the density of water (a constant value), C_p is the heat capacity of water (a constant value), c is a constant, and F the volumetric flow rate in and out of the kettle.

(b) If the heat capacity of the hot-plate material is assumed uniform and given as C_{p2} and its mass and effective lateral area for heat transfer to the atmosphere are given as m and A_c, respectively, obtain a second modeling equation that adequately describes the dynamics of T_2, the hot-plate temperature, in response to changes in Q, the rate of heat input. It may be assumed that the atmospheric temperature is T_a, a constant; the heat transfer coefficient may also be taken as a constant value equal to h.

Solution

Part a.   The assumptions are as follows:

- ρ and C_p are constant over the operating range.
- $T_{\text{water}} = T_1$ (ideal mixing inside the kettle).

- The instantaneous change in hot-plate temperature and the energy input from plate to kettle is directly proportional to $T_2 - T_1$.
- Heat losses to the atmosphere are negligible.
- The heat capacity of heating plate is neglected.

The generalized heat balance equation is

$$\sum n_{if}(H_{if} - H_{ir}) + Q = \sum n_i(H_i - H_{ir}) + \underbrace{0}_{\text{no reaction}} + \sum \frac{d(\bar{n}_i \overline{H}_i)}{dt}$$

Because we have only one component (water), the above equation becomes

$$n_f(H_f - H_r) + Q = n(H - H_{ir}) + \frac{d(\bar{n}\overline{H})}{dt}$$

Now, if there is no change in phase

$$H_f - H_r = \int_{T_r}^{T_0} C_{pM} \, dT$$

Note: Here, C_{pM} is the molar heat capacity, (i.e., the heat capacity per mole of water).

As it has been assumed that the value of C_p is constant within the operating range, we get

$$H_f - H_r = C_{pM}(T_0 - T_r)$$

Similarly,

$$H - H_r = C_{pM}(T_1 - T_r)$$

The heater duty (the heating plate rate of heat supply, where the heating plate is at temperature T_2) is given by

$$Q = c(T_2 - T_1)$$

The last term on the right-hand side of the heat balance equation can be changed in terms of temperature as follows:

$$\frac{d(\bar{n}\overline{H})}{dt} = \bar{n}\frac{d\overline{H}}{dt} = \bar{n}\frac{d(\overline{H} - H_r)}{dt}$$

and

$$(\overline{H} - H_r) = \int_{T_r}^{T_1} C_{pM} \, dT = C_{pM}(T_1 - T_r)$$

Thus, we get

$$\bar{n}\frac{d\left(\overline{H}-H_r\right)}{dt}=C_{\mathrm{pM}}\frac{d(T_1-T_r)}{dt}=C_{\mathrm{pM}}\frac{dT_1}{dt}$$

Thus, our heat balance equation becomes

$$n_fC_{\mathrm{pM}}(T_0-T_r)+c(T_2-T_1)=nC_{\mathrm{pM}}(T_1-T_r)+\bar{n}C_{\mathrm{pM}}\frac{dT_1}{dt}$$

Because there is no chemical reaction,

$$n=n_f$$

Thus, we get

$$nC_{\mathrm{pM}}(T_0-T_r)+c(T_2-T_1)=nC_{\mathrm{pM}}(T_1-T_r)+\bar{n}C_{\mathrm{pM}}\frac{dT_1}{dt}$$

On rearrangement, we get

$$nC_{\mathrm{pM}}(T_0-T_1)+c(T_2-T_1)=\bar{n}C_{\mathrm{pM}}\frac{dT_1}{dt}$$

Let the mass flow rate of water be m^*; then,

$$nC_{\mathrm{pM}}=m^*C_p$$

Similarly, for the contents (if the mass content of the tank is $\bar{m}$), we can write

$$\bar{n}C_{\mathrm{pM}}=\bar{m}C_p\ (C_p\text{ is per unit mass})$$

Thus, the heat balance equation becomes

$$m^*C_p(T_0-T_1)+c(T_2-T_1)=\bar{m}C_p\frac{dT_1}{dt}$$

Obviously

$$m^*=F\rho\quad\text{and}\quad\bar{m}=V\rho$$

where F is the volumetric flow rate, ρ is the density of water, and V is the volume of the tank contents.

Then, the heat balance equation becomes

$$F\rho C_p(T_0-T_1)+c(T_2-T_1)=V\rho C_p\frac{dT_1}{dt}$$

which on rearrangement, gives

$$V\rho C_p\frac{dT_1}{dt}=c(T_2-T_1)+F\rho C_pT_0-F\rho C_pT_1$$

with initial condition, at $t=0$, $T_1=T_1(0)$.

Part b. Take into consideration the heating capacity of the heating plate. There is no flow of material; hence,

$$\sum n_{if} H_{if} = \sum n_i H_i = 0$$

and we get

$$\breve{Q} = Q - c(T_2 - T_1) - A_c h(T_2 - T_a)$$

where Q is the heat coming into the plate, $c(T_2 - T_1)$ is the heat lost to the water, and $A_c h(T_2 - T_a)$ is the heat lost to the surrounding air, We then have

$$\breve{Q}\frac{d(\breve{n}\breve{H})}{dt} = mC_{p2}\frac{dT_2}{dt} \quad \text{(as shown in part a)}$$

Thus, the equation for the plate becomes

$$mC_{p2}\frac{dT_2}{dt} = Q - c(T_2 - T_1) - A_c h(T_2 - T_a)$$

with initial condition at $t = 0$, $T_2 - T_{2(0)}$.

Solved Example 5.3

The water heater as shown in Figure 5.63 is a well-mixed, constant-volume (V_l) tank through which fluid flows at a constant mass flow rate w (or $\rho_l F$ whereas F is the volumetric flow rate and ρ_l is the liquid density); the specific heat capacity of the fluid is C_{pl}. Because the incoming fluid temperature T_i is subject to fluctuations, an electric coil to which a simple proportional con-

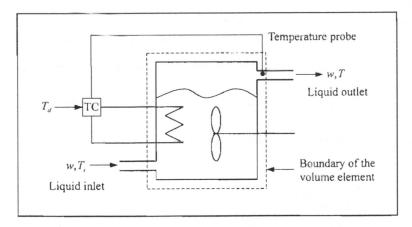

Figure 5.63 Schematic representation of the Solved Example 5.3.

troller is used to regulate the temperature of the liquid in the heater. To facilitate the design of this controller, it is desired to obtain a theoretical process model.

The electrical coil is made of a metal whose specific heat capacity is C_p, with a surface area A and mass m_c; the overall heat transfer coefficient is given as U. The temperature controller is provided with information about the tank temperature T, through a thermocouple probe; it is designed to supply energy to the coil at a rate of

$$Q_c = K_c(T_d - T)$$

where K_c is a predetermined constant and T_d is the desired tank temperature.

Develop a theoretical model for the water heater. You may wish to make use of the following assumptions:

1. The agitation is assumed perfect, so that the temperature within the heater may be considered uniform.
2. The coil temperature T_c is also assumed uniform at any instant, but different from the boiler liquid temperature T.
3. The physical properties of all the components of the process are assumed constant.
4. There are no heat losses to the atmosphere.

Cast your model in terms of deviation variables, and present in the state-space form.

Solution

A reasonable theoretical model for the water heater based on the given assumptions is obtained from the heat balances for the tank and for the coil.

The overall tank heat balance (using similar principles as in Example 5.2) is

$$V_l \rho_l C_{pl} \frac{dT}{dt} = UA(T_c - T) + w C_{pl}(T_i - T^*) - w C_{pl}(T - T^*)$$

where T^* is some reference temperature. The above equation simplifies to give

$$V_l \rho_l C_{pl} \frac{dT}{dt} = UA(T_c - T) + w C_{pl}(T_i - T)$$

with initial condition at $t = 0$, $T = T(0)$.

The overall heat balance for the coil is as follows. Using principles similar to those used in Example 5.2, we get

$$m_c C_p \frac{dT_c}{dt} = K_c(T_d - T) - UA(T_c - T)$$

with initial condition at $t = 0$, $T_c = T_c(0)$. The two differential equations are coupled and must be solved simultaneously.

Let us define the following deviation (or hat variables):

$$\hat{T} = T - T_{SS}$$

$$\hat{T}_c = T_c - T_{cSS}$$

$$\hat{T}_i = T_i - T_{iSS}$$

$$\hat{T}_d = T_d - T_{dSS}$$

Also, let us define the following parameters:

$$\alpha_1 = \frac{UA}{\rho_l V_l C_{pl}}$$

$$\alpha_2 = \frac{w}{\rho_l V_l}$$

$$\beta_1 = \frac{UA}{m_c C_{pc}}$$

$$\beta_2 = \frac{K_c}{m_c C_{pc}}$$

Then, using the above-defined deviation variables and parameters, the differential equations for the tank and coil become

$$\frac{d\hat{T}}{dt} = -(\alpha_1 + \alpha_2)\hat{T} + \alpha_1 \hat{T}_c + \alpha_2 \hat{T}_i$$

$$\frac{d\hat{T}_c}{dt} = (\beta_1 - \beta_2)\hat{T} - \beta_1 \hat{T}_c + \beta_2 \hat{T}_d$$

with initial conditions at $t = 0$, $\hat{T} = 0$ and $\hat{T}_c = 0$. Because the tank temperature is the sole measured variable, we have the complete water heater model in terms of deviation variables in state-space form.

Solved Example 5.4

The process shown in Figure 5.64 is a continuous stirred mixing tank used to produce F_B L/min of brine solution of mass concentration C_B g/liter. The raw materials are fresh water, supplied at a flow rate of F_w L/min, and a highly concentrated brine solution (mass concentration C_{Bf} g/L) supplied at a flow rate of F_{Bf} L/min. The volume of the material in the tank is V L, the liquid level in the tank is h m, and the tank's cross-sectional area $A(\text{m}^2)$ is assumed to be constant.

  (a) Assuming that the tank is well mixed, so that the brine concentration in the tank is uniformly equal to C_B, and that the flow rate out of the tank is proportional to the square root of the liquid level, obtain a mathematical model for this process.
  (b) If the process is redesigned to operate at constant volume, obtain a new process model and compare it with the one obtained in part (a).

Solution

Part a. An appropriate mathematical model may be obtained for this process by carrying out material balances on salt and on water as follows.

The salt material balance is

$$\frac{d}{dt}(VC_B) = F_{Bf}C_{Bf} - F_BC_B$$

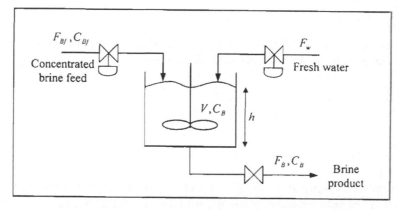

Figure 5.64 Schematic diagram for Solved Example 5.4

and because,

$$V = Ah, \quad A = \text{constant}$$
$$F_B = k\sqrt{h}$$

we get

$$A\frac{d}{dt}(hC_B) = F_{Bf}C_{Bf} - k\sqrt{h}C_B$$

The above differential equation can be rewritten as

$$A\left(C_B\frac{dh}{dt} + h\frac{dC_B}{dt}\right) = F_{Bf}C_{Bf} - C_B k\sqrt{h}$$

The water material balance is obtained as follows. Assume that mixing ξ g of salt with ϕ L of water does not alter appreciably the volume of the resulting mixture. This implies that F_{Bf} L of brine contains F_{Bf} L of water:

$$\frac{d}{dt}(\rho_W V) = \rho_W(F_{Bf} + F_w) - \rho_W F_B$$

where ρ_W is the constant density of water. Thus, we get

$$A\frac{dh}{dt} = F_{Bf} + F_w - k\sqrt{h} \tag{5.89}$$

Using the above differential equation to substitute the expression for dh/dt in the salt material balance, we get

$$A\left[C_B\left(\frac{F_{Bf} + F_w - k\sqrt{h}}{A}\right) + h\frac{dC_B}{dt}\right] = F_{Bf}C_{Bf} - C_B k\sqrt{h}$$

On rearrangement, we get

$$\frac{dC_B}{dt} = \frac{1}{Ah}\left[-C_B(F_{Bf} + F_w) + F_{Bf}C_{Bf}\right] \tag{5.90}$$

with initial conditions at $t = 0$, $C_B = C_B(0)$ and $h = h(0)$ $\hspace{1cm}$ (5.91)

Thus, Eq. (5.89) along with Eq. (5.90) and initial conditions (5.91) give the mathematical model of this system.

Notes about this model:

- It is a set of two coupled nonlinear ODEs, even though the coupling is "oneway," C_B is affected by changed in h, but h does not depend on C_B.

- At steady state, Eq. (5.90) indicates that

$$C_{B_{ss}} = \frac{F_{Bf}C_{Bf}}{F_{Bf} + F_w}$$

which is consistent with well-known mixing rules.

Part b. By redesigning the process to operate at constant volume, h becomes constant and the process model simplifies to only one equation:

$$\frac{dC_B}{dt} = \alpha\left[-C_B\left(F_{Bf} + F_w\right) + F_{Bf}C_{Bf}\right] \tag{5.92}$$

where

$$\alpha = \frac{1}{Ah} = \text{constant}$$

If the constant volume is achieved by fixing F_{Bf} and F_w, and manipulating only C_{Bf}, then observe that Eq. (5.92) will now be a linear ODE. If, on the other hand, constant volume is achieved by allowing both F_{Bf} and F_w to vary but in such a way that $F_{Bf} + F_w = F_B$, then Eq. (5.92) will be nonlinear.

PROBLEMS

Problem 5.1

Linearize the following nonlinear functions:

(a) $f(x) = \dfrac{\alpha x}{1 + (\alpha - 1)x}$, where α is a constant.

(b) $f(T) = e^{A/T+B}$, where A and B are constants.

(c) $f(v) = K(v)^{0.8}$, where K is a constant.

(d) $f(h) = K(h)^{\frac{3}{2}}$, where K is a constant.

Problem 5.2

Linearize the following ODEs, which describe a nonisothermal CSTR with constant volume. The input variables are T_f, T_J, C_{Af}, and F.

$$V\frac{dC_A}{dt} = F\left(C_{Af} - C_A\right) - VkC_A$$

$$V\rho C_p\frac{dT}{dt} = F\rho C_p\left(T_f - T\right) - \lambda VkC_A - UA(T - T_J)$$

where

$$k = k_0 e^{-E/RT}$$

Problem 5.3

Solve the following differential equations using Laplace transforms:

(a) $\dfrac{d^2 x}{dt^2} + 3\dfrac{dx}{dt} + x = 1$ with $x(0) = \left.\dfrac{dx}{dt}\right|_{t=0} = 0$

(b) $\dfrac{d^2 q}{dt^2} + \dfrac{dq}{dt} = t^2 + 2t$ with $q(0) = 0$ and $\left.\dfrac{dq}{dt}\right|_{t=0} = -2$

Problem 5.4

Obtain $y(t)$ for the following,

(a) $y(s) = \dfrac{s^2 + 2s}{s^4}$

(b) $y(s) = \dfrac{2s}{(s-1)^3}$

Problem 5.5

The function $f(t)$ has the Laplace transformation

$$\hat{f}(s) = \frac{1 - 2e^{-s} + e^{-2s}}{s^2}$$

Obtain the function $f(t)$ and plot its graph (its variation with t).

Problem 5.6

Given a system with the transfer function

$$\frac{y(s)}{x(s)} = \frac{T_1 s + 1}{T_2 s + 1}$$

find $y(t)$ if $x(t)$ is a unit-step function. If $T_1/T_2 = 5$, sketch $y(t)$ versus t/T_2. Show the numerical minimum, maximum, and ultimate values that may occur during the transient.

Problem 5.7

Simplified equations can describe a first-order, irreversible, exothermic reaction in a CSTR containing a heating coil to illustrate many of the principles of process dynamics. However, these equations neglect the possibility of dynamic effects in the heating coil. Assuming plug flow in the coil and no wall capacitance or resistance of the wall, derive an appropriate set of equations that includes the coil dynamics. Linearize the equations and obtain the transfer functions relating changes in the hot fluid inlet temperature and velocity to the reactor composition. Discuss the effect of this modification on the nature of characteristic equation and the stability of the system.

Problem 5.8

Derive a transfer function relating the tube outlet temperature to the shell inlet temperature for a two-tube-pass, single-shell-pass heat exchanger.

Problem 5.9

When chemists undertake a laboratory study of a new reaction, they often take data in a batch reactor at conditions corresponding to complete conversion or thermodynamic equilibrium. If the process economics appear promising, a pilot plant might be constructed to study the reaction in a continuous system. Normally, data are gathered at several steady-state operating conditions in an attempt to ascertain the most profitable operating region.

   Are any of these basic laboratory data useful if we are interested in establishing the dynamic characteristics of a process? If you were in charge of the whole project, what kind of experiments would you recommend? How would you try to sell your approach to top management? How would you expect the cost of your experimental program to compare to the conventional approach?

Problem 5.10

A simple model for a pair of exothermic parallel reactions, $A \rightarrow B$ and $A \rightarrow C$, in a CSTR containing a cooling coil might be written as

$$V \frac{dC_A}{dt} = q(C_{1f} - C_1') - k_1 V C_A^2 - k_2 V C_A$$

$$V \frac{dC_B}{dt} = -q C_B + k_1 V C_A^2$$

$$V C_p \rho \frac{dT}{dt} = q C_p \rho (T_f - T) + (-\Delta H_1) k_1 V C_A^2 + (-\Delta H_2) k_2 V C_A$$
$$- \frac{U A_c K q_c}{1 + K q_c} (T - T_c)$$

(a) See if you can list the assumptions implied by these equations.

(b) Describe a procedure for calculating the steady-state compositions and temperature in the reactor.

(c) Linearize these equations around the steady-state operating point.

(d) Calculate the characteristic roots of the linearized equations.

Problem 5.11

You have measured the frequency response of an industrial furnace and found that the transfer function relating the temperature of the effluent stream to the pressure supplied to a pneumatic motor valve on the fuel line could be represented by

$$\frac{\tilde{T}}{\tilde{P}} = \frac{40e^{-20s}}{(900s + 1)(25s + 1)} \quad \frac{°C}{psi}$$

Also, the response of the outlet temperature to feed temperature change is

$$\frac{\tilde{T}}{\tilde{T}_f} = \frac{e^{-600s}}{(60s + 1)}$$

where the time constants are given in seconds.

If a fast-acting temperature-measuring device is available, which has a gain of 0.2 psi/°C, select the gains for various kinds of pneumatic controllers that use air pressure for the input and output signals. Then, calculate the response of the closed-loop systems to a 20°C step change in the feed temperature.

Problem 5.12

A step change of magnitude 3 is introduced into the following transfer function

$$\frac{y(s)}{x(s)} = \frac{10}{2s^2 + 0.3s + 0.5}$$

Determine the overshoot and frequency of the oscillation.

Problem 5.13

The flow rate F of a manipulated stream through a control valve with equal percentage trim is given by

$$F = C_V \alpha^{x-1}$$

where F is flow in gallons per minute and C_V and α are constants set by the valve size and type. The control valve stem position x (fraction wide open) is set by the output signal β of an analog electronic feedback controller whose signal range is 4–20 mA. The valve cannot be moved instantaneously. It is approximately a first-order system:

$$\tau_v \frac{dx}{dt} + x = \frac{\beta - 4}{16}$$

The effect of the flow of the manipulated variable on the process temperature T is given by

$$\tau_p \frac{dT}{dt} + T = K_p F$$

Derive one linear ordinary differential equation that gives the dynamic dependence of process temperature on controller output signal β.

Problem 5.14

An isothermal, first-order, liquid-phase, reversible reaction is carried out in a constant-volume, perfectly mixed continuous reactor:

$$A \underset{k_2}{\overset{k_1}{\Longleftrightarrow}} B$$

The concentration of product B is zero in the feed, and in the reactor, it is C_B. The feed rate is equal to F.

(a) Derive a mathematical model describing the dynamic behavior of the system.

(b) Derive the steady-state relationship between C_A and C_{Af}. Show that the conversion of A and the yield of B decrease as k_2 increases.

(c) Assuming that the reactor is at this steady-state concentration and that a step change is made in C_{Af} to $C_{Af} + \Delta C_{Af}$, find the analytical solution that gives the dynamic response of $C_A(t)$.

Problem 5.15

Process liquid is continuously fed into a perfectly mixed tank in which it is heated by a steam coil. The feed rate F is 50,000 lb_m/h of material with a constant density ρ of 50 lb_m/ft^3 and heat capacity C_p of 0.5 Btu/lb_m °F. Holdup in the tank V is constant at 4000 lb_m. The inlet feed temperature T_f is 80°F.

Steam is added at a rate S lb_m/h that heats the process liquid up to temperature T. At the initial steady state, T is 190°F. The latent heat of vaporization λ_S of the steam is 900 Btu/lb_m.

(a) Derive a mathematical model of the system and prove that process temperature is described dynamically by the ODE

$$\tau\frac{dT}{dt} + T = K_1 T_f + K_2 S$$

where

$$\tau = \frac{V}{F}$$

$$K_1 = 1$$

$$K_2 = \frac{\lambda_S}{C_p F}$$

(b) Solve for the steady-state value of steam flow S_{SS}.
(c) Suppose a proportional feedback controller is used to adjust the steam flow rate,

$$S = S_{SS} + K_c(190 - T)$$

Solve analytically for the dynamic change in $T(t)$ for a step change in inlet feed temperature from 80°F down to 50°F. What will be the final values of T and S at the new steady state for a K_c of 100 $lb_m/h/°F$?

Problem 5.16

(a) Design of a water supply tank: You are required to design a cylindrical tank to supply a chemical plant with a constant water feed of 50 m³/h.
(b) Design of feedback control loop: The feed to this tank is suffering from some fluctuations with time, whereas the feed to the plant needs to be constant at 50 m³/h. Introduce the necessary feedback control loop using a PI (proportional–integral) controller. Derive the unsteady-state model equations for the closed-loop system (including the dynamics of controller, measuring instrument (DPC = differential pressure cell) and final control element (pneumatic control valve). Obtain the transfer functions (and their constants) for the DPC and control valve from the literature. Put the equations in linear form using Taylor series expansion, then use Laplace transformation to put the equations in the form

of transfer functions. Find the optimal controller settings (gains) using the following techniques:

(a)   The $\frac{1}{4}$ decay ratio criterion
(b)   Cohen and Coon reaction curve method

(c)   Evaluation of controller performance: Using the nonlinear model for the closed loop and solving the equations numerically, compare the performance of the nonlinear system with the controller settings obtained using the above two methods.

Problem 5.17

You are required to design two cylindrical tanks in series to supply a chemical plant with a constant water feed of 35 m^3/h. The feed to the first tank keeps fluctuating, whereas the feed to the plant needs to be constant at 35 m^3/hr. Design a suitable control loop (PI control action) for this system. Draw appropriate block diagrams for both the open- (uncontrolled) and closed- (controlled) loop cases. Find the optimal control settings (gains) using the $\frac{1}{4}$ decay ratio criterion, and the Cohen and Coon criterion and compare the results of both methods.

6

Heterogeneous Systems

Most real chemical and biochemical systems are heterogeneous. What do we mean by heterogeneous? We mean that the system is formed of more than one phase, with strong interaction between the various phases.

- Distillation is a heterogeneous system formed of at least a gas and a liquid. Distillation is not possible except when at least two phases are present (packed-bed distillation columns are formed of three phases, taking into consideration the solid packing phase).
- Absorption is another gas–liquid system (two phases). Packed-bed absorbers have three phases when taking into consideration the solid packing of the column.
- Adsorption is a solid gas (or solid–liquid) two-phase system.
- Fermentation is at least a two-phase system (the fermentation liquid mixture and the solid micro-organisms catalyzing the fermentation process). Fermentation can also be a three-phase system because aerobic fermentation involves gaseous oxygen which is bubbled through the fermentor. Immobilized packed-bed aerobic fermentors are formed of four phases; two of them are solid (the micro-organism and the solid carrier), one is liquid, and one is a gaseous phase.
- Gas–solid catalytic systems are two-phase systems involving a solid catalyst with reactants and products in the gas phase.

It is easily noticed that, in the earlier parts of this book, when dealing with one-phase systems, the emphasis was mainly on reacting systems. The reason was simply that for one-phase systems, the nonreacting cases are almost trivial (e.g., mixers and splitters).

However, in the present chapter dealing with heterogeneous systems, the nonreacting systems are as nontrivial as the reacting systems.

6.1 MATERIAL BALANCE FOR HETEROGENEOUS SYSTEMS

6.1.1 Generalized Mass Balance Equations

Let us first remember the generalized mass balance equations for a one phase system. (see Fig. 6.1). The generalized mass balance relation is given by

$$\sum_{k=1}^{K} n_{i_k} = \sum_{l=1}^{L} n_{if_l} + \sum_{j=1}^{N} \sigma_{ij} r_j \tag{6.1}$$

where r_j is the overall generalized rate of reaction for reaction j and σ_{ij} is the stoichiometric number of component i in reaction j.

The design equation for a lumped system is given by,

$$\sum_{k=1}^{K} n_{i_k} = \sum_{l=1}^{L} n_{if_l} + V \sum_{j=1}^{N} \sigma_{ij} r_j' \tag{6.2}$$

where r_j' is the generalized rate of the jth reaction per unit volume of reactor.

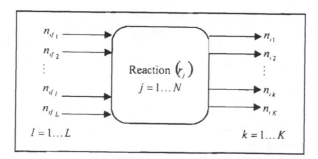

Figure 6.1 Material balance for homogeneous systems.

Let us, for simplicity, start by the simple single input–single output, single reaction mass balance which can be represented as shown in Figure 6.2. Now, we will move from the homogeneous one-phase system (as shown in Fig. 6.2) to the corresponding heterogeneous system.

6.1.2 Two-Phase Systems

If the system is formed of two phases, we can draw the mass balance schematic diagram as in Figure 6.3. The interaction between the two phases is usually some kind of mass transfer from phase I to phase II or vice-versa (also, it can be in one direction for one component and in the opposite direction for another component). The easiest way is to make the balance in one direction, as shown in Figure 6.4, and the sign of the driving force for each component determines the direction of mass transfer for the specific component. Therefore, the mass balance equations can be written as

$$n_i + \mathrm{RM}_i = n_{if} + \sigma_{iI} r_I \tag{6.3}$$

and

$$\bar{n}_i - \mathrm{RM}_i = \bar{n}_{if} + \sigma_{iII} r_{II} \tag{6.4}$$

Note that Eqs. (6.3) and (6.4) are coupled through the term RM_i, the overall rate of mass transfer of component i.

Let us take a simpler case with no reaction for which Eqs. (6.3) and (6.4) will become

$$n_i + \mathrm{RM}_i = n_{if}$$

and

$$\bar{n}_i - \mathrm{RM}_i = \bar{n}_{if}$$

For constant flow rates, we can write

$$q_1 C_i + \mathrm{RM}_i = q_1 C_{if}$$

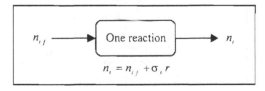

Figure 6.2 Mass balance for single-input, single-output, and single reaction system.

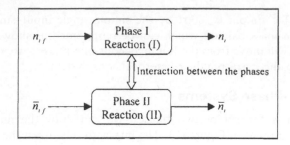

Figure 6.3   Interaction between two phases.

and,

$$q_{II}\overline{C}_i - RM_i = q_{II}\overline{C}_{if}$$

where q_I and q_{II} are the constant volumetric flow rates in phases I and II, respectively.

Then, RM_i should be expressed in terms of C_i and $\overline{C}_i$ as follows:

$$RM_i = a_m K_{g_i}\left(C_i - \overline{C}_i\right)$$

where a_m is the total area for mass transfer between the two phases and K_{g_i} is the mass transfer coefficient (for component i).
    Therefore, we have

$$q_I C_i + a_m K_{g_i}\left(C_i - \overline{C}_i\right) = q_I C_{if} \qquad\qquad (6.5)$$

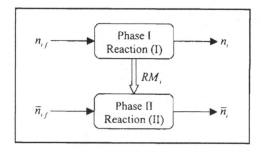

Figure 6.4   Interaction between different phases with the interaction in one direction.

and

$$q_{11}\overline{C}_i - a_m K_{g_i}(C_i - \overline{C}_i) = q_{11}\overline{C}_{if} \tag{6.6}$$

Therefore, for given values of q_1, q_{11}, a_m, k_{g_i}, C_{if}, and $\overline{C}_{if}$, we can compute the values of C_i and $\overline{C}_i$ from the Eqs. (6.5) and (6.6).

The countercurrent case: For countercurrent flow streams, we have the situation as shown in Figure 6.6. The two equations remain the same:

$$q_1 C_i + a_m K_{g_i}(C_i - \overline{C}_i) = q_1 C_{if} \tag{6.5}$$

and

$$q_{11}\overline{C}_i - a_m K_{g_i}(C_i - \overline{C}_i) = q_{11}\overline{C}_{if} \tag{6.6}$$

The difference between cocurrent and countercurrent flows does not appear in a single lumped stage. It appears in a sequence of stages (or in a distributed system) as shown later.

6.1.3 The Equilibrium Case

For a system in which the contact between the two phases is long and/or the rate of mass transfer is very high, the concentrations of the different components in the two phases reach a state of equilibrium (a state where no further mass transfer is possible). More details regarding equilibrium states have been covered in thermodynamics courses. The equilibrium relations relate the concentrations in the two phases, such that for example,

$$\overline{C}_i = f_i(C_i) \tag{6.7}$$

In certain narrow regions of concentration, we can use linear relations like the following:

$$\overline{C}_i = K_i C_i \tag{6.8}$$

where K_i is the equilibrium constant for component i. In this case, we neither know nor need the rates of mass transfer. The simple and systematic approach is to add Eqs. (6.5) and (6.6) for both the cocurrent case and the

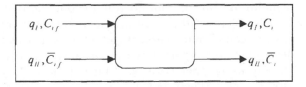

Figure 6.5 Cocurrent flow of streams.

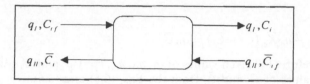

Figure 6.6 Counter-current flow of streams.

countercurrent case (as the mass balance equations are same for both flow cases).

Cocurrent case (Fig. 6.5):   Adding Eqs. (6.5) and (6.6) gives

$$q_1 C_i + q_{11}\overline{C}_i = q_1 C_{if} + q_{11}\overline{C}_{if} \tag{6.9}$$

This is a single algebraic equation with two unknowns (C_i and $\overline{C}_i$), but the equilibrium relation (6.7) can be used to make the equation solvable:

$$q_1 C_i + q_{11} f_i(C_i) = q_1 C_{if} + q_{11}\overline{C}_{if} \tag{6.10}$$

and as soon as Eq. (6.10) is solved and the value of C_i is obtained, it is straightforward to obtain the value of $\overline{C}_i$ from the equilibrium relation (6.7).

Countercurrent case (Fig. 6.6):   Adding Eqs. (6.5) and (6.6) gives

$$q_1 C_i + q_{11}\overline{C}_i = q_1 C_{if} + q_{11}\overline{C}_{if} \tag{6.11}$$

Again, we can replace $\overline{C}_i$ using relation (6.7) to get

$$q_1 C_i + q_{11} f_i(C_i) = q_1 C_{if} + q_{11}\overline{C}_{if} \tag{6.12}$$

As evident, there is no difference in the equations [compare Eqs. (6.10) and (6.12)] for the cocurrent and countercurrent cases. There will be evident differences when we have more than one stage of cocurrent or countercurrent operations, or a distributed system.

6.1.4 Stage Efficiency

For many mass transfer processes (e.g., distillation, absorption, extraction, etc.) which are multistage (in contradistinction to continuous, e.g., packed-bed columns), each stage (whether a cocurrent or countercurrent process) can be considered at equilibrium. These stages are usually called "the ideal stages." The deviation of the system from this "ideality" or "equilibrium" is compensated for by using what is called "stage efficiency", which varies between 1.0 (for an ideal stage) and 0 (for a "useless stage"). Usually, stage efficiencies are in the range 0.6–0.8, but they differ widely for different processes, components, and designs (specially retention time) for a particular stage.

However, in some cases, the concept of "stage efficiency" can cause problems and unacceptable inaccuracies may arise. The rate of mass transfer between the phases should be used in both the mass balance and the design equations, except when we are sure that equilibrium is established between the two phases.

It is clear from the above that the main concepts used in the mass balance for the one-phase system can be simply extended to heterogeneous system by writing mass balance equations for each phase and taking the interaction (mass transfer) between the phases into account. The same applies when there are reactions in both phases. Only the rate of reaction terms must be included in the mass balances of each phase, as shown earlier for homogeneous (one phase) systems.

6.1.5 Generalized Mass Balance for Two-Phase Systems

For the sake of generality, let us write the most general mass balance equations for a two-phase system, where each phase has multiple inputs and multiple outputs, with each phase undergoing multiple reactions within its boundaries.

Figure 6.7 presents the generalized mass balance for a heterogeneous system; i represents all the components in the two phases (reactants, products and inerts).

Now the mass balance equations can be written as follows:

Phase I

$$\sum_{k=1}^{K} n_{i_k} + RM_i = \sum_{l=1}^{L} n_{if_l} + \sum_{j=1}^{N} \sigma_{ij} r_j \qquad (6.13)$$

where

RM_i = overall rate of mass transfer of component i from phase I to phase II
r_j = overall generalized rate of reaction for reaction j in phase I
σ_{ij} = stoichiometric number of component i in reaction j in phase I
K = total number of output streams for phase I
L = total number of input streams for phase I
N = total number of reactions in phase I

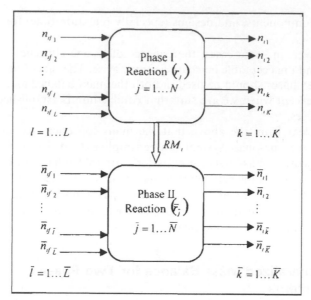

Figure 6.7 A two-phase heterogeneous system with multiple inputs–multiple output, and multiple reactions in each phase and mass transfer between the two phases.

Phase II

$$\sum_{\bar{k}=1}^{\overline{K}} \bar{n}_{i_{\bar{k}}} - \mathrm{RM}_i = \sum_{\bar{l}=1}^{\overline{L}} \bar{n}_{i f_i} + \sum_{\bar{j}=1}^{\overline{N}} \bar{\sigma}_{i\bar{j}} \bar{r}_{\bar{j}} \tag{6.14}$$

where

$\bar{r}_{\bar{j}}$ = overall generalized rate of reaction for reaction $\bar{j}$ in phase II
$\bar{\sigma}_{i\bar{j}}$ = stoichiometric number of component i in reaction $\bar{j}$ in phase II
$\overline{K}$ = total number of output streams for phase II
$\overline{L}$ = total number of input streams for phase II
$\overline{N}$ = total number of reactions in phase II

To change the above mass balance equations into design equations, we just replace r_j by Vr'_j and $\bar{r}_{\bar{j}}$ by $V\bar{r}'_{\bar{j}}$, where r'_j and $\bar{r}'_{\bar{j}}$ are the rates of reaction per unit volume of reaction mixture (or per unit mass of catalyst for catalytic reactions, etc.). More details about the design equations is given in the following sections. Note that these terms can also be $V_I\, r'_j$ (where V_I is the volume of phase I and r'_j is the rate of reaction per unit volume of phase I) and $V_{II}\bar{r}'_{\bar{j}}$ (where V_{II} is the volume of phase II and $\bar{r}'_{\bar{j}}$ is the rate of reaction per unit volume of phase II).

6.2 DESIGN EQUATIONS (STEADY-STATE MODELS) FOR ISOTHERMAL, HETEROGENEOUS LUMPED SYSTEMS

For simplicity and clarity, let us consider a two-phase system. Each phase has single input and single output with a single reaction taking place in each phase (see Fig. 6.8.)

The mass balance equations can be written as

$$n_i + \text{RM}_i = n_{if} + \sigma_i r \quad \text{(for phase I)} \tag{6.15}$$

and

$$\bar{n}_i - \text{RM}_i = \bar{n}_i + \bar{\sigma}_i \bar{r} \quad \text{(for phase II)} \tag{6.16}$$

where

n_i = molar flow rate of component i out of phase I

$\bar{n}_i$ = molar flow rate of component i out of phase II

n_{if} = molar flow rate of component i fed to phase I

$\bar{n}_{if}$ = molar flow rate of component i fed to phase II

r = generalized rate of the single reaction in phase I

σ_i = stoichiometric number of component i in the reaction in phase I

$\bar{r}$ = generalized rate of the single reaction in phase II

$\bar{\sigma}_i$ = stoichiometric number of component i in the reaction in phase II

RM_i = overall mass transfer rate of component i from phase I to phase II

Now, in order to turn these mass balance equations into design equations, we just turn all of the rate processes (r, $\bar{r}$, and RM_i) into rates per unit volume of the process unit (or the specific phase). For example, let us consider the following definitions for the rates:

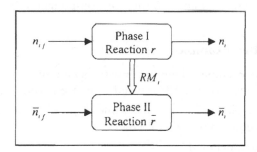

Figure 6.8 A two-phase heterogeneous system with single input–single output and single reaction in each phase.

r' = rate of reaction in phase I per unit volume of the process unit
$\bar{r}'$ = rate of reaction in phase II per unit volume of the process unit
RM_i' = is the rate of mass transfer of component i from phase I to phase II per unit volume of the process unit
V = volume of the process unit ($= V_I + V_{II}$)

Thus, Eqs. (6.15) and (6.16) can be rewritten as

$$n_i + V RM_i' = n_{if} + V \sigma_i r' \quad \text{(for phase I)} \tag{6.17}$$

and

$$\bar{n}_i - V RM_i' = \bar{n}_i + V \bar{\sigma}_i \bar{r}' \quad \text{(for phase II)} \tag{6.18}$$

For example, consider the case of first-order irreversible reactions in both phases with constant flow rates (q_I in phase I and q_{II} in phase II) and concentrations in phase I are C_is and in phase II are $\bar{C}_i$s. The rate of reactions are given by

$$r' = kC_A, (A \rightarrow B \text{ in phase I})$$

and

$$\bar{r}' = \bar{k}\,\bar{C}_B, (B \rightarrow C \text{ in phase II})$$

Note that $i = 1, 2, 3, \ldots A, \ldots B, \ldots$. Thus, Eqs. (6.17) and (6.18) become

$$q_1 C_i + V a_m' K_{g_i}(C_i - \bar{C}_i) = q_1 C_{if} + V \sigma_i k C_A \tag{6.19}$$

$$q_{11} \bar{C}_i - V a_m' K_{g_i}(C_i - \bar{C}_i) = q_{11} \bar{C}_{if} + V \bar{\sigma}_i \bar{k}\, \bar{C}_B \tag{6.20}$$

where a_m' is the area of mass transfer per unit volume of the process unit and K_{g_i} is the coefficient of mass transfer of component i between the two phases.

Simple Illustrative Example

Let us consider an example where the flow rates and rates of reactions are as shown in Figure 6.9. Note that C_{A_f} is fed to phase I while C_{B_f} and C_{C_f} are equal to zero. For phase II, $\bar{C}_{B_f}$ is fed while $\bar{C}_{A_f}$ and $\bar{C}_{C_f}$ are equal to zero.

The design equations are as follows:

Phase I: For component A,

$$q_1 C_A + V a_m' K_{g_A}(C_A - \bar{C}_A) = q_1 C_{A_f} - V k C_A \tag{6.21}$$

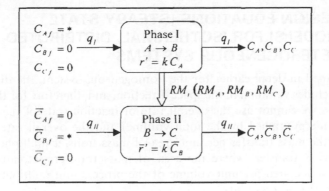

Figure 6.9 Mass flow diagram for a two-phase system.

For component B,

$$q_1 C_B + V a'_m K_{g_B}(C_B - \overline{C}_B) = \underbrace{q_1 C_{B_f}}_{\text{equal to zero}} + VkC_A \tag{6.22}$$

For component C,

$$q_1 C_C + V a'_m K_{g_C}(C_C - \overline{C}_C) = \underbrace{q_1 C_{C_f}}_{\text{equal to zero}} \tag{6.23}$$

Phase II: For component A,

$$q_{11}\overline{C}_A - V a'_m K_{g_A}(C_A - \overline{C}_A) = \underbrace{q_{11}\overline{C}_{A_f}}_{\text{equal to zero}} \tag{6.24}$$

For component B,

$$q_{11}\overline{C}_B - V a'_m K_{g_B}(C_B - \overline{C}_B) = q_{11}\overline{C}_{B_f} - V\overline{k}\,\overline{C}_B \tag{6.25}$$

For component C,

$$q_{11}\overline{C}_C - V a'_m K_{g_C}(C_C - \overline{C}_C) = \underbrace{q_{11}\overline{C}_{C_f}}_{\text{equal to zero}} + V\overline{k}\,\overline{C}_B \tag{6.26}$$

The above six equations are the design equations for this two-phase isothermal system when both phases are lumped systems.

6.3 DESIGN EQUATIONS (STEADY-STATE MODELS) FOR ISOTHERMAL, DISTRIBUTED HETEROGENEOUS SYSTEMS

As explained in detail earlier for the homogeneous system, the distributed system includes variation in the space direction, and, therefore for the design equations, we cannot use the overall rate of reaction (rate of reaction per unit volume multiplied by the total volume) and the overall rate of mass transfer (the mass transfer per unit area of mass transfer multiplied by the area of mass transfer, where the area of mass transfer is treated as the multiple of the area per unit volume of the process unit multiplied by the volume of the process unit).

In the case of distributed systems, we need to take the balance on an element; then, we take the limit when the size of element goes to zero, and so forth, as was detailed earlier for a distributed homogeneous system. As an illustration, let us consider a case with a single reaction

$$A \rightarrow B$$

taking place in phase I (no reaction is taking place in phase II) and we consider that mass transfer between the two phases takes place (see Fig. 6.10).

Thus, the balance for phase I (the design equation) is

$$n_i + \Delta n_i + \Delta V \mathrm{RM}'_i = n_i + \Delta V \sigma_i r' \tag{6.27}$$

and for phase II, it will be (for the cocurrent case),

$$\bar{n}_i + \Delta \bar{n}_i - \Delta V \mathrm{RM}'_i = \bar{n}_i + 0 \tag{6.28}$$

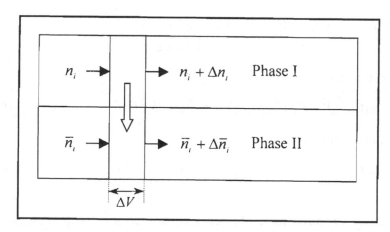

Figure 6.10   A distributed heterogeneous system.

because there is no reaction in this phase.

Arranging Eq. (6.27) gives

$$\frac{dn_i}{dV} + RM_i' = \sigma_i r'$$ (6.29)

and rearranging Eq. (6.28) gives

$$\frac{d\bar{n}_i}{dV} - RM_i' = 0$$ (6.30)

Both are differential equations with the initial conditions (for the cocurrent case) at $V = 0$, $n_i = n_{if}$ and $\bar{n}_i = \bar{n}_{if}$.

If the flow rate is constant and the reaction is first order ($r' = kC_A$), we can write Eqs. (6.29) and (6.30) in the following form:

$$q\frac{dC_i}{dV} + a_m' K_{g_i}(C_i - \bar{C}_i) = \sigma_i k C_A$$ (6.31)

and,

$$q\frac{d\bar{C}_i}{dV} - a_m' K_{g_i}(C_i - \bar{C}_i) = 0$$ (6.32)

and the initial conditions (for the cocurrent case) are at $V = 0$, $C_i = C_{if}$ and $\bar{C}_i = \bar{C}_{if}$, where, $i = A$ and B. Specifically, for the two components A and B, we can write the following:

Phase I

$$q\frac{dC_A}{dV} = -a_m' K_{g_A}(C_A - \bar{C}_A) - kC_A$$ (6.33)

and

$$q\frac{dC_B}{dV} = -a_m' K_{g_B}(C_B - \bar{C}_B) + kC_A$$ (6.34)

Phase II

$$\bar{q}\frac{d\bar{C}_A}{dV} = a_m' K_{g_A}(C_A - \bar{C}_A)$$ (6.35)

and

$$\bar{q}\frac{d\bar{C}_B}{dV} = a_m' K_{g_B}(C_B - \bar{C}_B)$$ (6.36)

with the initial conditions (for the cocurrent case) at $V = 0$, $C_A = C_{A_f}$, $C_B = C_{B_f}$, $\bar{C}_A = \bar{C}_{A_f}$ and $\bar{C}_B = \bar{C}_{B_f}$.

Generalized Form

Now, we develop the very *general equation*, when there are N (the counter is j) reactions taking place in phase 1 and $\overline{N}$ (the counter is $\bar{j}$) reactions taking place in phase II. (Refer to Fig. 6.11.)

Then, the design equations for the *cocurrent case* are as follows:

Phase I

$$n_i + \Delta n_i + \Delta V \mathrm{RM}_i' = n_i + \Delta V \sum_{j=1}^{N} \sigma_{ij} r_j'$$

which, after the usual manipulation, becomes

$$\frac{dn_i}{dV} + \mathrm{RM}_i' = \sum_{j=1}^{N} \sigma_{ij} r_j' \tag{6.37}$$

Phase II

Similarly for phase II, the design equation is,

$$\frac{d\overline{n}_i}{dV} - \mathrm{RM}_i' = \sum_{\bar{j}=1}^{\overline{N}} \overline{\sigma}_{ij} \overline{r}_{\bar{j}}' \tag{6.38}$$

The initial conditions are (for cocurrent case), at $V = 0$, $n_i = n_{if}$ and $\overline{n}_i = \overline{n}_{if}$.

Note that for distributed systems, the multiple-input problem corresponds to adding an idle-stage (with no reaction or mass transfer) mass balance at the input for each phase, as shown in Figure 6.12, which is a very simple problem. The same applies to multiple outputs.

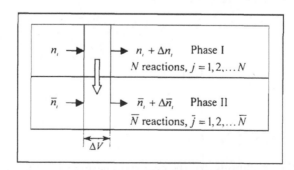

Figure 6.11　Mass flow for the heterogeneous distributed system.

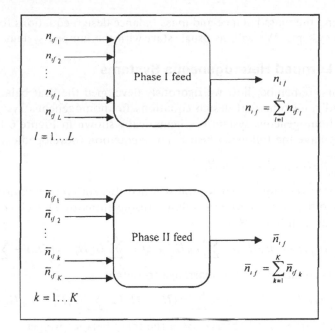

Figure 6.12 Multiple inputs can be combined together as a single feed.

There are cases in which one of the phases is distributed and the other phase is lumped. In such cases, the distributed phase will contribute its mass and/or heat transfer term to the lumped phase through an integral, as shown later with regard to the modeling of a bubbling fluidized bed.

For the *countercurrent case*, the equation of phase II will simply be

$$-\frac{d\bar{n}_i}{dV} - \mathrm{RM}_i' = \sum_{j=1}^{\overline{N}} \bar{\sigma}_{ij} \bar{r}_j' \tag{6.39}$$

with the boundary conditions at $V = V_t, \bar{n}_i = \bar{n}_{if}$, where V_t is the total volume of the process unit.

6.4 NONISOTHERMAL HETEROGENEOUS SYSTEMS

Based on the previous chapters, it should be quite easy and straightforward for the reader to use the heat balance and heat balance design equations developed earlier for homogeneous systems and the principles we used for

developing the mass balance and mass balance design equations for heterogeneous systems. We will, as usual, start with the lumped system.

6.4.1 Lumped Heterogeneous Systems

Let us first remember how we rigorously developed the heat balance equations and nonisothermal design equations for homogeneous systems. The
lumped homogeneous system is schematically shown in Figure 6.13.

We have the following heat balance equation (developed earlier):

$$\sum n_{if}H_{if} + Q = \sum n_i H_i \tag{6.40}$$

and then we remember that we transformed this enthalpy equation into an
enthalpy difference equation as follows (using enthalpy of each component
at reference condition H_{i_r}):

$$\sum (n_{if}H_{if} - n_{if}H_{i_r}) + \sum n_{if}H_{i_r} + Q = \sum (n_i H_i - n_i H_{i_r}) + \sum n_i H_{i_r}$$

From this equation, we can rearrange to obtain

$$\sum n_{if}(H_{if} - H_{i_r}) + Q = \sum n_i(H_i - H_{i_r}) + \sum (n_i H_{i_r} - n_{if}H_{i_r}) \tag{6.41}$$

As we did earlier in Chapter 2, from the mass balance we get

$$n_i = n_{if} + \sigma_i r$$

which can be rewritten as,

$$n_i - n_{if} = \sigma_i r$$

Now, we can write,

$$\sum (n_i - n_{if})H_{i_r} = \sum \sigma_i r H_{i_r} \tag{6.42}$$

Thus, relation (6.42) can be rewritten as $r\sum \sigma_i H_{i_r}$, and this can be finally
rewritten as $r\Delta H_r$. Thus the heat balance equation becomes, as shown in
Chapter 2,

$$\sum n_{if}(H_{if} - H_{i_r}) + Q = \sum n_i(H_i - H_{i_r}) + r\Delta H_r \tag{6.43}$$

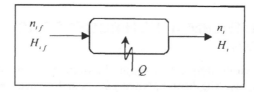

Figure 6.13 Heat balance for a homogeneous system.

In order to convert the equation into nonisothermal (heat balance) design equation, we just do the single step we did several times before: We replace r with $r'V$, where r' is the rate of reaction per unit volume and V is the volume of the reactor. Thus, Eq. (6.43) becomes

$$\sum n_{if}(H_{if} - H_{i_r}) + Q = \sum n_i(H_i - H_{i_r}) + Vr'\Delta H_r \qquad (6.44)$$

Note that the basis of r' and V depends on the type of process. For example, if we are dealing with a gas–solid catalytic system, we will usually define r' as per unit mass of the catalyst and replace V with W_S (the weight of the catalyst).

We do not need to repeat the above for multiple reactions (N reactions); Eq. (6.44) will become

$$\sum n_{if}(H_{if} - H_{i_r}) + Q = \sum n_i(H_i - H_{i_r}) + V \sum_{j=1}^{N} r'_j(\Delta H_{rj})$$

where (ΔH_{rj}) is the heat of reaction for reaction j.

The Heterogeneous System

For the heterogeneous system, the problem is very simple. We just write the above equation for each phase, taking into account $\overline{Q}$ as the heat transfer between the two phases. For a nonadiabatic system, the heat added from outside, $Q_{external}$ (which is the Q in Eq. (6.44)], will be added to the phase receiving it, or distributed between the two phases if it is added to both phases (this will depend very much on the configuration and knowledge of the physical system as will be shown with the nonadiabatic bubbling fluidized-bed catalytic reactor example).

Now, let us consider an adiabatic two-phase system with each phase having one reaction, as shown schematically in Figure 6.14. Here, $\overline{Q}$ is the heat transfer between the two phases. For phase I, the heat balance equation is

$$\sum n_{if}(H_{if} - H_{i_r}) - \overline{Q} = \sum n_i(H_i - H_{i_r}) + r_1\Delta H_{r_1} \qquad (6.45)$$

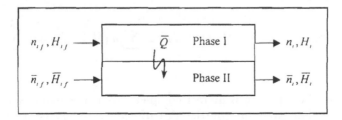

Figure 6.14   Heat balance for a two-phase lumped system.

and for phase II, it is

$$\sum \bar{n}_{if}(\bar{H}_{if} - \bar{H}_{i_r}) + \bar{Q} = \sum \bar{n}_i(\bar{H}_i - \bar{H}_{i_r}) + r_{11}\Delta H_{r_{11}} \tag{6.46}$$

To turn these heat balance equations into the nonisothermal heat balance design equations, we just define the rate of reaction per unit volume (or per unit mass of the catalyst depending on the system) and the heat transfer is defined per unit volume of the process unit (or per unit length, whichever is more convenient). Thus, Eqs. (6.45) and (6.46) become as follows:

For phase I,

$$\sum n_{if}(H_{if} - H_{i_r}) - VQ' = \sum n_i(H_i - H_{i_r}) + Vr'\Delta H_{r_1} \tag{6.47}$$

and for phase II, it is

$$\sum \bar{n}_{if}(\bar{H}_{if} - \bar{H}_{i_r}) + VQ' = \sum \bar{n}_i(\bar{H}_i - \bar{H}_{i_r}) + V\bar{r}'\Delta H_{r_1} \tag{6.48}$$

The rate of heat transfer per unit volume of the process unit Q' is given by

$$Q' = a'_h h(T - \bar{T}) \tag{6.49}$$

where a'_h is the area of heat transfer per unit volume of the system, h is the heat transfer coefficient between the two phases, $\bar{T}$ is the temperature of phase II, and T is the temperature of phase I.

Obviously, for multiple reactions in each phase, the change in the equations is straightforward, as earlier. For N reactions (counter j) in phase I and $\bar{N}$ reactions (counter $\bar{j}$) in phase II, the equations become as follows:

For phase I,

$$\sum n_{if}(H_{if} - H_{i_r}) - Va'_h h(T - \bar{T}) = \sum n_i(H_i - H_{i_r}) + V\sum_{j=1}^{N} r_j'\Delta H_{r_j} \tag{6.50}$$

For phase II, it is

$$\sum \bar{n}_{if}(\bar{H}_{if} - \bar{H}_{i_r}) + Va'_h h(T - \bar{T}) = \sum \bar{n}_i(\bar{H}_i - \bar{H}_{i_r}) + V\sum_{\bar{j}=1}^{\bar{N}} \bar{r}_{\bar{j}}'\Delta \bar{H}_{r_{\bar{j}}} \tag{6.51}$$

Now, the whole picture is almost complete; what is remaining is the heat balance design equation for the distributed two-phase system and the dynamic terms in heterogeneous systems.

6.4.2   Distributed Systems (see Fig. 6.15)

For phase I with one reaction,

$$\sum n_i H_i - \Delta V \sum \mathrm{RM}'_i H_i - \Delta V Q' = \sum (n_i H_i + \Delta(n_i H_i)) \qquad (6.52)$$

Equation (6.52) becomes

$$\sum \frac{d(n_i H_i)}{dV} = -Q' - \sum \mathrm{RM}'_i H_i$$

which can be rewritten as

$$\sum n_i \frac{dH_i}{dV} + \sum H_i \frac{dn_i}{dV} = -Q' - \sum \mathrm{RM}'_i H_i \qquad (6.53)$$

From the mass balance design equation [Eq. 6.37], when written for a single reaction we get

$$\frac{dn_i}{dV} + \mathrm{RM}'_i = \sigma_i r' \qquad (6.54)$$

Substituting for dn_i/dV from Eq. (6.54) into Eq. (6.53) gives

$$\sum n_i \frac{dH_i}{dV} + \sum H_i [\sigma_i r' - \mathrm{RM}'_i] = -Q' - \sum \mathrm{RM}'_i H_i$$

which gives

$$\sum n_i \frac{dH_i}{dV} + r' \Delta H = -Q' \qquad (6.55)$$

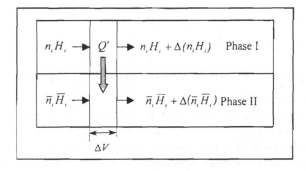

Figure 6.15   Heat balance for a heterogeneous distributed system.

For phase II, it will be (for cocurrent operation)

$$\sum \bar{n}_i \frac{d\overline{H}_i}{dV} + \bar{r}'\Delta\overline{H} = Q'$$ (6.56)

If the change is only sensible heat, we can use an average constant C_P for the mixture. Then, as shown earlier for the homogeneous system, we can write Eqs. (6.55) and (6.56) as,

$$q\rho C_P \frac{dT}{dV} + r'\Delta H = -a'_h h(T - \overline{T}) \quad \text{for Phase I}$$ (6.57)

and

$$\bar{q}\,\bar{\rho}\overline{C}_P \frac{d\overline{T}}{dV} + \bar{r}'\Delta\overline{H} = a'_h h(T - \overline{T}) \quad \text{for Phase II}$$ (6.58)

with the initial conditions (for the cocurrent case) at $V = 0$, $T = T_f$ and $\overline{T} = \overline{T}_f$.

For the countercurrent case,

$$q\rho C_P \frac{dT}{dV} + r'\Delta H = -a'_h h(T - \overline{T}) \quad \text{for Phase I}$$ (6.59)

Note that the equation for phase I for countercurrent case is the same as for cocurrent case. For phase II the equation is:

$$-\bar{q}\,\bar{\rho}\overline{C}_P \frac{d\overline{T}}{dV} + \bar{r}'\Delta\overline{H} = a'_h h(T - \overline{T}) \quad \text{for Phase II}$$ (6.60)

with the following two-point split boundary conditions:

At $\quad V = 0, T = T_f$.

At $\quad V = V_t, \overline{T} = \overline{T}_f$.

6.4.3 Dynamic Terms for Heterogeneous Systems

The dynamic terms for heterogeneous systems will be exactly the same as for the homogeneous systems (refer to Chapters 2 and 3), but repeated for both phases. The reader should take this as an exercise by just repeating the same principles of formulating the dynamic terms for both mass and heat and for both lumped and distributed systems. For illustration, see the dynamic examples later in this chapter.

6.5 EXAMPLES OF HETEROGENEOUS SYSTEMS

6.5.1 Absorption Column (High-Dimensional Lumped, Steady-State, and Equilibrium Stages System)

Let us consider the absorption column (tower), multistage (multitrays), and all stages as equilibrium stages (ideal stages). Component A is absorbed from the gas phase to the liquid phase.

Y_j = mole fraction of component A in the gas phase leaving the jth tray

n_j = molar flow rate of component A in the gas phase leaving the jth tray

X_j = mole fraction of component A in the liquid phase leaving the jth tray

$\bar{n}_j$ = molar flow rate of component A in the liquid phase leaving the jth tray

where,

$$n_j = n_{t_j} Y_j$$

where n_{t_j} is the total molar flow rate of the gas phase leaving the jth tray

$$\bar{n}_j = \bar{n}_{t_j} X_j$$

where $\bar{n}_{t_j}$ is the total molar flow rate of the liquid phase leaving the jth tray

The molar flow streams at tray j are shown in Figure 6.16. The molar flow balance on the jth tray gives

$$\bar{n}_{t_{j-1}} X_{j-1} + n_{t_{j+1}} Y_{j+1} = \bar{n}_{t_j} X_j + n_{t_j} Y_j \qquad (6.61)$$

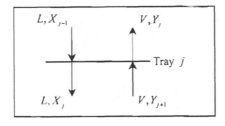

Figure 6.16 Molar flow across the jth tray.

For simplicity, we assume that the total molar flow rates of liquid and gas phases remain constant; that is,

$$\bar{n}_{l_j} = L \quad \text{and} \quad n_{t_j} = V$$

Based on the above-stated assumption, Eq. (6.61) becomes

$$L X_{j-1} + V Y_{j+1} = L X_j + V Y_j \tag{6.62}$$

For the equilibrium stage, we can write, in general,

$$Y_j = F(X_j)$$

where $F(X_j)$ is a function in terms of X_j only. For a linear case, we can write

$$Y_j = a X_j + b$$

On using the above-mentioned linear relation in Eq. (6.62), we get

$$L X_{j-1} + V (a X_{j+1} + b) = L X_j + V (a X_j + b)$$

which gives

$$L X_{j-1} + (V a) X_{j+1} = (L + V a) X_j$$

Further rearrangement gives

$$L X_{j-1} - \underbrace{(L + V a)}_{\alpha} X_j + \underbrace{(V a)}_{\beta} X_{j+1} = 0$$

Thus, we get

$$L X_{j-1} - \alpha X_j + \beta X_{j+1} = 0 \tag{6.63}$$

Figure 6.17 is the schematic diagram of the absorption column with the molar flow rates shown for each tray.

Note: The numbering can be reversed in order; it is up to the reader.

Looking at Eq. (6.63), for the first tray ($j = 1$) we get

$$L X_0 - \alpha X_1 + \beta X_2 = 0$$

which can be rewritten as

$$-\alpha X_1 + \beta X_2 + 0 X_3 + 0 X_4 + \cdots + 0 X_N = -L X_0$$

For $j = 2$,

$$L X_1 - \alpha X_2 + \beta X_3 + 0 X_4 + 0 X_5 + \cdots + 0 X_N = 0$$

For $j = 3$,

$$0 X_1 + L X_2 - \alpha X_3 + \beta X_4 + 0 X_5 + \cdots + 0 X_N = 0$$

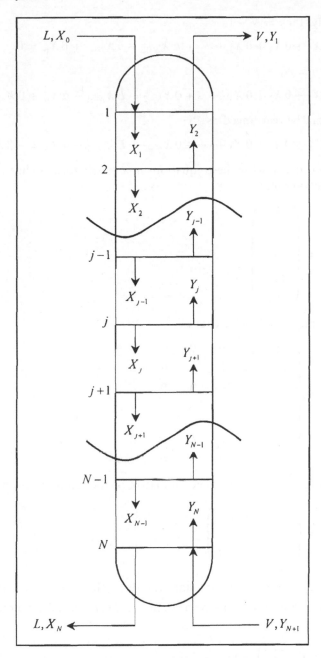

Figure 6.17 Absorption column.

Similarly, for $j = N - 1$,

$$0 X_1 + 0 X_2 + 0 X_3 + \cdots + L X_{N-2} - \alpha X_{N-1} + \beta X_N = 0$$

and for $j = N$,

$$0 X_1 + 0 X_2 + 0 X_3 + \cdots + 0 X_{N-2} + L X_{N-1} - \alpha X_N + \beta X_{N+1} = 0$$

which can be rearranged as

$$0 X_1 + 0 X_2 + 0 X_3 + \cdots + 0 X_{N-2} + L X_{N-1} - \alpha X_N = -\beta X_{N+1}$$

In order to write these equations in a matrix form, define the vector of the state variables

$$\underline{X} = \begin{pmatrix} X_1 \\ X_2 \\ \vdots \\ X_{N-1} \\ X_N \end{pmatrix}$$

and the matrix of coefficients as

$$\underline{A} = \begin{pmatrix} -\alpha & \beta & 0 & 0 & 0 & 0 \\ L & -\alpha & \beta & 0 & 0 & 0 \\ 0 & L & -\alpha & \beta & 0 & 0 \\ \vdots & \vdots & \ddots & \ddots & \ddots & \vdots \\ 0 & 0 & 0 & L & -\alpha & \beta \\ 0 & 0 & 0 & 0 & L & -\alpha \end{pmatrix}$$

The above matrix is a tridiagonal matrix of the following form:

$$\underline{A} = \begin{pmatrix} -\alpha & \beta & 0 \\ L & \ddots & \beta \\ 0 & L & -\alpha \end{pmatrix}$$

Note: X_0 and X_{N+1} are not state variables, they are input variables.

Define the input vector matrix as (input vector $\underline{I}$ is usually denoted by $\underline{m}$), so we get

$$\underline{I} = \begin{pmatrix} X_0 \\ X_{N+1} \end{pmatrix} = \underline{m}$$

The input coefficient matrix is given by

$$
\underline{B} = \begin{pmatrix} -L & 0 \\ 0 & 0 \\ 0 & 0 \\ \vdots & \vdots \\ 0 & -\beta \end{pmatrix}
$$

Thus, the mass balance equations for the absorption column in matrix form can be written as,

$$\underline{A}\,\underline{X} = \underline{B}\,\underline{m} \tag{6.64}$$

and its solution is given by (see Appendix A)

$$\underline{X} = \underline{A}^{-1}\underline{B}\,\underline{m} \tag{6.65}$$

Degrees of Freedom

Example: If all parameters and inputs are defined, then A, B, and m and m are defined. In such a case, we have N unknowns $(X_1, X_2, X_3, \ldots, X_N)$ and N equations; thus, the degrees of freedom is equal to zero, and the problem is solvable.

If the equilibrium relation is *not linear*,

$$L\,X_{j-1} + V\,F(X_{j+1}) = L\,X_j + V\,F(X_j)$$

where $F(X_j)$ is a function in terms of X_j only. In this case, the formulation gives a set of nonlinear equations to be solved numerically (iteratively, using, for example, the multidimensional Newton–Raphson or other modified gradient techniques; see Appendix B).

Nonequilibrium Stages

The overall molar balance we used will not be adequate because the equilibrium relation between the two phases cannot be used. In the present case of nonequilibrium stages, we must introduce the mass transfer rate between the two phases, as shown in Figure 6.18. In the figure, the term RMT is the rate of mass transfer of the component A from the vapor phase. It can be expressed as,

$$\mathrm{RMT} = K_{g,}\,\bar{a}_{m_i} \times (\text{driving force})$$

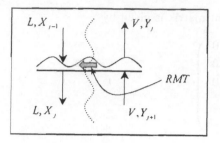

Figure 6.18 Mass transfer on a nonequilibrium stage.

where K_{gj} is the mass transfer coefficient and $\bar{a}_{m_j}$ is the area of mass transfer for tray j. Assuming the tray to be perfectly mixed with respect to both liquid and gas phases, we get

$$\text{RMT} = K_{g_j} \bar{a}_{m_j} \left(Y_j - \left(a\, X_j + b \right) \right)$$

where

$$\left(a\, X_j + b \right) = Y_j^*$$

which is the gas-phase mole fraction at equilibrium with X_j.

The molar balance for the gas phase is given by

$$V\, Y_{j+1} = V\, Y_j + K_{gj} \bar{a}_{m_j} \left(Y_j - \left(a\, X_j + b \right) \right)$$

On rearrangement, we get

$$\underbrace{\left(V + K_{gj} \bar{a}_{m_j} \right)}_{\alpha_j} Y_j - V\, Y_{j+1} - \underbrace{\left(K_{gj} \bar{a}_{m_j}\, a \right)}_{\beta_j} X_j - \underbrace{K_{gj} \bar{a}_{m_j}\, b}_{\gamma_j} = 0$$

thus giving

$$\alpha_j\, Y_j - V\, Y_{j+1} - \beta_j\, X_j - \gamma_j = 0 \qquad (6.66)$$

where

$$\alpha_j = V + K_{gj} \bar{a}_{mj}$$

$$\beta_j = K_{gj} \bar{a}_{mj}\, a$$

$$\gamma_j = K_{gj} \bar{a}_{mj}\, b$$

Using Eq. (6.66), we have the following:

For $j = 1$,

$$\alpha_1\, Y_1 - V\, Y_2 - \beta_1\, X_1 - \gamma_1 = 0$$

For $j = 2$,

$$0\,Y_1 + \alpha_2\,Y_2 - V\,Y_3 - \beta_2\,X_2 - \gamma_2 = 0$$

For $j = N - 1$,

$$0\,Y_1 + 0\,Y_2 + 0\,Y_3 + \cdots + 0\,Y_{N-2} + \alpha_{N-1}\,Y_{N-1} - V\,Y_N - \beta_{N-1}\,X_{N-1}$$
$$- \gamma_{N-1} = 0$$

For $j = N$,

$$0\,Y_1 + 0\,Y_2 + 0\,Y_3 + \cdots + 0\,Y_{N-1} + \alpha_N\,Y_N - V\,Y_{N+1} - \beta_N\,X_N$$
$$- \gamma_N = 0$$

The above equation can be rearranged as

$$0\,Y_1 + 0\,Y_2 + 0\,Y_3 + \cdots + 0\,Y_{N-1} + \alpha_N\,Y_N - \beta_N\,X_N - \gamma_N = V\,Y_{N+1}$$

In order to write these equations in a matrix form, we define the matrix of coefficients as

$$\underline{A}_Y = \begin{pmatrix} \alpha_1 & -V & 0 & 0 & 0 \\ 0 & \alpha_2 & -V & 0 & 0 \\ \vdots & \vdots & \ddots & \ddots & \vdots \\ 0 & 0 & 0 & \alpha_{N-1} & -V \\ 0 & 0 & 0 & 0 & \alpha_N \end{pmatrix}$$

The above matrix is a bidiagonal matrix of the form,

$$\underline{A}_Y = \begin{pmatrix} \alpha_1 & -V & 0 \\ & \ddots & -V \\ 0 & & \alpha_n \end{pmatrix}$$

and vectors of the state variables are,

$$\underline{X} = \begin{pmatrix} X_1 \\ X_2 \\ \vdots \\ X_N \end{pmatrix} \quad \text{and} \quad \underline{Y} = \begin{pmatrix} Y_1 \\ Y_2 \\ \vdots \\ Y_N \end{pmatrix}$$

Also, we get

$$\underline{B}_X = \begin{pmatrix} -\beta_1 & & & 0 \\ & -\beta_2 & & \\ & & \ddots & \\ 0 & & & -\beta_N \end{pmatrix} \quad \text{and} \quad \underline{\gamma} = \begin{pmatrix} \gamma_1 \\ \gamma_2 \\ \vdots \\ \gamma_N \end{pmatrix}$$

with

$$\underline{B}_Y = \begin{pmatrix} 0 & 0 \\ 0 & 0 \\ \vdots & \vdots \\ 0 & V \end{pmatrix} \quad \text{and} \quad \underline{m}_Y = \begin{pmatrix} 0 \\ Y_{N+1} \end{pmatrix}$$

Thus, we get

$$\underline{A}_Y \underline{Y} + \underline{\beta}_X \underline{X} - \underline{\gamma}_Y = \underline{B}_Y \underline{m}_Y \tag{6.67}$$

Similarly, for the molar balance on liquid phase, we get

$$L X_{j-1} + K_{gj} \bar{a}_{m_j} (Y_j - (a X_j + b)) = L X_j \quad j = 1, 2, \ldots, N \tag{6.68}$$

Putting Eq. (6.68) in matrix form gives

$$\underline{A}_X \underline{X} + \underline{\beta}_Y Y - \underline{\gamma}_X = \underline{B}_X \underline{m}_X \tag{6.69}$$

The reader should define $\underline{A}_X, \underline{B}_X, \underline{\gamma}_X$, and m_X.

The matrix Eqs. (6.67) and (6.69) can be solved simultaneously to get the solution, through the following simple matrix manipulation:

Matrix manipulation

From Eq. (6.69), we get

$$\underline{X} = \underline{A}_X^{-1} [\underline{\gamma}_X + \underline{B}_X \underline{m}_X - \underline{\beta}_Y \underline{Y}] \tag{6.70}$$

Substituting Eq. (6.70) into Eq. (6.67), we get

$$\underline{A}_Y \underline{Y} + \underline{\beta}_X \left\{ \underline{A}_X^{-1} \left[\underline{\gamma}_X + \underline{B}_X \underline{m}_X - \underline{\beta}_Y \underline{Y} \right] \right\} - \underline{\gamma}_Y = \underline{B}_Y \underline{m}_Y$$

Some simple manipulation gives

$$\underbrace{\left[\underline{A}_Y - \underline{\beta}_X \underline{A}_X^{-1} \underline{\beta}_Y \right]}_{\underline{C}} Y = \underbrace{\underline{B}_Y \underline{m}_Y + \underline{\gamma}_Y - \underline{\beta}_X \left[\underline{A}_X^{-1} \left[\underline{\gamma}_X + \underline{B}_X \underline{m}_X \right] \right]}_{\underline{M}}$$

where

$$\underline{C} = \underline{A}_Y - \underline{\beta}_X \underline{A}_X^{-1} \underline{\beta}_Y$$
$$\underline{M} = \underline{B}_Y \underline{m}_Y + \underline{\gamma}_Y - \underline{\beta}_X \{ \underline{A}_X^{-1} [\underline{\gamma}_X + \underline{B}_X \underline{m}_X] \}$$

Thus, we get

$$\underline{C} \underline{Y} = \underline{M} \tag{6.71}$$

Matrix Eq. (6.71) can be easily solved for vector $\underline{Y}$ as

$$\underline{Y} = \underline{C}^{-1} \underline{M} \tag{6.72}$$

Thus, we obtain the vector $\underline{Y}$ from matrix Eq. (6.72), and using the solved vector $\underline{Y}$ with matrix Eq. (6.70), we can compute the vector $\underline{X}$.

6.5.2 Packed-Bed Absorption Tower

Packed-bed absorption towers are distributed systems and Figure 6.19 shows the molar flow rates across a small element Δl of one of these towers. We consider the absorption of component A from the gas (vapor) phase to the liquid phase. The rate of mass transfer from the vapor phase to the liquid phase is given by the relation

$$\text{RMT} = a_m K_g (Y - (a X + b)) \tag{6.73}$$

where a_m is the specific area (mass transfer area per unit volume of column) and $(a X + b) = Y^*$ is the gas-phase mole fraction at equilibrium with X.

Molar balance on gas phase over a small element Δl gives

$$V Y = V (Y + \Delta Y) + A \Delta l\, a_m K_g (Y - (a X + b))$$

where V is the total molar flow rate of the gas phase (assumed constant), Y is the mole fraction of component A in the gas phase, and X is the mole

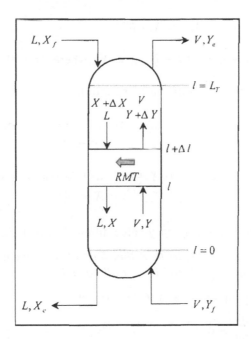

Figure 6.19 Schematic diagram of a packed-bed absorption tower.

fraction of component A in the liquid phase. By simple manipulation and rearrangement, we get

$$V \frac{\Delta Y}{\Delta l} = -A \, a_m \, K_g \, (Y - (a X + b))$$

Taking the limit value as $\Delta l \to 0$, we get

$$V \frac{dY}{dl} = -A \, a_m \, K_g \, (Y - (a X + b)) \tag{6.74}$$

with initial condition, at $l = 0$, $Y = Y_f$.

Molar balance on the liquid phase over a small element Δl gives

$$L X = L (X + \Delta X) + A \, \Delta l \, a_m \, K_g \, (Y - (a X + b))$$

where L is the total molar flow rate of the liquid phase (assumed constant). By simple manipulation and rearrangement, we get

$$L \frac{\Delta X}{\Delta l} = -A \, a_m \, K_g \, (Y - (a X + b))$$

Taking the limit value as $\Delta l \to 0$, we get

$$L \frac{dX}{dl} = -A \, a_m \, K_g \, (Y - (a X + b)) \tag{6.75}$$

with boundary condition at $l = H_t$, $X = X_f$, where, H_t is the total height of the absorption column.

Note that the set of differential equations

$$V \frac{dY}{dl} = -A \, a_m \, K_g \, (Y - (a X + b)) \quad \text{with initial condition } l = 0,$$
$$Y = Y_f$$

$$L \frac{dX}{dl} = -A \, a_m \, K_g \, (Y - (a X + b)) \quad \text{with boundary condition } l = H_t,$$
$$X = X_f$$

form a set of two-point boundary-value differential equations.

However because of the simplicity of this illustrative problem we can actually reduce it to one equation. How? Is it at all possible?

Subtracting Eq. (6.74) and (6.75), we get

$$V \frac{dY}{dl} - L \frac{dX}{dl} = 0$$

It can be rewritten as

$$\frac{d(V\,Y - L\,X)}{dl} = 0$$

On integration, we get

$$V\,Y - L\,X = C_1 \tag{6.76}$$

at $l = 0$, $Y = Y_f$ and $X = X_e$, where X_e is the mole fraction of component A in the liquid phase at the exit (at the bottom). Thus, using the initial condition in Eq. (6.76), we get

$$V\,Y_f - L\,X_e = C_1$$

So, we have obtained the value of the constant of integration C_1.

Now, on substituting the value of C_1 back in Eq. (6.76), we get

$$V\,Y - L\,X = V\,Y_f - L\,X_e$$

Thus,

$$V\,Y = L\,(X - X_e) + V\,Y_f$$

On rearrangement, we get

$$Y = \frac{L}{V}\,(X - X_e) + Y_f \tag{6.77}$$

On substituting the value of Y from Eq. (6.77) into Eq. (6.75), we get

$$L\,\frac{dX}{dl} = -A\,a_m\,K_g\left(\frac{L}{V}\,(X - X_e) + Y_f - (a\,X + b)\right) \tag{6.78}$$

with boundary condition at $l = L$, $X = X_f$.

Equation (6.78) is a single equation which, when solved for every l, gives the value of X; we can use this value of X to obtain the corresponding Y by simple substitution into the algebraic Eq. (6.77). However, the problem is still some kind of a two-point boundary-value differential equation because X_e at $l = 0$ (unknown) is on the right-hand side of Eq. (6.78).

The same relation [Eq. (6.77)] can be obtained by mass balance. Mass balance over the shown boundary in Figure 6.20 gives the following relation:

$$L\,X + V\,Y_f = L\,X_e + V\,Y$$

This gives

$$V\,Y = L\,(X - X_e) + V\,Y_f$$

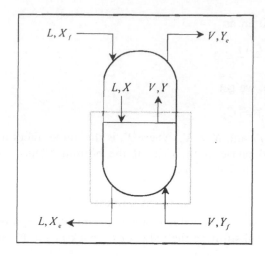

Figure 6.20 Mass balance over the region shown by the dashed lines.

thus giving

$$Y = \frac{L}{V}(X - X_e) + Y_f$$

The above equation is the same as the one obtained by subtracting the two differential equations [Eq. (6.77)].

6.5.3 Diffusion and Reaction in a Porous Structure (Porous Catalyst Pellet)

Consider a spherical particle and a simple reaction

$$A \rightarrow B$$

with rate of reaction equal to

$$r_A = kC_A \frac{\text{mol}}{\text{g catalyst} \times \text{min}}$$

where k is the reaction rate constant and C_A is the concentration of component A (in mol/cm³).

As the considered particle is spherical in shape, it is symmetrical around the center. The concentration profile inside the catalyst pellet is shown in Figure 6.21. Molar balance on component A over the element Δr (here, r is the radial position from the center of the pellet [in cm]) gives

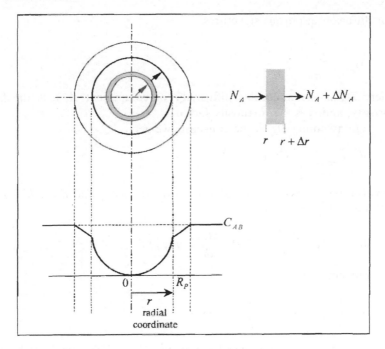

Figure 6.21 Elemental molar balance inside a catalyst pellet.

$$4\pi r^2 N_A = 4\pi(r + \Delta r)^2(N_A + \Delta N_A) + \left(4\pi r^2 \Delta r\right)\rho_C k C_A$$

where N_A is the diffusion flux of component A and ρ_C is the catalyst density, which can be written as

$$r^2 N_A = \left(r^2 + 2r\Delta r + (\Delta r)^2\right)(N_A + \Delta N_A) + r^2 \Delta r \rho_C k C_A$$

Further simplification and neglecting higher powers of Δr and $\Delta r \Delta N_A$ gives

$$0 = r^2 \Delta N_A + 2r N_A \Delta r + r^2 \Delta r \rho_C k C_A$$

On dividing the equation with Δr and taking the limits as $\Delta r \to 0$ and $\Delta N_A \to 0$, we get

$$0 = r^2 \frac{dN_A}{dr} + 2r N_A + r^2 \rho_C k C_A \tag{6.79}$$

On using the simple Fick's law,

$$N_A = -D_{e_A} \frac{dC_A}{dr} \tag{6.80}$$

For diffusion in porous structures,

$$D_{e_A} = \frac{D_A \varepsilon}{\tau}$$

where D_A is the molecular diffusivity of component A, ε is the pellet's porosity, and, τ is the tortuosity factor.

On assuming D_{e_A} to be constant, we get

$$\frac{dN_A}{dr} = -D_{e_A} \frac{d^2 C_A}{dr^2} \tag{6.81}$$

Substituting the expressions of N_A from Eq. (6.80) and dN_A/dr from Eq. (6.81) into Eq. (6.79) gives

$$0 = -r^2 D_{e_A} \frac{d^2 C_A}{dr^2} - 2r D_{e_A} \frac{dC_A}{dr} + r^2 \rho_C k C_A$$

Rearrangement gives

$$D_{e_A} \left(\frac{d^2 C_A}{dr^2} + \frac{2}{r} \frac{dC_A}{dr} \right) = \rho_C k C_A \tag{6.82}$$

Dividing and multiplying the left-hand side of Eq. (6.82) by R_P^2 (where R_P is the pellet's radius) gives

$$\frac{D_{e_A}}{R_P^2} \left(\frac{d^2 C_A}{d(r^2/R_P^2)} + \frac{2}{(r/R_P)} \frac{dC_A}{d(r/R_P)} \right) = \rho_C k C_A \tag{6.83}$$

Defining the dimensionless terms

$$\omega = \frac{r}{R_P}$$

and

$$x_A = \frac{C_A}{C_{A_{ref}}}$$

Eq. (6.83) can be written as

$$\frac{D_{e_A}}{R_P^2} \left(\frac{d^2 x_A}{d\omega^2} + \frac{2}{\omega} \frac{dx_A}{d\omega} \right) = \rho_C k x_A \tag{6.84}$$

which is

$$\nabla^2 x_A = \phi^2 x_A \tag{6.85}$$

where

$$\nabla^2 = \frac{d}{d\omega^2} + \frac{a}{\omega}\frac{d}{d\omega}$$

$$\phi^2 = \frac{\rho_C k R_P^2}{D_{e_A}} \text{(Thiele modulus)}$$

The value of a depends on the shape of the particle as follows:

$a = 2$ for the sphere

$a = 1$ for the cylinder

$a = 0$ for the slab

Boundary Conditions

At the center, due to symmetry, we have that at $\omega = 0$,

$$\frac{dx_A}{d\omega} = 0$$

At the surface ($r = R_P$), the mass transfer at the surface of catalyst pellet is shown in Figure 6.22. Balance at this boundary condition gives

$$4\pi R_P^2 k_{g_1}\left(C_{1_B} - C_A\big|_{r=R_P}\right) + 4\pi R_P^2 N_A\big|_{r=R_P} = 0$$

which simplifies to give

$$k_{g_A}\left(C_{A_B} - C_A\big|_{r=R_P}\right) = -N_A\big|_{r=R_P}$$

Since from Fick's law,

$$N_A = -D_{e_A}\frac{dC_A}{dr}$$

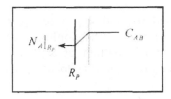

Figure 6.22 Mass transfer across the boundary at the surface of a catalyst pellet.

we obtain

$$k_{g_A}\left(C_{A_B} - C_A\big|_{r=R_P}\right) = \left(\frac{D_{e_A}}{R_P}\right)\frac{dC_A}{d(r/R_P)}\bigg|_{r=R_P}$$

We define

$$x_{A_B} = \frac{C_{A_B}}{C_{\text{ref}}}$$

and get the equation

$$\frac{dx_A}{d\omega}\bigg|_{\omega=1.0} = \underbrace{\left(\frac{R_P k_{g_A}}{D_{e_A}}\right)}_{\text{Sh}_A}\left(x_{A_B} - x_A\big|_{\omega=1.0}\right)$$

where

$$\left(\frac{R_P k_{g_A}}{D_{e_A}}\right) = \text{Sh}_A$$

the Sherwood number for component A. Thus, the second boundary condition at $\omega = 1.0$ is

$$\frac{dx_A}{d\omega} = \text{Sh}_A\left(x_{A_B} - x_A\right)$$

The Limiting Case

When the external mass transfer resistance is negligible (k_{g_A} is large, leading to $\text{Sh}_A \to \infty$), we can write the boundary condition at $\omega = 1.0$ as

$$\frac{1}{\text{Sh}_A}\frac{dx_A}{d\omega} = x_{A_B} - x_A$$

As $\text{Sh}_A \to \infty$ we get

$$x_A\big|_{\omega=1.0} = x_{A_B} \quad \text{(corresponding to negligible external mass transfer resistance)}$$

For both cases of limited Sh_A and $\text{Sh}_A \to \infty$, we get a two-point boundary-value differential equation. For the nonlinear cases, it has to be solved iteratively (we can use Fox's method, as explained for the axial dispersion model in Chapter 4 or orthogonal collocation techniques as explained in Appendix E).

As the above discussed case is linear, it can be solved analytically. How?

Analytical solution

The second-order differential equation for the catalyst pellet is given by

$$\frac{d^2x_A}{d\omega^2} + \frac{2}{\omega}\frac{dx_A}{d\omega} = \phi^2 x_A$$

It can be rewritten as

$$\omega\frac{d^2x_A}{d\omega^2} + 2\frac{dx_A}{d\omega} = \phi^2 x_A \omega \tag{6.86}$$

We define the following new variable:

$$y = x_A\omega \tag{6.87}$$

Differentiation of Eq. (6.87) gives

$$\frac{dy}{d\omega} = \omega\frac{dx_A}{d\omega} + x_A \tag{6.88}$$

The second differential of Eq. (6.88) gives

$$\frac{d^2y}{d\omega^2} = \omega\frac{d^2x_A}{d\omega^2} + \frac{dx_A}{d\omega} + \frac{dx_A}{d\omega} = \omega\frac{d^2x_A}{d\omega^2} + 2\frac{dx_A}{d\omega} \tag{6.89}$$

From Eqs. (6.86) and (6.89), we get

$$\frac{d^2y}{d\omega^2} = \phi^2 y \tag{6.90}$$

We can write Eq. (6.90) in the operator form as

$$D^2 y - \phi^2 y = 0$$

The characteristic equation for the above equation is

$$\lambda^2 - \phi^2 = 0$$

which gives the eigenvalues as

$$\lambda_1 = \phi \quad\text{and}\quad \lambda_2 = -\phi$$

Thus, the solution of Eq. (6.90) is given by

$$y = C_1 e^{\phi\omega} + C_2 e^{-\phi\omega} \tag{6.91}$$

C_1 and C_2 can be calculated using the boundary conditions and the complete analytical solution can be obtained.

The reader is advised to carry out the necessary calculations to get the complete analytical solution as an exercise.

For nonlinear cases, the nonlinear two-point boundary-value differential equation(s) of the catalyst pellet can be solved using the Fox's iterative method explained for the axial dispersion model in Chapter 4 or the orthogonal collocation technique as explained in Appendix E.

6.6 DYNAMIC CASES

6.6.1 The Multitray Absorption Tower

Consider the equilibrium tray (jth tray) as shown schematically in Figure 6.23. The holdup (V_t) on the tray will have two parts: V_L and V_G. So we can write

$$V_t = V_L + V_G$$

Thus, we can write the dynamic molar balance equation as

$$V_L \frac{dX_j}{dt} + V_G \frac{dY_j}{dt} = (LX_{j-1} + VY_{j+1}) - (LX_j + VY_j) \tag{6.92}$$

For equilibrium,

$$Y_j = aX_j + b$$

Thus, Eq. (6.92) becomes

$$V_L \frac{dX_j}{dt} + (V_G a) \frac{dX_j}{dt} = (LX_{j-1} + V(aX_{j+1} + b)) - (LX_j + V(aX_j + b))$$

It can be further simplified to get

$$\underbrace{(V_L + V_G a)}_{\alpha} \frac{dX_j}{dt} = (LX_{j-1} + V(aX_{j+1} + b)) - (LX_j + V(aX_j + b))$$

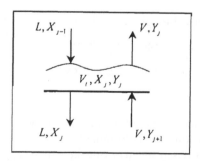

Figure 6.23 Molar flow on the tray.

The reader can arrange the above equation into matrix form, as we did earlier for the steady-state case; it becomes a matrix differential equation:

$$\alpha \frac{dX}{dt} = \underline{A}\,\underline{X} + \underline{B}\,\underline{m} \tag{6.93}$$

with initial condition at $t = 0$, $\underline{X} = \underline{X}_0$, where $\underline{X}_0$ is the initial conditions vector.

If the equilibrium relation is not used, then we carry out an unsteady molar balance for each phase with mass transfer between the two phases. The reader should do that as an exercise.

6.6.2 Dynamic Model for the Catalyst Pellet

For the same parameters and reaction as the steady-state case discussed earlier, unsteady-state molar balance on component A gives

$$4\pi r^2 N_A = 4\pi (r + \Delta r)^2 (N_A + \Delta N_A) + (4\pi r^2 \Delta r)\rho_c k C_A$$
$$+ \left(4\pi r^2 \Delta r \varepsilon\right)\frac{\partial C_A}{\partial t} + \left(4\pi r^2 \Delta r (1 - \varepsilon)\right)\rho_S \frac{\partial C_{AS}}{\partial t}$$

where, C_{A_s} is the concentration of the adsorbed (chemisorbed) A on the surface of the catalyst and is expressed in terms of gmol/g catalyst.

On some rearrangement, we get

$$r^2 N_A = \left(r^2 + 2r\Delta r + (\Delta r)^2\right)(N_A + \Delta N_A) + r^2 \Delta r \rho_c k C_A$$
$$+ \left(r^2 \Delta r \varepsilon\right)\frac{\partial C_A}{\partial t} + \left(r^2 \Delta r (1 - \varepsilon)\right)\rho_S \frac{\partial C_{AS}}{\partial t}$$

Further simplification and neglecting higher powers of Δr and $\Delta r \Delta N_A$ gives

$$0 = r^2 \Delta N_A + 2r N_A \Delta r + r^2 \Delta r \rho_c k C_A + \left(r^2 \Delta r \varepsilon\right)\frac{\partial C_A}{\partial t}$$
$$+ \left(r^2 \Delta r (1 - \varepsilon)\right)\rho_S \frac{\partial C_{AS}}{\partial t}$$

On dividing the equation by Δr and taking the limits as $\Delta r \to 0$ and $\Delta N_A \to 0$, we get

$$0 = r^2 \frac{\partial N_A}{\partial r} + 2r N_A + r^2 \rho_c k C_A + \left(r^2 \varepsilon\right)\frac{\partial C_A}{\partial t} + \left(r^2 (1 - \varepsilon)\right)\rho_S \frac{\partial C_{AS}}{\partial t}$$

Dividing the entire equation by r^2 gives

$$0 = \frac{\partial N_A}{\partial r} + \frac{2}{r} N_A + \rho_c k C_A + \varepsilon \frac{\partial C_A}{\partial t} + (1 - \varepsilon)\rho_S \frac{\partial C_{AS}}{\partial t} \tag{6.94}$$

On using the simple Fick's law,

$$N_A = -D_{e_A} \frac{\partial C_A}{\partial r} \tag{6.95}$$

Assuming D_{e_A} to be constant, we get

$$\frac{\partial N_A}{\partial r} = -D_{e_A} \frac{\partial^2 C_A}{\partial r^2} \tag{6.96}$$

Substituting Eqs. (6.95) and (6.96) into Eq. (6.94) and after some manipulations, we obtain

$$D_{e_A}\left(\frac{\partial^2 C_A}{\partial r^2} + \frac{2}{r}\frac{\partial C_A}{\partial r}\right) - \rho_C k C_A = \varepsilon \frac{\partial C_A}{\partial t} + (1 - \varepsilon)\rho_S \frac{\partial C_{AS}}{\partial t} \tag{6.97}$$

If the chemisorption is at equilibrium and the chemisorption isotherm is assumed to be linear, then the relation between C_A and C_{AS} is given by

$$C_{AS} = (K_A \overline{C}_m) C_A \tag{6.98}$$

where K_A is the equilibrium chemisorption constant and $\overline{C}_m$ is the concentration of active sites on the catalyst surface. Thus,

$$D_{e_A}\left(\frac{\partial^2 C_A}{\partial r^2} + \frac{2}{r}\frac{\partial C_A}{\partial r}\right) - \rho_C k C_A = \varepsilon \frac{\partial C_A}{\partial t} + (1 - \varepsilon)\rho_S K_A \overline{C}_m \frac{\partial C_A}{\partial t}$$

Defining the dimensionless variables

$$\omega = \frac{r}{R_P}$$

and

$$x_A = \frac{C_A}{C_{A_{ref}}}$$

we get

$$\left(\underbrace{(1 - \varepsilon)\rho_S K_A \overline{C}_m}_{\alpha_S} + \underbrace{\varepsilon}_{\alpha_g}\right)\frac{\partial C_A}{\partial t} = \frac{D_{e_A}}{R_P^2}\left(\frac{\partial^2 C_A}{\partial (r/R_P)^2} + \frac{2}{\frac{r}{R_P}}\frac{\partial C_A}{\partial (r/R_P)}\right) - \rho_C k C_A$$

where, $(1 - \varepsilon)\rho_S K_A \overline{C}_m = \alpha_S$ is the catalyst solid surface capacitance and $\varepsilon = \alpha_g$ is the catalyst pore capacitance (usually negligible). Thus, we obtain

$$\frac{R_P^2}{D_{e_A}}(\alpha_S + \alpha_g)\frac{\partial x_A}{\partial t} = \left(\frac{\partial^2 x_A}{\partial \omega^2} + \frac{2}{\omega}\frac{\partial x_A}{\partial \omega}\right) - \left(\frac{\rho_C k R_P^2}{D_{e_A}}\right)x_A \tag{6.99}$$

which can be written as

$$(Le_S + Le_g)\frac{\partial x_A}{\partial t} = \nabla^2 x_A - \phi^2 x_A \tag{6.100}$$

where

$$\nabla^2 = \frac{d}{d\omega^2} + \frac{a}{\omega}\frac{d}{d\omega}, \quad \phi^2 = \frac{\rho_C k R_P^2}{D_{e_A}} \text{ (Thiele modulus)}$$

and

$$Le_S = \frac{\alpha_S R_P^2}{D_{e_A}}$$

is the Lewis number for the solid catalyst surface and

$$Le_g = \frac{\alpha_g R_P^2}{D_{e_A}}$$

is the Lewis number for the catalyst gas voidage. The boundary conditions and initial condition are as discussed earlier: at $t = 0$, $x_A(\omega) = x_{A_o}(\omega)$. At the center, due to symmetry, we have, at $\omega = 0$, $dx_A/d\omega = 0$. At the surface $(r = R_P)$, at $\omega = 1.0$,

$$\frac{dx_A}{d\omega} = Sh_A(x_{A_B} - x_A)$$

For the limiting case, when $Le_S >>> Le_g$, we can neglect Le_g and call $Le_S = Le$; Eq. (6.100) becomes

$$Le\frac{\partial x_A}{\partial t} = \nabla^2 x_A - \phi^2 x_A \tag{6.101}$$

with the same initial and boundary conditions as given earlier.

6.7 MATHEMATICAL MODELING AND SIMULATION OF FLUIDIZED-BED REACTORS

When a flow of gas stream is passed through a bed of fine powder put in a tube (as shown in Fig. 6.24), we can observe the following:

1. In a certain range of flow, we will have a fixed bed (i.e., flow through porous medium).
2. The minimum fluidization condition is a condition at which ΔP across the bed is equal to the bed weight; a slight expansion occurs in this case (Fig. 6.24). The bed properties will be very similar to liquid properties (e.g., it can be transferred between two containers).

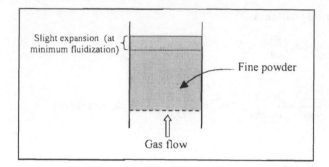

Figure 6.24 Fine powder entrained in a tube starting to fluidize due to gas flow.

3. Freely bubbling fluidized bed: With higher flow rate of gas than the minimum fluidization limit, we will reach a case in which there will be three phases: solid, gas in contact with solid, and gas in bubbles.

The two-phase theory of fluidization is applied: "Almost all the gas in excess of that necessary for minimum fluidization will appear as gas bubbles" (Fig. 6.25).

6.7.1 Advantages of Freely Bubbling Fluidized Beds

1. Perfect mixing of solids due to the presence of bubbles (isothermality).
2. Good heat transfer characteristics (high heat transfer coefficient).
3. Intraparticle as well as external mass and heat transfer resistances are negligible (so it can be neglected in modeling). (See Fig. 6.26.)
4. All of the advantages of minimum fluidization conditions.

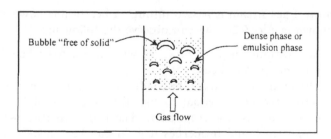

Figure 6.25 Fluidized bed.

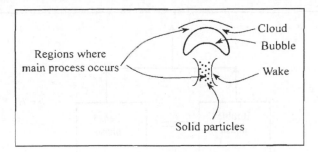

Figure 6.26 Diagram showing the main mass and heat transfer regions around the bubble.

6.7.2 Disadvantages of Fluidized Beds

1. Bypassing of bubbles through the bed.
2. Bubble explosion on surface causes entrainment.
3. Difficult mechanical design [e.g., industrial fluid catalytic cracking (FCC) unit has to handle 7000 tons of solid].

Notes:

- Bypassing of bubbles is compensated partially by diffusion between dense and bubble phases
- There is a solid exchange among the wake, cloud, and the dense phase. It is accounted for in three-phase models.
- Although the dense phase is perfectly mixed, the bubble phase is almost in plug flow condition.
- It is possible to break the bubbles using baffles or redistributors. Stirrers are not recommended because of vortex formation.

6.7.3 Mathematical Formulation (Steady State)

For illustration, consider a simple reaction

$$A \rightarrow D$$

Refer to Figure 6.27, where $C_A \neq f_1(h)$, which means that it is perfectly mixed condition, and $C_{A_B} = f_2(h)$, which means that it is plug flow condition. Here, C_A is the concentration of component A in the dense phase and C_{A_B} is the concentration of component A in the bubble phase.

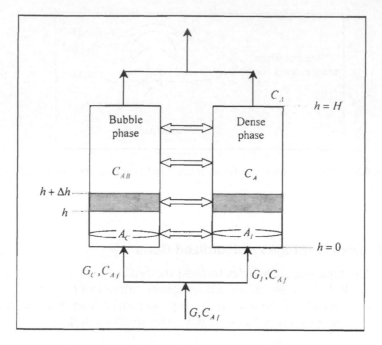

Figure 6.27 Schematic diagram of a fluidized bed.

Modeling of the Dense Phase

Molar balance on the dense phase gives

$$n_{Af} + \text{Exchange with bubble phase} = n_A - Vr$$

For constant volumetric flow rate G_I, we can write the above equation as

$$G_I C_{Af} + \int_0^H K_{gA} a (C_{AB} - C_A) \cdot 1_C \, dh = G_I C_A + A_I H \rho_b k C_A \qquad (6.102)$$

where

$$\rho_b = \text{bulk density of solid at minimum fluidization conditions}$$

$$a = \frac{\text{cm}^2 \text{ external surface area of bubbles}}{\text{cm}^3 \text{ volume of bubble phase}}$$

$$K_{gA} = \text{mass transfer coefficient between bubble and dense phase}$$

Modeling of the Bubble Phase

Molar balance on an element of the bubble phase gives (it is the plug flow mode and assumed to have negligible rate of reaction; see Fig. 6.28),

$$G_C C_{AB} = G_C(C_{AB} + \Delta C_{AB}) + K_{gA} a(C_{AB} - C_A) A_C \Delta h$$

With some manipulation and taking the limits as $\Delta C_{A_B} \to 0$ and $\Delta h \to 0$, we get

$$G_C \frac{dC_{AB}}{dh} = - \underbrace{K_{gA} a}_{Q_E} A_C(C_{AB} - C_A) \qquad (6.103)$$

with initial condition at $h = 0$, $\quad C_{AB} = C_{A_f}$. Thus, the balance gives the following two equations:

Dense phase (an integral equation)

$$G_I \left(C_{A_I} - C_A \right) + Q_E A_C \int_0^H (C_{AB} - C_A) \, dh = A_I \, H \, \rho_b \, k \, C_A \qquad (6.102)$$

Bubble phase (a differential equation which can be solved analytically)

$$G_C \frac{dC_{AB}}{dh} = -Q_E A_C (C_{AB} - C_A) \qquad (6.103)$$

at $h = 0$, $\quad C_{AB} = C_{A_f}$.

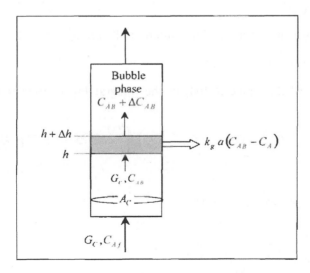

Figure 6.28 Mass flow in bubble phase.

Analytical Solution of Differential Equation (6.103)

By separation of variables, we get

$$\frac{dC_{AB}}{\left(C_{AB} - \underbrace{C_A}_{\text{constant}}\right)} = -\left(\frac{Q_E A_C}{G_C}\right) dh$$

On integration, we get

$$\ln(C_{AB} - C_A) = -\left(\frac{Q_E A_C}{G_C}\right) h + C_1$$

To get the value of constant of integration C_1, we use the initial condition at $h = 0$,

$$C_1 = \ln\left(C_{A_f} - C_A\right)$$

Thus, we get

$$\ln\left(\frac{C_{AB} - C_A}{C_{A_f} - C_A}\right) = -\underbrace{\left(\frac{Q_E A_C}{G_C}\right)}_{\alpha} h$$

Rearrangement gives

$$\frac{C_{AB} - C_A}{C_{A_f} - C_A} = e^{-\alpha h}$$

Finally, we get the solution of the differential equation as

$$C_{AB} - C_A = \left(C_{A_f} - C_A\right) e^{-\alpha h} \tag{6.104}$$

Substitution of this value (6.104) in the integral Eq. (6.102) of the dense phase gives

$$G_I\left(C_{A_f} - C_A\right) + Q_E A_C\left(C_{A_f} - C_A\right) \int_0^H e^{-\alpha h}\, dh = A_I H \rho_h k C_A$$

Because

$$\int_0^H e^{-\alpha h}\, dh = \frac{1}{\alpha}\left(1 - e^{-\alpha H}\right)$$

and

$$\alpha = \frac{Q_E A_C}{G_C}$$

we get

$$G_I(C_{Af} - C_A) + G_C(C_{Af} - C_A)(1 - e^{-\alpha H}) = A_I H \rho_b k C_A$$

In a more simplified and organized form

$$\underbrace{(G_I + G_C(1 - e^{-\alpha H}))}_{\overline{G}}(C_{Af} - C_A) = A_I H \rho_b k C_A \qquad (6.105)$$

where

$$\overline{G} = G_I + G_C(1 - e^{-\alpha H})$$

the modified volumetric flow rate.

Notes

1. Analogy between the above equations and the continuous-stirred tank reactor (CSTR) model can be easily realized, and from this analogy, we can define the modified flow rate with the following physical significance:

   - At very high Q_E, $\overline{G} = (G_I + G_C) = G$; thus, we approach CSTR.
   - At $Q_E \cong 0$, $G = G_I$; thus, we have complete segregation.

2. Output concentration of A is calculated by the relation $G_I C_A + G_C C_{A_B}|_{h=H} = G C_{A_{out}}$

Heat Balance Design Equations for the Fluidized-Bed Reactors (see Fig. 6.29)

For a reaction

$$A \to D$$

where $r = kC_A$ and $k = k_0 e^{-E/RT}$

Heat balance in the dense phase gives

$$G_I(T - T_f)\rho C_P = A_I H \rho_b k C_A(-\Delta H)$$
$$+ \int_0^H h_{Ba} a A_C(T_B - T) \, dh - U A_J(T - T_J) \qquad (6.106)$$

Note that the heat supply/removal coil is considered to be with the dense-phase balance, where

T_B = bubble phase temperature (varies with height)
T_J = jacket temperature
T = dense phase temperature (constant with height)

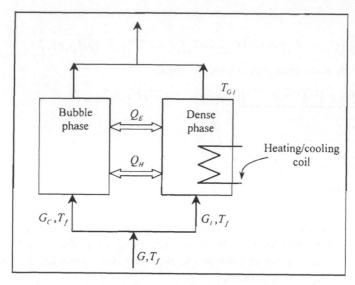

Figure 6.29   Heat transfer in the fluidized bed.

U = heat transfer coefficient between the jacket and dense phase
A_J = jacket area available for heat transfer

Heat balance in the bubble phase gives (it is assumed to be in plug flow mode and with negligible rate of reaction), (see Fig. 6.30)

$$G_C \, \rho \, C_P \, T_B = G_C \, \rho \, C_P \, (T_B + \Delta T_B) + h_B \, a \, A_C \, \Delta h \, (T_B - T)$$

which, after some rearrangements and taking the limits as $\Delta T_B \to 0$ and $\Delta h \to 0$, gives

$$G_C \, \rho \, C_P \, \frac{dT_B}{dh} = h_B \, a \, A_C \, (T - T_B) \tag{6.107}$$

Analytical Solution of Eq. (6.107)

On separating the variables of differential Eq. (6.107), we get

$$\frac{dT_B}{T - T_B} = \beta \, dh$$

where

$$\beta = \frac{h_B \, a \, A_C}{G_C \, \rho \, C_P}$$

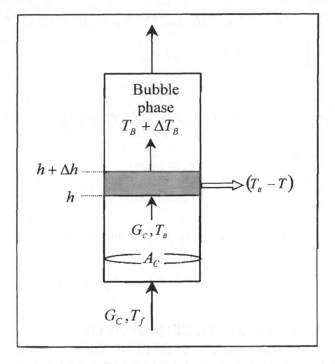

Figure 6.30 Heat flow in the bubble phase.

On integration, we get

$$-\ln(T - T_B) = \beta h + C_1$$

The integration constant C_1 can be calculated using the initial condition at $h = 0$, $T_B = T_f$. Thus, we get

$$C_1 = -\ln(T - T_f)$$

On substituting the value of constant of integration, we get

$$\ln\left(\frac{T - T_B}{T - T_f}\right) = -\beta h$$

On rearrangement, we get

$$\frac{T - T_B}{T - T_f} = e^{-\beta h}$$

We can use the above relation in the heat balance for the dense phase (6.106) in order to calculate the heat transfer integral; we get

$$G_I \rho C_P (T - T_f) = A_I H \rho_b k_0 e^{-E/RT} C_A (-\Delta H) + h_B a A_C (T - T_f)$$
$$\times \int_0^H e^{-\beta h} dh - UA_J (T - T_J)$$

Because

$$\int_0^H e^{-\beta h} \, dh = \frac{1}{\beta} \left(1 - e^{-\beta H} \right) = \frac{G_C \rho C_P}{h_B a A_C} \left(1 - e^{-\beta H} \right)$$

Thus, we can write

$$\left[G_I \rho C_P + G_C \rho C_P \left(1 - e^{-\beta H} \right) \right] (T - T_f) = A_I H \rho_b k_0 e^{-E/RT} C_A (-\Delta H)$$
$$- UA_J (T - T_J)$$

For a fluidized bed without chemical reaction, the heat transfer equation will be

$$\rho C_P \left[G_I + G_C \left(1 - e^{-\beta H} \right) \right] (T - T_f) = -UA_J (T - T_J) \qquad (6.108)$$

6.8 UNSTEADY-STATE BEHAVIOR OF HETEROGENEOUS SYSTEMS: APPLICATION TO FLUIDIZED-BED CATALYTIC REACTORS

In the mass balance equation, an accumulation term will be added as follows:

$$\left(G_I + G_C \left(1 - e^{-(Q_E A_C / G_C) H} \right) \right) (C_{Af} - C_A) = A_I H \rho_b k C_A$$
$$+ \text{Accumulation or depletion with time} \qquad (6.109)$$

where

> Accumulation or depletion with time = The unsteady state dynamic term

and we can write the holdup of any component (e.g., component A) as

$$\bar{n}_A = A_I H \varepsilon C_A + A_I H \rho_b C_{AS} \qquad (6.110)$$

where

> C_A is the gas-phase concentration of component A
> C_{AS} = solid phase (chemisorbed) concentration of component A
> $\bar{n}_A$ = number of moles of component A in the fluidized bed reactor (component A holdup)

$$\rho_b = \frac{\text{g catalyst}}{\text{cm}^3 \text{ bed}} = \rho_S(1 - \varepsilon)$$

$$\rho_S = \frac{\text{g catalyst}}{\text{cm}^3 \text{ catalyst}}$$

On differentiation of $\bar{n}_A$ [in Eq. (6.110)], we get

$$\frac{d\bar{n}_A}{dt} = A_I H\left(\varepsilon \frac{dC_A}{dt} + \rho_b \frac{dC_{AS}}{dt}\right) = \text{Accumulation or depletion with time}$$

(6.111)

Substituting Eq. (6.111) into Eq. (6.109) gives

$$\left(G_I + G_C\left(1 - e^{-(Q_E A_C/G_C)H}\right)\right)\left(C_{Af} - C_A\right) = A_I H \rho_b k C_A$$
$$+ A_I H\left(\varepsilon \frac{dC_A}{dt} + \rho_b \frac{dC_{AS}}{dt}\right)$$

(6.112)

Thus, we get a differential equation in which two variables (C_A and C_{AS}) are functions of time. How do we handle putting C_{AS} (the chemisorbed concentration of A) in terms of C_A (the gas phase concentration of A)? The simplest approach is the one given as follows:

Chemisorption mechanism

We give some detail regarding the relation between C_A and C_{AS}.

Reactions:

$$A \rightarrow D$$
$$A + S \Leftrightarrow AS \text{ at equilibrium}$$
$$AS \rightarrow D + S$$

where S is the active site

Thus, we can write

$$K_A = \frac{C_{AS}}{C_A C'_S}$$

where,

$$C'_S = \bar{C}_m - C_{AS}, \text{ where } C'_S \text{ is the concentration of free active sites.}$$

and, $\bar{C}_m$ is the total concentration of active sites. Thus, so we get,

$$K_A C_A\left(\bar{C}_m - C_{AS}\right) = C_{AS}$$

Thus,

$$C_{AS} = \frac{K_A \bar{C}_m C_A}{1 + K_A C_A}$$

This is a Langmuir isotherm. Considering the low concentration of C_A (i.e., linear relation between C_A and C_{AS}), we get

$$C_{AS} = (K_A \overline{C}_m) C_A \text{ (as earlier)} \tag{6.113}$$

On differentiation of Eq. (6.113) and assuming that K_A and $\overline{C}_m$ are constant, we get

$$\frac{dC_{AS}}{dt} = (K_A \overline{C}_m) \frac{dC_A}{dt} \tag{6.114}$$

Thus, the final mass balance equation derived earlier [Eq. (6.112)] becomes

$$
\begin{aligned}
&\left(G_I + G_C \left(1 - e^{-(Q_E A_C / G_C H)} \right) \right) (C_{Af} - C_A) = A_I H \rho_b k C_A \\
&+ A_I H \left(\varepsilon + K_A \overline{C}_m \rho_b \right) \frac{dC_A}{dt}
\end{aligned}
\tag{6.115}
$$

Note: It is always difficult to find the values of K_A and $\overline{C}_m$, except for very common processes.

Utilizing the *nonlinear isotherm* will complicate this dynamic term as follows. Because

$$C_{AS} = \frac{K_A \overline{C}_m C_A}{1 + K_A C_A}$$

differentiating gives

$$\frac{dC_{AS}}{dt} = K_A \overline{C}_m \left((1 + K_A C_A)^{-1} \frac{dC_A}{dt} + C_A(-1)(1 + K_A C_A)^{-2} K_A \frac{dC_A}{dt} \right)$$

Thus, we get

$$\frac{dC_{AS}}{dt} = K_A \overline{C}_m \left(\frac{1}{1 + K_A C_A} - \frac{K_A C_A}{(1 + K_A C_A)^2} \right) \frac{dC_A}{dt} \tag{6.116}$$

For the nonisothermal case, the heat balance equation is

$$\gamma(T - T_f) = A_I H k_0 e^{-E/RT} \rho_b C_A(-\Delta H) - U A_J (T - T_J) - \frac{d\overline{Q}}{dt} \tag{6.117}$$

where

$$\overline{Q} = A_I H \varepsilon \rho_g C_{P_g} T_g + A_I H \rho_b C_{PS} T_S \tag{6.118}$$

is the heat content of the fluidized bed reactor and

$$\gamma = \left(G_I \rho C_P + G_C \rho C_P \left(1 - e^{-\beta H} \right) \right)$$

Assuming negligible heat transfer resistances between the gas and solid, we get

$$T_g = T_S = T$$

On differentiating Eq. (6.118), we get

$$\frac{d\overline{Q}}{dt} = A_I H \left(\underbrace{\varepsilon \rho_g C_{P_g}}_{\text{negligible}} + \rho_b C_{PS} \right) \frac{dT}{dt}$$

Thus, we get

$$\frac{d\overline{Q}}{dt} = A_I H \rho_b C_P S \frac{dT}{dt} \tag{6.119}$$

The expression for $d\overline{Q}/dt$ from Eq. (6.119) can be substituted in Eq. (6.117) to get the final unsteady heat balance equation

$$\gamma(T - T_f) = A_I H k_0 e^{-E/RT} \rho_b C_A(-\Delta H) - UA_J(T - T_J)$$
$$- A_I H \rho_b C_{PS} \frac{dT}{dt} \tag{6.120}$$

Note: For the case of the linear isotherm, we can define a Lewis number for the process as follows:

$$\frac{\text{Heat capacitance term}}{\text{Mass capacitance term}} = \frac{A_I H \rho_b C_{PS}}{A_I H(\varepsilon + K_A \overline{C}_m \rho_b)} = \frac{\rho_b C_{PS}}{(\varepsilon + K_A \overline{C}_m \rho_b)}$$
$$= \text{Lewis number (Le)}$$

6.9 EXAMPLE: SIMULATION OF A BUBBLING FLUIDIZED-BED CATALYTIC REACTOR

A consecutive reaction

$$A \rightarrow B \rightarrow C$$

is taking place in a bubbling fluidized-bed reactor with both reactions being exothermic and of first order. The intermediate product B is the desired product.

Figure 6.31 shows the a schematic representation of this two-phase fluidized-bed reactor with a simple proportional control. It should be noted that the proportional control is based on the exit temperature (the average between the dense-phase and the bubble-phase temperatures), which is the measured variable, and the steam flow to the feed heater is the manipulated variable.

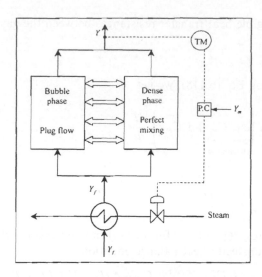

Figure 6.31 Simulation model for two-phase fluidized-bed reactor with proportional control.

As we have shown in the previous sections, the dynamics of the dense phase are described by nonlinear ordinary differential equations (ODEs) on time having an integral term for the mass and heat transfer between the dense and the bubble phases. The bubble phase is described by pseudo-steady state linear ODEs on height. The linear ODEs of the bubble phase are solved analytically and the solution is used to evaluate the integrals in dense-phase equations. The reader should do this as practice to reach the dimensionless model equations given next.

The dynamic equations describing this fluidized bed after the above-described manipulations are the following nonlinear ODEs in terms of dimensionless variables and parameters:

$$\frac{1}{\text{Le}_A}\frac{dX_A}{d\tau} = \overline{B}(X_{Af} - X_A) - \alpha_1 e^{-(\gamma_1/Y)}X_A$$

$$\frac{1}{\text{Le}_B}\frac{dX_B}{d\tau} = \overline{B}(X_{Bf} - X_B) + \alpha_1 e^{-(\gamma_1/Y)}X_A - \alpha_2 e^{-(\gamma_2/Y)}X_B$$

$$\frac{dY}{d\tau} = \overline{B}(Y_f - Y) + \alpha_1\beta_1 e^{-(\gamma_1/Y)}X_A + \alpha_2\beta_2 e^{-(\gamma_2/Y)}X_B$$

where X_j is the dimensionless dense-phase concentration of the component j ($j = A, B, C$) and Y is the dimensionless dense-phase temperature. The dimensionless feed temperature (after neglecting the dynamics of the feed heater) is given by

$$\overline{Y}_f = Y_f + K(Y_m - Y)$$

The base values for Y_f is Y_{f0} and $\overline{B}$ is the reciprocal of the effective residence time of the bed given by

$$\overline{B} = \frac{G_I + G_C(1 - e^{-a})}{A_I H}$$

The yield of the reactor is given by

$$y = \frac{G_I X_B + G_C \overline{X}_{BH}}{(G_I + G_C)X_{Af}}$$

where

$$\overline{X}_{BH} = X_B + \left(X_{Bf} - X_B\right)e^{-a}$$

and

$$a = \frac{Q_E H A_C}{G_C}$$

The dimensionless parameters are defined as follows:

α_i = dimensionless pre-exponential factor for the reaction i
$\quad = \rho_S(1 - \varepsilon)k_{oi}$
β_i = dimensionless exothermicity factor for the reaction i
$\quad = \dfrac{(-\Delta H_i)C_{ref}}{\rho_f C_{Pf} T_{ref}}$
γ_i = dimensionless activation energy for the reaction i
$\quad = \dfrac{E_i}{RT_{ref}}$
τ = normalized time (time/heat capacity of the system)
Le_j = Lewis number of component j
$\quad = \dfrac{C_{PS}}{K_j C_m C_{Pf} \rho_f}$

where

ρ_S = solid catalyst density
C_{PS} = specific heat of the catalyst
ε = voidage occupied by the gas in the dense phase
$-\Delta H_i$ = heat of reaction for reaction i
k_{0_i} = pre-exponential factor for reaction i

C_{ref} = reference concentration

T_{ref} = reference temperature

ρ_f = density of the gas

C_{Pf} = specific heat of the gas

E_i = activation energy for reaction i

$\overline{K}_j$ = chemisorption equilibrium constant for component j

$\overline{C}_m$ = concentration of active sites in the catalyst

The values of parameters used in this simulation are tabulated in Table 6.1.

Figure 6.32 shows that the desired steady state (corresponding to the highest yield of intermediate desired product B) corresponds to a middle unstable steady state. In order to stabilize this desired steady state, the use of a simple proportional controller is proposed and the behavior is studied for different values of the proportional gain (K), as shown in Fig. 6.32B. It can

Table 6.1 Data Used for the Simulation

Bed cross-section area, A	3000 cm^2
Superficial gas velocity (based on unit cross section of the whole bed), U_2	10.0 cm/s
Minimum fluidization velocity (based on unit cross section of the whole bed), U_{mf}	0.875 cm/s
Voidage of the dense phase	0.4
Bed height, H	100 cm
Normalized pre-exponential factor for the reaction $A \rightarrow B$	10^8
Normalized pre-exponential factor for the reaction $B \rightarrow C$	10^{11}
Dimensionless overall thermicity factor for the reaction $A \rightarrow B$	0.4
Dimensionless overall thermicity factor for the reaction $B \rightarrow C$	0.6
Dimensionless activation energy for the reaction $A \rightarrow B$	18.0
Dimensionless activation energy for the reaction $B \rightarrow C$	27.0
Lewis number of component A, Le$_A$	1.0
Lewis number of component B, Le$_B$	2.2
Feed concentration of component A, X_{Af}	1.0
Feed concentration of component B, X_{Bf}	0.0
Feed concentration of component C, X_{Cf}	0.0
Dimensionless feed temperature to the reactor (base value), Y_f	0.553420765035
Set point for the controller, Y_m	0.929551298359

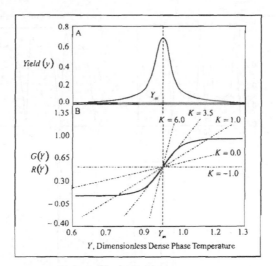

Figure 6.32 Yield and van Heerden diagrams for the data in Table 6.1. (A) Yield (*y*) of *B* versus dimensionless dense-phase temperature (*Y*), where Y_m is *Y* at the maximum yield of *B*; (B) van Heerden diagram showing the heat generation function of the system $G(Y)$ and heat removal line $R(Y)$ for the different values of *K* versus the dimensionless dense-phase temperature *Y*.

be observed that a *K* value of at least 3.5 is necessary to stabilize this unstable saddle-type steady state.

The dynamic behavior of this three-dimensional system is quite rich and complicated. The reader can find the detailed study in the paper published by one of the authors of this book (1). A sample of the results is shown in Fig. 6.33, in which the complex oscillations of the state variables can be observed.

6.10   A DISTRIBUTED PARAMETER DIFFUSION–REACTION MODEL FOR THE ALCOHOLIC FERMENTATION PROCESS

The following is an example for the development of a distributed parameter model for the yeast floc in the alcoholic fermentation process. The model takes into consideration the external mass transfer resistances, the mass transfer resistance through the cellular membrane, and the diffusion resistances inside the floc. The two-point boundary-value differential equations for the membrane are manipulated analytically, whereas the nonlinear two-point boundary-value differential equations of diffusion and reaction inside

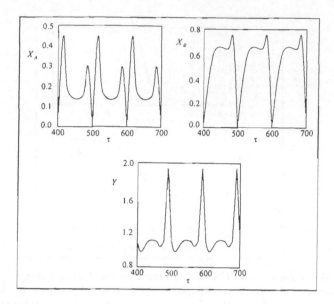

Figure 6.33 Classical dynamic characteristics (time traces) for a case with $K = 3.083691$.

the floc are approximated using the orthogonal collocation technique (Appendix E).

The evaluation of the necessary diffusion coefficients have involved a relatively large number of assumptions because of the present limited knowledge regarding the complex process of diffusion and biochemical reactions in these systems.

A comparison between the model and an experimental laboratory batch fermentor as well as an industrial fed-batch fermentor is also presented to the reader. The model is shown to simulate reasonably well the experimental results, with the largest deviation being for the concentration of yeast.

A heterogeneous model is being developed by taking three mass transfer resistances in series for both ethanol and sugar; these are as follows:

1. External mass transfer resistances that depend on the physical properties of the bulk fluid (extracellular fluid) and the degree of mixing as well as temperature.
2. Diffusion through the cell membrane: A process that is not fully understood and which is affected by many factors in a complex way; for example, it is well known that ethanol increases the fluidity of biological membranes; in addition, ethanol causes a

change in phospholipid composition and a decrease of the lipid to protein ratio of the membrane. At steady state, the transport of sugar through biological membranes is considered a carrier-mediated process, which consists of diffusion and biochemical reactions in the membrane. Glucose diffusion versus concentration gradient has been beautifully simulated by Kernevez for the glucose pump formed of artificial membranes (2). However, for the present unsteady-state model, all bioreactions are assumed to take place inside the cell, with the assumption that no reaction is taking place in the membrane. Therefore, Fickian diffusion of sugar is considered through the membrane with the appropriate diffusion coefficient.

3. Intracellular diffusion, which is dependent on the intracellular conditions. A reasonably rigorous description of this diffusion process requires a distributed diffusion–reaction model. A more rigorous description, which is not considered in this undergraduate book, requires a more structured diffusion–reaction model.

6.10.1 Background on the Problems Associated with the Heterogeneous Modeling of Alcoholic Fermentation Processes

The rigorous heterogeneous modeling of alcoholic fermentation process requires a large amount of fundamental information which is not completely available in the literature now. However, rational mathematical models based on the available knowledge regarding this process represent considerable improvement over the usual pseudohomogeneous models. Such models also help to direct experimental investigations in a more organized manner.

The alcoholic fermentation process is quite a complex process and involves the following main processes.

Product Inhibition

The ethanol inhibition effect is of the non-competitive type, where ethanol concentration affects only the maximum specific growth rate. This dependence of the maximum specific rate of growth of ethanol concentration involves two main types of dependence: linear dependence and nonlinear dependence. Jobses and Roels [3] have shown that inhibition kinetics can be approximated by a linear relation between the specific growth rate and the ethanol concentration up to 50 g/L ethanol. Above this level, deviation from linearity is observed.

As discussed by Elnashaie and Ibrahim [4], many investigators reported that produced ethanol (intracellular ethanol) is more toxic than added ethanol (extracellular ethanol) and an intermediate value of $K_p = 35$ has been observed, which lies between $K_p = 105.2$ g/L for added ethanol and $K_p = 3.04$ g/L for produced ethanol. This value is believed to represent K_p for ethanol inhibition when using the heterogeneous model, which distinguishes between intracellular and extracellular fluids.

Sugar Inhibition

It is well known that a high concentration of sugar inhibits the growth of yeasts. However, the inhibitory effect of sugar is negligible below a sugar concentration of 100 g/L [5]. The inhibitory effect of sugar is not included in the present model; however, it can easily be added to the kinetic rate equation without any extra complications.

Cell Inhibition

In alcoholic fermentation, most kinetic models express cell growth rate as a linear function of cell concentration. However, the experimental results of Cysewski and Wilke [6] showed that at a high cell concentration, the linear relationship is incorrect.

Ethanol Diffusion

The dispute in the literature regarding the diffusion of ethanol has been discussed by Elnashaie and Ibrahim [4]. The slightly more detailed diffusion reaction model for the alcoholic fermentation process presented in this section of the book will help to throw more light on this dispute than the model used by Elnashaie and Ibrahim [4]. However, we are certainly still quite far from a completely rigorous representation of this rather complex diffusion–reaction problem.

Flocculation of the Yeast

The flocculation of the yeast affects the mass transfer rates and, thus, the rate of fermentation considerably through the change of the exposed surface area for mass transfer per unit mass of the micro-organism. The flocculation process has been investigated experimentally by many researchers; however, the theoretical basis for understanding the flocculation process is severely lacking.

6.10.2 Development of the Model

The model equations are developed for the flocs with three mass transfer resistances in series for sugar and ethanol. The present floc model is more sophisticated than the one developed earlier by Elnahsaie and Ibrahim [4], whereas for the extracellular fluid, the model does not differ from the that used by Elnashaie and Ibrahim [4].

Single-Floc Model

The floc is formed of a number of cells. This number depends on many factors, including the liquid-phase composition and the flocculating tendency of the specific strain of the yeast. The structure of the floc is quite complex and branched, giving rise to what may be called a fractal structure. In this undergraduate book, we use the more classical approach of reducing the complex structure of the floc to an equivalent sphere. The geometric dimensions of the floc fractal structure can be used instead. However, there are certain difficulties associated with the determination of the fractal dimension f and the constant of proportionality α (needed for the calculation of the geometric dimension of the fractal structure), which are beyond the scope of this undergraduate book. Figure 6.34 shows the idealized equivalent sphere used in this model, where the internal floc (the equivalent sphere) is considered a substrate sink where sugar reacts to produce ethanol.

Diffusion through the cell membrane (the membrane of the floc equivalent sphere) is assumed to obey Fick's law with membrane diffusion coefficients that take into account the complex nature of the membrane. Carrier-

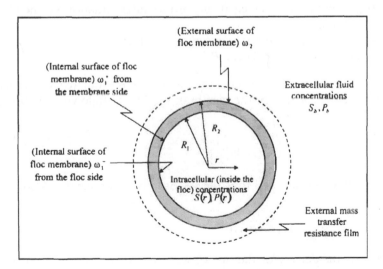

Figure 6.34 Schematic presentation of the proposed model of an equivalent sphere for the microbial floc.

mediated transport of sugar through the membrane is not considered because of the unsteady-state nature of the model and the fact that the model assumes that all of the bioreactions take place inside the cell, with no reactions taking place in the membrane. The external mass transfer resistances are assumed to be lumped in one hypothetical external mass transfer film, with the mass transfer coefficient depending on the extracellular conditions.

Single-Floc Equations

A schematic diagram showing the details of the proposed model for the equivalent sphere representing the microbial floc is presented in Figure 6.34. A number of simplifying assumptions are implicit in the lengthy manipulation of the model equations.

Mass Balance of Ethanol Within the Membrane

If we assume that no reaction is taking place in the membrane, then the diffusion equation becomes

$$\frac{dN_p}{d\omega} = \frac{d}{d\omega}\left[\frac{-Dp_m\omega^2 dP}{R_1 d\omega}\right] = 0 \qquad (6.121)$$

where

N_p = flux of intracellular ethanol (g/cm^2 s)
Dp_m = diffusion coefficient of intracellular ethanol in membrane (cm^2/s)
P = intracellular ethanol concentration (g/cm^3)
$\omega = r/R_1$ (see Fig. 6.34 for R_1)

The diffusivity of ethanol through the membrane is estimated as a function of ethanol mole fraction in the membrane x_1 by the following relation [7]:

$$Dp_m = \frac{\bar{\alpha}_{22}}{\left[x_1\mu_1^{1/3} + (1 - x_1)\mu_2^{1/3}\right]^3} \qquad (6.122)$$

where α_{22} is a constant and μ_i is the viscosity of component i. Also,

$$x_1 = \frac{P}{\left[\rho_w R_M + P(1 - R_M)\right]} \qquad (6.123)$$

where ρ_w is the density of the floc (g wet weight/cm^3 wet volume) = 1.23 g/cm^3 and R_M is the ratio of the molecular weight of ethanol and the membrane lipid.

Substitution of Eqs. (6.122) and (6.123) into Eq. (6.121) and integration gives

$$f(P) = \alpha_{22}\left[AP + \frac{B}{d}\ln(c + dP) - \frac{A_1}{d(c + dP)} - \frac{A_2}{2d(c + dP)^2}\right] = C_1 - \frac{C_2}{\omega}$$

(6.124)

where,

$$A = \frac{b^3}{d^3}$$

$$B = \left(\frac{3b^2}{d^3}\right)(ad - cb)$$

$$A_1 = \left(\frac{3b}{d^3}\right)(ad - cb)^2$$

$$A_2 = \left(\frac{1}{d^3}\right)(ad - cb)^3$$

$$a = R_M \rho_w$$

$$b = 1 - R_M$$

$$c = \left(\mu_{lipid}\right)^{1/3} R_M \rho_w$$

$$d = \left(\mu_{ethanol}\right)^{1/3} - R_M \left(\mu_{lipid}\right)^{1/3}$$

$$\alpha_{22} = \bar{\alpha}_{22}/R_1$$

and C_1 and C_2 are constants of integration to be fitted to boundary conditions.

Boundary Condition at the External Surface of the Membrane

The external surface of the membrane corresponds to $r = R_2$ (i.e., $\omega_2 = R_2/R_1$).

The boundary condition at this point is given by

$$\frac{dP}{d\omega}\bigg|_{\omega_2} = \left(\frac{K_{gp}R_1}{Dp_{m(\omega_2)}}\right)(P_b - P_{\omega_2})$$

(6.125)

where K_{gi} is the external mass transfer coefficient for component i (cm/s) and P_b is the extracellular ethanol concentration (g/cm^3). From Eq. (6.121), we obtain

$$\frac{dP}{d\omega}\bigg|_{\omega_2} = \frac{C_2}{Dp_{m(\omega_2)}\omega_2^2}$$

(6.126)

Thus, from Eqs. (6.125) and (6.126), we can express the constant of integration C_2 in terms of the concentration at the point ω_2 (i.e., R_2/R_1):

$$C_2 = \left(\frac{R_2^2}{R_1}\right) K_{gp}(p_b - p_{\omega_2}) \tag{6.127}$$

Equation (6.124) at point ω_2 can be written as

$$
f(P_{\omega_2}) = \alpha_{22}\left[AP_{\omega_2} + \frac{B}{d}\ln(c + dP_{\omega_2}) - \frac{A_1}{d(c + dP_{\omega_2})} - \frac{A_2}{2d(c + dP_{\omega_2})^2}\right]
$$
$$
= C_1 - \frac{C_2}{\omega_2}
$$
$$
= C_1 - R_2 K_{gp}(P_b - P_{\omega_2}) \tag{6.128}
$$

Therefore, C_1 can also be written in terms of P_{ω_2}, using the above relation, as follows:

$$C_1 = f(P_{\omega_2}) + R_2 K_{gp}(P_b - P_{\omega_2}) \tag{6.129}$$

At the point $\omega_1^+ (R_1^+/R_1 = 1.0^+)$ (i.e., R_1^+ is R_1 at the membrane side as distinguished from R_1, which is R_1 at the floc side). Equation (6.124) can be rewritten using Eqs. (6.127) and (6.129) in the following form:

$$f\left(P_{\omega_1^+}\right) = C_1 - \frac{C_2}{\omega_1} = f(P_{\omega_2}) + R_2 K_{gp}(P_b - P_{\omega_2}) - \left(\frac{R_2^2}{R_1}\right)K_{gp}(P_b - P_{\omega_2})$$

Thus,

$$f\left(P_{\omega_1^+}\right) = f(P_{\omega_2}) + R_2 K_{gp}\left(1 - \frac{R_2}{R_1}\right)(P_b - P_{\omega_2}) \tag{6.130}$$

At the interface between the membrane and the floc (the point ω_1^-), we have,

$$Dp_f \left.\frac{dP}{d\omega}\right|_{\omega_1^-} = Dp_m \left.\frac{dP}{d\omega}\right|_{\omega_1^+} \tag{6.131}$$

Similar to Eq. (6.126), we have

$$\left.\frac{dP_m}{d\omega}\right|_{\omega_1^+} = \left(\frac{C_2}{\omega_1^+}\right) = C_2 \tag{6.132}$$

Thus, from Eqs. (6.131), (6.132), and (6.127), we get

$$Dp_f \left.\frac{dP}{d\omega}\right|_{\omega_1^-} = \left(\frac{R_2^2}{R_1}\right)K_{gp}(P_b - P_{\omega_2}) \tag{6.133}$$

If we use the one internal collocation point approximation for the diffusion reaction equation inside the floc as shown in the next section, then the left-hand side of Eq. (6.133) can be approximated using the orthogonal collocation formula (refer to Appendix E for orthogonal collocation method), and, thus, Eq. (6.133) becomes

$$Dp_{f_{\omega_1}}\left(A_{21}P_c + A_{22}P_{\omega_1^-}\right) = \left(\frac{R_2^2}{R_1}\right)K_{gp}\left(P_b - P_{\omega_2}\right) \tag{6.134}$$

where A_{11} and A_{22} are collocation constants for the spherical shape ($A_{21} = -3.5$, $A_{22} = 3.5$) and c indicates the internal collocation point for the floc.

Ethanol Mass Balance Inside the Floc

For unsteady-state mass balance inside the floc, considering that all of the reactions producing ethanol and consuming sugar is taking place inside the floc, we obtain the following parabolic partial differential equations:

$$\frac{\partial P}{\partial t} = \frac{\bar{\mu}_m \rho_d Y_P \mu'}{Y_C} + \left(\frac{1}{r^2}\right)\left(\frac{\partial}{\partial r}\left[Dp_f r^2 \frac{\partial P}{\partial r}\right]\right) \tag{6.135}$$

where

  Y_P = yield factor for ethanol (g ethanol produced/g sugar consumed)
  Y_C = yield factor for yeast (g yeast produced/g sugar consumed)
  ρ_d = dry density of floc (g dry wt/cm³ wet vol.) = 0.2 g cm³
  $\bar{\mu}_m$ = maximum specific growth rate (s⁻¹)

Putting the above equation in dimensionless coordinate. $\omega \in (0, 1)$, $\omega = r/R_1$, we obtain the following equation:

$$R_1^2 \frac{\partial P}{\partial t} = \frac{\bar{\mu}_m \rho_d Y_P \mu' R_1^2}{Y_C} + \left(\frac{1}{\omega^2}\right)\left(\frac{\partial}{\partial \omega}\left[Dp_f \omega^2 \frac{\partial P}{\partial \omega}\right]\right) \tag{6.136}$$

Applying the orthogonal collocation technique (refer to Appendix E for orthogonal collocation method) using one internal collocation point at ω_c, at the internal collocation point we obtain the following differential equation:

$$R_1^2 \frac{dP_c}{dt} = B_{11}\left[Dp_{f_c} P_c\right] + B_{12}\left[Dp_{f_{\omega_1^-}} P_{\omega_1^-}\right] + \Phi_S^2 Y_P \mu'[S_c P_c] \tag{6.137}$$

where $B_{11} = -10.5$, $B_{12} = 10.5$, S_c is the sugar concentration at the internal collocation point (g/cm³), and

$$\Phi_S^2 = \frac{\bar{\mu}_m \rho_d R_1^2}{Ds_f Y_C}$$

Equations (6.130), (6.134), and (6.137) are two algebraic equations and one differential equation in four variables, namely $P_{\omega_1^+}$, $P_{\omega_1^-}$, P_{ω_1}, and P_c; thus, one equation is missing. This extra equation can be furnished by the equilibrium relation between $P_{\omega_1^+}$ and P_{ω_1}, which has the following form:

$$P_{\omega_1^-} = K_1 P_{\omega_1^+} \tag{6.138}$$

where K is the equilibrium partition constant for the floc side of the membrane

Equations (6.130), (6.134), (6.137), and (6.138) can now be solved to obtain four variables at every time during the fermentation process, provided that at these times, the concentration of sugar at the collocation point is known (the necessary equations are derived in the next section) and the bulk (extracellular) concentrations are known (the extracellular equations are exactly the same as in Ref. 4; however, for the benefit of the reader, we will present those equations again in a later section). We will also discuss the solution algorithm to be employed to solve this complex set of equations.

Mass Balance for Sugar Inside the Membrane

We assume that no reaction is taking place inside of the membrane and also assume a pseudo-steady-state for the membrane. Furthermore, we assume constant sugar diffusivity through the membrane. Based on these assumptions, the sugar material balance inside the membrane becomes

$$\nabla^2 S = 0 \tag{6.139}$$

where

$$\nabla^2 = \frac{d^2}{d\omega^2} + \left(\frac{2}{\omega}\right)\left(\frac{d}{d\omega}\right) \quad \text{and} \quad \omega = \frac{r}{R_1}$$

The boundary condition at the outer surface of the membrane ω_2 (i.e., R_2/R_1) is given by

$$\left.\frac{dS}{d\omega}\right|_{\omega_2} = Sh_S\left(S_b - S_{\omega_2}\right) \tag{6.140}$$

where

$$Sh_S = \frac{Kg_s R_1}{Ds_m}$$

and Kg_s is the external mass transfer coefficient for sugar (cm/s) and Ds_m is the diffusion coefficient of sugar in membrane phase (cm²/s).

The boundary condition at the inner surface of the membrane ω_1 (i.e., $R_1/R_1 = 1.0$) is given by

$$Ds_f \left.\frac{dS}{d\omega}\right|_{\omega_1^-} = Ds_m \left.\frac{dS}{d\omega}\right|_{\omega_1^-} \tag{6.141}$$

The general equation of Eq. (6.139) is given by

$$S = C_1' - \frac{C_2'}{\omega} \tag{6.142}$$

After some manipulation, we obtain S_{ω_2} in terms of $S_{\omega_1}^+$ as follows,

$$S_{\omega_2} = \frac{S_{\omega_1^+} + Sh_{Sm}S_b(1 - R_1/R_2)}{1 + Sh_{Sm}(1 - R_1/R_2)} \tag{6.143}$$

where

$$Sh_{Sm} = Sh_S\left(\frac{R_2}{R_1}\right)^2$$

The boundary condition in Eq. (6.141) coupled with the floc differential equation after simplifying it using the one-internal orthogonal collocation point approximation, gives the following relation among $S_{\omega_1}^-$, $S_{\omega_1}^+$, and the concentration of sugar at the internal collocation point S_c:

$$S_{\omega_1^-} = \frac{Sh_{Sf}\left(S_b - S_{\omega_1^+}\right)}{A_{22}[1 + Sh_{Sm}(1 - R_1/R_2)]} - \left(\frac{A_{21}}{A_{22}}\right)S_c \tag{6.144}$$

where

$$Sh_{Sf} = Sh_{Sm}\left(\frac{Ds_m}{Ds_f}\right)$$

The change of sugar concentration at the internal collocation point is given by the following nonlinear differential equation:

$$\left(\frac{R_1^2}{Ds_f}\right)\frac{dS_c}{dt} = B_{11}S_c + B_{12}S_{\omega_1^-} - \Phi_S^2\mu'[S_cP_c] \tag{6.145}$$

where

$$\Phi_S^2 = \frac{\bar{\mu}_m P_d R_1^2}{Ds_f Y_C}$$

$$\mu' = \frac{K_p S(1 - X/X_m)^N}{(K_p + P)(K_S + S)} + \frac{K_p'}{K_p' + P}$$

K_p is the inhibition constant (g/cm^3), K_p' is the rate constant (g/cm^3), X is the biomass concentration (g dry wt/cm^3), and X_m is the maximum biomass concentration (g dry wt/cm^3). In addition, we assume the validity of the equilibrium relation between $S_{\omega_1^-}$ and $S_{\omega_1^+}$ written as

$$S_{\omega_1^-} = K_2 S_{\omega_1^+} \tag{6.146}$$

where K_2 is the equilibrium partition constant for the floc side of the membrane.

The Extracellular Balance Equation

For the batch fermentor case, the extracellular mass balance equations are the same as those in Ref. 4; these are as follows:

For sugar,

$$\frac{dS_b}{dt} = \left[\frac{-3XKg_S}{(\rho_d - X)R_2}\right](S_b - S_{\omega_2}) \tag{6.147}$$

For ethanol,

$$\frac{dP_b}{dt} = \left[\frac{-3XKg_p}{(\rho_d - X)R_2}\right](P_{\omega_2} - P_b) \tag{6.148}$$

For micro-organisms,

$$\frac{dX}{dt} = \bar{\mu}_m\mu'X \tag{6.149}$$

where, Kg_i is the external mass transfer coefficient for component i (cm/s).

Equations (6.130), (6.134), (6.137), (6.138), and (6.143)–(6.149) are the model equations that are to be solved together with the physical properties, external mass transfer, and diffusivities correlations to obtain the change of sugar, ethanol, and micro-organism concentrations with time.

Note: The physical properties, external mass transfer, and diffusivity correlations can be obtained from Ref. 4.

6.10.3   Solution Algorithm

For the given initial bulk conditions at $t = 0, P_{b0}$, S_{b0}, and the initial conditions for P_c and S_c the solution algorithm is as follows:

1. Equations (6.130), (6.134), and (6.138) together with the relations for physical properties and diffusivities [7] are solved to obtain the values of P_{ω_2}, $P_{\omega_1^+}$, and $P_{\omega_1^-}$.
2. Equations (6.143), (6.144), and (6.146) together with the relations for physical properties and diffusivities [7] are solved to obtain the values of S_{ω_2}, $S_{\omega_1^+}$, and $S_{\omega_1^-}$.
3. The differential Eqs. (6.137) and (6.145) for the concentration at the internal collocation point are integrated using a fourth-order

Runge–Kutta routine with automatic step size (to ensure accuracy) to obtain the values of P_c and S_c for the next time step.

4. The differential Eqs. (6.147), (6.148), and (6.149) are integrated using the same routine to obtain the values of P_b, S_b, and X for the next time step.

5. Repeat steps 1–4 until you reach the final time of fermentation.

6.10.4 Comparison Between the Model and Experimental/Industrial Data

The model has been solved using the algorithm discussed in Section 6.10.3, the physical parameters estimations are given in Ref. 7, and the kinetic and physical parameters are given in Ref. 4. The thickness of the membrane is taken as 10^{-6} cm for the two simulated cases. The results of the present model are compared with the experimental and industrial results.

Figure 6.35 shows an example of the results of the experimental intracellular ethanol concentration profile versus time together with the profile predicted by the above-discussed distributed parameter model (DPM) at the internal collocation point. The results of the present DPM model give the nonmonotonic shape of the intracellular ethanol concentration and the profile as a whole follows the same experimental trend.

Figure 6.36 shows the profile of the extracellular sugar concentration. The results of the DPM are very close to the experimental results.

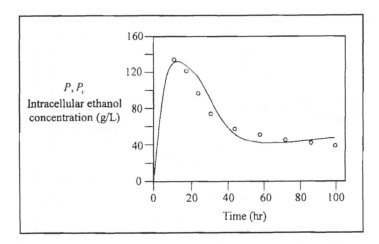

Figure 6.35 Intracellular ethanol concentration versus time ($\bigcirc$: experimental; —: DPM for an experimental batch fermentor).

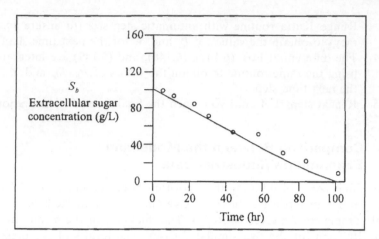

Figure 6.36 Extracellular sugar concentration versus time (○: experimental; —: DPM for an experimental batch fermentor).

Figure 6.37 shows the extracellular ethanol concentration profiles where the DPM predicts ethanol concentrations that are very close to the experimental results.

This discussion of the development of a DPM of the fermentation process gives the reader a good overview and the necessary understanding

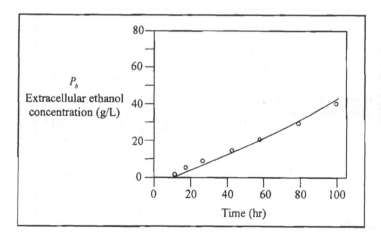

Figure 6.37 Extracellular ethanol concentration versus time (○: experimental; —: DPM for an experimental batch fermentor).

for the development of mathematical models for industrial biochemical processes.

REFERENCES

1. Elnashaie, S. S. E. H., Harraz, H. M., and Abashar, M. E. Homoclinical chaos and the period-adding route to complex non-chaotic attractors in fluidized bed catalytic reactors. *Chaos Solitons Fractals* 12, 1761–1792, 2001.
2. Kernevez, J. P. Control, optimization and parameter identification in immobilized enzyme systems. In *Proceedings of the International Symposium on Analysis and Control of Immobilized Enzyme Systems.* (Thomas, D. and Kernevez, J. P, eds.). North Holland/American Elsevier, 1976, pp. 199–225.
3. Jobses, I. M. L and Roels, J. A. The inhibition of the maximum specific growth and fermentation rate of *Zymomonas mobilis* by ethanol. *Biotechnol. Bioeng.* 28(4), 554–563, 1986.
4. Elnashaie, S. S. E. H. and Ibrahim, G. Heterogeneous modeling for the alcoholic fermentation process. *Appl. Biochem. Biotechnol.* 19(1), 71–101, 1988.
5. Ciftci, T., Constantinides, A., and Wang, S. S. Optimization of conditions and cell feeding procedures for alcohol fermentation. *Biotechnol. Bioeng.* 25(8), 2007–2023, 1983.
6. Cysewski, G. R. and Wilke, C. R. Process design and economic studies of alternative fermentation methods for the production of ethanol. *Biotechnol. Bioeng.*, 20(9), 1421–1444, 1978.
7. Elnashaie, S. S. E. H. and Ibrahim, G. A distributed parameter diffusion-reaction model for the alcoholic fermentation process. *Appl. Biochemi. Biotechnol.* 30(3), 339–358, 1991.
8. Garhyan, P. and Elnashaie, S. S. E. H. Exploitation of static/dynamic bifurcation and chaotic behavior of fermentor for higher productivity of fuel ethanol. AIChE Annual Meeting, 2001.

PROBLEMS

Problem 6.1

Derive the steady-state mass balance equations for a one-component two-plate absorption tower with ideal (equilibrium) stages; that is, the concentrations of the transferable component in the liquid and gas leaving each plate are at equilibrium according to the linearized equilibrium relation

$$Y_j = aX_j + b$$

where $a = 0.72$ and $b = 0.0$, and Y_j and X_j are weight fractions of the solute (the transferable component) to inerts in the gas and liquid phases. The mass flow rate of the gas and liquid (based on inerts) are $G = 66.7$ lbs inerts/min and $L = 40.8$ inerts/min. Solve for the steady-state plate composition when the liquid feed is pure and the gas feed has a concentration of 0.3 lb solute/lb inerts.

What is the effect of relaxing the assumption of equilibrium stages. Derive the steady-state mass balance equations for this case of nonideal stages and express the solution vectors in terms of the different coefficient matrices and input vectors.

Chapter 6

Problem 6.2

Flooded condensers are sometimes used in distillation columns. The liquid level runs up in the condenser, covering some of the tubes. Thus, a variable amount of heat transfer area is available to condense the vapor. The column pressure can be controlled by changing the distillate (or reflux) draw-off rate.

Write equations describing the dynamics of the condenser.

Problem 6.3

Derive the unsteady-state mass balance equations for a countercurrent absorption column with N plates, where a single component is absorbed from the vapor phase into the liquid phase as shown in Figure P6.3. Assume ideal stages and that the liquid and gas on each plate are perfectly mixed and that the liquid and gas molar flow rate from one plate to the other are constant (dilute system). Assume linear equilibrium relations. Put the resulting equation in a matrix form. Deduce the steady-state matrix equation from the unsteady-state matrix equation.

Problem 6.4

A mixture of two immiscible liquids is fed to a decanter. The heavier liquid A settles to the bottom of the tank. The lighter liquid B forms a layer on the top. The two interfaces are detected by floats and are controlled by manipulating the two flows F_A and F_B (m^3/h) according to the relations

$$F_A = K_A h_A \quad \text{and} \quad F_B = K_B (h_A + h_B)$$

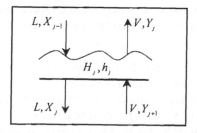

Figure P6.3   Schematic diagram for Problem 6.3.

where h_A is the height of the interface above the bottom of the tank and h_B is the height of the light liquid (B) above the interface. The controllers increase or decrease the flows as the level rise or fall.

The total feed rate is W_o (kg/h). The weight fraction of liquid A in the feed is X_A, and the two densities ρ_A and ρ_B (kg/m^3) are constant. Write the equations describing the dynamic behavior of the system; also express the steady-state height of both phases in terms of other parameters.

Problem 6.5

For the partial oxidation of ethylene to ethylene oxide,

$$C_2H_4 + \tfrac{1}{2}O_2 \rightarrow C_2H_4O$$

the catalyst consists of silver supported on alumina, and although it is reasonably specific, appreciable amounts of CO_2 and H_2O are also formed. Over the range of interest, the yield of ethylene oxide is relatively constant, so that for present purposes, we may regard the reaction stoichiometry as

$$C_2H_4 + 1.5O_2 \rightarrow 0.6C_2H_4O + 0.8CO_2 + 0.8H_2O$$

The rate of reaction may be expressed as

$$r = 1.23 \times 10^6 e^{-9684/T} P_{C_2H_4}^{0.323} P_{O_2}^{0.658}$$

where the partial pressures are expressed in atmospheres, temperature is expressed in degrees Kelvin, and the rate of reaction r is expressed in pound moles per pound of catalyst per hour.

If $\tfrac{1}{8}$-in. catalyst pellets are packed in 1-in.-inner-diameter tubes, which, in turn, are immersed in a liquid bath that maintains the tube walls at 240°F, consider the effects of varying the feed temperature and of diluting the feed with N_2 to moderate the thermal effects accompanying the reaction. Consider inlet temperatures from 350°F to 480°F and N_2/C_2H_4 ratios from 0 to 5.0.

Your analysis will be governed by the following constraints:

(a) If the temperature at any point in the reactor exceeds 550°F, the conditions will be inappropriate in that explosions may occur in this regime.

(b) If the temperature decreases as the mixture moves down the reactor, the reaction must be regarded as self-extinguishing.

(c) If the pressure drop exceeds 14 atm, the calculations must be terminated.

Determine the range of satisfactory performance and the resultant yields for various reactor lengths for the following operating specifications:

1. Inlet gas pressure = 15 atm
2. Inlet C_2H_4/O_2 ratio (moles) = 4:1.
3. Superficial mass velocity = 9000 lb/h ft^2.
4. Bulk density of catalyst = 81.9 lb/ft^3.

External heat and mass transfer effects are to be neglected in your analysis, but you should estimate the potential magnitudes of these effects.

Problem 6.6

Before ethylene feedstocks produced by thermal cracking can be used chemically for most applications, it is necessary to remove the traces of acetylene present in such streams. This purification can be accomplished by selective hydrogenation of acetylene to ethylene. The process involves adding sufficient hydrogen to the feedstock so that the mole ratio of hydrogen to acetylene exceeds unity. Using a palladium-on-alumina catalyst under typical reaction conditions (25 atm, 50–200°C), it is possible to achieve extremely high selectivity for the acetylene hydrogenation reaction. As long as acetylene is present, it is selectively adsorbed and hydrogenated. However, once it disappears, hydrogenation of ethylene takes place. The competitive reactions may be written as

$$C_2H_2 + H_2 \rightarrow C_2H_4 \qquad \text{(desired)}$$
$$C_2H_4 + H_2 \rightarrow C_2H_6 \qquad \text{(undesired)}$$

The intrinsic rate expressions for these reactions are both first order in hydrogen and zero order in acetylene or ethylene. If there are diffusional limitations on the acetylene hydrogenation reaction, the acetylene concentration will go to zero at some point within the core of the catalyst pellet. Beyond this point within the central core of the catalyst, the undesired hydrogenation of ethylene takes place to the exclusion of the acetylene hydrogenation reaction.

(a) In the light of the above facts, what do the principles enunciated in this chapter have to say about the manner in which the reactor should be operated and the manner in which the catalyst should be fabricated?

(b) It is often observed that the catalysts used for this purpose in commercial installations do not achieve maximum selectivity until they have been on stream for several days. How do you explain this observation?

Problem 6.7

Show that the equilibrium relationship $y = mx + b$ reduces to the equation $Y = mX$ if we let $Y = y - b/(1 - m)$ and $X = x - b/(1 - m)$.

Problem 6.8

A dynamic model for a plate absorber was developed in the chapter for the case of unit-plate efficiency (ideal stages). In order to make the model more realistic, we could assume that the plate efficiency is constant throughout the column and is given by the relation,

$$E = \frac{y_n - y_{n-1}}{y_n^* - y_{n-1}}$$

In this expression y_n^* is the vapor composition in equilibrium with liquid mixture of composition x_n, so that the equilibrium relationship becomes

$$y_n^* = mx_n + b$$

Develop a dynamic model of a plate absorption column using these assumptions.

Problem 6.9

Consider the dynamics of an isothermal CSTR followed by a simple (single stage) separating unit (e.g., an extractor, crystallizer, or a settler). The reaction is reversible, $A \Leftrightarrow B$, and the effluent stream from the separator, which is rich in unreacted material, is recycled. It is assumed that the reaction rate is first order, the equilibrium relationship for the separator is linear, and the rate of mass transfer between the phases in the separator could be written in terms of mass transfer coefficient and a linear driving force. Making certain that you define all your terms carefully, show that the dynamic model for the plant can be put into the following form:

Reactor

$$V\frac{dC_A}{dt} = qC_{Af} + QC_{A2} - (q + Q)C_A - V(k_1C_A - k_2C_B)$$

$$V\frac{dC_B}{dt} = QC_{B2} - (q + Q)C_B + V(k_1C_A - k_2C_B)$$

Lean phase in separator

$$H\frac{dC_{A2}}{dt} = (q + Q)C_A - QC_{A2} - (k_{gA}a_v)(C_{A2} - m_1C_{A1})$$

$$H\frac{dC_{B2}}{dt} = (q + Q)C_B - QC_{B2} - (k_{gB}a_v)(C_{B2} - m_2C_{B1})$$

Rich phase in separator

$$h\frac{dC_{A1}}{dt} = (k_{gA}a_v)(C_{A2} - m_1 C_{A1}) - qC_{A1}$$

$$h\frac{dC_{B1}}{dt} = (k_{gB}a_v)(C_{B2} - m_2 C_{B1}) - qC_{B1}$$

Problem 6.10

If we consider a plate gas absorption unit containing only two trays for a case where the equilibrium relationship is linear, we can write a dynamic model for this unit as

$$H\frac{dx_1}{dt} = (L_1 + Vm)x_1 + L_2 x_2 - Vmx_0$$

$$H\frac{dx_2}{dt} = Vmx_1 - (L_2 + Vm)x_2$$

where x_0 is the liquid composition in equilibrium with the vapor stream entering the bottom of the column. For the simplest case, we might hope that the liquid flow rate was constant, so that $L_1 = L_2 = L$. However, there are situations in which the liquid rate is observed to be time dependent, and we often model the tray hydraulics with the simple equations

$$\tau\frac{dL_1}{dt} = -L_1 + L_2$$

$$\tau\frac{dL_2}{dt} = -L_2 + L_3$$

where L_3 is the liquid rate entering the top plate.
    First consider the case of constant flow rates:

(a) Find the steady-state solutions of the material balance expressions.
(b) Linearize the equations around this steady-state operating point.
(c) Determine the characteristic roots of the linearized equations.

Next, try to reproduce parts (a)–(c) for the case of variable flow rates. At what point does the procedure fail? Why?

Problem 6.11

Derive a dynamic model that can be used to describe an adsorption or ion-exchange column. If you desire, consider a particular case where benzene is being removed from an airstream by passing through a bed packed with silica gel. Carefully define your terms and list your assumptions.

What is the meaning of steady-state operation in this kind of unit? How does the model change as you change one or more of the assumptions?

Problem 6.12

Show that the dynamic model for a double-pipe, countercurrent heat exchanger can have the same form as the model of a packed absorber. Discuss the assumptions inherent in both the heat exchanger and absorber models which might lead to significant differences in the kinds of equations used to describe each system.

Problem 6.13

Derive a dynamic model for a packed-bed extraction unit that includes an axial dispersion term for the dispersed phase. Give an appropriate set of boundary conditions for the model.

Problem 6.14

In the contact process for manufacturing sulfuric acid, sulfur is burned to SO_2, the SO_2 is reacted with oxygen over a catalyst to produce SO_3, and the SO_3 gas is absorbed in a packed tower by concentrated sulfuric acid. The particular concentration of acid fed to the absorber is a critical design variable, for the heat effects can be so large that there might be significant amount of mist formation. Derive a dynamic model describing an SO_3 absorber. Carefully define your terms and list your assumptions. How would you decide whether you could neglect the accumulation of energy in the packing?

Problem 6.15

Heat regenerators are encountered in a number of large-scale industrial processes, such as open-hearth furnaces, liquefaction of a vapor, and the separation of its components in the liquid state. A hot gas, possibly leaving a reactor, is passed through a checkwork of bricks, or even a bed of stones, so that the heat is removed from the gas and stored in the solid. Then, a cold gas is passed through this bed, normally in the opposite direction, so that the gas is preheated before it enters the reactor. Derive a dynamic model for the system. Carefully define your terms and list your assumptions.

Problem 6.16

The catalytic dehydrogenation of butane is important both in the manufacture of butadiene and as one step in the synthetic manufacture of gasoline. It has been reported that the principal reaction and important secondary reaction are as follows:

$$C_4H_{10} \rightarrow C_4H_8 + H_2 \quad \text{(principal reaction)}$$
$$C_4H_8 \rightarrow C_4H_6 + H_2 \quad \text{(secondary reaction)}$$

In the high-temperature range of interest, an appreciable amount of the reactions occurred by pyrolysis in the homogeneous phase, as well as by the catalytic path. Also, it was observed that a large number of other reactions were taking place, including the following:

(a) The dealkylation or cracking of butane to form methane, ethane, ethylene and propylene
(b) The dealkylation of butanes to form methane, ethane, propane, ethylene, propylene and coke
(c) The dimerization of butadiene to form 4-vinyl cyclohexane-1
(d) The decomposition of butadiene to form hydrogen, methane, ethylene, acetylene, and coke

Obviously, a complete determination of the reaction kinetics, both homogeneous and catalytic, of this process would be extremely difficult. Hence, for gasoline manufacture, it was suggested to use the following simplified reaction model:

$$C_4H_{10} \rightarrow C_4H_8 + H_2$$
$$C_4H_{10} \rightarrow 0.1C_4H_8 + 0.1H_2 + 1.8 \quad \text{(dealkylation products)}$$
$$C_4H_8 \rightarrow 0.1H_2 + 1.8 \quad \text{(dealkylation products)}$$

with the rate equations

$$r_A = \frac{C(P_A - P_R P_S/K)}{(1 + K_A P_A + K_{Rs} P_{Rs})^2} \quad \text{(main dehydrogenation reaction)}$$
$$r_B = kP_A$$
$$r_C = kP_R$$

where,

r_A = rate of reaction (moles/kg catalyst h)
C = overall rate constant (moles/kg catalyst h atm)

K = overall gas phase equilibrium constant (atm)

K_A = effective adsorption equilibrium constant of butane (1/atm)

K_{Rs} = effective average adsorption constant of hydrogen and butene

$P_{Rs} = \frac{1}{2}(P_R + P_s)$ = average partial pressure of hydrogen and butene

r_B, r_C = rate of cracking reactions (moles/kg catalyst h)

P_A, P_R = partial pressures of butane and butene

P_S = partial pressure of hydrogen

k = cracking reaction rate constant

Using this simplified kinetic model, develop a dynamic model for a nonisothermal fixed-bed, catalytic reactor. Include the possibility of feeding steam to the bed to act as a dilutent. Carefully define your terms and list your assumptions.

Problem 6.17

The reaction-rate expressions for heterogeneous, catalytic reactions are based on the assumption that the rates of the individual steps taking place in the overall reaction are equal; that is, an equality exists among the rate of diffusion of reactants through a stagnant film of gas surrounding the catalyst particle, the rate of adsorption of reactants on the surface of the catalyst, the rate of surface reaction; the rate of desorption of the products from the surface, and the rate of diffusion of products across the stagnant film back into the bulk of the gas stream flowing through the catalyst bed. A further modification must be made if the catalyst particle is porous and if the rate of diffusion into the pores is important.

Normally, the equations we obtain based on this assumption are so complicated that they are unmanageable. Hence, we often attempt to simplify the approach by assuming that one of the steps is rate controlling and that the others are at equilibrium. The final form of the rate expression, as well as the values of the unknown constants, is usually obtained by comparing the model predictions with experimental data from isothermal reactors operated at steady-state conditions.

Discuss, in detail, the applicability of these rate expressions in dynamic models.

Problem 6.18

Fluid catalytic cracking is one of the most important processes in the petroleum refining industry. The process cracks gas oil to produce high-octane-number gasoline. The Type IV industrial design consists of two bubbling fluidized beds with continuous catalyst circulation between the two vessels;

the reactor and the regenerator. In the reactor gas oil cracks to produce high-octane-number gasoline; overcracking also occurs, giving rise to light gases and deposition of carbon on the catalyst that causes catalyst deactivation. The simplest representation of the reaction scheme is the following consecutive reaction network using pseudocomponents,

$$A \rightarrow B \rightarrow C + \text{Light gases}$$

where, A is the gas oil, B is the gasoline, and C is the carbon.

The cracking reactions are endothermic. In the regenerator, the carbon deposited on the catalyst is burned off using air and, thus, the catalyst is regenerated. This regeneration reaction is highly exothermic. The catalyst is recirculating continuously between the reactor and the regenerator.

(a) Develop a dynamic model for this process.
(b) Is multiplicity of the steady states possible for this process? Explain your answer.
(c) Discuss the possible sources of instability for this process and suggest/discuss suitable control loops to stabilize the process.
(d) Suggest a simple PI feedback control loop to "improve" the performance of your unit. Explain the procedure for the design of this control loop.

Problem 6.19

A first-order reversible reaction

$$A \Leftrightarrow B$$

takes place in a nonisothermal porous spherical catalyst pellet. The forward reaction rate is given by

$$r_f = k_f C_A \quad \text{and} \quad k_f = k_{f0} e^{-(E_f/RT)}$$

The backward reaction rate is given by

$$r_b = k_b C_A \quad \text{and} \quad k_b = k_{b0} e^{-(E_h/RT)}$$

The equilibrium constant of the reaction is independent of temperature and is equal to 16.5. The dimensionless activation energy for the forward reaction is equal to 18, the net dimensionless exothermicity factor is equal to 0.7, and the external mass and heat transfer resistances are negligible.

Use bulk phase gas concentration and temperature as the reference values, such that your $X_{AB} = 1.0$, $X_{BB} = 0.0$ and $Y_B = 1.0$.

(a) Use a value of the Thiele modulus for the forward reaction that gives multiple steady states (if possible).

(b) Compute the effectiveness factor(s) for the steady state(s) using the orthogonal collocation technique.
(c) Use both integral and derivative formulas for the calculation of the effectiveness factor(s) and comment on your results.

Problem 6.20

Jobses and Roels (3) proposed a four-dimensional model to simulate the oscillatory behavior of ethanol fermentation using *Zymomonas mobilis*. Four state variables in the model are X (micro-organisms), e (an internal key component related to the rate of growth of micro-organism), S (substrate; i.e., sugar), and P (product; i.e., ethanol). The rates of formation of these variables obtained in batch experiments are

$$r_e = \left(k_1 - k_2 C_P + k_3 C_P^2\right)\left(\frac{C_S C_e}{K_S + C_S}\right)$$

$$r_X = \frac{C_S C_e}{K_S + C_S}$$

$$r_S = \left(\frac{-1}{Y_{SX}}\right)\left(\frac{C_S C_e}{K_S + C_S}\right) - m_S C_X$$

$$r_P = \left(\frac{1}{Y_{PX}}\right)\left(\frac{C_S C_e}{K_S + C_S}\right) + m_P C_X$$

Parameter values are as follows:

$k_1(h^{-1}) = 16.0,$ $k_2(m^3/kg\ h) = 4.97 \times 10^{-1},$ $k_3(m^6/kg^2\ h) = 3.83 \times 10^{-3}.$
$m_S(kg/kg\ b) = 2.16,$ $m_P(kg/kg\ h) = 1.1,$ $Y_{SX}(kg/kg) = 2.44498 \times 10^{-2}.$
$Y_{PX}(kg/kg) = 5.26315 \times 10^{-2},$ $K_S(kg/m^3) = 0.5$

Initial conditions are as follows:

$$C_e(0) = 0.06098658\ kg/m^3, \quad C_X(0) = 1.1340651\ kg/m^3,$$
$$C_P(0) = 58.24388\ kg/m^3 \qquad C_S(0) = 45.89339\ kg/m^3$$

(a) Formulate a dynamic model for a continuous-stirred tank fermentor (CSTF) of an active volume V_F. Assume that the inlet and outlet flow rates remain constant and are equal to q.
(b) Using the given values of the parameters, solve the dynamic model and comment on the dependence of dynamic behavior on the dilution rate (in the range 0.0–0.1 h^{-1}) and inlet substrate feed concentration (in the range 140–200 kg/m^3). Assume that the feed to CSTF is pure sugar.

Problem 6.21

As the ethanol produced in fermentation acts as an inhibitor for the process, continuous ethanol removal is generally used to improve the productivity/ yield of the fermentation process. Garhyan and Elnashaie (8) modeled a continuous-membrane fermentor with in situ removal of ethanol produced using a sweep liquid as shown in Figure P6.21. The rate of ethanol removal is considered to be proportional to the ethanol concentration gradient across the membrane and area of permeation.

Extend the four-dimensional model of Problem 6.20 to model this continuous-membrane fermentor. List all of the assumptions you make and justify them.

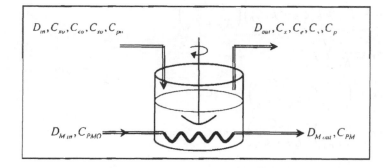

Figure P6.21   Schematic diagram for Problem 6.21

7

Practical Relevance of Bifurcation, Instability, and Chaos in Chemical and Biochemical Systems

Multiplicity phenomenon in chemically reactive systems was first observed in 1918 (1) and in the Russian literature in 1940 (2). However, it was not until the 1950s that the great interest in the investigation of this phenomenon started, inspired by the Minnesota school of Amundson and Aris (3 5) and their students (6–12). The Prague school also had a notable contribution to the field (13–16). Since then, this phenomenon has been high on the agenda of chemical engineering research, in general, and chemical reaction engineering research, in particular. The fascination with this phenomenon has also spread widely to the biological and biochemical engineering literature where it is referred to as "short-term memory" (17,18). The mathematical literature has also caught up with the fascination of this phenomenon, where it is treated in a more general and abstract terms under "bifurcation theory" (19).

A major breakthrough with regard to the understanding of this phenomenon in the field of chemical reaction engineering was achieved by Ray and co-workers (20,21) when, in one stroke, they uncovered a large variety of possible bifurcation behaviours in nonadiabatic continuous-stirred tank reactors (CSTRs). In addition to the usual hysteresis-type bifurcation, Uppal et al. (21) uncovered different types of bifurcation diagrams, the most important of which is the "isola," which is a closed curve disconnected from the rest of the continuum of steady states.

515

Isolas were also found by Elnashaie et al. (22–24) for enzyme systems where the rate of reaction depends nonmonotonically on two of the system's state variables (substrate concentration and pH), a situation which was shown to be applicable to the acetylcholinesterase enzyme system.

Later development in the singularity theory, especially the pioneering work of Golubitsky and Schaeffer (19), provided a powerful tool for the analysis of bifurcation behavior of chemically reactive systems. These techniques have been used extensively, elegantly, and successfully by Luss and his co-workers (7–11) to uncover a large number of possible types of bifurcation. They were also able to apply the technique successfully to complex reaction networks as well as distributed parameter systems.

Many laboratory experiments were also designed to confirm the existence of bifurcation behavior in chemically reactive systems (25–29) as well as in enzyme systems (18).

Multiplicity (or bifurcation) behavior was found to occur in other systems such as distillation (30), absorption with chemical reaction (31), polymerization of olefins in fluidized beds (32), char combustion (33,34), heating of wires (35), and, recently, in a number of processes used for the manufacturing and processing of electronic components (36,37).

Although the literature is rich in theoretical investigations of the bifurcation behavior and laboratory experimental work for the verification of the existence of the bifurcation behavior, it is extremely poor with regard to the investigation of this phenomenon for industrial systems. In fact, very few articles have been published that address the question of whether this phenomenon is important industrially or is only of theoretical and intellectual interest. Research in this field has certainly raised the intellectual level of chemical engineering and helped tremendously in the development of a more advanced and rigorous approach to the modeling of chemical engineering processes. Moreover, in addition to the intellectual benefits of bifurcation research to the engineering level of thinking and simulation of difficult processes, it seems that the phenomenon is also of great importance and relevance to certain industrial units. Elnashaie et al. (38) found, using heterogeneous models, that an industrial TVA ammonia converter is operating at the multiplicity region. The source of multiplicity in this case is the countercurrent flow of reactants in the catalyst bed and the cooling tubes and heat exchanger. Multiplicity of the steady states in TVA ammonia converters had been observed much earlier by using a simple pseudohomogeneous model (39,40). The fluid catalytic cracking (FCC) units for the conversion of heavy gas oil to gasoline and light hydrocarbons not only exhibit a complex bifurcation behavior but also operate mostly in the middle unstable steady state (41–44). For these systems, the stable high- and low-temperature steady states both give very low gasoline yield.

Despite the extensive interest in the bifurcation behavior of chemical reactors manifested in the chemical engineering literature, the industrial interest in this phenomenon is extremely limited. It seems that the industrial philosophy is to avoid these troublesome regions of operating and design parameters which are characterized by instabilities, quenching, ignition, and so forth, where design and control can be quite difficult. This conservative philosophy was quite justifiable before the great advances achieved during the last decades in the development of rigorous models for chemical reactors and the revolutionary advancement achieved in computer power and digital computer control. The present state of affairs indicates that industrial philosophy should change more and more into the direction of exploring the possible opportunities available with regard to higher conversion, higher yield, and selectivity in this region of operating and design parameters. In fact, in the last two decades, many outstanding academic researchers have demonstrated the possibility of exploiting these instabilities for higher conversions, selectivities, and yields (45-56). In the Russian literature, Matros and co-workers (57-59) demonstrated the advantage of operating catalytic reactors deliberately under unsteady-state conditions to achieve higher performance. Industrial enterprises in the West are showing great interest in the concept. Although this deliberate unsteady-state operation is not directly related to the bifurcation phenomenon and its associated instabilities, it, nevertheless, demonstrates a definite change in the conservative industrial attitude.

7.1 SOURCES OF MULTIPLICITY

Catalytic reactors are exceptionally rich in bifurcation and instability problems. These can come from many sources, as summarized in the following subsections.

7.1.1 Isothermal Multiplicity (or Concentration Multiplicity)

This is probably the most intrinsic source of multiplicity in catalytic reactors because it results from the nonmonotonic dependence of the intrinsic rate of reaction on the concentration of reactants and products. Although a decade ago nonmonotonic kinetics of catalytic reactions were considered the exceptional cases, nowadays it is clear that nonmonotonic kinetics in catalytic reactions is much more widespread than previously thought. Readers can learn more about various examples from a long list of catalytic reactions exhibiting nonmonotonic kinetics (60-65). However, nonmonotonic kinetics alone will not produce multiplicity. It has to be coupled with some diffusion process, either a mass transfer resistance between the catalyst pellet surface and the bulk gas or within the pores of the pellets. For if the flow conditions and the catalyst pellets size are such that diffusional resistances between the bulk gas phase and the catalytic active centers are negligible (the system in this case is described by a pseudohomogeneous model) and the bulk gas phase is in plug flow, then multiplicity is not possible. However, for such a pseudohomogeneous reactor, if the bulk phase flow conditions are not in plug flow, then multiplicity of the steady states is possible (66,67). The range of deviations from plug flow which gives multiplicity of the steady states corresponds to shallow beds with small gas flow rates, a situation not applicable to most industrial reactors.

A word of caution is necessary. Isothermal multiplicity resulting from nonmonotonic kinetics occurs only when the nonmonotonic kinetic dependence of the rate of reaction on species concentration is sharp enough. For flat nonmonotonic behavior, multiplicity can occur only for bulk-phase concentrations, which are too high to be considered of much practical relevance.

7.1.2 Thermal Multiplicity

This is the most widespread and extensively investigated type of multiplicity. It is associated with exothermic reactions. In fact, this type of multiplicity results from some sort of nonmonotonic behavior associated with the change of the rate of reaction under the simultaneous variation of reactants concentration and temperature accompanying the reactions taking place within the boundaries of the system (the reactor). For the case of exothermic reactions, as the reaction proceeds, the reactants deplete, which tends to cause a decrease in the rate of reaction, while heat release increases the temperature and thus causes the rate of reaction to increase through the Arrhenius dependence of the rate of reaction upon temperature. These two conflicting effects on the rate of reaction associated with the conversion of reactants lead to a nonmonotonic dependence of the rate of reaction on the reactant concentration. This, in turn, leads to the possibility of multiple steady states. However, similar to the isothermal case, multiplicity will not occur in fixed-bed catalytic reactors due to this effect alone; there has to be some diffusional mechanism coupled with the reaction in order to give rise to multiplicity of the steady states. Therefore, for a fixed-bed catalytic reactor where the flow conditions are such that external mass and heat transfer resistance between the surface of the catalyst pellets and the bulk gas phase are negligible, the catalyst pellet size, pore structure, and conductivity are such that intraparticle heat and mass transfer resistances are also negligible (a system that can be described by a pseudohomogeneous model), the bulk gas phase is in plug flow, and the system is adiabatic or cooled cocurrently, then multiplicity of the steady states is usually not possible. If the bulk flow conditions is not in plug flow, then multiplicity is possible. However, as

indicated earlier, this situation is very unlikely in industrial fixed-bed catalytic reactors. Therefore, the main source of multiplicity in fixed-bed catalytic reactors is through the coupling between the exothermic reaction and
the catalyst pellet mass and heat transfer resistances.

Isothermal (concentration) multiplicity and thermal multiplicity may
coexist in certain systems, when the kinetics are nonmonotonic, the reaction
is exothermic, and the reactor system is nonisothermal.

7.1.3 Multiplicity Due to the Reactor Configuration

Industrial fixed-bed catalytic reactors have a wide range of different configurations. The configuration of the reactor may give rise to multiplicity of the
steady states when other sources are not sufficient to produce the phenomenon. The most well known is the case of catalytic reactors where the gas
phase is in plug flow and all diffusional resistances are negligible; however,
the reaction is exothermic and is counter-currently cooled (68). One of the
typical examples for this case is the TVA-type ammonia converter.

7.2 SIMPLE QUANTITATIVE DISCUSSION OF THE MULTIPLICITY PHENOMENON

Consider three very simple lumped systems, each one described by similar
algebraic equations:

1. CSTR (continuous-stirred tank reactor)
2. Nonporous catalyst pellet
3. Cell with permeable membrane containing an enzyme

A reaction is taking place in each system with a rate of reaction $r = f(C_A)$,
where $f(C_A)$ is a nonmonotonic function.

Figure 7.1 shows that for a certain combination of parameters, the
supply line $\alpha(C_{Af} - C_A)$ intersects the consumption curve $f(C_A)$ in three
points, giving rise to multiple steady states.

For all the three cases in Table 7.1 [see Eqs. (7.1)–(7.3)]

$$\alpha(C_{Af} - C_a) = f(C_A)$$

where

$$\alpha = \frac{q}{V} \quad \text{(for CSTR)}$$

$$\alpha = \frac{a_p K_g}{W_p} \quad \text{(for catalyst pellet)}$$

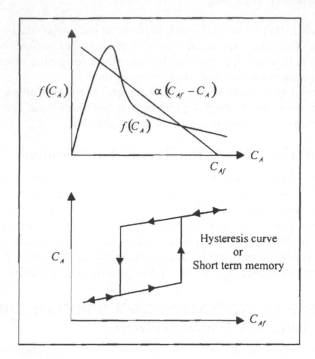

Figure 7.1 Hysteresis curve for concentration multiplicity.

$$\alpha = \frac{a_C P_A}{E V_C} \qquad \text{(for cell)}$$

7.3 BIFURCATION AND STABILITY

In the multiplicity region of operating and design parameters, the stability characteristics of the system are quite different from those in the uniqueness region of parameters for the same system. It is also important to note that for systems showing multiplicity behavior over a certain region of parameters, even when the system is operating in the region of unique steady state, the existence of the multiplicity region will have its implications on the behavior of the system in the face of external disturbances that may move the system into this multiplicity region.

7.3.1 Steady-State Analysis

The full appreciation of the stability characteristics of any system requires dynamic modeling and analysis of the system. Detailed dynamic modeling

Table 7.1 Multiplicity of the Steady States for Three Chemical/Biochemical Systems

CSTR	Nonporous catalyst pellet	cell containing an enzyme

CSTR

$$q(C_{Af} - C_A) = Vf(C_A) \qquad (7.1)$$

where

q = volumetric flow rate
V = reactor volume
C_{Af} = feed concentration
C_A = effluent concentration of reactant
$f(C_A)$ = reaction rate based on unit volume of reaction mixture

Nonporous catalyst pellet

$$a_p K_g(C_{Af} - C_A) = W_p f(C_A) \qquad (7.2)$$

where

a_p = surface area of pellet
W_p = mass of pellet
K_g = mass transfer coefficient
C_{Af} = concentration of reactant in bulk fluid
C_A = concentration of reactant on pellet surface
$f(C_A)$ = reaction rate based on unit volume of reaction mixture

cell containing an enzyme

$$a_C P_a(C_{Af} - C_A) = \bar{E} V_C f(C_A) \qquad (7.3)$$

where

a_C = surface area of cell
V_C = volume of cell
P_A = permeability of cell membrane
$\bar{E}$ = weight of enzyme per unit volume cell
C_{Af} = concentration of reactant in bulk fluid
C_A = concentration of reactant inside the cell
$f(C_A)$ = reaction rate based on unit mass of enzyme

and analysis is beyond the scope of this book; however, a considerable amount of insight into the stability and dynamic characteristics of the system can be extracted from steady-state analysis as shown in this section. In Section 7.3.2, a simple and brief introduction to the dynamical side of the picture is given.

A simple and almost obvious illustration of this point is that when it is found from steady-state analysis that the system is operating in the multiplicity region, then global stability for any of the steady states is not possible. Every steady state will have its region of asymptotic stability (RAS) or, in the more modern terminology of dynamical systems theory, a basin of attraction. This fact has very important implications for the dynamic behavior, stability, and control of the system. Also, it will be shown that the initial direction of the dynamic trajectory can be predicted from steady-state arguments. Of course, there remains qualitative and quantitative dynamic questions that need to be answered and which can only be answered through dynamic modeling and analysis of the system.

To illustrate the above points, let us consider a simple homogeneous CSTR, where a consecutive exothermic reaction

$$A \xrightarrow{k_1} B \xrightarrow{k_2} C$$

is taking place, and the reactor is at steady-state conditions, as shown in Figure 7.2.

The mass and heat balance equations for the model are given in dimensionless form as follows:

$$X_{Af} = X_A + \alpha_1 e^{-y_1/y} X_A \tag{7.4}$$

$$X_{Bf} = X_B + \alpha_2 e^{-y_2/y} X_B - \alpha_1 e^{-y_1/y} X_A \tag{7.5}$$

$$y - y_f = \alpha_1 \beta_1 e^{-y_1/y} X_A + \alpha_2 \beta_2 e^{-y_2/y} X_B + \bar{K}_c(y_c - y) \tag{7.6}$$

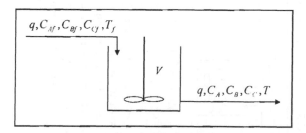

Figure 7.2   Schematic diagram of a CSTR.

where the following dimensionless variables are used:

$$X_A = \frac{C_A}{C_{ref}}$$

$$X_{Af} = \frac{C_{Af}}{C_{ref}}$$

$$X_B = \frac{C_B}{C_{ref}}$$

$$X_{Bf} = \frac{C_{Bf}}{C_{ref}}$$

$$y = \frac{T}{T_{ref}}$$

$$y_c = \frac{T_c}{T_{ref}}$$

The following dimensionless parameters are also used:

$$\alpha_i = \frac{Vk_{i0}}{q}$$

$$\gamma_i = \frac{E_i}{R_G T_{ref}}$$

$$\beta_i = \frac{(\Delta H_i) C_{ref}}{\rho C_p T_{ref}}$$

$$\bar{K}_c = \frac{UA_H}{q\rho C_p}$$

Note:

$$i = 1, 2$$

For the special case of adiabatic operation, we set

$$\bar{K}_c = 0.0 \tag{7.7}$$

in Eq. (7.6).

The equations of the adiabatic case also represent the case of non-porous catalyst pellets with external mass and heat transfer resistances and negligible intraparticle heat transfer resistance but with different meanings to the parameters.

Equations (7.4)–(7.6) can be solved simultaneously for a certain set of parameters in order to obtain the values X_A, X_B, and y (the concentrations and temperature at the exit of the reactor for the CSTR case (and the surface of the catalyst pellet for the nonporous catalyst pellet case). However, it is possible to reduce Eqs. (7.4)–(7.6) to a single nonlinear equation in y, together with two explicit linear equations for the computation of X_A and X_B once y has been determined. The single nonlinear equation (for $X_{Bf} = 0$) can be written as

$$R(y) = (y - y_f) - \bar{K}_c(y_c - y)$$

$$= \frac{\alpha_1 \beta_1 e^{-y_1/y} X_{Af}}{1 + \alpha_1 e^{-y_1/y}} + \frac{\alpha_2 \beta_2 e^{-y_2/y} \alpha_1 e^{-y_1/y} X_{Af}}{\left(1 + \alpha_1 e^{-y_1/y}\right)\left(1 + \alpha_2 e^{-y_2/y}\right)} \equiv G(y) \qquad (7.8)$$

The right-hand side is proportional to the heat generation and will be termed the heat generation function, $G(y)$, whereas the left-hand side is proportional to the heat removal due to the flow and the cooling jacket and will be termed the heat removal function:

$$R(y) = \left(1 + \bar{K}_c\right)y - \left(y_f + \bar{K}_c y_c\right) \qquad (7.9)$$

Equation (7.8) can be solved using any of the standard methods (bisectional, Newton–Raphson, etc.); however, it is more instructive to solve it graphically by plotting $G(y)$ and $R(y)$ versus y, as shown in Figure 7.3 for a case with parameters corresponding to three steady states.

The slope of the heat removal line is

$$b = 1 + \bar{K}_c$$

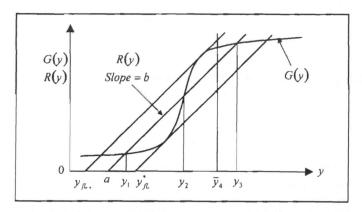

Figure 7.3 Schematic diagram for the heat generation function $G(y)$ and the heat removal function $R(y)$, for a case with maximum number of steady states equal to 3.

and the intersection with the horizontal axis is at

$$a = \frac{y_f + \bar{K}_c y_c}{1 + \bar{K}_c} \tag{7.10}$$

For the adiabatic case with $\bar{K}_c = 0$, we get

$$b_{ad} = 1 \tag{7.11}$$

and

$$a_{ad} = y_f \tag{7.12}$$

For the case shown in Figure 7.3, three steady states are possible: y_1, y_2, and y_3. Some information regarding the stability of the three steady states and the dynamic behavior can be obtained from this static diagram. The steady-state temperatures y_1, y_2, and y_3 correspond to points where the heat generation and heat removal are equal (steady states). If the reactor is disturbed to a point $\bar{y}_4$ that is not a steady state, then it is easy from this static diagram to determine the direction of temperature change, because at this point, it is clear that the heat generation is higher than the heat removal and, therefore, the temperature will increase. Of course, the full behavior of the temperature–time trajectory can only be determined through the dynamic model of the system.

With regard to this limited steady-state test of stability, let us examine y_3 using the above argument. Suppose that we disturb y_3 by an infinitesimal amount δy, which is positive; in this case, we notice from Figure 7.3 that the heat removal is higher than the heat generation and the system cools down toward y_3. If δy is negative, the heat generation is higher than the heat removal and, therefore, the system will heat up toward y_3. This indicates that y_3 is stable. However, this is only a necessary condition for stability, but it is not sufficient. Other stability conditions should be checked through the computation of the eigenvalues of the dynamic model (69).

If we do the same experiment for y_1, we find that it also satisfies the necessary condition for stability. However, if we check the intermediate temperature y_2, we find that if δy is positive, then the heat generation is higher than the heat removal and, therefore, the system will heat up away from y_2. On the other hand, if δy is negative, the heat removal is higher than the heat generation and, therefore, the system will cool down away from y_2, computing the eigenvalues of the dynamic model is not necessary because violation of the necessary condition for stability is a sufficient condition for instability.

The shape of the bifurcation diagram can be easily concluded from this simple heat generation–heat removal diagram. Let us take, for simplicity, the adiabatic case with y_f as a bifurcation parameter. If we plot y versus the bifurcation parameter y_f, we obtain the S-shaped hysteresis curve shown in Figure 7.4. The points $y_{fL.}$ and y_{fL}^* are the two limit points which are the boundaries for the multiplicity region; thus, multiplicity of the steady states exists for values of y_f lying in the region

$$y_{fL.} \leq y_f \leq y_{fL}^* \tag{7.13}$$

The curves shown in Figures 7.3 and 7.4 are not the only possible cases; another case with another set of parameters is shown in Figure 7.5. In this case, five steady states are possible. With regard to static stability, y_1 is stable, y_2 is unstable, y_3 is stable, y_4 is unstable, and y_5 is stable. The bifurcation diagram for this case is shown in Figure 7.6. In this case, the diagram can be divided into the following regions:

Unique low-temperature steady state for the region

$$y_f \leq y_{fL1} \tag{7.14}$$

Three steady states for the region

$$y_{fL1} \leq y_f \leq y_{fL2}$$

Five steady states in the region

$$y_{fL2} \leq y_f \leq y_{fL3}$$

Three steady states in the region

$$y_{fL3} \leq y_f \leq y_{fL4}$$

Unique high-temperature steady state in the region

$$y_f \geq y_{fL4}$$

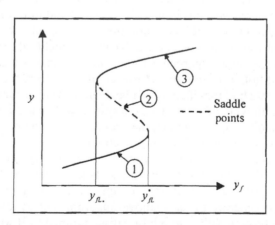

Figure 7.4 Bifurcation diagram for a case with three steady sates (two limit points).

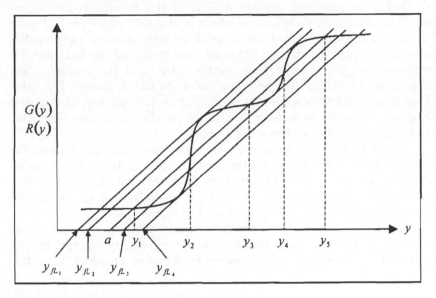

Figure 7.5 Heat generation and heat removal functions for a case with five steady states.

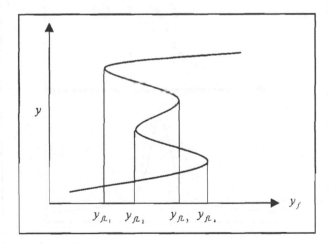

Figure 7.6 Schematic bifurcation diagram with five steady states (four limit points).

For some cases, the maximum yield of the desired product corresponds to the middle steady state, which is unstable, as shown in Figure 7.7. In these cases, efficient adiabatic operation is not possible and nonadiabatic operation is mandatory. However, the choice of the heat transfer coefficient (U), the area of heat transfer (A_H), and the cooling jacket temperature are critical for the stable operation of the system. The value of the dimensionless heat transfer coefficient ($\bar{K}_c$) should exceed a critical value ($\bar{K}_{c,\text{crit}}$) in order to stabilize the unstable middle steady state. The value of ($\bar{K}_{c,\text{crit}}$) corresponds to line 4 in Figure 7.7.

The increase in $\bar{K}_c$ increases the slope of the removal line because the slope is given by ($1 + \bar{K}_c$). If a bifurcation diagram is drawn for this nonadiabatic case with $\bar{K}_c$ as the bifurcation parameter and the jacket cooling temperature is the temperature of the middle steady state y_m, we get the pitchfork-type bifurcation diagram shown in Figure 7.8.

In Figure 7.8, for $\bar{K}_c = 0$ the three steady states y_H, y_m, and y_L are those of the adiabatic case. For $\bar{K}_c < \bar{K}_{c,\text{crit}}$, y_H and y_L change, but y_m remains the same because it is the jacket temperature. However, y_m in this

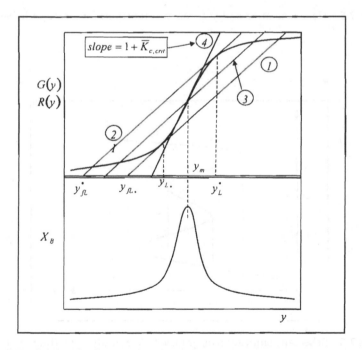

Figure 7.7 Schematic diagram for the case where the maximum yield of B corresponds to a middle unstable steady state.

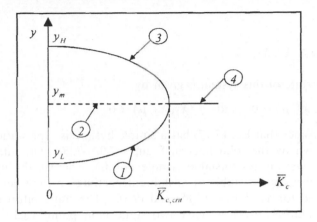

Figure 7.8 Pitchfork bifurcation diagram $(y - \bar{K}_c)$.

region is a middle unstable steady state. At $\bar{K}_{c,\text{crit}}$, the multiplicity disappears, giving rise to a unique steady state y_m. From a static point of view, this steady state is stable. However, dynamically it may be unstable for a certain range of $\bar{K}_c > \bar{K}_{c,\text{crit}}$, giving rise to limit cycle behavior in this region (69). If the cooling jacket is not y_m, we get "imperfect pitchfork" (43,44,77,78).

7.3.2 Dynamic Analysis

The steady states which are unstable using the above-discussed static analysis are always unstable. However, steady states that are stable from a static point of view may prove to be unstable when the full dynamic analysis is performed. In other words branch 2 in Figure 7.8 is always unstable, whereas branches 1, 3, and 4 can be stable or unstable depending on the dynamic stability analysis of the system. As mentioned earlier, the analysis for the CSTR presented here is mathematically equivalent to that of a catalyst pellet using lumped models or a distributed model made discrete by a technique such as the orthogonal collocation technique (see Appendix E). However, in the latter case, the system dimensionality will increase considerably, with n dimensions for each state variable, where n is the number of internal collocation points.

For simplicity of presentation, we consider a two-dimensional system with one bifurcation parameter:

$$\frac{dX_1}{dt} = f_1(X_1, X_2, \mu) \tag{7.15}$$

and

$$\frac{dX_2}{dt} = f_2(X_1, X_2, \mu) \qquad (7.16)$$

The steady state of this system is given by

$$f_1(X_1, X_2, \mu) = 0 \quad \text{and} \quad f_2(X_1, X_2, \mu) = 0 \qquad (7.17)$$

We will consider that Eq. (7.17) has a simple hysteresis-type static bifurcation as shown by the solid lines in Figures 7.9A–7.9C. The intermediate static branch is always unstable (saddle points), whereas the upper and lower branches can be stable or unstable depending on the eigenvalues of the linearized forms of Eqs. (7.15) and (7.16). The static bifurcation diagrams in Figures 7.9A–7.9C have two static limit points each; these are usually called saddle-node bifurcation points

The stability characteristics of the steady-state points can be determined from the eigenvalue analysis of the linearized version of Eqs. (7.15) and (7.16) which will have the following form:

$$\frac{d\hat{x}_1}{dt} = g_{11}\hat{x}_1 + g_{12}\hat{x}_2 \qquad (7.18)$$

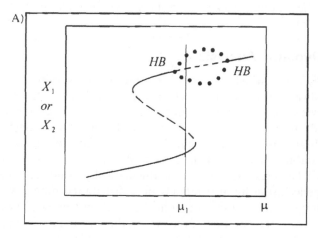

Figure 7.9 Bifurcation diagram for Eqs. (7.15)–(7.17). (A) A case with two Hopf bifurcation points; (B) a case with two Hopf bifurcation points and one periodic limit point; (C) a case with one Hopf bifurcation point, two periodic limit points, and one homoclinical orbit (infinite period bifurcation point). Solid curve: stable branch of the bifurcation diagram; dashed curve: saddle points; filled circles: stable limit cycles; open circles: unstable limit cycles; HB = Hopf bifurcation point; **PLP** = periodic limit point.

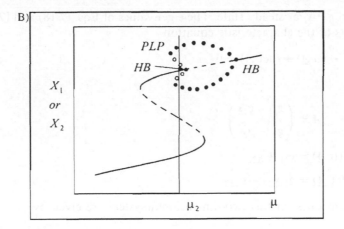

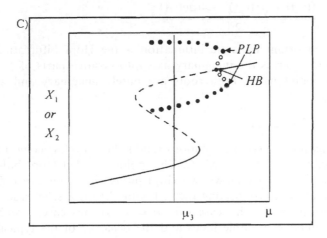

Figure 7.9 Continued

$$\frac{d\hat{x}_2}{dt} = g_{21}\hat{x}_1 + g_{22}\hat{x}_2 \tag{7.19}$$

where

$$\hat{x}_i = x_i - x_{iss}$$

and

$$g_{ij} = \left.\frac{\partial f_i}{\partial x_j}\right|_{ss}$$

where $x_{iss} = x_i$ at steady state. The eigenvalues of Eqs. (7.18) and (7.19) are the roots of the characteristic equation

$$\lambda^2 - (\text{tr } \underline{A}) + (\det \underline{A}) = 0 \tag{7.20}$$

where

$$\underline{A} = \begin{pmatrix} g_{11} & g_{12} \\ g_{21} & g_{22} \end{pmatrix}$$

$$(\text{tr } \underline{A}) = g_{11} + g_{22} \tag{7.21}$$

$$(\det \underline{A}) = g_{11}g_{22} - g_{12}g_{21}$$

The eigenvalues for this two-dimensional system are given by

$$\lambda_{1,2} = \frac{(\text{tr } \underline{A}) \pm \sqrt{(\text{tr } \underline{A})^2 - 4(\det \underline{A})}}{2} \tag{7.22}$$

The most important dynamic bifurcation is the Hopf bifurcation point, when λ_1 and λ_2 cross the imaginary axis into positive parts of λ_1 and λ_2. This is the point where both roots are purely imaginary and at which $\text{tr } \underline{A} = 0$, giving

$$\lambda_{1,2} = \pm i\sqrt{(\det \underline{A})} \tag{7.23}$$

At this point, periodic solutions (stable limit cycles) come into existence, as shown in Figure 7.9A–C. This point is called Hopf bifurcation. The case in Figure 7.9A shows two Hopf bifurcation points with a branch of stable limit cycles connecting them. Figure 7.10 is a schematic diagram of the phase plane for this case with $\mu = \mu_1$. In this case, a stable limit cycle surrounds an unstable focus and the behavior of the typical trajectories are as shown. The case in Figure 7.9B has two Hopf bifurcation points in addition to a periodic limit point (PLP) and a branch of unstable limit cycles in addition to the stable limit cycles branch.

Figure 7.11 shows the phase plane for this case when $\mu = \mu_2$. In this case, there is an unstable limit cycle surrounding a stable focus and the unstable limit cycle is surrounded by a stable limit cycle. The behavior of the typical trajectories is as shown in Figure 7.11.

The case of Figure 7.9C has one Hopf bifurcation point and one periodic limit point, and the stable limit cycle terminates at a homoclinical orbit (infinite period bifurcation). For $\mu = \mu_3$, we get a case of an unstable steady state surrounded by a stable limit cycle similar to the case in Figure 7.10. However, in this case, as μ decreases below μ_3, the limit cycle grows until we reach a limit cycle that passes through the static saddle point, as

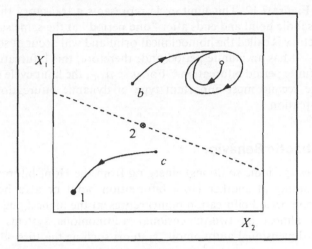

Figure 7.10 Phase plane for $\mu = \mu_1$ in Figure 7.9A. Solid curve: stable limit cycle trajectories; $\otimes$: unstable saddle; $\bullet$: stable steady state (node or focus); $\bigcirc$: unstable steady state (node or focus); dashed curve: separatix.

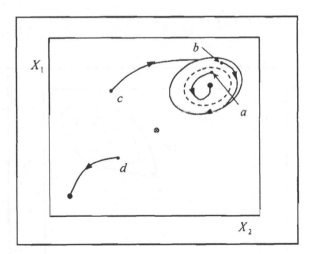

Figure 7.11 Phase plane for $\mu = \mu_2$ in Figure 7.9B. Solid curve: stable limit cycle trajectories; dashed curve: unstable limit cycle; $\otimes$: unstable saddle; $\bullet$: stable steady state (node or focus).

shown in Figure 7.12. This limit cycle represents a trajectory that starts at the static saddle point and ends after "one period" at the same saddle point. This trajectory is called the homoclinical orbit and will occur at some critical value μ_{HC}. It has an infinite period and, therefore, this bifurcation point is called "infinite period bifurcation." For $\mu < \mu_{HC}$, the limit cycle disappears. This is the second most important type of dynamic bifurcation after the Hopf bifurcation.

7.3.3 Chaotic Behavior

Limit cycles (periodic solutions) emerging from the Hopf bifurcation point and terminating at another Hopf bifurcation point or at a homoclinical orbit (infinite period bifurcation point) represent the highest degree of complexity in almost all two-dimensional autonomous systems. However, for higher-dimensional autonomous systems such as the three-dimensional system,

$$\frac{dX_1}{dt} = f_1(X_1, X_2, X_3, \mu) \tag{7.24}$$

$$\frac{dX_2}{dt} = f_2(X_1, X_2, X_3, \mu) \tag{7.25}$$

$$\frac{dX_3}{dt} = f_3(X_1, X_2, X_3, \mu) \tag{7.26}$$

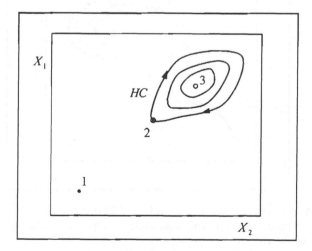

Figure 7.12 Homoclinical orbit.

or the nonautonomous two-dimensional system such as the sinusoidal forced two-dimensional system,

$$\frac{dX_1}{dt} = f_1(X_1, X_2, \mu) + A \sin(\omega t) \tag{7.27}$$

$$\frac{dX_2}{dt} = f_2(X_1, X_2, \mu) \tag{7.28}$$

Higher degrees of dynamic complexity are possible, including period doubling, quasiperiodicity (torus), and chaos. Phase-plane plots are not the best means of investigating these complex dynamics for in such cases (which are at least three-dimensional); the three (or more)-dimensional phase plane can be quite complex, as shown in Figures 7.13 and 7.14 for two of the most well-known attractors, the Lorenz strange attractor (70) and the Rössler strange attractor (71–73). Instead, stroboscopic maps for forced systems (nonautonomous) and Poincaré maps for autonomous systems are better suited for the investigation of these types of complex dynamic behavior.

The equations for the Lorenz model are

$$\frac{dX}{dt} = \sigma(Y - X)$$

$$\frac{dY}{dt} = rX - Y - XZ \tag{7.29}$$

$$\frac{dZ}{dt} = -bZ + XY$$

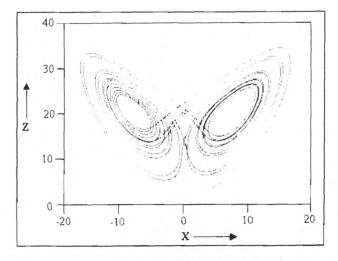

Figure 7.13 Lorenz strange attractor projected on the two-dimensional xz plane ($\sigma = 10$, $b = 8/3$, $r = 28$).

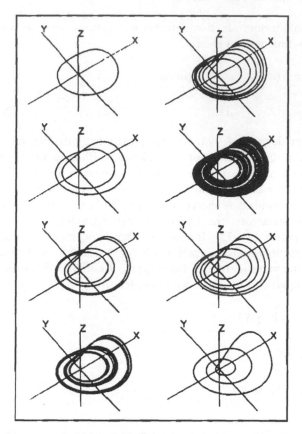

Figure 7.14 Final trajectories of the Rössler attractor [73] for different values of the parameter a. Left row, top to bottom: limit cycle, $a = 0.3$; period 2 limit cycle $a = 0.35$; period 4, $a = 0.375$; four-band chaotic attractor, $a = 0.386$; Right row, top to bottom: period 6, $a = 0.3904$; single-band chaos, $a = 0.398$; period 5, $a = 0.4$; period 3, $a = 0.411$. In all cases $b = 2$ and $c = 4$. (From Ref. 71.)

The stroboscopic and Poincaré maps are different from the phase plane in that they plot the variables on the trajectory at specific chosen and repeated time intervals. For example, for the forced two-dimensional system, these points are taken at every forcing period. For the Poincaré map, the interval of strobing is not as obvious as in the case of the forced system and many techniques can be applied. Different planes can be used in order to get a deeper insight into the nature of strange attractors in these cases. A periodic solution (limit cycle) on the phase plane will appear as one point on the stroboscopic (or Poincaré) map. When period doubling takes place, period 2 will appear as two points on the map, period 4 will appear as four points, and so on. Quasiperiodicity (torus), which looks complicated on the phase plane, will appear as an invariant close circle with discrete points on the map. When chaos takes place, a complicated collection of points appear on the stroboscopic map. The shapes formed have fractal dimensions and are usually called strange attractors.

The equations for the Rössler model are

$$\frac{dX}{dt} = -(Y - Z)$$

$$\frac{dY}{dt} = X + aY \qquad\qquad (7.30)$$

$$\frac{dZ}{dt} = b + Z(X - c)$$

A strange attractor resulting from such a deterministic model is called "deterministic chaos" to emphasize the fact that it is not a random or stochastic variation.

This is the minimum information necessary for the interested reader to get an appreciation of the complex bifurcation, instability, and chaos associated with chemical and biochemical processes. Although industrial practice in petrochemical and petroleum refining and other chemical and biochemical industries does not appreciate the importance of these phenomena and their implications on the design, optimization, and control of catalytic and biocatalytic processes, it is easy to recognize that these phenomena are important because bifurcation, instability, and chaos in these systems are generally due to nonlinearity and, specifically, nonmonotonicity, which is widespread in catalytic reactors either as a result of exothermicity or as a result of the nonmonotonic dependence of the rate of reaction on the concentration of the reactant species. It is expected that in the near future and through healthy scientific interaction between industry and academia, these important phenomena will be better appreciated by industry and that more academicians will turn their attention to the investigation of these phenomena in industrial systems.

For the readers interested in this field, it is useful to recommend the following further reading in bifurcation, instability and chaos:

1. The work of Uppal et al. (20,21) and that of Ray (46) on the bifurcation behavior of continuous-stirred tank reactors.
2. The work of Teymour and Ray (74–76) on bifurcation, instability, and chaos of polymerization reactors.

3. The work of Elnashaie and coworkers on the bifurcation and instability of industrial fluid catalytic cracking (FCC) units (42–44,77).
4. The book by Elnashaie and Elshishni on the chaotic behavior of gas–solid catalytic systems (78).
5. The interesting review article by Razon and Schmitz (79) on the bifurcation, instability, and chaos for the oxidation of carbon monoxide on platinum.
6. The two articles by Wicke and Onken (80) and Razon et al. (81) give different views on the almost similar dynamics observed during the oxidation of carbon monoxide on platinum. Wicke and Onken (80) analyze it as statistical fluctuations, whereas Razon et al. (81) analyze it as chaos. Later, Wicke changed his views and analyzed the phenomenon as chaos (82).
7. The work of the Minnesota group on the chaotic behavior of sinusoidally forced two-dimensional reacting systems is useful from the points of view of the analysis of the system as well as the development of suitable efficient numerical techniques for the investigation of these systems (82–85).
8. The book by Marek and Schreiber (86) is important for chemical engineers who want to get into this exciting field. The book covers a wide spectrum of subjects, including the following: differential equations, maps, and asymptotic behavior; transition from order to chaos; numerical methods for studies of parametric dependences, bifurcations, and chaos; chaotic dynamics in experiments; forced and coupled chemical oscillators; and chaos in distributed systems. It also contains two useful appendices, the first dealing with normal forms and their bifurcation diagrams, and the second, a computer program for construction of solution and bifurcation diagrams.
9. The software package "Auto 97" (87) is useful in computing bifurcation diagrams using the efficient continuation technique for both static and periodic branches of the bifurcation diagram.
10. The software package "Dynamics" (88) is useful in computing Floquette multipliers and Lyapunov exponents.
11. Elnashaie et al. (77) presented a detailed investigation of bifurcation, instability, and chaos in fluidized-bed catalytic reactor for both the unforced (autonomous) and forced (nonautonomous) cases.

The compilation of articles on chaos in different fields published by Cvitanovic (89) is also useful as an introduction to the field.

The dynamic behavior of fixed-bed reactors has not been extensively investigated in the literature. The only reaction which received close attention was the CO oxidation over platinum catalysts. The investigations revealed interesting and complex dynamic behavior and have shown the possibility of oscillatory behavior as well as chaotic behavior (79–81). It is easy to speculate that more emphasis on the study of the dynamic behavior of catalytic reactions will reveal complex dynamics like those discovered for the CO oxidation over a Pt catalyst, for most phenomena are due to the nonmonotonicity of the rate process, which is widespread in catalytic systems.

There are many other interesting and complex dynamic phenomena in addition to oscillation and chaos, which have been observed but not followed in depth both theoretically and experimentally; an example is the wrong directional behavior of catalytic fixed-bed reactors, for which the dynamic response to input disturbances is opposite to that suggested by the steady-state response (90,91). This behavior is most probably connected to the instability problems in these catalytic reactors, as shown crudely by Elnashaie and Cresswell (90).

It is needless to say that the above briefly discussed phenomena (bifurcation, instability, chaos, and wrong directional responses) may lead to unexpected pitfalls in the design, operation, and control of industrial fixed-bed catalytic reactors and that they merit extensive theoretical and experimental research, as well as research directed toward industrial units.

REFERENCES

1. Liljenroth, F. G. Starting and stability phenomena of ammonia-oxidation and similar reactions. *Chem. Met. Eng.* 19, 287–293, 1918.
2. Frank-Kamenetskii, D. A. *Diffusion and Heat Transfer in Chemical Kinetics.* Plenum Press, New York, 1969.
3. Aris, R, Amundson, and N. R. An analysis of chemical-reactor stability and control. I. The possibility of local control, with perfect or imperfect control mechanisms. *Chem. Eng. Sci.* 7, 121–131. 1958.
4. Aris, R. Chemical reactors and some bifurcation phenomena. In *Bifurcation Theory and Applications in Scientific Disciplines* (Gurrel, O. and Rossler, O. E., eds.) New York Academy Press, New York. 1979, pp. 314–331.
5. Aris, R., and Varma, A. (eds). *The Mathematical Understanding of Chemical Engineering Systems. Selected Papers of N. R. Amundson.* Pergamon Press, Oxford, 1980.
6. Balakotaiah, V., and Luss, D. Analysis of the multiplicity patterns of a CSTR. *Chem. Eng. Commun.* 13, 111–132, 1981.
7. Balakotaiah, V., and Luss D. Structure of the steady-state solutions of lumped-parameter chemically reacting systems. *Chem. Eng. Sci.* 37, 1611–1623, 1982.

8. Balakotaiah, V., and Luss, D. Exact steady-state multiplicity criteria for two consecutive or parallel reactions in lumped-parameter systems. *Chem. Eng. Sci.* 37, 433–445, 1982.

9. Balakotaiah, V., and Luss, D. Multiplicity features of reacting systems. Dependence of the steady-state of a CSTR on the residence time. *Chem. Eng. Sci.* 38, 1709–1721, 1983.

10. Balakotaiah, V., and Luss, D. Multiplicity criteria for multireaction networks. *AIChE J.* 29, 552–560, 1983.

11. Balakotaiah, V., and Luss, D. Global analysis of the multiplicity features of multireaction lumped-parameter systems. *Chem. Eng. Sci.* 39, 865–881, 1984.

12. Lee, C. K., Morbidelli, M., and Varma, A. Steady state multiplicity behavior of an isothermal axial dispersion fixed-bed reactor with nonuniformly active catalyst. *Chem. Eng. Sci.* 42, 1595–1608, 1987.

13. Hlavacek, V., Kubicek, M., and Marek, M. Modelling of chemical reactors. XIV. Analysis of nonstationary heat and mass transfer in a porous catalyst particle. 1. *J. Catal.* 15, 17-30, 1969.

14. Hlavacek, V., Kubicek, M., and Marek, M. Modelling of chemical reactors. XV. Analysis of nonstationary heat and mass transfer in a porous catalyst particle. 2. *J. Catal.* 15, 31–42, 1969.

15. Hlavacek, V., Kubicek, M., and Visnak, K. Modeling of chemical reactors XXVI. Multiplicity and stability analysis of a continuous stirred tank reactor with exothermic consecutive reactions $A \to B \to C$. *Chem. Eng. Sci.* 27, 719–742, 1972.

16. Hlavacek, V., and Van Rompay, P. On the birth and death of isolas. *Chem. Eng. Sci.* 36(10), 1730–1731, 1981.

17. Thomas, D. *Proceedings of International Symposium on Analysis and Control of Immobilized Enzyme Systems, Compiegne* (Thomas, D. and Kernevez, J., eds.). Elsevier, Amsterdam, 1975.

18. Friboulet, A., David, A., and Thomas, D. Excitability, memory, and oscillations in artificial acetylcholinesterase membranes. *J. Membr. Sci.* 8, 33–39, 1981.

19. Golubitsky, M., and Schaeffer, D. G. *Singularities and Groups in Bifurcation Theory, Volume I.* Applied Mathematical Science, Vol. 51, Springer-Verlag, New York, 1985.

20. Uppal, A., Ray, W. H., and Poore, A. B. Dynamic behavior of continuous stirred tank reactors. *Chem. Eng. Sci* 29(4), 967–985, 1974.

21. Uppal, A., Ray, W. H., and Poore, A. B. The classification of the dynamic behavior of continuous stirred tank reactors. Influence of reactor residence time. *Chem. Eng. Sci.* 31(3), 205–214, 1976.

22. Elnashaie, S. S. E. H., Elrifaie, M. A., Ibrahim, G., and Badra, G. The effect of hydrogen ion on the steady-state multiplicity of substrate-inhibited enzymic reactions. II. Transient behavior. *Appl. Biochem. Biotechnol.* 8(6), 467–479, 1983.

23. Elnashaie, S. S. E. H., El-Rifai, M. A., and Ibrahim, G. The effect of hydrogen ion on the steady-state multiplicity of substrate-inhibited enzymic

reactions. I. Steady-state considerations. *Appl. Biochem. Biotechnol.* 8(4), 275–288, 1983.

24. Elnashaie, S. S. E. H., Ibrahim, G., and Elshishini, S. S. The effect of hydrogen ions on the steady state multiplicity of substrate-inhibited enzymic reactions. III. Asymmetrical steady states in enzyme membranes. *Appl. Biochem. Biotechnol.* 9(5–6), 455–474, 1984.

25. Root, R. B., and Schmitz, R. A. Experimental study of steady state multiplicity in a loop reactor. *AIChE J.* 15(5), 670–679, 1969.

26. Chang, M., and Schmitz, R. A. Oscillatory states in a stirred reactor. *Chem. Eng. Sci* 30(1), 21–34, 1975.

27. Hlavacek, V., Mikus, O., Jira, E., and Pour, V. Experimental investigation of temperature profiles, parametric sensitivity, and multiple steady state in deactivated fixed bed reactors. *Chem. Eng. Sci* 35(1–2), 258–263, 1980.

28. Harold, M. P., and Luss, D. An experimental study of steady-state multiplicity features of two parallel catalytic reactions. *Chem. Eng. Sci.* 40(1), 39–52, 1985.

29. Marb, C. M., and Vortmeyer, D. Multiple steady states of a crossflow moving bed reactor: theory and experiment. *Chem. Eng. Sci.* 43(4), 811–819, 1988.

30. Widagdo, S., Seider, W. D., and Sebastian, D. H. Bifurcation analysis in heterogeneous azeotropic distillation. *AIChE J.* 35(9), 1457–1464, 1989.

31. White, D., Jr., and Johns, L. E. Temperature rise multiplicity in the absorption of chemically reactive gases. *Chem. Eng. Sci.* 40(8), 1598–1601, 1985.

32. Choi, K. Y. and Ray, W. The dynamic behavior of fluidized bed reactors for solid catalyzed gas phase olefin polymerization. *Chem. Eng. Sci* 40(12), 2261–2279, 1985.

33. Hsuen, H. K. D., and Sotirchos, S. V. Multiplicity analysis of intraparticle char combustion. *Chem. Eng. Sci* 44(11), 2639–2651, 1989.

34. Hsuen, H. K. D., and Sotirchos, S. V. Multiplicity analysis of char combustion with homogeneous carbon monoxide oxidation. *Chem. Eng. Sci.* 44(11), 2653–2665, 1989.

35. Luss, D. A note on steady state multiplicity of an electrically heated wire. *Chem. Eng. Sci.* 33(3), 409, 1978.

36. Kushner, M. J. A kinetic study of plasma-etching process. I. A model for the etching of silicon and silicon dioxide in carbon fluoride (CnFm)/hydrogen and carbon fluoride (CnFm)/oxygen plasmas. *J. Appl. Phys.* 53(4), 2923–2938, 1982.

37. Jensen, K. F. Chemical vapor deposition. In *Chemical Engineering Materials Process.* Advances in Chemistry Series No. 221, 1989.

38. Elnashaie, S. S. E. H., Mahfouz, A. T., and Elshishini, S. S. Digital simulation of an industrial ammonia reactor. *Chem. Eng. Process* 23(3), 165–177, 1988.

39. Baddour, R. F., Brian, P. L. T., Logcais, B. A., and Eymery, J. P. Steady-state simulation of an ammonia synthesis reactor. *Chem. Eng. Sci* 20(4), 281–292, 1965.

40. Brian, P. L. T., Baddour, R. F., and Eymery, J. P. Transient behavior of an ammonia synthesis reactor. *Chem. Eng. Sci* 20(4), 297–310, 1965.

41. Iscol, L. Automatic Control Conference, 1970, p. 602.

42. Elnashaie, S. S. E. H., and El-Hennawi, I. M. Multiplicity of the steady state in fluidized bed reactors. IV. Fluid catalytic cracking. *Chem. Eng. Sci.* 34(9), 1113–1121, 1979.

43. Elshishini, S. S., and Elnashaie, S. S. E. H. Digital simulation of industrial fluid catalytic cracking units: bifurcation and its implications. *Chem. Eng. Sci.* 45(2), 553–559, 1990.

44. Elshishini, S. S., and Elnashaie, S. S. E. H. Digital simulations of industrial fluid catalytic cracking units—II. Effect of charge stock composition on bifurcation and gasoline yield. *Chem. Eng. Sci.* 45(9), 2959–2964, 1990.

45. Douglas, J. M., and Rippin, D. W. T. Unsteady-state process operation. *Chem. Eng. Sci.* 21(4), 305–315, 1966.

46. Ray, W. H. Periodic operation of polymerization reactors. *Ind. Eng. Chem., Process Des. Dev.*, 7(3), 422–426, 1968.

47. Gaitonde, N. Y., and Douglas, J. M. Use of positive feedback control systems to improve reactor performance. *AIChE J.* 15(6), 902–910, 1969.

48. Bailey, J. E., Horn, F. J. M., and Lin, R. C. Cycle operation of reaction systems. Effects of heat and mass transfer resistance. *AIChE J.* 17(4), 819–825, 1971.

49. Miertschin, G. N., and Jackson, R. Optimal switching policies for fixed bed catalytic reactors in series. *Can. J. Chem. Eng.* 49(4), 514–521, 1971.

50. Lindberg, R. C., and Schmitz, R. A. Multiplicity of states with surface reaction on a blunt object in a convective system. *Chem. Eng. Sci.* 25(5), 901–904, 1970.

51. Dorawala, T. G., and Douglas, J. M. Complex reactions in oscillating reactors. *AIChE J.* 17(4), 974–981, 1971.

52. Renken, A. Use of periodic operation to improve the performance of continuous stirred tank reactors. *Chem. Eng. Sci.* 27(11), 1925–1932, 1972.

53. Bailey, J. E. Periodic operation of chemical reactors. Review. *Chem. Eng. Commun.* 1(3), 111–124, 1974.

54. Bruns, D. D., Bailey, J. E., and Luss, D. Steady-state multiplicity and stability of enzymic reaction systems. *Biotechnol. Bioeng.* 15(6), 1131–1145, 1973.

55. Lee, C. K., and Bailey, J. E. Diffusion waves and selectivity modifications in cyclic operation of a porous catalyst. *Chem. Eng. Sci.* 29(5), 1157–1163, 1974.

56. Cinar, A., Deng, J., Meerkov, S. M., and Shu, X. Vibrational control of an exothermic reaction in a CSTR: Theory and experiments. *AIChE J.* 33(3), 353–365. 1987.

57. Matros, Yu. Sh. Reactors with fixed catalyst beds. *Kinet. Katal.* 22(2), 505–512, 1981.

58. Matros, Yu. Sh. Unsteady State Processes in Catalytic Reactors. Naukya, Novosibirsk, USSR 1982 (in Russian).

59. Boreskov, G. K., and Matros, Yu. Sh. Fixed catalyst bed reactors operated in steady- and unsteady-state conditions. *Recent Adv. Eng. Anal. Chem. React. Syst.* 142–155, 1984.

60. Satterfield, C. N., and Roberts, G. W. Kinetics of thiophene hydrogenolysis on a cobalt molybdate catalyst. *AIChE J.* 14(1), 159–164, 1968.

61. Nicholas, D. M., and Shah, Y. T. Carbon monoxide oxidation over a platinum-porous fiber glass supported catalyst. Ind. Eng. Chem. Prod. Res. Dev. 15(1), 35–40, 1976.

62. Peloso, A., Moresi, M., Mustachi, C., and Soracco, B. Kinetics of the dehydrogenation of ethanol to acetaldehyde on unsupported catalysts. Can. J. Chem. Eng. 57(2), 159–164, 1979.

63. Yue, P. L., and Olaofe, O. Kinetic analysis of the catalytic dehydration of alcohols over zeolites. Chem. Eng. Res. Deve 62(2), 81–91, 1984.

64. Yue, P. L., and Birk, R. H. Catalytic Dehydration of Ethanol in a Fluidized Bed Reactor. Institute of the Engineers Symposium Series No. 87. 1984, pp. 487–494.

65. Das, G., and Biswas, A. K. Initial rate study of the vapor-phase condensation of aniline to diphenylamine. Can. J. Chem. Eng. 64(3), 473–477, 1986.

66. Elnashaie, S., Marek, M., and Ray, W. H. Effect of catalyst deactivation on the performance of dispersed catalytic reactors. Chem. React. Eng. Proc. Eur. Symp. B5, 39–48, 1972.

67. Ray, W. H., Marek, M., and Elnashaie, S. S. E. H. Effect of heat and mass dispersion on tubular reactor performance. Chem. Eng. Sci. 27(8), 1527–1536, 1972.

68. Luss, D., and Medellin, P. Steady state multiplicity and stability in a counter-currently cooled tubular reactor. Chem. React. Eng., Proc. Eur. Symp., B4, 47–56, 1972.

69. Elnashaie, S. S. E. H. Multiplicity of the steady state in fluidized bed reactors. III. Yield of the consecutive reaction A → B → C. Chem. Eng. Sci 32(3), 295–301, 1977.

70. Sparrow, C. Bifurcations in the Lorenz Equations. Lecture Notes in Applied Mathematics, Springer-Verlag, Berlin, 1982.

71. Thompson, J. M. T., and Stewart, H. B. Non-linear Dynamics and Chaos. Geometrical Methods for Engineers and Scientists. Wiley, New York, 1986.

72. Rossler, O. E. Phys. Lett. 57A, 397, 1976.

73. Rossler, O. E. Phys. Lett. 60A, 392, 1977.

74. Teymour, F., and Ray, W. H. The dynamic behavior of continuous solution polymerization reactors—IV. Dynamic stability and bifurcation analysis of an experimental reactor. Chem. Eng. Sci. 44(9), 1967–1982, 1989.

75. Teymour, F., and Ray, W. H. The dynamic behavior of continuous polymerization reactors. V. Experimental investigation of limit-cycle behavior for vinyl acetate polymerization. Chem. Eng. Sci. 47(15–16), 4121–4132, 1992.

76. Teymour, F., and Ray, W. H. The dynamic behavior of continuous polymerization reactors. VI. Complex dynamics in full-scale reactors. Chem. Eng. Sci. 47(15–16), 4133–4140, 1992.

77. Elnashaie, S. S., Abashar, M. E., and Teymour, F. A. Bifurcation, instability and chaos in fluidized bed catalytic reactors with consecutive exothermic chemical reactions. Chaos Solitons Fractals 3(1), 1–33, 1993.

78. Elnashaie, S. S. E. H., and Elshishni, S. S. *Dynamic Modelling. Bifurcation and Chaotic Behavior of Gas–Solid Catalytic Reactors.* Gordon and Breach Publishers London, 1996.
79. Razon, L. F., and Schmitz, R. A. Intrinsically unstable behavior during the oxidation of carbon monoxide on platinum. *Catal. Rev. Sci. Eng.* 28(1), 89–164, 1986.
80. Wicke, E., and Onken, H. U. Statistical fluctuations of conversion and temperature in an adiabatic fixed-bed reactor for carbon monoxide oxidation. *Chem. Eng. Sci.* 41(6), 1681–1687, 1986.
81. Razon, L. F., Chang, S. M., and Schmitz, R. A. Chaos during the oxidation of carbon monoxide on platinum-experiments and analysis. *Chem. Eng. Sci.* 41(6), 1561–1576, 1986.
82. Wicke, E., and Onken, H. U. Periodicity and chaos in a catalytic packed bed reactor for carbon monoxide oxidation. *Chem. Eng. Sci.* 43(8), 2289–2294, 1988.
83. Kevrekidis, I. G., Schmidt, L. D., and Aris, R. On the dynamics of periodically forced chemical reactors. *Chem. Eng. Commun.* 30(6), 323–330, 1984.
84. Kevrekidis, I. G., Aris, R., and Schmidt, L. D. The stirred tank forced. *Chem. Eng. Sci.* 41(6), 1549–1560, 1986.
85. Cordonier, G. A., Schmidt, L. D., and Aris, R. Forced oscillations of chemical reactors with multiple steady states. *Chem. Eng. Sci.* 45(7), 1659–1675, 1990.
86. Marek, M., and Schreiber, I. *Chaotic Behavior of Deterministic Dissipative Systems.* Cambridge University Press, Cambridge, 1991.
87. Doedel, E. J., Champneys, A. R., Fairgrieve, T. F., Kuznetsov, Y. A., Sandstede, B., and Wang, X. J. AUTO97: Continuation and Bifurcation Software for Ordinary Differential Equations, 1997.
88. Yorke, J. A. *Dynamics: Program for IBM PC Clones.* Institute for Science and Technology, University of Maryland, College Park, 1988.
89. Cvitanovic, P. *Universality in Chaos*, 2nd ed. Adam Hilger, New York, 1989.
90. Elnashaie, S. S. E., and Cresswell, D. L. Dynamic behavior and stability of adiabatic fixed bed reactors. *Chem. Eng. Sci* 29(9), 1889–1900, 1974.
91. Pinjala, V., Chen, Y. C., and Luss, D. Wrong-way behavior of packed-bed reactors: II. Impact of thermal dispersion. *AIChE J.* 34(10), 1663–1672, 1988.

8

Novel Designs for Industrial Chemical/Biochemical Systems

The material explained in this book is necessary for the development of novel configurations that can revolutionize certain critical processes, especially in the field of clean fuel production.

The rigorous mathematical modeling approach explained in full detail in previous chapters can be used to examine alternative novel configurations before (or without) the expensive pilot-plant and semicommercial units stages. Two examples are given in this concluding chapter of the book and the student, under the supervision of the course instructor, should get training in using the mathematical modeling and computer simulation approach to develop the novel configurations.

8.1 NOVEL REFORMING PROCESS FOR THE EFFICIENT PRODUCTION OF THE ULTRACLEAN FUEL HYDROGEN FROM HYDROCARBONS AND WASTE MATERIALS

8.1.1 Introduction

In the United States, solid waste materials are disposed off in landfills at a staggering annual rate of 200 million tons of municipal solid waste, 300 million scrap tires, and 15 million tons of plastic waste. Another major environmental problem results from the use of hydrocarbons as fuel and for industrial applications, generating greenhouse gases and other pollutants. There is a strategic need for the United States and other countries to develop alternative sources of clean energy in order to mitigate dependence on foreign crude oil importation. Chemical engineering students should recognize that recently there has been strong recognition that hydrogen offers significant advantages as the ultraclean fuel of the future. The main process for hydrogen production is the fixed-bed catalytic steam reforming of natural gas, which is inefficient, highly polluting, and suffers from catalyst deactivation, especially when using higher hydrocarbons as feedstock. The objective of this exercise for the student is to use the knowledge learned in this book to develop a new process which converts a wide range of hydrocarbons (e.g., natural gas, diesel, gasoline, etc.) as well as solid-waste materials into hydrogen and other useful chemicals. The students are assisted by providing them with the proposed novel configuration, which is basically a fast circulating membrane-fluidized bed integrated to a novel dry reforming process. This process addresses most of the limitations of the conventional steam reforming process for the production of hydrogen/syngas, resulting in an efficient/compact novel reformer for a wide range of feedstocks. This proposed process will offer the following environmental benefits:

1. Efficient production of an ultraclean fuel, H_2, from a wide range of feedstocks
2. Mitigation of the solid-waste problem by conversion into clean fuel
3. Conservation of natural resources
4. A novel processing scheme which has minimum environmental impact

Furthermore, the proposed process will provide several economic and strategic merits, as it will produce hydrogen at prices that are comparable to or even less expensive than hydrocarbon sources. The reader is requested to use a computer-modeling approach to develop this ultraclean and efficient process for the conversion of solid wastes and a wide range of hydrocarbons into hydrogen and other useful chemicals. The reader should check other processes in the literature and should become aware of the fact that this proposed novel process is much more efficient, flexible, and cleaner than the modern integrated bubbling fluidized-bed membrane steam reformer (BFBMST) for natural gas (e.g., Ref. 1).

8.1.2   Literature Review

A preliminary literature review is given here as a starter for the reader. However, the reader should update this literature review.

Waste Materials Pretreatment

Waste materials such as municipal solid waste, scrap tires, and waste plastics have traditionally been placed in sanitary landfills. However, with landfill space rapidly decreasing in the United States and worldwide, an alternative disposal method for these waste materials becomes imperative. The recycling of solid wastes is a challenging problem, with both economic and environmental constraints. Recently, two broad approaches have been attempted to reclaim solid wastes. The first approach relies on thermal or catalytic conversion of waste materials into fuel and valuable chemical feedstocks. Examples of this approach include gasification, pyrolysis, depolymerization, and liquefaction. The second approach relies on the physical recovery of valuable ingredients in the waste materials.

In spite of the experimental viability of conversion methods such as depolymerization and liquefaction, severe technical and economic limitations that prevented their commercial feasibility have been encountered (2,3). Feedstock characteristics and variability constituted a major challenge.

Catalytic Steam Reforming of Hydrocarbons

Nickel-supported catalysts are very efficient for steam reforming of hydrocarbons to hydrogen/syngas. The main reforming reactions are reversible and highly endothermic as follows:

$$CH_4 + H_2O \Leftrightarrow CO + 3H_2 \qquad (\Delta H_1 = 206\,kJ/mol)$$

$$CO + H_2O \Leftrightarrow CO_2 + H_2 \qquad (\Delta H_2 = -40\,kJ/mol)$$

$$CH_4 + 2H_2O \Leftrightarrow CO_2 + 4H_2 \qquad (\Delta H_3 = 196\,kJ/mol)$$

Elnashaie and Elshishini (4) have shown that the rate of steam reforming is nonmonotonic with respect to steam. Steam reforming reactions have a strong tendency to carbon formation, causing deactivation of the catalyst as follows:

$$CH_4 \rightarrow C + 2H_2 \qquad (\Delta H = 75\,kJ/mol)$$

$$2CO \rightarrow C + CO_2 \qquad (\Delta H = -171\,kJ/mol).$$

This tendency increases with the increase in temperature and the percentage of higher hydrocarbons in the feed, and it decreases with the increase in steam/hydrocarbon ratio.

Fixed-Bed Reformers: The Present Generation of Reformers

Main Characteristics

Catalyst Tubes: Fixed-bed catalytic reactors with hundreds of tubes (e.g., 100–900) having diameters in the range 7–12 cm and length in the range 10–15 m. Large catalyst particles to avoid excessive pressure drop, typically Rashig rings with dimensions $1.6 \times 0.6 \times 1.6$ cm. High steam/methane (S/M) ratio 3:6, to avoid carbon formation and catalyst deactivation. S/M ratio needed increases as the percentage of higher hydrocarbons in the feed increases. Typical feed flow rate/catalyst tube: 3.5–4.5 kmol/h/catalyst tube. Typical natural gas feed composition: 70–80% CH_4, 20–25% higher hydrocarbons $+ H_2 + CO_2$ $+N_2$. Typical feed temperature to catalyst tubes: 730–770 K (850–930°F); typical feed pressure: 2400–2900 kPa (24–29 bar).

Furnace: Huge furnace (top/side fired) to supply the endothermic heat needed for the steam reforming reactions. Typical dimensions: $21.8 \times 35.5 \times 13.7$ m.

Typical Furnace Fuel Feed

1. Flow rate (in kmol fuel/kmol natural gas) to catalyst tubes: 0.4–0.9
2. Excess air above stoichiometry: 10–15%
3. Fuel/air feed temperature: 320–580°K
4. Fuel composition: 10–20% CH_4, 65–75% H_2 (note the large consumption of hydrogen as fuel in furnace); the rest is CO_2 and N_2.

Main Previous Attempts to Develop an Efficient Compact Steam Reformer

Over the past decade, significant progress has been made toward overcoming some of the limitations of conventional fixed-bed reforming systems. These trials include the following: fixed-bed with hydrogen permselective membranes (5–7); fixed bed with/without hydrogen permselective membranes using methanol feed (8); microchannel reformer for hydrogen production from natural gas (9); catalytic oxidative steam reforming (10,11); and bubbling fluidized-bed reformer with/without hydrogen permselective membranes (1,12,13). The most successful milestone to date involves the bubbling fluidized-bed membrane steam reformer (BFBMSR) for natural gas (1), which is more efficient than the classical fixed-bed configuration. The present proposed configuration presents a much more efficient and flexible process than the BFBMSR.

The Basic Characteristics of BFBMSRs

1. Using a powdered catalyst with $\eta = 1.0$ (100–1000 times higher rate of reaction than conventional fixed-bed steam reformers).
2. Hydrogen-selective membranes "break" the thermodynamic barrier, giving higher conversions at lower temperatures.
3. In situ burning of natural gas, achieving efficient heat transfer with no large furnace. The hydrogen productivity (moles hydrogen/h/cm^3 of catalyst bed) for these BFBMSRs is 0.43–0.48 as compared with 0.06–0.12 for the fixed bed, which is more than a 300–800% increase in hydrogen productivity.
4. The operating temperature: For the fixed-bed configuration, it is in the range 980–1100 K (1305–1520°F), whereas for the BFBMSR, it is in the range 780–850 K (944–1070°F), about 20% reduction in operating temperature.

Hydrogen Fuel (14,15)

Hydrogen is rightly often called the perfect fuel. Its major reserve on earth (water) is inexhaustible. Steam reforming not only extracts the hydrogen from the hydrocarbon but also extracts the hydrogen from water. It can be used directly by fuel cells to produce electricity very efficiently (> 50%) and with zero emissions. Ultralow emissions are also achievable when hydrogen is combusted with air to power an engine. Operating fuel cells on natural gas (or other hydrocarbon source) will require a clean/efficient fuel processor to provide hydrogen to the fuel cell. Total U.S. hydrogen consumption in 1997 was approximately 8.5 billion standard cubic feet per day (bscfpd) (= 20.53 million kg H_2/day), with 40% dedicated to petroleum refining and 59% to chemical manufacturing. The U.S. hydrogen production is increasing at a rate of 5–10% per year. In most industries, hydrogen is produced using the classical inefficient fixed-bed reformers.

Hydrogen Production and CO_2 Sequestration Costs (16,17)

The proposed novel configuration will not only have an appreciable positive impact on the environment but will also achieve considerable economic advantages by making clean hydrogen energy economically competitive. The delivered cost of H_2 transportation fuel in the New York City/New Jersey areas varies considerably from different data sources. It is reported to be in the wide range of $15–40/GJ (= $0.054–

0.144/kWh = $1.96–5.23/equivalent gallon gasoline). In some other regions of the United States (and using other sources of data), the cost of hydrogen production is $10/GJ (= $0.036/kWh = $1.306/equivalent gallon gasoline). The following is hydrogen cost (including reformer capital cost + operating cost, purification, etc.) from another data source: Production cost = $10/GJ (= $0.036/kWh = $1.306/equivalent gallon gasoline) and delivered cost = $14/GJ (= $0.05/kWh = $1.83/equivalent gallon gasoline). The strategic objective should be to develop processes to produce hydrogen from fossil fuels, waste materials, and renewable energy sources to meet both central and distributed production goals of $6–8 per million BTUs ($0.82–1.09/kg H_2, $0.9–1.2/gallon equivalent gasoline). The CO_2 removal from the exit gases during the production of hydrogen or from other sources is an important step in decreasing the emission of this greenhouse gas which contributes strongly to global warming. The added cost of CO_2 separation is approximately $0.65/kg H_2 (= $00–0.17/kWh). The cost of removing CO_2 from power-plant flue gases is most typically $30–60/ton CO_2. The suggested novel CO_2 reforming which is an integral part of the proposed novel configuration will increase the H_2 productivity, improve the economics of the process, and reduce the CO_2 emission considerably.

8.1.3   Limitations of Current Reforming Technologies

Catalytic Steam Reforming Diffusional Limitations

The effectiveness factor η of the catalyst pellets, expressing the fraction of the intrinsic rate of reaction being exploited in the fixed-bed configuration, is extremely small ($\eta = 10^{-2}–10^{-3}$). This is due to the large catalyst pellet sizes used to avoid excessive pressure drop along the length of the reactor (18–20m). This severe limitation can be "broken" by using a fluidized-bed reactor with fine catalyst particles (effectiveness factor = 1.0); thus the full intrinsic activity of the catalyst is utilized.

The Thermodynamic Equilibrium Barrier for Reversible Steam Reforming

The reversible, endothermic steam reforming reactions are thermodynamically limited, dictating a very high temperature for high conversion. This thermodynamic equilibrium barrier can be "broken" through the continuous removal of one (or more) of the products. The most promising technique is the hydrogen removal using hydrogen permselective membranes. However, it is also possible to "break" this thermodynamic equilibrium barrier by the continuous removal of CO_2 (21,22) using CaO as a CO_2 acceptor according to the reaction

$$CaO(s) + CO_2(g) \Leftrightarrow CaCO_3(s) \ (725°C)$$

This is an exothermic reaction; therefore, it supplies part of the endothermic heat necessary for reforming. The CaO acceptor can be recovered through the reaction

$$CaCO_3(s) \Leftrightarrow CaO(s) + CO_2(g)$$

The above reaction is endothermic and needs high temperature of about 975°C.

Either of the two techniques can be used or a third hybrid technique utilizing both of them.

Carbon Formation and Catalyst Deactivation (23,24)

The reforming reactions (specially at high temperatures and for higher hydrocarbons) have a strong tendency to deposit carbon on the catalyst, which deactivates it. The usual technique for the solution of this problem is to increase the steam/hydrocarbon ratio in the feed. For the proposed configuration, the use of membranes for the removal of hydrogen and/or CO_2 "breaks" the thermodynamic barrier of the reversible reforming reactions, achieving high equilibrium conversion at low temperatures. The circulating nature of the proposed fast fluidized bed allows the continuous regeneration of the catalyst in its "way back" to the bed. It is also interesting to note that the endothermic carbon formation and the exothermic burning of carbon to regenerate the catalyst results in a net heat supply which is higher than oxidative reforming. For methane, this heat is accompanied by the production of two moles of hydrogen per mole of methane, and the excess heat can be used to carry out steam reforming simultaneously. A design challenge using this technology is to maximize hydrogen production under autothermic conditions.

Supply of Heat Necessary for the Endothermic Steam Reforming Reactions

The need to use high temperatures and the tendency for carbon formation, which deactivates the catalyst, limits the range of hydrocarbon to be used as feedstocks in the present generation of reformers. The heat transfer from the furnace to the catalyst in the tubes is not very efficient, causing an excessive amount of heat dissipation and environmental pollution. This problem is addressed in the suggested novel configuration through simultaneous oxidative reforming (and more efficiently through carbon formation and carbon burning as described above). CaO acceptor, and continuous regeneration/ circulation of the catalyst.

Materials (Membrane) Limitations

The idea of breaking the thermodynamic equilibrium limitations through the use of a selective membrane in not completely new. However, it became practically feasible in the last decade due to the impressive advancement in material science and the development of a new generation of highly selective inorganic membranes (5,6,25). The reader is requested to review the advancement in this field to choose the most suitable membranes.

Hydrodynamic Limitations

Whenever the diffusional limitation is "broken" through the use of fine catalyst powder in a bubbling fluidized bed, a new limitation arises related to the hydrodynamics of the system. In the bubbling fluidized bed, it is not possible to fully exploit the very intrinsic kinetics of the powdered catalyst. Fast fluidization (transport) reactor configuration offers excellent potential to "break" this limitation.

Catalyst Attrition and Entrainment

Catalytic attrition and entrainment represent a problem in bubbling fluidized-bed reactor, limiting the range of flow rate and dictates the use of cyclones. In the suggested fast fluidization (transport) reformer, the solid movement is exploited to the maximum limit in an integrated circulating configuration.

8.1.4 Main Characteristics of the Suggested Novel Ultraclean/Efficient Reforming Process Configuration

The proposed novel reformer will utilize five main catalytic processes, namely steam reforming, dry reforming, oxidative steam reforming, and carbon formation/burning in addition to a number of other noncatalytic processes.

Catalytic Processes

1. *Steam reforming*: Supported nickel catalysts (e.g., 32 mol% Ni/α-alumina, 3 mol% Ni/MgO or Mg-Ca oxide) are very efficient for steam reforming of hydrocarbons to hydrogen/syngas.
2. *Dry reforming*: The main reaction is endothermic:

$$CO_2 + CH_4 \Leftrightarrow 2CO + 2H_2 \quad (\Delta H = 247\,kJ/mol)$$

Suitable catalysts include Ni-, Rh-, and Ru-supported catalysts and ZrO_2-supported Pt catalysts promoted with cerium.

3. *Oxidative steam reforming*: The main reaction is highly exothermic:

$$CH_4 + \tfrac{1}{2}O_2 \rightarrow CO + 2H_2 \quad (\Delta H = -208 \text{ kJ/mol, exothermic})$$

Main suitable catalysts are nickel-based catalysts, titanates-based perovskite oxides, cobalt-containing catalysts.

4. *Carbon formation:*

$$CH_4 \rightarrow C + 2H_2 \quad (\Delta H = 74.8 \text{ kJ/mol, endothermic})$$

5. *Carbon burning:*

$$C + O_2 \rightarrow CO_2 \quad (\Delta H = -393.5 \text{ kJ/mol, exothermic})$$

Other Noncatalytic Processes

The main other noncatalytic reaction is homogeneous combustion of hydrocarbons. In addition, when CaO is utilized as a CO_2 acceptor, two additional noncatalytic reactions are involved, as explained earlier.

8.1.5 Components of the Suggested Novel Ultraclean Process for the Production of the Ultraclean Fuel Hydrogen

The exercise involves the optimization of the process not only with regard to the design parameters and operating conditions, but also with regard to the configuration itself. A simplified diagram of the suggested integrated novel configuration is shown in Figure 8.1. A brief description and discussion of the different components and their functions and interactions follows. They will be the basis for the modeling exercise for the reader.

The suggested novel configuration consists of the following (see Fig. 8.1):

1. Fast-circulating fluidized bed (transport reactor). The circulating fluidized bed (transport reactor) achieves a number of advantages over the bubbling fluidized bed:

   (a) "Breaks" the hydrodynamic limitation "born" after the break of the diffusional limitations by using fine catalyst particles, as discussed earlier

   (b) Gives near plug flow conditions which are beneficial for conversion and hydrogen selectivity

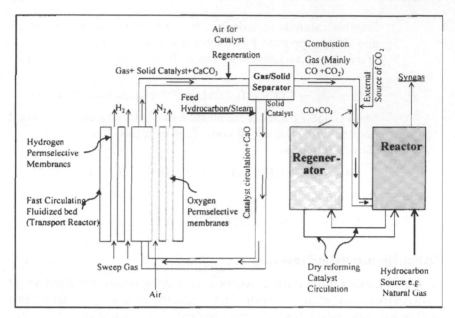

Figure 8.1 Preliminary suggestion for the novel hydrogen production configuration.

(c) Allows distribution of the oxidizing agent, as detailed later

(d) Allows the flexibility of using different configuration for the sweep gas (26) in the membrane side

(e) Allows the continuous regeneration of catalyst by burning the reformed carbon.

(f) Allows the continuous use and regeneration of CaO for the continuous removal of CO_2

2. Hydrogen permselective membranes. Hydrogen permselective membranes are used in the reformer to break the thermodynamic equilibrium barrier of the reforming reactions and to produce pure hydrogen in the membrane side.

3. Carbon dioxide acceptors. Addition of CaO to the catalyst circulating bed to remove CO_2, as described earlier.

4. Oxygen permselective membranes. Oxygen selective membranes are used to supply the oxygen along the height of the reformer for the oxidative reforming reaction.

5. The catalyst mixture in the reformer. A mixture of mainly a steam reforming catalyst with an optimum percentage of oxidative reforming catalyst will be used in the reformer in order to

minimize the homogeneous combustion and maximize hydrogen production and heat integration.

6. Gas–solid separation unit. The catalyst from the reformer will be separated from the gases in a gas–solid separator and the regenerated solid catalyst recycled to the reformer. The hydrocarbon/steam feed will be fed in the downer pipe used for recirculating the regenerated catalyst to make use of the residence time in this pipe for the reforming reactions.

7. The novel reactor–regenerator dry reformer. The output gas from the steam reformer will be very poor in hydrogen and quite rich in CO_2 due to hydrogen removal by membranes, regeneration of catalyst, and $CaCO_3$. It can be used together with a suitable hydrocarbon in a dry reforming process (27,28). A promising approach suggested in the present exercise is to use a novel fluidized-bed reactor-regenerator system that utilizes the carbon formed on the catalyst in the reactor to supply the necessary heat for the endothermic dry reforming reaction through the regeneration of the catalyst in the regenerator and recycling the hot regenerated catalyst to the reactor [Similar in a sense to the fluid catalytic cracking (FCC) process]. This part of the unit can also be used separately as a stand-alone unit for the sequestration of CO_2 produced from different sources (e.g., power plants), thus contributing to the national/international efforts to control global warming.

8. Feedstocks. To simplify the exercise, natural gas is to be used as the feedstock. However, other higher hydrocarbons can also be used.

8.1.6 Main Tasks for the Exercise

A number of extensive literature surveys will be carried out by the reader to produce the necessary data.

Task 1: Kinetics, Catalysis, and Hydrodynamics

1. Collection of steam reforming kinetics. The kinetics of the steam reforming of natural gas (as well as higher hydrocarbons) on a nickel catalyst should be obtained from the literature (29,30).

2. Collection of oxidative steam reforming kinetic data. Rate equations should be obtained from the literature for the most promising oxidative reforming catalysts, including Ni-based catalysts, Titanates-based perovskite oxides, and cobalt-containing catalysts. Careful consideration will be given to the calorimetry of

the reactions under oxidative conditions. Mixtures of reforming and oxidative reforming catalysts will also be tested.
3. Collection of dry reforming kinetic data. Rate equations should be obtained from the literature (31,32).
4. Carbon formation and catalyst regeneration. Rate equations should be obtained from the literature (33).
5. CO_2 acceptor and regeneration. Rate equations should be obtained from the literature (33).
6. Hydrodynamics of gas–solid flow and separation. Necessary data should be obtained from the literature and hydrodynamic models of different degree of sophistication/rigor be developed.

Task 2: Separation

The permeation and selectivity of different commercial hydrogen and oxygen selective membranes as well as membranes developed by the students should (if possible) be determined using permeation cells and the rate of permeation equations will be developed.

Task 3: Mathematical Modeling, Computer Simulation, and Optimization of the Process

Rigorous static/dynamic models should be developed for the different parts of the novel configuration. The models of the different parts should be integrated for the entire system. The overall model should then be utilized for the prediction of system performance, optimization of the design and operating conditions, as well as the development and design of optimal control strategies. This task will utilize all of the data and rate equations collected. Because a number of reactions are exothermic and recirculation is widely used in this novel configuration, the steady state and dynamic models should be used to investigate the possible instability problems associated with this configuration. The ultimate goal should be to develop a comprehensive model of the process to be used as a CAD/CAM tool for analysis, design, and optimization of this novel process (34).

> *Important note:* It is stated everywhere that all kinetics and permeation parameters as well as other parameters and rate equations are to be obtained from the literature. However, whenever the reader has the experimental facilities to determine any of these parameters, he/she is encouraged to obtain these parameters. This will be useful and add to the educational benefits of this mathematical modeling and computer simulation exercise.

8.2 A NOVEL FERMENTOR

This exercise involves not only novel configuration but, more essentially, also a novel mode of operation of a fermentor.

8.2.1 Introduction

Both "end-of-pipe" as well as "in-process" modifications are not sufficient means for achieving long-term economic growth while sustaining a clean environment. It is essential to develop a new generation of technologies that achieve pollution avoidance/prevention and produce cleaner fuels utilizing abundant waste. Cellulosic biomass is an attractive feedstock for the production of fuel ethanol by fermentation, because it is renewable and available domestically in great abundance at low cost (in some cases, it has a negative price). Cellulosic biomass includes tree leaves, forest products (particularly those fast-growing energy trees), agricultural residues (e.g., corn cobs, corn stalks, rice straws, etc.), waste streams from agricultural processing, municipal waste, yard and wood waste, waste from paper mills, and so forth. It is very important for the United States and other countries to decrease their dependence on hydrocarbons and introduce cleaner fuels to the energy and transportation sectors. The main aim of this exercise is to use mathematical and computer modeling to develop an efficient and environmentally friendly process for the production of the cleaner fuel ethanol through fermentation of the sugars resulting from the hydrolysis of biomass and cellulosic waste and energy crops (e.g., switch grass). The main bottleneck in this process is associated with the fermentation step, because the sugars produced from the hydrolysis step are difficult to ferment.

In order to "break" this bottleneck an integrated multidisciplinary system as explained in this book should be adopted (35). It consists of the following nonlinearly interacting means:

1. Use of genetically engineered mutated micro-organisms capable of efficiently fermenting these difficult sugars (36,37).
2. The use of advanced fermentor configuration (immobilized packed bed) with ethanol-selective membranes for the continuous removal of ethanol from the fermentation mixture.
3. The exploitation of the oscillatory and chaotic behavior of the system at high sugar concentrations, for higher sugar conversion and ethanol productivity (38).
4. Control of chaos for the generation of new attractors with higher ethanol productivity.

5. Extensive utilization of reliable mathematical and computer modeling to minimize the cost of development of novel efficient and environmentally friendly technologies capable of achieving Maximum Production–Minimum Pollution (MPMP).

The synergy of the above means to radically improve the performance of this process is expected to give considerable increases in ethanol productivity.

Mathematical modeling should be exploited in order to reach the optimal designs and operating mode/conditions.

Both the continuous-stirred tank configuration with/without ethanol pervaporation membrane separation as well as the packed-bed immobilized fermentors with their different configurations for the membrane case (cocurrent, countercurrent, countercurrent with partial blinding and mixed flow) are parts of this exercise for the reader. The kinetics of genetically engineered mutated micro-organisms should be used to simulate the fermentation of sugars resulting from the hydrolysis of different types of biomass and/or cellulosic waste. The biokinetic data should be obtained for the different strains of micro-organisms and different feedstocks through extensive literature survey by the reader. The effect of the membrane removal of ethanol on the biokinetics should also be investigated. The permeation and diffusion parameters should be obtained from the literature. Reliable bioreaction, permeation, and diffusion rate equations for use in the mathematical models should be obtained from the literature. Reliable fermentor models should be developed and used to find the optimal design and operating condition.

8.2.2 Basic Research Description

The development and use of innovative technologies is a must in order to develop new generation of environmentally benign technologies. This important strategic aim requires an orchestrated effort utilizing fundamental knowledge, maximum utilization of mathematical and computer modeling (as discussed in this book), and extensive use of computing resources coupled to optimally planned experimentation. This approach will be able to minimize the cost and systemize the approach and procedure for the development of efficient and environmentally friendly novel processes and products.

This exercise will, of course, concentrate on the mathematical and computer modeling part of this strategy.

Continuous Ethanol Removal

It is important for efficient production to use a suitable technique for continuous removal of this inhibitory product, ethanol. There are many techniques for the continuous removal of ethanol from the fermentation process.

Membranes

Membranes have been used in various configurations with integrated reaction/recovery schemes. *Pervaporation* is most probably the most promising technique for the efficient continuous removal of ethanol from the fermentation mixture for the efficient breaking of the ethanol inhibition barrier (39,40).

Perstraction is a similar technique; however, in this case, a solvent is used to remove product away from the membrane surface. In situations in which the solvent is highly toxic, even in dissolved amounts (41), the membrane can act to prevent the solubilization of solvent into the aqueous phase containing the biocatalyst. Alternatively, in situations in which contact rather than dissolved toxicity is a concern (42), the membrane can be used to prevent direct contact between the solvent an the biocatalyst. Supported liquid membranes can also be used in perstraction (43).

Hollow fibers

The economics of Hollow Fiber Ethanol Fermentation (HFEF) was examined (44) and the process was found to be as competitive as conventional fermentation processes even at the high cost ($4/ft^2) of hollow fibers at the time of that study.

Continuous extractive fermentation of high-strength glucose feeds to ethanol

Extractive fermentation using olel alcohol was used successfully to produce 90–95% conversion for higher sugar concentrations (400–450 g/L) (45).

Optimal Fermentor Configuration

Different fermentor configurations can be used, including batch, fed-batch, continuous-stirred tank, packed bed, and so forth. Immobilization of the micro-organism allows the use of high flow rates without facing the problem of micro-organisms washout. However, immobilization adds an additional mass transfer resistance. The choice of the best configuration being a conceptual problem is rather difficult; in this exercise, different configurations should be examined using different mutated micro-organisms and different feedstocks. Earlier studies by Elnashaie and co-workers (46,47) suggest that in many cases, the optimal configuration is the countercurrent configuration with partial blinding for a membrane packed-bed immobilized fermentor.

Immobilization of the micro-organisms also increases the yield of ethanol per gram of sugar converted (48).

Exploitation of Oscillatory and Chaotic Behavior at High Sugar Concentrations

It is interesting to note that in the fermentation process oscillations are evident (49,50). Other investigators observed and reported these continuous oscillations (51,52). Elnashaie and co-workers (35,38,46,47) analyzed some of these oscillations and found that in certain regions of operating and design parameters, the behavior is not periodic, but chaotic. They also discovered that the route to chaotic behavior is through the period-doubling mechanism. Control of chaos through external disturbances (53,54) can be used to control the fermentor in usually unattainable regions with higher ethanol yield and productivity.

Bruce and co-workers (55) assumed that the inhibitory effect of ethanol is the cause of oscillations and argued that ethanol removal using extractive fermentation stops the oscillations. Same preliminary results carried out by the authors of this book, shown in Figures 8.2 and 8.3, support this conclusion of Bruce and co-workers (55).

It is interesting also to note that periodic attractors may give higher ethanol yield than the corresponding steady states. The higher productivity

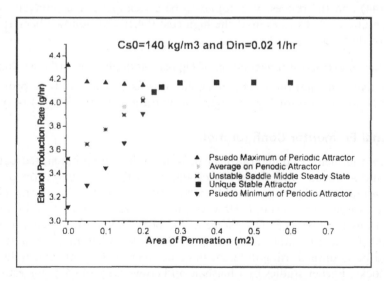

Figure 8.2 Effect of membrane permeation area on stability and ethanol production rate. A case with periodic attractor when area of permeation is $0\,m^2$.

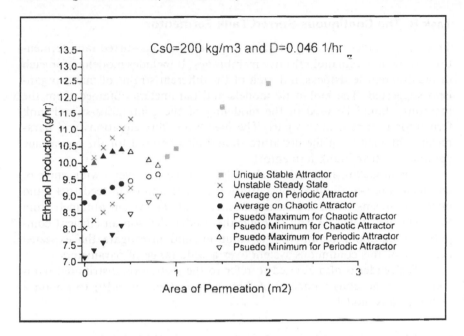

Figure 8.3 Effect of membrane permeation area on stability and ethanol production rate. A case with chaotic attractor when area of permeation is $0\,\mathrm{m}^2$.

of the periodic attractor over the corresponding steady state is shown in Figure 8.2, whereas the higher productivity of the chaotic attractor over the corresponding steady state is shown in Figure 8.3. This effect will be of considerable importance for the above-discussed unclassical approach to optimization and control of these bioreactors.

8.2.3 Tasks for the Exercise

Task A. Kinetics of Sugars Fermentation to Ethanol and Ethanol Membrane Removal

The objective of this part is to collect the data regarding the kinetics of sugar fermentation to ethanol using different sugars (resulting from the hydrolysis of biomass and cellulosic waste) using different strains of micro-organisms and genetically engineered mutated micro-organisms. The feedstocks are mainly the difficult-to-ferment sugars resulting from the hydrolysis of cellulosic waste and energy crops. Unstructured and structured kinetic models should be used.

Task B. The Continuous-Stirred Tank Fermentor

This part includes modeling (46,53) for the continuous-stirred tank fermentor with/without ethanol-selective membranes. It includes modeling for each of the different feedstocks and each of the different strains of micro-organisms suggested. The biokinetic models and parameters obtained from the literature should be used in the modeling of the continuous-stirred tank fermentor proposed in this part. The membrane flux equations and parameters obtained from the literature should also be used for the membrane continuous-stirred tank fermentor.

   The modeling and analysis should be carried out over a wide range of parameters (especially feed concentrations) for the different feedstocks and micro-organisms. With the basic parameters of the system obtained from the literature, it will be wise to build the model, develop an efficient computer simulation program for its solution, and investigate the expected behavior of this continuous system over a wide range of parameters.

   The reader is also advised to refer to the section on distributed parameter heterogeneous modeling of fermentation (Section 6.10) to obtain a very rigorous model.

Task C. Immobilized Packed-Bed Fermentors With/Without Membranes

In this part, the more advanced configuration of immobilized packed-bed fermentor with/without membranes is modeled and simulated (Figure 8.4). The micro-organisms in this configuration is immobilized using different

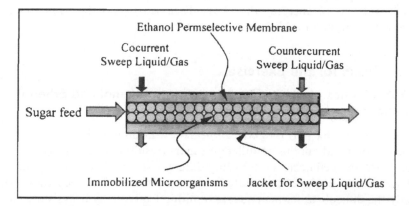

Figure 8.4 Schematic simplified diagram of the immobilized packed-bed membrane fermentor (cocurrent and countercurrent configurations).

types of carrier (e.g., calcium alginate) with different diffusivities. The diffusion coefficients of the beads will be obtained from the literature.

Rigorous heterogeneous steady-state and dynamic models for this immobilized packed-bed fermentor with/without membranes should be developed and used to analyze and optimize this novel configuration (refer to the section in Chapter 6 regarding distributed modeling of fermentation).

For more senior readers, the dynamic model should be used not only to simulate the dynamic behavior of this heterogeneous distributed system but also to investigate the possible spatio-temporal chaos associated with it.

Important note: As for the previous exercise, it is stated everywhere that all biokinetics and permeation parameters as well as other parameters are to be obtained from the literature. However, whenever the reader has the experimental facilities to determine any of these parameters, he/she is encouraged to obtain these values. This will be useful and add to the educational benefits of this mathematical modeling and computer simulation exercise.

REFERENCES

1. Adris, A., Grace, J., Lim, C., and Elnashaie, S. S. E. H., Fluidized bed reaction system for system/hydrocarbon gas reforming to produce hydrogen. U.S. Patent 5,326,550, June 5, 1994.
2. Shelley, M. D., and El-Halwagi, M. M. Techno-economic feasibility and flow sheet synthesis of scrap tire/plastic waste liquefaction. *J. Elastomers Plastics* 31(3), 232–254, 1999.
3. Warren, A., and El-Halwagi, M. M. An economic study on the co-generation of liquid fuel and hydrogen from coal and municipal solid waste. *Fuel Process. Technol.* 49, 157–166, 1996.
4. Elnashaie, S. S. E. H., and Elshishini, S. S. *Modeling, Simulation and Optimization of Industrial Fixed Bed Catalytic Reactors.* Gordon & Breach Science, London, 1993.
5. Sammells, A. F., Barton, A. F., Peterson, D. R., Harford, S. T., Mackey, R., Van Calcar, P. H., Mundshau, M. V., and Schultz, J. B. Methane conversion to syngas in mixed conducting membrane reactors. 4th International Conference on Catalysis in Membrane Reactors, 2000.
6. Sammels, A. F., Schwartz, M., Mackay, R. A., Barton T. F., and Peterson, D. R. Catalytic membrane reactors for spontaneous synthesis gas production. *Catal. Today* 56, 325, 2000.
7. Dyer, P. N., and Chen, C. M. Engineering development of ceramic membrane reactor system for converting natural gas to H_2 and syngas for liquid transportation fuel. Proceedings of the 2000 Hydrogen Program Review, 2000.
8. Amphlett, J. C., Holland, R. M., Mann, R. F., and Pepply, B. A. Modeling a steady state catalytic methanol steam reformer for reactor development and

design. AMSE Symposia Series: Emerging Energy Technology, ASME, 1996, pp. 118–124.

9.   Makel, D. Low cost microchannel reformer for hydrogen production from natural gas. California Energy Commission (CEG), Energy Innovations Small Grant (EISG) Program, 1999.

10. Hayakawa, T., Andersen, A. G., Shimizu, M., Suzuki, K., and Takehira, K. Partial oxidation of methane to synthesis gas over some tibanates based perovskite oxides. *Catal. Lett.* 22, 307–317, 1993.

11. Theron, J. N., Dry, M. E., Steen, E. V., and Fletcher, J. C. G. Internal and external transport effects during the oxidative reforming of methane on a commercial steam reforming catalyst. *Studies Surface Sci. Catal.* 107, 455–460, 1997.

12. Elnashaie, S. S. E. H., and Adris, A. Fluidized beds steam reformer for methane. Proceedings of the IV International Fluidization Conference, 1989.

13. Adris, A., Elnashaie, S. S. E. H., and Hughes, R. Fluidized bed membrane steam reforming of methane. *Can. J. Chem. Eng.* 69, 1061–1070, 1991.

14. Ogden, J. Hydrogen: A low polluting energy carrier for the next century. *Pollut. Preven. Rev.* 8, 4, 1998.

15. Ogden, J. M. Hydrogen energy systems studies. Proceedings of the 1999 US DOE Hydrogen Program Review, 1999.

16. Ogden, J. A technical and economical assessment of hydrogen energy systems with CO_2 sequestration. 12th World Hydrogen Energy Conference, 1998.

17. Ogden, J. Hydrogen systems and CO_2 sequestration. 9th National Hydrogen Association Meeting, 1998.

18. Soliman, M. A., Elnashaie, S. S. E. H., Al-Ubaid, A. S., and Adris, A. Simulation of steam reformers for methane. *Chem. Eng. Sci.* 43, 1801–1806, 1988.

19. Elnashaie, S. S. E. H., Adris, A., Soliman, M. A., and Al-Ubaid, A. S. Digital simulation of industrial steam reformers. *Can. J. Chem. Eng.* 70, 786–793, 1992.

20. Alhabdan, F. M., Abashar, M. A., and Elnashaie, S. S. E. H. A flexible software package for industrial steam reformers and methanators based on rigorous heterogeneous models. *Math. Computer Model.* 16, 77–86, 1992.

21. Han, C., and Harrison, D. P. Simultaneous shift and carbon dioxide separation for the direct production of hydrogen. *Chem. Eng. Sci.* 49. 5875–5883, 1994.

22. Brun-Tsekhovoi, A. R., Zadorin, A. R., Katsobashvilli, Y. R., and Kourdyumov, S. S. The process of catalytic steam reforming of hydrocarbons in the presence of carbon dioxide acceptor. Hydrogen Energy Progress VII, Proceedings of the 7th World Hydrogen Energy Conference, 1988, pp. 885–900.

23. Demicheli, M. C., Duprez, D., Barbier, J., Ferretti, O., and Ponzi, E. Deactivation of steam-reforming catalysts by coke formation. *J. Cataly.* 145, 437–449, 1994.

24. Barbier, J. Coking of reforming catalysts. *Studies Surface Sci. Cataly.* 34, 1–19, 1987.

25. Collins, J. P., and Way, J. D. Preparation and characterization of a composite palladium–ceramic membrane. *Ind. Eng. Chem. Res.* 32, 3006, 1993.

26. Elnashaie, S. S. E. H., Moustafa, T. M., Alsoudani, T., and Elshishini, S. S. Modeling and basic characteristics of novel integrated dehydrogenation–hydrogenation membrane catalytic reactors. *Computers Chem. Eng.* 24, 1293–1300, 2000.

27. Gadalla, A. M., and Sommer, M. E. Carbon dioxide reforming of methane on nickel catalysts. *Chem. Eng. Sci.* 44, 2825, 1989.

28. Bradford, M. C. J., and Vannice, M. A. CO_2 reforming of CH_4. *Catal. Rev. Sci. Eng.* 41, 1–42, 1999.

29. Chen, D., Lodeng, R., Omdahl, K., Anundskas, A., Olsvik, O., and Holmen, A. A model for reforming on Ni catalyst with carbon formation and deactivation. *Studies Surface Sci. Cataly.* 139, 93–100, 2001.

30. Caton, N., Villacampa, J. I., Royo, C., Romeo, E., and Monzon, A. Hydrogen production by catalytic cracking of methane using $Ni-Al_2O_3$ catalysts. Influence of the operating conditions. *Studies Surface Sci. Catal.* 139, 391–398, 2001.

31. Wurzel, T., Olsbye, U., and Mleczko, L. Kinetic and reaction engineering studies of dry reforming of methane over a $Ni/La/Al_2O_3$ catalyst. *Ind. Eng. Chem. Res.* 36(12), 5180–5188, 1997.

32. Ferreira-Aparicio, P., Marquez-Alvarez, C., Rodriguez-Ramos, I., Schuurman, Y., Guerrero-Ruiz, A., and Mirodatos, C. A transient kinetic study of the carbon dioxide reforming of methane over supported Ru catalysts. *J. Catal.* 184(10), 202–212, 1999.

33. Arandes, J. M., Abajo, I., Fernandez, I., Lopez, D., and Bilbao, J. Kinetics of gaseous product formation in the coke combustion of a fluidized catalytic cracking catalyst. *Ind. Eng. Chem. Res.* 38(9), 3255–3260, 1999.

34. Natarajan, T. T., and El-Kaddah, N. Finite element analysis of electromagnetically driven flow in sub-mold stirring of steel billets and slabs. *ISIJ Int.* 38(7), 680–689, 1998.

35. Garhyan, P., and Elnashaie, S. S. E. H. Integrated multidisciplinary approach for efficient production of fuel ethanol from cellulosic waste. International Conference on Fiber Industry, 2002.

36. Ingram, L. O., Conway, T., and Alterhum, F. U.S. Patent 5,000,000, 1991.

37. Ho, N. W. Y. and Tsao, G. T. U.S. Patent 5,789,210, 1998.

38. Garhyan, P. and Elnashaie, S. S. E. H. Exploration and exploitation of bifurcation chaotic behavior for efficient ethanol fermentation. AIChE Annual Conference, 2002.

39. Yamasaki, A., Ogasawara, K., and Mizoguchi, K. Pervaporation of water/alcohol mixtures through the PVA membranes containing cyclodextrin oligomer. *J. Appl. Polym. Sci.* 54(7), 867–872, 1994.

40. Ruckenstein, E., and Liang, L. Pervaporation of ethanol–water mixtures through polyvinyl alcohol–polyacrylamide interpenetrating polymer network membranes unsupported and supported on polyethersulfone ultrafiltration membranes: A comparison. *J. Membr. Sci.* 110(1), 99–107, 1996.

41. Shukla, R., Kang, W., and Sirkar, K. K. Novel hollow fiber immobilization techniques for whole cells and advanced bireactors. *Appl. Biochem. Biotechnol.* 20(2), 571–586, 1989.

42. Steinmeyer, D. E., and Shuler, M. L. Mathematical modeling and simulations of membrane bioreactor extractive fermentations. *Biotechnol. Progr.* 6(5), 362–369, 1990.

43. Deblay, P., Minier, M., and Renon, H. Separation of L-valine from fermentation broths using a supported liquid membrane. *Biotechnol. Bioeng.* 35(2), 123–131, 1990.

44. Naser, S. F., and Fournier, R. L. A numerical evaluation of a hollow fiber extractive fermentor process for the production of ethanol. *Biotechnol. Bioeng.* 32(5), 628–638, 1988.

45. Daugulis, A. J., Axford, D. B., Ciszek, B., and Malinowski, J. J. Continuous fermentation of high-strength glucose feeds to ethanol. *Biotechnol. Lett.* 16(6), 637–642, 1994.

46. Abasaed, A. E., and Elnashaie, S. S.E .H. A novel configuration for packed bed membrane fermentors for maximizing ethanol productivity and concentration. *J. Membr. Sci.* 82(1–2), 75–82, 1993.

47. Elnashaie, S. S. E. H., Garhyan, P., Haddad, S., Ibrahim, G., Abasaeed, A., and Elshishini, S. S. Integrated multidisciplinary approach for efficient production of fuel alcohol. AIChE Annual Meeting, 2000.

48. Doran, P. M., and Bailey, J. E. Effects of immobilization on growth, fermentation properties, and macromolecular composition of *Saccharomyces cerevisiae* attached to gelatin. *Biotechnol. Bioeng.* 28(1), 73–87, 1986.

49. Jobses, I. M. L., Egberts, G. T. C., Luyben, K. C. A. M., and Roels, J. A. Fermentation kinetics of *Zymomonas mobilis* at high ethanol concentrations: oscillations in continuous cultures. *Biotechnol. Bioeng.* 28, 868–877, 1986.

50. Daugulis, A. J., McLellan, P. J., and Li, J. Experimental investigation and modeling of oscillatory behavior in the continuous culture of *Zymomonas mobilis*. *Biotechnol. Bioeng.* 56, 99–105, 1997.

51. Ghommidh, C., Vaija, J., Bolarinwa, S., and Navarro, J. M. Oscillatory behavior of *Zymomonas* in continuous cultures: A simple stochastic model. *Biotechnol. Lett.* 9, 659–664, 1989.

52. McLellan, P. J., Daugulis, A. J., and Li, J. The incidence of oscillatory behavior in the continuous fermentation of *Zymomonas mobilis*. *Biotechnol. Progr.* 15, 667–680, 1999.

53. Haddad, Sherif, M.Sc. thesis, Salford University, U.K., 1996.

54. Elnashaie, S. S. E. H., and Abashar, M. E. Chaotic behavior of periodically forced fluidized-bed catalytic reactors with consecutive exothermic chemical reactions. *Chem. Eng. Sci.* 49(15), 2483–2498, 1994.

55. Bruce, L. J., Axford, D. B., Ciszek,B., and Daugulis, A. J. Extractive fermentation by *Zymomonas mobilis* and the control of oscillatory behavior. Biotechnol. Lett. 13(4), 291–296, 1991.

Appendix A

Matrices and Matrix Algebra

The following facts make a working knowledge of matrix methods virtually indispensable for an engineer dealing with multivariable systems:

1. The convenience of representing systems of several equations in compact, vector-matrix form
2. How such representations facilitate further analysis

It is perhaps useful to briefly review the essential concepts here because most readers will have encountered matrix methods previously.

A.1 DEFINITION, ADDITION, SUBTRACTION, AND MULTIPLICATION

A *matrix* is an array of numbers arranged in rows and columns, written in the following notation:

$$\mathbf{A} = \begin{pmatrix} a_{11} & a_{12} & a_{13} \\ a_{21} & a_{22} & a_{23} \end{pmatrix} \tag{A.1}$$

Because the above matrix $\mathbf{A}$ contains two rows and three columns, it is called a 2×3 matrix. The elements $a_{11}, a_{12}, \ldots$ have the notation a_{ij}; that is, the element of the ith row and jth column of $\mathbf{A}$.

In general, a matrix with m rows and n columns is said to have dimension $m \times n$. In the special case when $n = m$, the matrix is said to be *square*; otherwise, the matrix is said to be *rectangular* or *nonsquare*.

As in the case of scalar quantities, matrices can also be added, subtracted, multiplied and so forth, but special rules of matrix algebra must be followed.

A.1.1 Addition/Subtraction

Only matrices having the same dimensions (i.e. the same number of rows and columns) can be added or subtracted. For example, given the matrices **A**, **B**, and **C**,

$$\mathbf{A} = \begin{pmatrix} a_{11} & a_{12} & a_{13} \\ a_{21} & a_{22} & a_{23} \end{pmatrix}, \quad \mathbf{B} = \begin{pmatrix} b_{11} & b_{12} & b_{13} \\ b_{21} & b_{22} & b_{23} \\ b_{31} & b_{32} & b_{33} \end{pmatrix}$$

$$\mathbf{C} = \begin{pmatrix} c_{11} & c_{12} & c_{13} \\ c_{21} & c_{22} & c_{23} \end{pmatrix} \tag{A.2}$$

only **A** and **C** can be added or subtracted. **B** cannot be added or subtracted from either **A** or **C** because its dimensions are different from both.

The sum of $\mathbf{A} + \mathbf{C} = \mathbf{D}$ can be written as

$$\begin{aligned} \mathbf{D} = \mathbf{A} + \mathbf{C} &= \begin{pmatrix} a_{11} & a_{12} & a_{13} \\ a_{21} & a_{22} & a_{23} \end{pmatrix} + \begin{pmatrix} c_{11} & c_{12} & c_{13} \\ c_{21} & c_{22} & c_{33} \end{pmatrix} \\ &= \begin{pmatrix} a_{11} + c_{11} & a_{12} + c_{12} & a_{13} + c_{13} \\ a_{21} + c_{21} & a_{22} + c_{22} & a_{23} + c_{23} \end{pmatrix} \end{aligned} \tag{A.3}$$

Thus, addition (or subtraction) of two matrices is accomplished by adding (or subtracting) the corresponding elements of the two matrices. These operations may, of course, be extended to any number of matrices of the same order. In particular, observe that adding k identical matrices **A** (equivalent to scalar multiplication of **A** by k) results in a matrix in which each element is merely the corresponding element of **A** multiplied by the scalar k. Thus, when a matrix is to be multiplied by a scalar, the required operation is done by multiplying each element of the matrix by the scalar. For example, for **A** given in relation (A.2),

$$k\mathbf{A} = k \begin{pmatrix} a_{11} & a_{12} & a_{13} \\ a_{21} & a_{22} & a_{23} \end{pmatrix} = \begin{pmatrix} k\,a_{11} & k\,a_{12} & k\,a_{13} \\ k\,a_{21} & k\,a_{22} & k\,a_{23} \end{pmatrix} \tag{A.4}$$

A.1.2 Multiplication

Matrix multiplication can be performed on two matrices **A** and **B** to form the product **AB** only if the number of columns of **A** is equal to the number of rows in **B**. If this condition is satisfied, the matrices **A** and **B** are said to be *conformable* in the order **AB**. Let us consider that **P** is an $m \times n$ matrix (m rows and n columns) whose elements p_{ij} are determined from the elements of **A** and **B** according to the following rule:

$$p_{ij} = \sum_{k=1}^{n} a_{ik}b_{kj} \quad i = 1, 2, \ldots, m; \quad j = 1, 2, \ldots, r \qquad (A.5)$$

For example, if **A** and **B** are defined by Eq. (A.2), then **A** is a 2×3 matrix, **B** a 3×3 matrix, and the product **P** = **AB**, a 2×3 matrix, is obtained as

$$\mathbf{P} = \begin{pmatrix} a_{11} & a_{12} & a_{13} \\ a_{21} & a_{22} & a_{23} \end{pmatrix} \begin{pmatrix} b_{11} & b_{12} & b_{13} \\ b_{21} & b_{22} & b_{23} \\ b_{31} & b_{32} & b_{33} \end{pmatrix}$$

Thus we get

$$\mathbf{P} = \begin{pmatrix} (a_{11}b_{11} + a_{12}b_{21} + a_{13}b_{31}) & (a_{11}b_{12} + a_{12}b_{22} + a_{13}b_{32}) & (a_{11}b_{13} + a_{12}b_{23} + a_{13}b_{33}) \\ (a_{21}b_{11} + a_{22}b_{21} + a_{23}b_{31}) & (a_{21}b_{12} + a_{22}b_{22} + a_{23}b_{32}) & (a_{21}b_{13} + a_{22}b_{23} + a_{23}b_{33}) \end{pmatrix}$$
$$(A.6)$$

Unlike in scalar algebra, the fact that **AB** exists does not imply that **BA** exists. So, in general,

$$\mathbf{AB} \neq \mathbf{BA} \qquad (A.7)$$

Note: In matrix algebra, therefore, unlike in scalar algebra, the order of multiplication is very important.

Those special matrices for which **AB** = **BA** are said to *commute* or be *permutable*.

A.2 TRANSPOSE OF A MATRIX

The transpose of the matrix **A**, denoted by $\mathbf{A}^T$, is the matrix formed by interchanging the rows with columns. For example, if

$$\mathbf{A} = \begin{pmatrix} a_{11} & a_{12} & a_{13} \\ a_{21} & a_{22} & a_{23} \end{pmatrix}$$

then

$$\mathbf{A}^T = \begin{pmatrix} a_{11} & a_{21} \\ a_{12} & a_{22} \\ a_{13} & a_{23} \end{pmatrix} \tag{A.8}$$

Thus, if $\mathbf{A}$ is an $m \times n$ matrix, then $\mathbf{A}^T$ is an $n \times m$ matrix. Observe, therefore, that both products $\mathbf{A}^T\mathbf{A}$ and $\mathbf{A}\mathbf{A}^T$ can be formed and that each will be a square matrix. However, we note, again, that, in general,

$$\mathbf{A}\mathbf{A}^T \neq \mathbf{A}^T\mathbf{A} \tag{A.9}$$

The transpose of a matrix made up of the product of the other matrices may be expressed by the relation

$$(\mathbf{ABCD})^T = \mathbf{D}^T\mathbf{C}^T\mathbf{B}^T\mathbf{A}^T \tag{A.10}$$

which is valid for any number of matrices, $\mathbf{A}$, $\mathbf{B}$, $\mathbf{C}$, $\mathbf{D}$, and so on.

A.3   SOME SCALAR PROPERTIES OF MATRICES

Associated with each matrix are several *scalar* quantities that are quite useful for characterizing matrices in general; the following are some of the most important.

A.3.1   Trace of a Matrix

If $\mathbf{A}$ is a square, $n \times n$ matrix, the trace of $\mathbf{A}$, sometimes denoted as $\mathrm{Tr}(\mathbf{A})$, is defined as the sum of the elements on the main diagonal; that is,

$$\mathrm{Tr}(\mathbf{A}) = \sum_{i=1}^{n} a_{ii} \tag{A.11}$$

A.3.2   Determinants and Cofactors

Of all the scalar properties of a matrix, none is perhaps more important than the *determinant*. By way of definition, the "absolute value" of a *square* matrix $\mathbf{A}$, denoted by $|\mathbf{A}|$, is called its determinant. For a simple 2×2 matrix

$$\mathbf{A} = \begin{pmatrix} a_{11} & a_{12} \\ a_{21} & a_{22} \end{pmatrix} \tag{A.12}$$

the determinant is defined by

$$|\mathbf{A}| = a_{11}a_{22} - a_{21}a_{12} \tag{A.13}$$

For higher-order matrices, the determinant is more conveniently defined in terms of *cofactors*.

A.3.3 Cofactors

The *cofactor* of the element a_{ij} is defined as the determinant of the matrix left when the ith row and the jth column are removed, multiplied by the factor $(-1)^{i+j}$. For example, let us consider the 3×3 matrix

$$\mathbf{A} = \begin{pmatrix} a_{11} & a_{12} & a_{13} \\ a_{21} & a_{22} & a_{23} \\ a_{31} & a_{32} & a_{33} \end{pmatrix} \qquad (A.14)$$

The cofactor, C_{12}, of element a_{12} is

$$C_{12} = (-1)^3 \begin{vmatrix} a_{21} & a_{23} \\ a_{31} & a_{33} \end{vmatrix} = -(a_{21}a_{33} - a_{31}a_{23}) \qquad (A.15)$$

Similarly, the cofactor C_{23} is

$$C_{23} = (-1)^5 \begin{vmatrix} a_{11} & a_{12} \\ a_{31} & a_{32} \end{vmatrix} = -(a_{11}a_{32} - a_{31}a_{12}) \qquad (A.16)$$

A.3.4 The General $n \times n$ Determinant

For the general $n \times n$ matrix $\mathbf{A}$ whose elements a_{ij} have corresponding cofactors C_{ij}, the determinant $|\mathbf{A}|$ is defined as

$$|\mathbf{A}| = \sum_{i=1}^{n} a_{ij}C_{ij} \quad \text{(for any column } j) \qquad (A.17)$$

or

$$|\mathbf{A}| = \sum_{j=1}^{n} a_{ij}C_{ij} \quad \text{(for any row } i) \qquad (A.18)$$

Equations (A.17) and (A.18) are known as the Laplace expansion formulas; they imply that to obtain $|\mathbf{A}|$, one must do the following:

1. Choose any row or column of $\mathbf{A}$.
2. Evaluate the n cofactors corresponding to the elements in the chosen row or column.
3. The sum of the n products of elements and cofactors will give the required determinant. (It is important to note that the same answer is obtained regardless of the row or column chosen.)

To illustrate, let us evaluate the determinant of the 3×3 matrix given by Eq. (A.14) by expanding in the first row. In this case, we have that

$$|A| = a_{11} C_{11} + a_{12} C_{12} + a_{13}$$

that is

$$\begin{vmatrix} a_{11} & a_{12} & a_{13} \\ a_{21} & a_{22} & a_{23} \\ a_{31} & a_{32} & a_{33} \end{vmatrix} = a_{11} \begin{vmatrix} a_{22} & a_{23} \\ a_{32} & a_{33} \end{vmatrix} - a_{12} \begin{vmatrix} a_{21} & a_{23} \\ a_{31} & a_{33} \end{vmatrix} + a_{13} \begin{vmatrix} a_{21} & a_{22} \\ a_{31} & a_{32} \end{vmatrix}$$

$$= a_{11} a_{22} a_{33} - a_{11} a_{32} a_{23} + a_{12} a_{31} a_{23} - a_{12} a_{21} a_{33}$$
$$+ a_{13} a_{21} a_{32} - a_{13} a_{31} a_{22} \qquad \text{(A.19)}$$

It is easily shown that the expansion by any other row or column yields the same result. The reader should try to prove this fact.

A.3.5   Properties of Determinants

The following are some important properties of determinants frequently used to simplify determinant evaluation.

1. Interchanging the rows with columns leaves the value of a determinant unchanged. The main implications of this property are twofold:

   (a) $|A| = |A^T|$.
   (b) Because rows and columns of a determinant are interchangeable, any statement that holds true for rows is also true for columns, and vice versa.

2. If any two rows (or columns) are interchanged, the sign of the determinant is reversed.

3. If any two rows (or columns) are identical, the determinant is zero.

4. Multiplying the elements of any one row (or column) by the same constant results in the value of determinant being multiplied by this constant factor; if, for example,

$$D_1 = \begin{vmatrix} a_{11} & a_{12} & a_{13} \\ a_{21} & a_{22} & a_{23} \\ a_{31} & a_{32} & a_{33} \end{vmatrix} \quad \text{and} \quad D_2 = \begin{vmatrix} ka_{11} & a_{12} & a_{13} \\ ka_{21} & a_{22} & a_{23} \\ ka_{31} & a_{32} & a_{33} \end{vmatrix}$$

Then, $D_2 = k D_1$.

5. If for any i and j ($i \neq j$) the elements of row (or column) i are related to the elements of row (or column) j by a constant multiplicative factor, the determinant is zero; for example,

$$D_1 = \begin{vmatrix} ka_{12} & a_{12} & a_{13} \\ ka_{22} & a_{22} & a_{23} \\ ka_{32} & a_{32} & a_{33} \end{vmatrix}$$

where we observe that column 1 is k times column 2; then, $D_1 = 0$.

6. Adding a constant multiple of the elements of one row (or column) to corresponding elements of another row (or column) leaves the value of the determinant unchanged; for example,

$$D_1 = \begin{vmatrix} a_{11} & a_{12} & a_{13} \\ a_{21} & a_{22} & a_{23} \\ a_{31} & a_{32} & a_{33} \end{vmatrix} \quad \text{and} \quad D_2 = \begin{vmatrix} a_{11} + ka_{12} & a_{12} & a_{13} \\ a_{21} + ka_{22} & a_{22} & a_{23} \\ a_{31} + ka_{32} & a_{32} & a_{33} \end{vmatrix}$$

Then, $D_2 = D_1$.

7. The value of a determinant that contains a whole row (or column) of zeros is zero.

8. The determinant of a *diagonal* matrix (a matrix with nonzero elements only on the main diagonal) is the product of the elements on the main diagonal. The determinant of a *triangular* matrix (a matrix for which all elements below the main diagonal *or* above the main diagonal are zero) is also equal to the product of the elements on the main diagonal.

A.3.6 Minors of a Matrix

Let $\mathbf{A}$ be a general $m \times n$ matrix with $m < n$. Many $m \times m$ determinants may be formed from the m rows and any m of the n columns of $\mathbf{A}$. These $m \times m$ determinants are called *minors* (of order m) of the matrix $\mathbf{A}$. We may similarly obtain several $(m - 1) \times (m - 1)$ determinants from the various combinations of $(m - 1)$ rows and $(m - 1)$ columns chosen from the m rows and n columns of $\mathbf{A}$. These determinants are also minors, but of order $(m - 1)$.

Minors may also be categorized as first minor, the next highest minor is the second minor, and so forth, until we get to the lowest-order minor, the one containing a single element. In the special case of a square $n \times n$ matrix, there is, of course, one and only one minor of order n, and there is no minor of higher order than this. Observe, therefore, that the first minor of a square matrix is what we have referred to as the *determinant* of the matrix. The second minor will be of order $(n - 1)$. Note, therefore, that the cofactors of the elements of a square matrix are intimately related to the second minors of the matrix.

A.3.7 Rank of a Matrix

The *rank* of a matrix is defined as the order of the highest nonvanishing minor of that matrix (or, equivalently, the order of the highest square submatrix having a nonzero determinant). A square $n \times n$ matrix is said to be of *full rank* if its rank is equal to n. The matrix is said to be rank deficient if the rank is less than n.

In several practical applications of matrix methods, the rank of the matrix involved provides valuable information about the nature of the problem at hand. For example, in the solution of the system of linear algebraic equations by matrix methods, the number of *independent* solutions that can be found is directly related to the rank of the matrix involved.

A.4 SOME SPECIAL MATRICES

A.4.1 The Diagonal Matrix

If all the elements of a square matrix are zero except the main diagonal from the top left-hand corner to the bottom right-hand corner, the matrix is said to be *diagonal*. Some examples are

$$\mathbf{A} = \begin{pmatrix} a_{11} & 0 \\ 0 & a_{22} \end{pmatrix} \quad \text{and} \quad \mathbf{B} = \begin{pmatrix} b_{11} & 0 & 0 \\ 0 & b_{22} & 0 \\ 0 & 0 & b_{33} \end{pmatrix}$$

A.4.2 The Triangular Matrix

A square matrix in which all the elements below the main diagonal are zero is called an *upper-triangular matrix*. By the same token, a *lower-triangular matrix* is one for which all the elements above the main diagonal are zero. Thus,

$$\begin{pmatrix} b_{11} & b_{12} & b_{13} \\ 0 & b_{22} & b_{23} \\ 0 & 0 & b_{33} \end{pmatrix} \quad \text{and} \quad \begin{pmatrix} b_{11} & 0 & 0 \\ b_{21} & b_{22} & 0 \\ b_{31} & b_{32} & b_{33} \end{pmatrix}$$

are, respectively, examples of upper- and lower-triangular matrices.

A.4.3 The Identity Matrix

A diagonal matrix in which the nonzero elements on the main diagonal are all unity is called the *identity matrix*, it is frequently denoted by $\mathbf{I}$:

$$\mathbf{I} = \begin{pmatrix} 1 & 0 & 0 & \cdots & 0 \\ 0 & 1 & 0 & \cdots & 0 \\ 0 & 0 & 1 & \cdots & 0 \\ \vdots & \vdots & \vdots & \ddots & 0 \\ 0 & 0 & 0 & 0 & 1 \end{pmatrix}$$

The significance of the identify matrix will be obvious later, but, for now, we note that it has the unique property that

$$\mathbf{IQ} = \mathbf{Q} = \mathbf{QI} \tag{A.20}$$

for any general matrix $\mathbf{Q}$ of conformable order with the multiplying identity matrix.

A.4.4 Orthogonal Matrix

If $\mathbf{A}^T\mathbf{A} = \mathbf{I}$, the identity matrix, then $\mathbf{A}$ is an *orthogonal* (or *orthonormal*, or *unitary*) matrix. For example, as the reader may easily verify,

$$\mathbf{A}\begin{pmatrix} \left(\frac{3}{5}\right) & \left(\frac{4}{5}\right) \\ \left(\frac{4}{5}\right) & -\left(\frac{3}{5}\right) \end{pmatrix}$$

is an orthogonal matrix.

A.4.5 Symmetric and Skew-Symmetric Matrices

When every pair of (real) elements that are symmetrically placed in a matrix with respect to the main diagonal are equal, the matrix is said to be a (real) *symmetric* matrix; that is, for a (real) symmetric matrix,

$$a_{ij} = a_{ji} \quad \text{and} \quad i \neq j \tag{A.21}$$

Some examples of symmetric matrices are

$$\mathbf{A} = \begin{pmatrix} 1 & 3 \\ 3 & 4 \end{pmatrix} \quad \text{and} \quad \mathbf{B} = \begin{pmatrix} 2 & 1 & 5 \\ 1 & 0 & -2 \\ 5 & -2 & 4 \end{pmatrix}$$

Thus, a symmetric matrix and its transpose are identical; that is

$$\mathbf{A} = \mathbf{A}^T \tag{A.22}$$

A *skew-symmetric* matrix is one for which

$$a_{ij} = -a_{ji} \tag{A.23}$$

The following are some examples of such matrices:

$$\mathbf{A} = \begin{pmatrix} 0 & -3 \\ 3 & 0 \end{pmatrix} \quad \text{and} \quad \mathbf{B} = \begin{pmatrix} 0 & 1 & 5 \\ -1 & 0 & -2 \\ -5 & 2 & 4 \end{pmatrix}$$

For skew-symmetric matrices,

$$\mathbf{A} = -\mathbf{A}^T \tag{A.24}$$

By noting that a_{ij} can be expressed as

$$a_{ij} = \tfrac{1}{2}\left(a_{ij} + a_{ij}\right) + \tfrac{1}{2}\left(a_{ij} - a_{ij}\right) \tag{A.25}$$

it is easily established that every square matrix $\mathbf{A}$ can be resolved into the sum of a symmetric matrix $\mathbf{A}^0$ and a skew-symmetric matrix $\mathbf{A}'$; that is,

$$\mathbf{A} = \mathbf{A}^0 + \mathbf{A}' \tag{A.26}$$

A.4.6 Hermitian Matrices

When symmetrically situated elements in a matrix of complex numbers are complex conjugates, that is,

$$a_{ji} = \bar{a}_{ij} \tag{A.27}$$

the matrix is said to be *Hermitian*. For example,

$$\begin{pmatrix} 3 & (2 - 3j) \\ (2 + 3j) & 5 \end{pmatrix}$$

is a Hermitian matrix. Note that the (real) symmetric matrix is a special form of a Hermitian matrix in which the coefficients of j are absent.

A.4.7 Singular Matrices

A matrix whose determinant is zero is said to be *singular*. A matrix whose determinant is not zero is said to be *nonsingular*.

A.5 INVERSE OF A MATRIX

The matrix counterpart of division is the inverse operation. Thus, corresponding to the reciprocal of a number in scalar algebra is the inverse of the matrix A, denoted by A^{-1} (read as "A inverse"). A matrix for which both AA^{-1} and $A^{-1}A$ give I, the identity matrix, i.e.,

$$AA^{-1} = A^{-1}A = I \tag{A.28}$$

The inverse matrix A^{-1} does not exist for all matrices, it exists only if the following hold:

1. A is square.
2. Its determinant is not zero (i.e., the matrix is nonsingular).

The inverse of A is defined as

$$A^{-1} = \frac{C^T}{|A|} = \frac{\text{adj}(A)}{|A|} \tag{A.29}$$

where C is the matrix of cofactors; that is,

$$C = \begin{pmatrix} C_{11} & C_{12} & \cdots & C_{1n} \\ C_{21} & C_{22} & \cdots & C_{2n} \\ \vdots & \vdots & \ddots & \vdots \\ C_{n1} & C_{n2} & \cdots & C_{nn} \end{pmatrix}$$

as defined previously, C_{ij} is the cofactor of the element a_{ij} of the matrix A. C^T, the transpose of this matrix of cofactors, is termed the *adjoint* of A and often abbreviated as $\text{adj}(A)$.

Clearly because $|A| = 0$ for singular matrices, the matrix inverse as expressed in Eq. (A.29) does not exist for such matrices.

The following should be noted:

1. The inverse of a matrix composed of a product of other matrices may be written in the form

   $$(ABCD)^{-1} = D^{-1}C^{-1}B^{-1}A^{-1}$$

2. For the simple 2×2 matrix, the general definition in Eq. (A.29) simplifies to a form easy to remember. Thus, given

   $$A = \begin{pmatrix} a_{11} & a_{12} \\ a_{21} & a_{22} \end{pmatrix}$$

   we have that $|A| = a_{11}a_{22} - a_{12}a_{21}$, and it is easily shown that

   $$A^{-1} = \frac{1}{|A|} \begin{pmatrix} a_{22} & -a_{12} \\ -a_{21} & a_{11} \end{pmatrix} \tag{A.30}$$

A.6   LINEAR ALGEBRAIC EQUATIONS

Much of the motivation behind developing matrix operations is due to their particular usefulness in handling sets of linear algebraic equations; for example, the following three linear equations:

$$a_{11}x_1 + a_{12}x_2 + a_{13}x_3 = b_1$$

$$a_{21}x_1 + a_{22}x_2 + a_{23}x_3 = b_2 \qquad (A.31)$$

$$a_{31}x_1 + a_{32}x_2 + a_{33}x_3 = b_3$$

The above equations contain the unknowns x_1, x_2, and x_3. They can be represented by the single matrix equation.

$$\mathbf{Ax} = \mathbf{b} \qquad (A.32)$$

where

$$\mathbf{A} = \begin{pmatrix} a_{11} & a_{12} & a_{13} \\ a_{21} & a_{22} & a_{23} \\ a_{31} & a_{32} & a_{33} \end{pmatrix}, \qquad \mathbf{x} = \begin{pmatrix} x_1 \\ x_2 \\ x_3 \end{pmatrix}, \qquad \mathbf{b} = \begin{pmatrix} b_1 \\ b_2 \\ b_3 \end{pmatrix} \qquad (A.33)$$

The solution to these equations can be obtained very simply by premultiplying Eq. (A.32) by $\mathbf{A}^{-1}$ to yield

$$\mathbf{A}^{-1}\mathbf{Ax} = \mathbf{A}^{-1}\mathbf{b}$$

which becomes

$$\mathbf{x} = \mathbf{A}^{-1}\mathbf{b} \qquad (A.34)$$

upon recalling the property of a matrix inverse that $\mathbf{A}^{-1}\mathbf{A} = \mathbf{I}$ and recalling Eq. (A.20) for the property of the identity matrix.

There are several standard, computer-oriented numerical procedures for performing the operation $\mathbf{A}^{-1}$ to produce the solution given in Eq. (A.34). Indeed, the availability of these computer subroutines makes the use of matrix techniques very attractive from a practical viewpoint.

It should be noted that the solution equation (A.34) exists *only* if $\mathbf{A}$ is nonsingular, as only then will $\mathbf{A}^{-1}$ exist. When $\mathbf{A}$ is singular, $\mathbf{A}^{-1}$ ceases to exist and Eq. (A.34) will be meaningless. The singularity of $\mathbf{A}$, of course, implies that $|\mathbf{A}| = 0$, and from property 5 given for determinants, this indicates that some of the rows and columns of $\mathbf{A}$ are *linearly dependent*. Thus, the singularity of $\mathbf{A}$ means that only a subset of the equations we are trying to solve are truly linearly independent. As such, linearly independent solutions can be found only for this subset. It can be shown that the actual number of such linearly independent solutions is equal to the rank of $\mathbf{A}$.

Solved Example A.1

To illustrate the use of matrices in the solution of linear algebraic equations, let us consider the following set of equations:

$$2x_1 + 2x_2 + 3x_3 = 1$$

$$4x_1 + 5x_2 + 6x_3 = 3$$

$$7x_1 + 8x_2 + 9x_3 = 5$$

which in matrix notation is given by

$$\mathbf{Ax} = \mathbf{b}$$

where $\mathbf{A}$, $\mathbf{b}$, and $\mathbf{x}$ are

$$\mathbf{A} = \begin{pmatrix} 2 & 2 & 3 \\ 4 & 5 & 6 \\ 7 & 8 & 9 \end{pmatrix}, \qquad \mathbf{b} = \begin{pmatrix} 1 \\ 3 \\ 5 \end{pmatrix}, \qquad \mathbf{x} = \begin{pmatrix} x_1 \\ x_2 \\ x_3 \end{pmatrix}$$

Notice that $\mathbf{A}$ is simply a (3×3) matrix whose inverse can be computed as

$$\mathbf{A}^{-1} = \begin{pmatrix} 1 & -2 & 1 \\ -2 & 1 & 0 \\ 1 & \frac{2}{3} & -\frac{2}{3} \end{pmatrix}$$

Thus, from Eq. (A.34), the solution to this set of algebraic equations is obtained as

$$\mathbf{x} = \mathbf{A}^{-1}\mathbf{b} = \begin{pmatrix} 1 & -2 & 1 \\ -2 & 1 & 0 \\ 1 & \frac{2}{3} & -\frac{2}{3} \end{pmatrix} \begin{pmatrix} 1 \\ 3 \\ 5 \end{pmatrix} = \begin{pmatrix} 1-6+5 \\ -2+3+0 \\ 1+\frac{6}{3}-\frac{10}{3} \end{pmatrix} = \begin{pmatrix} 0 \\ 1 \\ -\frac{1}{3} \end{pmatrix}$$

Thus, $x_1 = 0$, $x_2 = 1$, and $x_3 = -\frac{1}{3}$ is the required solution.

A.7   EIGENVALUES AND EIGENVECTORS

The eigenvalue/eigenvector problem arises in the determination of the values of a constant λ for which the following set of n linear algebraic equations has nontrivial solutions:

$$\begin{aligned}
a_{11}x_1 &+ a_{12}x_2 + \cdots + a_{1n}x_n = \lambda x_1 \\
a_{21}x_1 &+ a_{22}x_2 + \cdots + a_{2n}x_n = \lambda x_2 \\
&\vdots \\
a_{n1}x_1 &+ a_{n2}x_2 + \cdots + a_{nn}x_n = \lambda x_n
\end{aligned} \tag{A.35}$$

This can be expressed as

$$\mathbf{Ax} = \lambda \mathbf{x}$$

or we can write

$$(\mathbf{A} - \lambda \mathbf{I})\mathbf{x} = 0 \tag{A.36}$$

Being a system of linear homogeneous equations, the solution to Eq. (A.35) or, equivalently, Eq. (A.36) will be nontrivial (i.e., $\mathbf{x} \neq 0$) if any only if

$$|\mathbf{A} - \lambda \mathbf{I}| = 0 \tag{A.37}$$

that is, $|\mathbf{A} - \lambda \mathbf{I}|$ is a singular matrix

Expanding this determinant results in a polynomial of order n in λ:

$$a_0 \lambda^n + a_1 \lambda^{n-1} + a_2 \lambda^{n-2} + \cdots + a_{n-1}\lambda + a_n = 0 \tag{A.38}$$

This equation [or its determinant from Eq. (A.37)] is called the *characteristic equation* of the $n \times n$ matrix $\mathbf{A}$, its n roots $(\lambda_1, \lambda_2, \lambda_3, \ldots, \lambda_n)$, which may be real or imaginary and may not be distinct, are the *eigenvalues* of the matrix.

Solved Example A.2

Let us illustrate the computation of eigenvalues by considering the matrix

$$\mathbf{A} = \begin{pmatrix} 1 & 2 \\ 3 & -4 \end{pmatrix}$$

In this case, the matrix $(\mathbf{A} - \lambda \mathbf{I})$ is obtained as

$$(\mathbf{A} - \lambda \mathbf{I}) = \begin{pmatrix} 1 & 2 \\ 3 & -4 \end{pmatrix} - \begin{pmatrix} \lambda & 0 \\ 0 & \lambda \end{pmatrix} = \begin{pmatrix} (1-\lambda) & 2 \\ 3 & (-4-\lambda) \end{pmatrix}$$

from which the characteristic equation is obtained as

$$|\mathbf{A} - \lambda \mathbf{I}| = -(1-\lambda)(4+\lambda) - 6 = \lambda^2 + 3\lambda - 10 = 0$$

which is a second-order polynomial (quadratic) with the following solutions:

$$\lambda_1, \lambda_2 = -5, 2$$

Thus, in this case, the eigenvalues are both real. Note that $\lambda_1 + \lambda_2 = -3$ and that $\lambda_1 \lambda_2 = -10$; also note that $\text{Tr}(\mathbf{A}) = 1 - 4 = -3$ and $|\mathbf{A}| = -10$. Thus, we see that $\lambda_1 + \lambda_2 = \text{Tr}(\mathbf{A})$ and $\lambda_1 \lambda_2 = |\mathbf{A}|$.

If we change $\mathbf{A}$ slightly to

$$\mathbf{A} = \begin{pmatrix} 1 & 2 \\ -3 & 4 \end{pmatrix}$$

then the characteristic equation becomes

$$|\mathbf{A} - \lambda\mathbf{I}| = \begin{pmatrix} 1-\lambda & 2 \\ -3 & 4-\lambda \end{pmatrix} = \lambda^2 - 5\lambda + 10 = 0$$

with roots

$$\lambda_1, \lambda_2 = \tfrac{1}{2}\left[5 \pm (\sqrt{15})j\right], \text{ where } j = \sqrt{-1}$$

which are *complex conjugates*. Note, again, that $\text{Tr}(\mathbf{A}) = 5 = \lambda_1 + \lambda_2$ and $|\mathbf{A}| = 10 = \lambda_1\lambda_2$.

Finally, if we make another small change in $\mathbf{A}$, that is,

$$\mathbf{A} = \begin{pmatrix} 1 & 2 \\ -3 & -1 \end{pmatrix}$$

then the characteristic equation is

$$|\mathbf{A} - \lambda\mathbf{I}| = \lambda^2 - 1 + 6 = 0$$

which has the solutions

$$\lambda_1, \lambda_2 = \pm\left(\sqrt{5}\right)j$$

so the eigenvalues are *purely imaginary* in this case. Once again, $\text{Tr}(\mathbf{A}) = 0 = \lambda_1 + \lambda_2$ and $|\mathbf{A}| = 5 = \lambda_1\lambda_2$.

The following are some general properties of eigenvalues worth noting:

1. The *sum* of the eigenvalues of a matrix is equal to the *trace* of that matrix:

$$\text{Tr}(\mathbf{A}) = \sum_{j=1}^{n} \lambda_j$$

2. The *product* of the eigenvalues of a matrix is equal to the determinant of that matrix:

$$|\mathbf{A}| = \prod_{j=1}^{n} \lambda_j$$

3. A singular matrix has at least one zero eigenvalue.
4. If the eigenvalues of $\mathbf{A}$ are $\lambda_1, \lambda_2, \lambda_3, \ldots, \lambda_n$, then the eigenvalues of $\mathbf{A}^{-1}$ are

$$1/\lambda_1, 1/\lambda_2, 1/\lambda_3, \ldots, 1/\lambda_n, \text{ respectively.}$$

5. The eigenvalues of a diagonal or a triangular matrix are identical to the elements of the main diagonal.
6. Given any nonsingular matrix $\mathbf{T}$, the matrices $\mathbf{A}$ and $\bar{\mathbf{A}} = \mathbf{TAT}^{-1}$ have identical eigenvalues. In this case, such matrices $\mathbf{A}$ and $\bar{\mathbf{A}}$ are called as similar matrices.

Eigenvalues are essential for the study of the stability of different systems.

Solved Example A.3

Let us illustrate the computation of eigenvectors by considering the following 2×2 matrix:

$$\mathbf{A} = \begin{pmatrix} 1 & 2 \\ 3 & -4 \end{pmatrix}$$

for which eigenvalues $\lambda_1 = -5$ and $\lambda_2 = 2$ were calculated in the previous example.

For λ_1,

$$(\mathbf{A} - \lambda_1 \mathbf{I}) = \begin{pmatrix} 6 & 2 \\ 3 & 1 \end{pmatrix}$$

and its adjoint is given by

$$\text{adj}(\mathbf{A} - \lambda_1 \mathbf{I}) = \begin{pmatrix} 1 & -2 \\ -3 & 6 \end{pmatrix}$$

We may now choose either column of the adjoint as the eigenvector $\mathbf{x}_1$, because one is a scalar multiple of the other. Let us choose

$$\mathbf{x}_1 = \begin{pmatrix} 1 \\ -3 \end{pmatrix}$$

as the eigenvector for $\lambda_1 = -5$. Because the norm of this vector is $\sqrt{1^2 + (-3)^2} = \sqrt{10}$, the *normalized eigenvector* corresponding to $\lambda_1 = -5$ will be

$$\bar{\mathbf{x}}_1 = \begin{pmatrix} \frac{1}{\sqrt{10}} \\ \frac{-3}{\sqrt{10}} \end{pmatrix}$$

Similarly, for $\lambda_2 = 2$,

$$(\mathbf{A} - \lambda_2 \mathbf{I}) = \begin{pmatrix} -1 & 2 \\ 3 & -6 \end{pmatrix}$$

and its adjoint is given by

$$\text{adj}(\mathbf{A} - \lambda_2 \mathbf{I}) = \begin{pmatrix} -6 & -2 \\ -3 & -1 \end{pmatrix}$$

Thus,

$$\mathbf{x}_2 = \begin{pmatrix} 2 \\ 1 \end{pmatrix}$$

can be chosen as the eigenvector for eigenvalue $\lambda_2 = 2$. The *normalized eigenvector* in this case is

$$\bar{\mathbf{x}}_2 = \begin{pmatrix} \frac{2}{\sqrt{5}} \\ \frac{1}{\sqrt{5}} \end{pmatrix}$$

The following are some useful properties of eigenvectors:

1. Eigenvectors associated with separate, distinct eigenvalues are *linearly independent*.
2. The eigenvectors $\mathbf{x}_j$ are solutions of a set of homogeneous linear equations and are thus determined only up to a scalar multiplier; that is, $\mathbf{x}_j$ and $k\mathbf{x}_j$ will both be a solution to the equation if k is a scalar constant. We obtain a normalized eigenvector $\bar{x}_j$ when the eigenvector $\mathbf{x}_j$ is scaled by the vector norm $\|\mathbf{x}_j\|$; that is

$$\bar{\mathbf{x}}_j = \frac{\mathbf{x}_j}{\|\mathbf{x}_j\|}$$

3. The normalized eigenvector $\bar{\mathbf{x}}_j$ is a valid eigenvector because it is a scalar multiple of $\mathbf{x}_j$. It enjoys the sometimes desirable property that it has unity norm; that is,

$$\|\bar{\mathbf{x}}_j\| = 1$$

4. The eigenvectors $\mathbf{x}_j$ may be calculated by resorting to methods of solution of homogeneous linear equations.

It can be shown that the eigenvector $\mathbf{x}_j$ associated with the eigenvalue λ_j is given by *any* nonzero column of $\mathrm{adj}(\mathbf{A} - \lambda_j \mathbf{I})$. It can also be shown that there will always be one and only one independent nonzero column of $\mathrm{adj}(\mathbf{A} - \lambda_j \mathbf{I})$.

A.8 SOLUTION OF LINEAR DIFFERENTIAL EQUATIONS AND THE MATRIX EXPONENTIAL

It is possible to use the results of matrix algebra to provide general solutions to sets of linear ordinary differential equations. Let us recall a dynamic system with control and disturbances neglected:

$$\frac{dx}{dt} = Ax \quad \text{with } x(0) = x_0 \tag{A.39}$$

Here, A is a constant $n \times n$ matrix with distinct eigenvalues, and $x(t)$ is a time-varying solution vector. Let us define a new set of n variables, $z(t)$, as

$$z(t) = M^{-1}x(t) \tag{A.40}$$

so that

$$x(t) = Mz(t) \tag{A.41}$$

Here, M is the modal matrix (having the property that matrix $M^{-1}AM$ is diagonal) formed of the eigenvectors of matrix A. Substituting Eq. (A.41) into Eq. (A.39) gives

$$M\frac{dz(t)}{dt} = AMz(t) \quad \text{with } z(0) = M^{-1}x_0 \tag{A.42}$$

Premultiplying both sides by M^{-1} gives

$$M^{-1}M\frac{dz(t)}{dt} = M^{-1}AMz(t)$$

or

$$\frac{dz(t)}{dt} = \Lambda z(t) \quad \text{with } z(0) = z_0 = M^{-1}x_0 \tag{A.43}$$

where $\Lambda = M^{-1}AM$

Because Λ is diagonal, these equations may now be written as

$$\frac{dz_1(t)}{dt} = \lambda_1 z_1(t) \quad \text{with } z_1(0) = z_{10}$$

$$\frac{dz_2(t)}{dt} = \lambda_2 z_2(t) \quad \text{with } z_2(0) = z_{20}$$

$$\vdots$$

$$\frac{dz_n(t)}{dt} = \lambda_n z_n(t) \quad \text{with } z_n(0) = z_{n0}$$

Thus, the transformation equation (A.41) has converted the original equation (A.39) into n completely decoupled equations, each of which has a solution of the form

$$z_j(t) = e^{\lambda_1 t}z_{j0}, \quad j = 1, 2, 3, \ldots, n$$

This may be expressed in matrix notation as

$$z(t) = e^{\Lambda t}z_0 \tag{A.44}$$

where

$$
e^{\Lambda t} = \begin{pmatrix}
e^{\lambda_1 t} & 0 & 0 & \cdots & 0 \\
0 & e^{\lambda_2 t} & 0 & \cdots & 0 \\
0 & 0 & e^{\lambda_2 t} & \cdots & 0 \\
\vdots & \vdots & \vdots & \ddots & \vdots \\
0 & 0 & 0 & 0 & e^{\lambda_n t}
\end{pmatrix} \tag{A.45}
$$

It is the *matrix exponential* for the diagonal matrix Λ. Equation (A.44) may be put back in terms of the original variables $x(t)$ by using Eq. (A.40):

$$
\mathbf{M}^{-1}\mathbf{x}(t) = e^{\Lambda t}\mathbf{M}^{-1}\mathbf{x}_0
$$

or

$$
\mathbf{x}(t) = \mathbf{M}e^{\Lambda t}\mathbf{M}^{-1}\mathbf{x}_0 \tag{A.46}
$$

This is often written as

$$
\mathbf{x}(t) = e^{\mathbf{A}t}\mathbf{x}_0 \tag{A.47}
$$

where the matrix $e^{\mathbf{A}t}$ is defined as

$$
e^{\mathbf{A}t} = \mathbf{M}e^{\Lambda t}\mathbf{M}^{-1} \tag{A.48}
$$

It is called the *matrix exponential* of $\mathbf{A}t$. It is very important to note that Eqs. (A.47) and (A.48) provide the general solution to any set of homogeneous equations (A.39) provided $\mathbf{A}$ has distinct eigenvalues.

Solved Example A.4

To illustrate the solution of linear constant-coefficient, homogeneous differential equations, let us consider the following systems:

$$
\frac{d\mathbf{x}}{dt} = \mathbf{A}\mathbf{x} \quad \text{with } \mathbf{x}(0) = \mathbf{x}_0 \tag{A.49}
$$

where

$$
\mathbf{A} = \begin{pmatrix} 1 & 2 \\ 3 & -4 \end{pmatrix} \quad \text{and} \quad \mathbf{x}_0 = \begin{pmatrix} 1 \\ 2 \end{pmatrix}
$$

Observe that $\mathbf{A}$ is the same matrix we have analyzed in previous examples, so we already know its eigenvalues (λ_1 and λ_2) and modal matrix $\mathbf{M}$. Thus, we have that

$$
\Lambda = \begin{pmatrix} -5 & 0 \\ 0 & 2 \end{pmatrix}, \quad \mathbf{M} = \begin{pmatrix} 1 & 2 \\ -3 & 1 \end{pmatrix}, \quad \mathbf{M}^{-1} = \begin{pmatrix} \frac{1}{7} & -\frac{2}{7} \\ \frac{3}{7} & \frac{1}{7} \end{pmatrix}
$$

Using Eq. (A.48), we get

$$e^{\mathbf{A}t} = \mathbf{M}e^{\Lambda t}\mathbf{M}^{-1} = \begin{pmatrix} 1 & 2 \\ -3 & 1 \end{pmatrix}\begin{pmatrix} e^{-5t} & 0 \\ 0 & e^{2t} \end{pmatrix}\begin{pmatrix} \frac{1}{7} & -\frac{2}{7} \\ \frac{3}{7} & \frac{1}{7} \end{pmatrix}$$

Thus, we get

$$e^{\mathbf{A}t} = \begin{pmatrix} \dfrac{e^{-5t} + 6e^{2t}}{7} & \dfrac{-2e^{-5t} + 2e^{2t}}{7} \\[3mm] \dfrac{-3e^{-5t} + 3e^{2t}}{7} & \dfrac{6e^{-5t} + e^{2t}}{7} \end{pmatrix}$$

Finally, we get

$$\mathbf{x}(t) = \begin{pmatrix} x_1(t) \\ x_2(t) \end{pmatrix} = \begin{pmatrix} \dfrac{-3e^{-5t} + 10e^{2t}}{7} \\[3mm] \dfrac{9e^{-5t} + 5e^{2t}}{7} \end{pmatrix}$$

The extension of these ideas provides solutions for the case of *nonhomogeneous linear differential equations*

$$\frac{d\mathbf{x}(t)}{dt} = \mathbf{A}\mathbf{x}(t) + \mathbf{B}\mathbf{u}(t) + \mathbf{\Gamma}\mathbf{d}(t) \quad \text{with } \mathbf{x}(t_0) = \mathbf{x}_0 \tag{A.50}$$

The method of obtaining the solution is straightforward. Here, $\mathbf{u}(t)$ and $\mathbf{d}(t)$ represent the control variables and disturbances, respectively. When the matrices $\mathbf{A}$, $\mathbf{B}$, and $\mathbf{\Gamma}$ are *constant* matrices, a derivation completely similar to the one just given leads to the solution

$$\mathbf{x}(t) = e^{\mathbf{A}(t-t_0)}\mathbf{x}_0 + \int_{t_0}^{t} e^{\mathbf{A}(t-\tau)}[\mathbf{B}\mathbf{u}(\tau) + \mathbf{\Gamma}\mathbf{d}(\tau)]\,d\tau \tag{A.51}$$

where the matrix exponentials $e^{\mathbf{A}(t-t_0)}$ and $e^{\mathbf{A}(t-\tau)}$ can be evaluated using Eq. (A.48).

Solved Example A.5

To illustrate the solution of nonhomogeneous equations using Eq. (A.51), let us suppose we have a system equation (A.50) with $\mathbf{A}$ the same as the previous example,

$$\mathbf{A} = \begin{pmatrix} 1 & 2 \\ 3 & -4 \end{pmatrix}$$

and

$$\mathbf{B} = \begin{pmatrix} 1 & 0 \\ 0 & 1 \end{pmatrix}, \quad \boldsymbol{\Gamma} = \begin{pmatrix} 1 \\ 0 \end{pmatrix}, \quad \mathbf{x}_0 = \begin{pmatrix} 1 \\ 2 \end{pmatrix},$$

$$\mathbf{u} = \begin{pmatrix} u_1 \\ u_2 \end{pmatrix}, \quad \mathbf{d} = d_1 \text{ (a scalar)}$$

In addition, assume that $u_1 = u_2 = 0$, $d_1 = 1$ (a constant disturbance), and $t_0 = 0$. In this case, Eq. (A.51) takes the form

$$\mathbf{x}(t) = e^{\mathbf{A}t}\mathbf{x}_0 + \int_0^t e^{\mathbf{A}(t-\tau)}\boldsymbol{\Gamma}\mathbf{d}(\tau)\, d\tau$$

Substituting for $e^{\mathbf{A}t}$ and $e^{\mathbf{A}t}\mathbf{x}_0$ calculated in the previous example, we obtain the solution as follows:

$$\mathbf{x}(t) = \begin{pmatrix} \dfrac{-3e^{-5t} + 10e^{2t}}{7} \\[2ex] \dfrac{9e^{-5t} + 5e^{2t}}{7} \end{pmatrix} + \int_0^t \begin{pmatrix} \dfrac{e^{-5(t-\tau)} + 6e^{2(t-\tau)}}{7} \\[2ex] \dfrac{-3e^{-5(t-\tau)} + 3e^{2(t-\tau)}}{7} \end{pmatrix} dt$$

or we can write

$$\mathbf{x}(t) = \begin{pmatrix} x_1(t) \\ x_2(t) \end{pmatrix} = \begin{pmatrix} -\dfrac{2}{5} - \dfrac{16}{35}e^{-5t} + \dfrac{13}{7}e^{2t} \\[2ex] -\dfrac{21}{70} + \dfrac{48}{35}e^{-5t} + \dfrac{13}{14}e^{2t} \end{pmatrix}$$

For the situation where the matrices $\mathbf{A}$, $\mathbf{B}$, and $\boldsymbol{\Gamma}$ can themselves be functions of time, the solution requires a bit more computation. In this case,

$$\mathbf{x}(t) = \boldsymbol{\Phi}(t, t_0)\mathbf{x}_0 + \boldsymbol{\Phi}(t, t_0)\int_{t_0}^t \boldsymbol{\Phi}^{-1}(\tau, t_0)[\mathbf{B}\mathbf{u}(\tau) + \boldsymbol{\Gamma}\mathbf{d}(\tau)]\, d\tau \qquad (A.52)$$

Here, $\boldsymbol{\Phi}(t, t_0)$ is an $n \times n$ time-varying matrix known as the *fundamental matrix solution*, which may be found from

$$\frac{d\boldsymbol{\Phi}(t, t_0)}{dt} = \mathbf{A}(t)\boldsymbol{\Phi}(t, t_0) \quad \text{with } \boldsymbol{\Phi}(t_0, t_0) = \mathbf{I} \qquad (A.53)$$

Because $\mathbf{A}(t)$, $\mathbf{B}(t)$, and $\boldsymbol{\Gamma}(t)$ can have arbitrary time dependence, the fundamental matrix solution is usually obtained by numerical methods. Nevertheless, once $\boldsymbol{\Phi}(t, t_0)$ is determined, the solution $\mathbf{x}(t)$ may be readily calculated for a variety of control and disturbance inputs $\mathbf{u}(t)$ and $\mathbf{d}(t)$ using Eq. (A.52).

REFERENCES

1. Kreyszig, E. *Advanced Engineering Mathematics*, 8th ed. Wiley, New York, 1998.
2. Amundson, N. R. *Mathematical Method in Chemical Engineering*. Prentice-Hall, Englewood Cliffs, NJ. 1966.
3. Wylie, C. R. *Advanced Engineering Mathematics*. Wiley, New York, 1966.
4. Duncan, W. J., Collar, A. R., and Frazer, R. Q. *Elementary Matrices*, Cambridge University Press, Cambridge, 1963.

Appendix B

Numerical Methods

Analytical solution methods are restricted to linear, low-dimensional problems. For nonlinear problems and/or high-dimensional systems, a numerical solution must be used.

B.1 SOLUTION OF ALGEBRAIC AND TRANSCENDENTAL EQUATIONS

To find the roots of an equation $f(x) = 0$, we start with a known approximate solution and apply any of the following methods.

B.1.1 Bisection Method

This method consists of locating the root of the equation $f(x) = 0$ between a and b. If $f(x)$ is continuous between a and b and $f(a)$ and $f(b)$ are of opposite signs, then there is a root between a and b. For example, let $f(a)$ be negative and $f(b)$ be positive. Then, the first approximation to the root is $x_1 = \frac{1}{2}(a + b)$.

If $f(x_1) = 0$, then x_1 is a root of $f(x) = 0$. Otherwise, the root lies between a and x_1, or x_1 and b according to whether $f(x_1)$ is positive or negative. Then, we bisect the interval as previously and continue the process until the root is found to the desired accuracy.

In Figure B.1, $f(x_1)$ is positive and $f(a)$ is negative; therefore, the root lies between a and x_1. Then, the second approximation to the root is

$x_2 = \frac{1}{2}(a + x_1)$. If $f(x_2)$ is negative, the root lies between x_1 and x_2. Then, the third approximation to the root is $x_3 = \frac{1}{2}(x_1 + x_2)$, and so on.

B.1.2 Newton–Raphson Method

Let x_0 be an approximation root of the equation $f(x) = 0$. If $x_1 = x_0 + h$ is the exact root, then $f(x_1) = 0$. Therefore, expanding $f(x_0 + h)$ by Taylor's series gives

$$f(x_0) + hf'(x_0) + \frac{h^2}{2!} f''(x_0) + \cdots = 0$$

Because h is very small, neglecting h^2 and higher powers of h, we get

$$f(x_0) + hf'(x_0) = 0 \quad \text{or} \quad h = -\frac{f(x_0)}{f'(x_0)} \tag{B.1}$$

Therefore, a closer approximation to the root is given by

$$x_1 = x_0 - \frac{f(x_0)}{f'(x_0)}$$

Similarly, starting with x_1, a still better approximation x_2 is given by

$$x_2 = x_1 - \frac{f(x_1)}{f'(x_1)}$$

In general,

$$x_{n+1} = x_n - \frac{f(x_n)}{f'(x_n)}$$

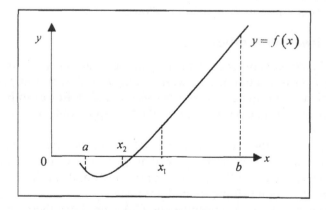

Figure B.1   Bisection method.

This technique is known as the Newton–Raphson formula or Newton's iteration formula.

Geometrical Interpretation of Newton–Raphson Technique (see Fig. B.2)

Let x_0 be a point near the root α of the equation $f(x) = 0$. Then, the equation of the tangent point $A_0[x_0, f(x_0)]$ is $y - f(x_0) = [f'(x_0)(x - x_0)]$. This tangent cuts the x axis at $x_1 = x_0 - f(x_0)/f'(x_0)$. This point is a first approximation to the root α. If A_1 is the point corresponding to x_1 on the curve, then the tangent at A_1 will cut the x axis at x_2, which is closer to α and is, therefore, a second approximation to the root. Repeating this process, we approach the root α quite rapidly.

B.2   SOLUTION OF LINEAR SIMULTANEOUS EQUATIONS

Simultaneous linear equations occur in various engineering problems. The reader knows that a given system of linear equations can be solved by Cramer's rule or by the matrix method. However, these methods become tedious for large systems. However, there exist other numerical methods of solution which are well suited for computing machines. The following is an example.

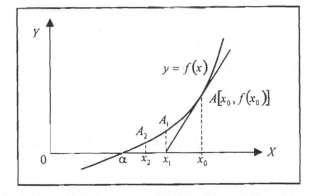

Figure B.2   Geometrical interpretation of the Newton–Raphson method.

Gauss Elimination Method

In this method, the unknowns are eliminated successively and the system is reduced to an upper-triangular system from which the unknowns are found by back substitution. The method is quite general and is well adapted for computer calculations. Here, we shall explain it by considering a system of three equations for sake of simplicity.

Consider the following equations,

$$a_1 x + b_1 y + c_1 z = d_1$$
$$a_2 x + b_2 y + c_2 z = d_2 \tag{B.2}$$
$$a_3 x + b_3 y + c_3 z = d_3$$

Step I. To eliminate x from second and third equations. Assuming $a_1 \neq 0$, we eliminate x from the second equation by subtracting a_2/a_1 times the first equation from the second equation. Similarly, we eliminate x from the third equation by eliminating (a_3/a_1) times the first equation form the third equation. We thus obtain the new system

$$a_1 x + b_1 y + c_1 z = d_1$$
$$b_2' y + c_2' z = d_2' \tag{B.3}$$
$$b_3' y + c_3' z = d_3'$$

Here, the first equation is called the *pivotal equation* and a_1 is called the *first pivot*.

Step II. To eliminate y from the third equation in Eq. (B.3). Assuming $b_2' \neq 0$, we eliminate y from the third equation of Eq. (B.3), by subtracting (b_3'/b_2') times the second equation from the third equation. Thus, we get the new system

$$a_1 x + b_1 y + c_1 z = d_1$$
$$b_2' y + c_2' z = d_2' \tag{B.4}$$
$$c_3'' z = d_3''$$

Here, the second equation is the *pivotal equation* and b_2' is the *new pivot*.

Step III. To evaluate the unknowns. The values of x, y, and z are found from the reduced system (B.4) by back-substitution.

Note: Clearly, the method will fail if any one of the pivots a_1, b_2', or c_3'' becomes zero. In such cases, we rewrite the equations in a different order so that the pivots are nonzero.

B.3 SOLUTION OF NONLINEAR SIMULTANEOUS EQUATIONS (MULTIDIMENSIONAL NEWTON–RAPHSON METHOD)

Consider the equations

$$f(x, y) = 0 \quad \text{and} \quad g(x, y) = 0 \tag{B.5}$$

If an initial approximation (x_0, y_0) to a solution has been found by the graphical method or otherwise, then a better approximation (x_1, y_1) can be obtained as follows:

Let

$$x_1 = x_0 + h \quad \text{and} \quad y_1 = y_0 + k$$

so that

$$f(x_0 + h, y_0 + k) = 0 \quad \text{and} \quad g(x_0 + h, y_0 + k) = 0 \tag{B.6}$$

Expanding each of the functions in Eq. (B.6) by the Taylor's series to first-degree terms, we get approximately

$$
\begin{aligned}
f_0 + h\frac{\partial f}{\partial x_0} + k\frac{\partial f}{\partial y_0} &= 0 \\
g_0 + h\frac{\partial g}{\partial x_0} + k\frac{\partial g}{\partial y_0} &= 0
\end{aligned}
\tag{B.7}
$$

where

$$f_0 = f(x_0, y_0), \quad \frac{\partial f}{\partial x_0} = \left(\frac{\partial f}{\partial x}\right)_{x_0, y_0} \quad \text{and so forth}$$

Solving Eqs. (B.7) for h and k, we get a new approximation to the root:

$$x_1 = x_0 + h \quad \text{and} \quad y_1 = y_0 + k$$

This process is repeated until we obtain the values of x and y with the desired accuracy.

Solved Example B.1

Solve the system of nonlinear equations

$$x^2 + y = 11 \quad \text{and} \quad y^2 + x = 7$$

Solution

An initial approximation to the solution is obtained from a rough graph of the given equation as $x_0 = 3.5$ and $y_0 = -1.8$. We have $f = x^2 + y - 11$ and $g = y^2 + x - 7$, so that

$$\frac{\partial f}{\partial x} = 2x, \quad \frac{\partial f}{\partial y} = 1, \quad \frac{\partial g}{\partial x} = 1, \quad \frac{\partial g}{\partial y} = 2y$$

Then, Newton–Raphson's equation (B.7) will be

$$7h + k = 0.55 \quad \text{and} \quad h - 3.6k = 0.26$$

Solving the above set of algebraic equations, we get

$$h = 0.0855 \quad \text{and} \quad k = -0.0485$$

Therefore, the better approximation to the root is

$$x_1 = x_0 + h = 3.5855 \quad \text{and} \quad y_1 = y_0 + k = -1.8485$$

Repeating the above process, replacing (x_0, y_0) by (x_1, y_1), we obtain

$$x_2 = 3.5844 \quad \text{and} \quad y = -1.8482$$

This procedure is to be repeated till x and y are located to the desired accuracy.

B.4 NUMERICAL SOLUTION OF NONLINEAR ORDINARY DIFFERENTIAL EQUATIONS

A number of numerical methods are available for the solution of nonlinear ordinary differential equations of the form

$$\frac{d\underline{Y}}{dx} = \underline{f}(x, \underline{Y}) \quad \text{with given } \underline{Y}(x_0) = \underline{Y}_0 \tag{B.8}$$

B.4.1 Euler's Method

Consider the equation

$$\frac{dy}{dx} = f(x, y) \quad \text{with } y(x_0) = y_0 \tag{B.9}$$

Its curve of solution through $P(x_0, y_0)$ is shown in Figure B.3. Now, we have to find the ordinate of any other point Q on this curve.

Let us divide LM into n subintervals, each of width h at $L_1, L_2, L_3, \ldots$ so that h is quite small. In the interval LL_1, we approximate the curve by the

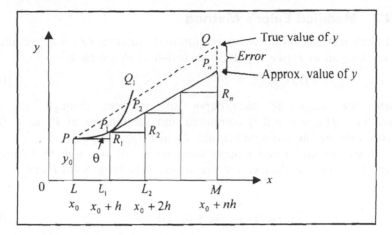

Figure B.3 Geometrical representation of Euler's method.

tangent at P. If the ordinate through L_1 meets this tangent in $P_1(x_0 + h. y_1)$, then

$$y_1 = L_1 P_1 = LP + R_1 P_1 = y_0 + PR_1 \tan \theta = y_0 + h\left(\frac{dy}{dx}\right)_P$$

$$= y_0 + hf(x_0, y_0)$$

Let $P_1 Q_1$ be the curve of the solution of Eq. (B.9) through P_1 and let its tangent at P_1 meet the ordinate through L_2 in $P_2(x_0 + 2h, y_2)$. Then

$$y_2 = y_1 + hf(x_0 + h, y_1) \tag{B.10}$$

Repeating this process n times, we finally reach an approximation MP_n of MQ given by

$$y_n = y_{n-1} + hf(x_0 + (n - 1)h, y_{n-1})$$

This is the Euler's method of finding an approximate solution of (B.9).

Note: In Euler's method, we approximate the curve to solution by the tangent in each interval (i.e., by a sequence of short lines). Unless h is very small, the error is bound to be quite significant. This sequence of lines may also deviate considerably from the curve of solution. Hence, there is a modification of this method, which is given in the next section.

B.4.2 Modified Euler's Method

In Euler's method, the curve of solution in the interval LL_1 is approximated by the tangent at P (see Fig. B.3), such that at P_1 we have

$$y_1 = y_0 + hf(x_0, y_0) \tag{B.11}$$

Then, the slope of the curve of solution through P_1 [i.e., $(dy/dx)_{P_1} = f(x_0 + h, y_1)$] is computed and the tangent at P_1 to P_1Q_1 is drawn, meeting the ordinate through L_2 in $P_2(x_0 + 2h, y_2)$.

Now, we find a better approximation $y_1^{(1)}$ of $y(x_0 + h)$ by taking the slope of the curve as the mean of the slopes of the tangents at P and P_1; that is,

$$y_1^{(1)} = y_0 + \frac{h}{2}[f(x_0, y_0) + f(x_0 + h, y_1)] \tag{B.12}$$

As the slope of the tangent at P_1 is not known, we use y_1 as computed from Eq. (B.11) by Euler's method and insert it on the right-hand side of Eq. (B.12) to obtain the first modified value of $y_1^{(1)}$. Equation (B.11) is, therefore, called the *predictor*, whereas Eq. (B.12) serves as the *corrector* of y_1.

Again, a corrector is applied and we find a still better value of $y_1^{(1)}$ corresponding to L_1:

$$y_1^{(2)} = y_0 + \frac{h}{2}\left[f(x_0, y_0) + f(x_0 + h, y_1^{(1)})\right]$$

We repeat this step until two consecutive values of y agree. This is taken as the starting point for the next interval L_1L_2.

Once y_1 is obtained to the desired degree of accuracy, y corresponding to L_2 is found from the *predictor*,

$$y_2 = y_1 + hf(x_0 + h, y_1)$$

and a better approximation, $y_2^{(1)}$, is obtained from the *corrector*,

$$y_2^{(1)} = y_1 + \frac{h}{2}[f(x_0 + h, y_1) + f(x_0 + 2h, y_2)]$$

We repeat this step until y_2 becomes stationary. Then, we proceed to calculate y_3 as above.

This is the modified Euler's method, which is a relatively simple predictor–corrector method.

Solved Example B.2

Using the modified Euler method, find an approximate value of y when $x = 0.3$ for the following differential equation:

$$\frac{dy}{dx} = x + y \quad \text{and} \quad y = 1 \quad \text{when } x = 0$$

Solution

Taking $h = 0.1$, the various calculations are arranged as shown in Table B.1. Hence,

$$y(0.3) = 1.4004 \quad \text{approximately.}$$

B.4.3 Runge–Kutta Method

The class of methods known as Runge–Kutta methods do not require the calculations of higher-order derivatives. These methods agree with Taylor's series solution up to the terms in h^r, where r differs from method to method and is called the *order of that method*. The fourth-order Runge–Kutta method is most commonly used and is often referred to as the "Runge–Kutta method" only.

Table B.1 Calculations of Solved Example B.2

x	$x + y = y'$	Mean slope	Old $y + 0.1$ (mean slope) = new y
0.0	$0 + 1$		$1.00 + 0.1(1.00) = 1.10$
0.1	$0.1 + 1.1$	$\frac{1}{2}(1 + 1.2)$	$1.00 + 0.1(1.10) = 1.11$
0.1	$0.1 + 1.11$	$\frac{1}{2}(1 + 1.21)$	$1.00 + 0.1(1.105) = 1.1105$
0.1	$0.1 + 1.1105$	$\frac{1}{2}(1 + 1.2105)$	$1.00 + 0.1(1.1052) = 1.1105$
0.1	1.2105		$1.1105 + 0.1(1.2105) = 1.2316$
0.2	$0.2 + 1.2316$	$\frac{1}{2}(1.2105 + 1.4316)$	$1.1105 + 0.1(1.3211) = 1.2426$
0.2	$0.2 + 1.426$	$\frac{1}{2}(1.2105 + 1.4426)$	$1.1105 + 0.1(1.3266) = 1.2432$
0.2	$0.2 + 1.2432$	$\frac{1}{2}(1.2105 + 1.4432)$	$1.1105 + 0.1(1.3268) = 1.2432$
0.2	1.4432	—	$1.2432 + 0.1(1.4432) = 1.3875$
0.3	$0.3 + 1.3875$	$\frac{1}{2}(1.4432 + 1.6875)$	$1.2432 + 0.1(1.5654) = 1.3997$
0.3	$0.3 + 1.3997$	$\frac{1}{2}(1.4432 + 1.6997)$	$1.2432 + 0.1(1.5715) = 1.4003$
0.3	$0.3 + 1.4003$	$\frac{1}{2}(1.4432 + 1.7003)$	$1.2432 + 0.1(1.5718) = 1.4004$
0.3	$0.3 + 1.4004$	$\frac{1}{2}(1.4432 + 1.7004)$	$1.2432 + 0.1(1.5718) = 1.4004$

Working Rule

Finding the increment k of y corresponding to an increment h of x by the Runge–Kutta method from

$$\frac{dy}{dx} = f(x, y), \qquad y(x_0) = y_0$$

is as follows:

Calculate successively

$$k_1 = hf(x_0, y_0)$$

$$k_2 = hf\left(x_0 + \frac{h}{2}, \; y_0 + \frac{k_1}{2}\right)$$

$$k_3 = hf\left(x_0 + \frac{h}{2}, \; y_0 + \frac{k_2}{2}\right)$$

and

$$k_4 = hf(x_0 + h, \; y_0 + k_3)$$

Finally, calculate

$$k = \tfrac{1}{6}[k_1 + 2k_2 + 2k_3 + k_4]$$

which gives the required approximate value $y_1 = y_0 + k$.

Notes:

1. k is the weighted mean of k_1, k_2, k_3, and k_4.
2. One of the advantages of these methods is that the operation is identical whether the differential equation is linear or nonlinear.

Solved Example B.3

Apply the Runge–Kutta method to find an approximate value of y for $x = 0.2$ in steps of 0.1 for the following differential equation:

$$\frac{dy}{dx} = x + y^2 \quad \text{and} \quad y = 1 \quad \text{when } x = 0$$

Solution

Here, we have taken $h = 0.1$ and carry out the calculations in two steps.

Step 1

$$x_0 = 0, \qquad y_0 = 1, \qquad h = 0.1$$

$$k_1 = hf(x_0, y_0) = 0.1f(0, 1) = 0.100$$

$$k_2 = hf\left(x_0 + \frac{h}{2}, \ y_0 + \frac{k_1}{2}\right) = 0.1f(0.05, 1.1) = 0.1152$$

$$k_3 = hf\left(x_0 + \frac{h}{2}, \ y_0 + \frac{k_2}{2}\right) = 0.1f(0.05, 1.1152) = 0.1168$$

and

$$k_4 = hf(x_0 + h, \ y_0 + k_3) = 0.1f(0.1, 1.1168) = 0.1347$$

Finally, we have

$$k = \tfrac{1}{6}[k_1 + 2k_2 + 2k_3 + k_4]$$

On substituting the values, we get

$$k = \tfrac{1}{6}[0.100 + 0.2304 + 0.2336 + 0.1347] = 0.1165$$

giving

$$y(0.1) = y_0 + k = 1.1165$$

Step 2

$$x_1 = x_0 + h = 0.1, \qquad y_1 = 1.165, \qquad h = 0.1$$

$$k_1 = hf(x_1, y_1) = 0.1f(0.1, 1.1165) = 0.1347$$

$$k_2 = hf\left(x_1 + \frac{h}{2}, y_1 + \frac{k_1}{2}\right) = 0.1f(0.15, 1.1838) = 0.1551$$

$$k_3 = hf\left(x_1 + \frac{h}{2}, y_1 + \frac{k_2}{2}\right) = 0.1f(0.15, 1.194) = 0.1576$$

and

$$k_4 = hf(x_1 + h, y_1 + k_3) = 0.1f(0.2, 1.1576) = 0.1823$$

Finally, we have

$$k = \tfrac{1}{6}[k_1 + 2k_2 + 2k_3 + k_4]$$

On substituting the values, we get

$$k = \tfrac{1}{6}[0.134 + 0.1551 + 0.1576 + 0.1823] = 0.1571$$

giving

$$y(0.2) = y_1 + k = 1.2736$$

Hence, the required approximate value of y is equal to 1.2736.

Predictor–Corrector Methods

In the methods explained so far, to solve a differential equation over an interval (x_1, x_{i+1}) only the value of y at the beginning of the interval was required. In the *predictor–corrector* methods, however, four prior values are required for finding the value of y at x_{i+1}. A predictor formula is used to predict the value of y at x_{i+1} and then a corrector formula is applied to improve this value. We now explain one such method.

Milne's Method
Given

$$\frac{dy}{dx} = f(x, y), \qquad y(x_0) = y_0$$

To find an approximate value of y for $x = x_0 + nh$ by Milne's method for the above-mentioned differential equation, we proceed as follows:
    The value $y_0 = y(x_0)$ being given, we compute

$$y_1 = y(x_0 + h), \qquad y_2 = y(x_0 + 2h), \qquad y_3 = y(x_0 + 3h)$$

by Picard's or Taylor's series method. Next, we calculate

$$f_0 = f(x_0, y_0), \qquad f_1 = f(x_0 + h, y_1), \qquad f_2 = f(x_0 + 2h, y_2),$$

$$f_3 = f(x_0 + 3h, y_3)$$

Then, to find $y_4 = y(x_0 + 4h)$, we substitute Newton's forward interpolation formula

$$f(x, y) = f_0 + n\Delta f_0 + \frac{n(n-1)}{2} \Delta^2 f_0 + \frac{n(n-1)(n-2)}{6} \Delta^3 f_0 + \cdots$$

in the relation

$$y_4 = y_0 + \int_{x_0}^{x_0+4h} f(x, y)\, dx$$

We get

$$y_4 = y_0 + \int_{x_0}^{x_0+4h} \left(f_0 + n\Delta f_0 + \frac{n(n-1)}{2} \Delta^2 f_0 + \frac{n(n-1)(n_2)}{6} \Delta^3 f_0 + \cdots \right) dx$$

Setting $x = x_0 + nh$ and $dx = h\,dn$, we get

$$y_4 = y_0 + h \int_0^4 \left(f_0 + n\Delta f_0 + \frac{n(n-1)}{2} \Delta^2 f_0 + \frac{n(n-1)(n-2)}{6} \Delta^3 f_0 + \cdots \right) dn$$

Thus, we get

$$y_4 = y_0 + h \left(4f_0 + 8\Delta f_0 + \frac{20}{3} \Delta^2 f_0 + \frac{8}{3} \Delta^3 f_0 + \cdots \right)$$

Neglecting fourth- and higher-order differences and expressing Δf_0, $\Delta^2 f_0$, and $\Delta^3 f_0$ in terms of the function values, we get

$$y_4 = y_0 + \frac{4h}{3} (2f_1 - f_2 + 2f_3)$$

which is called a *predictor*. Having found y_4, we obtain a first approximation to $f_4 = f(x_0 + 2h, y_4)$.

Then, a better value of y_4 is found by Simpson's rule as

$$y_4 = y_2 + \frac{h}{3}(f_2 + 4f_3 + f_4)$$

which is called a *corrector*. Then, an improved value of f_4 is computed and, again, the corrector is applied to find a still better value of y_4. We repeat this step until y_4 remains unchanged. Once y_4 and f_4 are obtained to the desired degree of accuracy, $y_5 = y(x_0 + 5h)$ is found from the *predictor* as

$$y_5 = y_1 + \frac{4h}{3} (2f_2 - f_3 + 2f_4)$$

and $f_5 = f(x_0 + 5h, y_5)$ is calculated. Then, a better approximation to the value of y_5 is obtained form the *corrector* as

$$y_5 = y_3 + \frac{h}{3}(f_3 + 4f_4 + f_5)$$

We repeat this step until y_5 becomes stationary and then we proceed to calculate y_6, and so on.

This is *Milne's predictor–corrector* method. To ensure greater accuracy, we must first improve the accuracy of the starting values and then subdivide the intervals.

Solved Example B.4

Apply Milne's method to find a solution of the differential equation $dy/dx = x - y^2$ in the range $0 \le x \le 1$ for the initial condition $y = 0$ at $x = 0$.

Solution

Using Picard's method, we have

$$y = y(0) + \int_0^x f(x, y)\, dx$$

where $f(x, y) = x - y^2$. To get the first approximation, we put $y = 0$ in $f(x, y)$, thus giving

$$y_1 = 0 + \int_0^x x\, dx = \frac{x^2}{2}$$

To find the second approximation, we put $y = x^2/2$ in $f(x, y)$, thus giving

$$y_2 = \int_0^x \left(x - \frac{x^4}{4} \right) dx = \frac{x^2}{2} - \frac{x^5}{20}$$

Similarly, the third approximation is

$$y_3 = \int_0^x \left[x - \left(\frac{x^2}{2} - \frac{x^5}{20} \right)^2 \right] dx = \frac{x^2}{2} - \frac{x^5}{20} + \frac{x^8}{160} - \frac{x^{11}}{4400} \qquad \text{(i)}$$

Now, let us determine the starting values of the Milne's method from (i) by choosing $h = 0.2$; we get

$$\begin{aligned}
x = 0.0, \quad & y_0 = 0.0000, \quad && f_0 = 0.0000 \\
x = 0.2, \quad & y_1 = 0.0200, \quad && f_1 = 0.1996 \\
x = 0.4, \quad & y_2 = 0.0795, \quad && f_2 = 0.3937 \\
x = 0.6, \quad & y_3 = 0.1762, \quad && f_3 = 0.5689
\end{aligned}$$

Using the *predictor*,

$$y_4 = y_0 + \frac{4h}{3} (2f_1 - f_2 + 2f_3)$$

we get

$$x = 0.8, \quad y_4 = 0.3049, \quad f_4 = 0.7070$$

and the *corrector* is given by

$$y_4 = y_2 + \frac{h}{3} (f_2 + 4f_3 + f_4)$$

which gives

$$y_4 = 0.3046, \quad f_4 = 0.7072 \qquad \text{(ii)}$$

Again using the *corrector*, $y_4 = 0.3046$, which is the same as in (ii).

Now using the *predictor*,

$$y_5 = y_1 + \frac{4h}{3}(2f_2 - f_3 + 2f_4)$$

we get

$$x = 1.0, \qquad y_5 = 0.4554, \qquad f_4 = 0.7926$$

and the *corrector*

$$y_5 = y_3 + \frac{h}{3}(f_3 + 4f_4 + f_5)$$

which gives

$$y_5 = 0.4555, \qquad f_5 = 0.7925$$

Again, using the *corrector*, $y_5 = 0.4555$, a value which is the same previously. Hence, $y(1) = 0.4555$.

B.5 SOLUTION OF SIMULTANEOUS FIRST-ORDER DIFFERENTIAL EQUATIONS

The simultaneous differential equations of the type

$$\frac{dy}{dx} = f(x, y, z) \qquad \text{and} \qquad \frac{dz}{dx} = \phi(x, y, z)$$

with initial condition

$$y(x_0) = y_0 \quad \text{and} \quad z(x_0) = z_0$$

can be solved by the methods discussed earlier, especially Picard's or Runge–Kutta methods. The method is best illustrated by the following example.

Solved Example B.5

Using Picard's method, find the approximate value of y and z corresponding to $x = 0.1$ given that

$$\frac{dy}{dx} = x + z \quad \text{and} \quad \frac{dz}{dx} = x - y^2$$

with $y(0) = 2$ and $z(0) = 1$.

Solution

Here,

$$x_0 = 0, \qquad y_0 = 2, \qquad z_0 = 1$$

Also,

$$\frac{dy}{dx} = f(x, y, z) = x + z \quad \text{and} \quad \frac{dz}{dx} = \phi(x, y, z) = x - y^2$$

Therefore,

$$y = y_0 + \int_{x_0}^{x} f(x, y, z)\, dx \quad \text{and} \quad z = z_0 + \int_{x_0}^{x} \phi(x, y, z)\, dx$$

Firstly approximations are

$$y_1 = y_0 + \int_{x_0}^{x} f(x, y_0, z_0)\, dx = 2 + \int_0^x (x + 1)\, dx = 2 + x + \frac{x^2}{2}$$

$$z_1 = z_0 + \int_{x_0}^{x} \phi(x, y_0, z_0)\, dx = 1 + \int_0^x (x - 4)\, dx = 1 - 4x + \frac{x^2}{2}$$

Second approximations are

$$y_2 = y_0 + \int_{x_0}^{x} f(x, y_1, z_1)\, dx = 2 + \int_0^x \left(x + 1 - 4x + \frac{x^2}{2} \right) dx$$

$$= 2 + x - \frac{3}{2} x^2 + \frac{x^3}{6}$$

$$z_2 = z_0 + \int_{x_0}^{x} \phi(x, y_1, z_1)\, dx = 1 + \int_0^x \left[x - \left(2 + x + \frac{x^2}{2} \right)^2 \right] dx$$

$$= 1 - 4x + \frac{3}{2} x^2 - x^3 - \frac{x^4}{4} - \frac{x^5}{20}$$

Third approximations are

$$y_3 = y_0 + \int_{x_0}^{x} f(x, y_2, z_2)\, dx = 2 + x - \frac{3}{2} x^2 - \frac{x^3}{2} - \frac{x^4}{.4} - \frac{x^5}{20} - \frac{x^6}{120}$$

$$z_3 = z_0 + \int_{x_0}^{x} \phi(x, y_2, z_2)\, dx = 1 - 4x + \frac{3}{2} x^2 + \frac{5}{3} x^3 + \frac{7}{12} x^4$$

$$- \frac{31}{60} x^5 + \frac{x^6}{12} - \frac{x^7}{252}$$

and so on.

When

$$x = 0.1$$

$y_1 = 2.015,$	$y_2 = 2.08517,$	$y_3 = 2.08447$
$z_1 = 0.605,$	$z_2 = 0.58397,$	$z_3 = 0.58672$

Then

$$y(0.1) = 2.0845 \quad \text{and} \quad z(0.1) = 0.5867$$

correct up to four decimal places.

B.6 SECOND-ORDER DIFFERENTIAL EQUATIONS

Consider the second-order differential equation

$$\frac{d^2y}{dx^2} = f\left(x, y, \frac{dy}{dx}\right)$$

By writing $dy/dx = z$, it can be reduced to two first-order differential equations:

$$\frac{dy}{dx} = z \quad \text{and} \quad \frac{dz}{dx} = f(x, y, z)$$

These equations can be solved as explained in the previous method (for simultaneous first-order differential equations).

Solved Example B.6

Using the Runge–Kutta method, solve the following second-order differential equation for $x = 0.2$, correct up to four decimal places:

$$\frac{d^2y}{dx^2} = x\left(\frac{dy}{dx}\right)^2 - y^2$$

with initial conditions $x = 0$, $y = 1$, and $dy/dx = 0$.

Solution

Let

$$\frac{dy}{dx} = z = f(x, y, z)$$

Then,

$$\frac{dz}{dx} = xz^2 - y^2 = \phi(x, y, z)$$

Table B.2 Calculations of Solved Example B.6

$k_1 = hf(x_0, y_0, z_0)$	$l_1 = h\phi(x_0, y_0, z_0)$
$\quad = 0.2(0) = 0$	$\quad = 0.2(-1) = -0.2$

$$k_2 = hf\left(x_0 + \frac{h}{2}, y_0 + \frac{k_1}{2}, z_0 + \frac{l_1}{2}\right) \qquad l_2 = h\phi\left(x_0 + \frac{h}{2}, y_0 + \frac{k_1}{2}, z_0 + \frac{l_1}{2}\right)$$

$$\quad = 0.2(-0.1) = -0.02 \qquad\qquad\qquad = 0.2(-0.999) = -0.1998$$

$$k_3 = h_f\left(x_0 + \frac{h}{2}, y_0 + \frac{k_2}{2}, z_0 + \frac{l_2}{2}\right) \qquad l_3 = h\phi\left(x_0 + \frac{h}{2}, y_0 + \frac{k_2}{2}, z_0 + \frac{l_2}{2}\right)$$

$$\quad = 0.2(-0.0999) = -0.02 \qquad\qquad\qquad = 0.2(-0.9791) = -0.1958$$

$$k_4 = hf(x_0 + h, y_0 + k_3, z_0 + l_3) \qquad l_4 = h\phi(x_0 + h, y_0 + k_3, z_0 + l_3)$$

$$\quad = 0.2(-0.1958) = -0.0392 \qquad\qquad = 0.2(-0.9527) = -0.1905$$

$$k = \tfrac{1}{6}(k_1 + 2k_2 + 2k_3 + k_4) \qquad l = \tfrac{1}{6}(l_1 + 2l_2 + 2l_3 + l_4)$$

$$\quad = -0.0199 \qquad\qquad\qquad\qquad = -0.1970$$

We have

$$x_0 = 0, \qquad y_0 = 1, \qquad z_0 = 0, \qquad h = 0.2$$

Using $k_1, k_2, k_3, \ldots$ for $f(x, y, z)$ and $l_1, l_2, l_3, \ldots$ for $\phi(x, y, z)$, Runge–Kutta formulas become (as shown in Table B.2).

Hence at $x = 0.2$,

$$y = y_0 + k = 1 - 0.0199 = 0.9801$$

$$y' = z = z_0 + l = 1 - 0.1970 = 0.803$$

Appendix C

Analytical Solution of Differential Equations

C.1 ANALYTICAL SOLUTION OF LINEAR ORDINARY DIFFERENTIAL EQUATIONS

C.1.1 Homogeneous

$$\frac{d^2y}{dt^2} + a\frac{dy}{dt} + by = 0 \tag{C.1}$$

The solution will be a linear combination (addition) of terms in the following form:

$$y = e^{\lambda t} \qquad \left(\text{e.g., } y = C_1 e^{\lambda_1 t} + C_2 e^{\lambda_2 t}\right) \tag{C.2}$$

From Eq. (C.2), we get

$$\frac{dy}{dt} = \lambda e^{\lambda t} \quad \text{and} \quad \frac{d^2y}{dt^2} = \lambda^2 e^{\lambda t} \tag{C.3}$$

Substituting Eqs. (C.2) and (C.3) in Eq. (C.1), we get

$$\lambda^2 e^{\lambda t} + a\lambda e^{\lambda t} + b e^{\lambda t} = 0$$

On simplifying, we get

$$e^{\lambda t}\left(\lambda^2 + a\lambda + b\right) = 0$$

which gives

$$\lambda^2 + a\lambda + b = 0$$

The solutions of the above quadratic equation is given by

$$\lambda_{1,2} = \frac{-a \pm \sqrt{a^2 - 4b}}{2} \tag{C.4}$$

So the solution is given by

$$y = C_1 e^{\lambda_1 t} + C_2 e^{\lambda_2 t}$$

where C_1 and C_2 are obtained from the initial or boundary conditions.

C.1.2   Types of Roots (λ_1, λ_2)

1. λ_1 and λ_2 are real and distinct.
2. When $a^2 - 4b = 0$, we get repeated roots

$$\lambda_{1,2} = -\frac{a}{2} = \lambda$$

Then, the solution is given by

$$y = (C_1 + C_2 t)e^{\lambda_2 t}$$

3. When $4b > a^2$, we get complex roots

$$\lambda_1 = -\frac{a}{2} + i\omega \quad \text{and} \quad \lambda_2 = -\frac{a}{2} - i\omega$$

where

$$i = \sqrt{-1}$$

$$\omega = \sqrt{b - \frac{a^2}{4}}$$

The solution is given by

$$y = C_1 e^{(-(a/2) + i\omega)t} + C_2 e^{(-(a/2) - i\omega)t}$$

On rearranging, we get

$$y = C_1 \left(e^{-(a/2)t} e^{i\omega t} \right) + C_2 \left(e^{-(a/2)t} e^{-i\omega t} \right)$$

Recalling the complex function analysis, for a complex number $z = x + iy$, we can write

$$e^z = e^x(\cos y + i \sin y)$$

So for our case, we can write

$$z_1 = -\frac{a}{2}t + i\omega t \quad \text{and} \quad z_2 = -\frac{a}{2}t - i\omega t$$

We can also write

$$e^{z_1} = e^{-(a/2)t}(\cos\omega t + i\sin\omega t) \equiv e^{\lambda_1 t}$$

and

$$e^{z_2} = e^{-(a/2)t}(\cos\omega t - i\sin\omega t) \equiv e^{\lambda_2 t}$$

Thus, the solution $y = C_1'e^{\lambda_1 t} + C_2'e^{\lambda_2 t}$ becomes

$$y = C_1'\left[e^{-(a/2)t}(\cos\omega t + i\sin\omega t)\right] + C_2'\left[e^{-(a/2)t}(\cos\omega t - i\sin\omega t)\right]$$

On simplification, we get

$$y = e^{-(a/2)t}\left[(C_1' + C_2')\cos\omega t + (C_1' - C_2')\sin\omega t\right]$$

which can be rewritten as

$$y = e^{-(a/2)t}[C_1\cos\omega t + C_2\sin\omega t] \tag{C.5}$$

Thus, Eq. (C.5) is the solution of the differential equation.

Special Cases

When $a > 0$, we have decaying oscillations, as shown in Figure C.1 (clearly when $a < 0$, we have oscillations with increasing amplitudes).

When $a = 0$, we have sustained oscillations, as shown in Figure C.2. The solution is of the form, $y = C_1\cos\omega t + C_2\sin\omega t$.

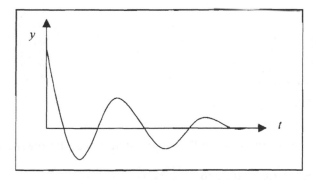

Figure C.1   Decaying oscillations.

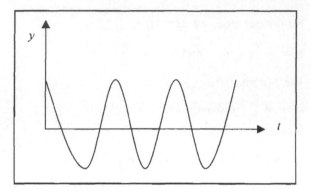

Figure C.2 Sustained oscillations.

C.2 THE TRANSFORMATION OF HIGHER-ORDER DIFFERENTIAL EQUATIONS INTO A SET OF FIRST-ORDER DIFFERENTIAL EQUATIONS

For example, let us consider the following second-order differential equation:

$$\frac{d^2y}{dt^2} + a\frac{dy}{dt} + by = 0$$

Let us define

$$x_1 = y$$

from which we can write

$$\frac{dx_1}{dt} = \frac{dy}{dt}$$

and

$$x_2 = \frac{dy}{dt}$$

from which we can write

$$\frac{dx_2}{dt} = \frac{d^2y}{dt^2}$$

Thus, our main second-order differential equation can be rewritten as

$$\frac{dx_2}{dt} + ax_2 + bx_1 = 0 \quad \text{and} \quad \frac{dx_1}{dt} = x_2 \tag{C.6}$$

The set of two first-order differential equations (C.6) can replace the original second-order differential equation. Thus, we have

$$\frac{dx_1}{dt} = 0x_1 + x_2$$

$$\frac{dx_2}{dt} = -bx_1 - ax_2$$

Let us define the following vectors and matrix:

$$\underline{X} = \begin{pmatrix} x_1 \\ x_2 \end{pmatrix}$$

(this is the state vector, i.e., vector of state variables).

We can also write

$$\frac{d\underline{X}}{dt} = \begin{pmatrix} \dfrac{dx_1}{dt} \\ \dfrac{dx_2}{dt} \end{pmatrix}$$

and

$$\underline{A} = \begin{pmatrix} 0 & 1 \\ -b & -a \end{pmatrix}$$

In matrix form, we can write

$$\frac{d\underline{X}}{dt} = \underline{A}\,\underline{X}$$

with initial conditions

$$\underline{X}(0) = \underline{X}_o = \begin{pmatrix} x_{1o} \\ x_{2o} \end{pmatrix}$$

C.3 NONHOMOGENEOUS DIFFERENTIAL EQUATIONS

$$\frac{d^2 y}{dt^2} + a\frac{dy}{dt} + by = f(t)$$

The solution of the above nonhomogeneous differential equation is of the form

$$y = y_h + y_p$$

where y_h is the solution of the homogeneous equation (complementary function) and y_p is the particular integral depending on $f(t)$. Some of the values of the particular integral depending on the nature of $f(t)$ is tabulated in Table C.1.

Table C.1   Choice of y_p

$f(t)$	Choice of y_p
$ke^{\gamma t}$	$Ce^{\gamma t}$
kt^n (where $n = 0, 1, 2, 3, \ldots$)	$k_n t^n + k_{n-1} t^{n-1} + \cdots + k_1 t^1 + k_0$
$k \cos \omega t$ or $k \sin \omega t$	$K_1 \cos \omega t + K_2 \sin \omega t$
$ke^{\alpha t} \cos \omega t$ or $ke^{\alpha t} \sin \omega t$	$e^{\alpha t}(K_1 \cos \omega t + K_2 \sin \omega t)$

C.4   EXAMPLES OF SOME SPECIAL CASES

Example C.1

$$\frac{d^2 y}{dt^2} + 4y = 8t^2$$

First, we calculate for the homogeneous solution y_h:

$$\frac{d^2 y}{dt^2} + 4y = 0$$

We get

$$\lambda^2 + 4 = 0$$

The solutions are

$$\lambda_1 = 2i \quad \text{and} \quad \lambda_2 = -2i$$

Thus, the solution is given by

$$y_h = C_1 \cos 2t + C_2 \sin 2t$$

Now, to calculate the particular integral y_p. From Table C.1,

$$y_p = k_2 t^2 + k_1 t + k_0$$

and we get

$$\frac{d^2 y_p}{dt^2} = 2k_2$$

As y_p must satisfy the differential equation, we get

$$2k_2 + 4(k_2 t^2 + k_1 t + k_0) = 8t^2$$

Reorganizing the equation with all the terms with same t^n together gives

$$(4k_2 - 8)t^2 + 4k_1 t + (2k_2 + 4k_0) = 0$$

For the above equation to be correct for all values of t, all of the coefficients of t must be equal to zero, so we get

$$4k_2 - 8 = 0 \Rightarrow k_2 = 2$$
$$4k_1 0 = 0 \Rightarrow k_1 = 0$$
$$2k_2 + 4k_0 = 0 \Rightarrow k_0 = -1$$

Thus

$$y_p = 2t^2 - 1$$

So the total solution is given by

$$y = y_h + y_p = C_1 \cos 2t + C_2 \sin 2t + 2t^2 - 1$$

Example C.2

$$\frac{d^2 y}{dt^2} - 3\frac{dy}{dt} + 2y = e^t$$

First, we calculate for the homogeneous solution y_h,

$$\frac{d^2 y}{dt^2} - 3\frac{dy}{dt} + 2y = 0$$

We get

$$\lambda^2 - 3\lambda + +2 = 0$$

The solutions are

$$\lambda_1 = 2 \quad \text{and} \quad \lambda_2 = 1$$

Thus, the solution is given by

$$y_h = C_1 e^t + C_2 e^{2t}$$

Now, to calculate the particular integral y_p. From Table C.1, we usually choose $y_p = Ce^t$, but now e^t is a part of y_h. So, in this case, we must choose

$$y_p = Cte^t$$

To obtain the value of C, we perform the following calculations:

$$\frac{dy_p}{dt} = C(te^t + e^t)$$

$$\frac{d^2y_p}{dt^2} = C(2e^t + te^t)$$

Substituting the above values into the differential equation gives

$$C(2e^t + te^t) - 3C(te^t + e^t) + 2Cte^t = e^t$$

On reorganizing, we get

$$e^t(2C - 3C - 1) + te^t(C - 3C + 2C) = 0$$

We finally get

$$C = -1$$

So the particular integral is equal to

$$y_p = -te^t$$

The total solution is given by

$$y = y_h + y_p = C_1e^t + C_2e^{2t} - te^t$$

Note: Notice not to have in y_p a term which is same as a term in y_h (not the same in functional form). The reader is advised to consult his mathematics book for details.

Appendix D

Table of Laplace Transforms of Some Common Functions

Solution of differential equations using Laplace transformation is given in Section 5.6.6.

Table D.1 Table of Laplace Transforms

	$f(t)$	$\hat{f}(s)$
1	1	$\dfrac{1}{s}$
2	t	$\dfrac{1}{s^2}$
3	$\dfrac{t^n - 1}{(n-1)!}$	$\dfrac{1}{s^n}$
4	$2\sqrt{\dfrac{t}{\pi}}$	$s^{-3/2}$
5	t^{k-1}	$\dfrac{\Gamma(k)}{s^k} \quad (k \geq 0)$
6	$t^{k-1}e^{-at}$	$\dfrac{\Gamma(k)}{(s+a)^k} \quad (k \geq 0)$
7a	e^{-at}	$\dfrac{1}{s+a}$

Table D.1   Continued

$f(t)$	$\hat{f}(s)$
7b $\quad \dfrac{1}{\tau}e^{-t/\tau}$	$\dfrac{1}{\tau s + 1}$
8 $\quad 1 - e^{-t/\tau}$	$\dfrac{1}{s(\tau s + 1)}$
9a $\quad te^{-at}$	$\dfrac{1}{(s + a)^2}$
9b $\quad \dfrac{t}{\tau^2}e^{-t/\tau}$	$\dfrac{1}{(\tau s + 1)^2}$
10 $\quad 1 - \left(1 + \dfrac{t}{\tau}\right)e^{-t/\tau}$	$\dfrac{1}{s(\tau s + 1)^2}$
11a $\quad \dfrac{1}{(n-1)!}t^{n-1}e^{-at}$	$\dfrac{1}{(s + a)^n}$ $\quad (n = 1, 2, \dots)$
11b $\quad \dfrac{1}{\tau^n(n-1)!}t^{n-1}e^{-t/\tau}$	$\dfrac{1}{(\tau s + 1)^n}$ $\quad (n = 1, 2, \dots)$
12 $\quad \dfrac{1}{(b-a)}\left(e^{-at} - e^{-bt}\right)$	$\dfrac{1}{(s + a)(s + b)}$
13 $\quad \dfrac{1}{(b-a)}\left(be^{-bt} - ae^{-at}\right)$	$\dfrac{s}{(s + a)(s + b)}$
14 $\quad \dfrac{1}{ab}\left[1 + \dfrac{1}{(a-b)}\left(be^{-at} - ae^{-bt}\right)\right]$	$\dfrac{1}{s(s + a)(s + b)}$
15 $\quad \sin bt$	$\dfrac{b}{s^2 + b^2}$
16 $\quad \sinh bt$	$\dfrac{b}{s^2 - b^2}$
17 $\quad \cos bt$	$\dfrac{s}{s^2 + b^2}$
18 $\quad \cosh bt$	$\dfrac{s}{s^2 - b^2}$
19 $\quad 1 - \cos bt$	$\dfrac{b^2}{s(s^2 + b^2)}$

	$f(t)$	$\hat{f}(s)$
20	$e^{-at}\sin bt$	$\dfrac{b}{(s+a)^2+b^2}$
21	$e^{-at}\cos bt$	$\dfrac{s+a}{(s+a)^2+b^2}$
22	$e^{-at}f(t)$	$\hat{f}(s+a)$
23	$f(t-b)$; with $f(t)=0$ for $t<0$	$e^{-bs}\hat{f}(s)$

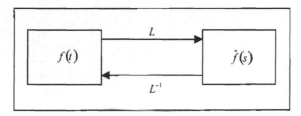

Figure D.1 Relation between s and t domains.

Appendix E

Orthogonal Collocation Technique

The orthogonal collocation technique is a simple numerical method which is easy to program and which converges rapidly. Therefore, it is useful for the solution of many types of second-order boundary-value problems (it transforms them to a set of algebraic equations satisfying the boundary conditions), as well as partial differential equations (it transforms them to a set of ordinary differential equations satisfying the boundary conditions). This method in its simplest form as presented in the appendix was developed by Villadsen and Stewart (1). The orthogonal collocation method has the advantage of ease of computation. This method is based on the choice of a suitable trial series to represent the solution. The coefficients of the trial series are determined by making the residual equation vanish at a set of points called "collocation points," in the solution domain (2).

E.1 APPLICATION OF THE SIMPLE ORTHOGONAL COLLOCATION TECHNIQUE

When the Fickian diffusion model is used, many reaction–diffusion problems in a porous catalyst pellet can be reduced to a two-point boundary-value differential equation of the form of Eq. (E.1). This is not necessary condition for the application of this simple orthogonal collocation technique. The technique, in principle, can be applied to any number of simultaneous two-point boundary-value differential equations

$$\nabla^2 Y + F(Y) = 0 \tag{E.1}$$

where

$$\nabla^2 = \frac{d^2}{d\omega^2} + \frac{a}{\omega}\frac{d}{d\omega} \tag{E.2}$$

where $a = 0$ for a slab, $a = 1$ for a cylinder, and $a = 2$ for a sphere, with the following boundary conditions

At $\omega = 1$,

$$\frac{dY}{d\omega} = \alpha(Y_B - Y) \tag{E.3a}$$

At $\omega = 0$,

$$\frac{dY}{d\omega} = 0 \tag{E.3b}$$

where Y_B is the value of Y at bulk conditions and α can be Nusselt or Sherwood numbers depending on whether we are using the heat or mass balance equations, respectively.

For interior collocation, the approximation function is chosen so that the boundary conditions are satisfied. A suitable function is the following:

$$Y^{(n)} = Y(1) + (1 - \omega^2) \sum_{i=1}^{n-1} a_1^{(n)} P_i(\omega^2) \tag{E.4}$$

in which $P_i(\omega^2)$ are polynomials of degree i in ω^2, that are to be specified, and also $a_i^{(n)}$ are undetermined constants. The even polynomials, $P_i(\omega^2)$, automatically satisfy the symmetry condition at the center of the pellet [Eq. (E3.b)].

A set of (n) equations is needed to determine the collocation constants a_i. We note that once $Y^{(n)}$ has been adjusted to satisfy Eq. (E.1) at n collocation points $\omega_1, \omega_2, \ldots, \omega_n$, the residual either vanishes everywhere or contains a polynomial factor of degree (n), whose zeros are the collocation points.

Villadsen and Stewart (1) pointed out that the equations can be solved in terms of the solution at the collocation points, instead of the coefficients $a_i^{(n)}$. This is more convenient and reduces the Laplacian and the first-order derivative to

$$\nabla^2 Y\big|_{\omega=\omega_1} = \sum_{j=1}^{n+1} B_{ij} Y(\omega_j) = \underline{B}\,\underline{Y} \qquad (E.5)$$

$$\frac{dY}{d\omega}\bigg|_{\omega=\omega_1} = \sum_{j=1}^{n+1} A_{ij} Y(\omega_j) = \underline{A}\,\underline{Y} \qquad (E.6)$$

respectively, where $i = 1, 2, 3, \ldots, n$ is the number of interior collocation points.

Integrals of the solution (which are useful for the calculation of the effectiveness factor) can also be calculated with high accuracy via the formula

$$\int_0^{1.0} F(Y)\omega^2\, d\omega = \sum_{i=1}^{n+1} \omega_i^{(n)} F(Y_i) = \underline{W}\,\underline{F(Y)} \qquad (E.7)$$

$\underline{A}$, $\underline{B}$, and $\underline{W}$ can be calculated as follows:

First, the values of $\underline{Y}$ can be expressed at the collocation points by

$$\underbrace{\begin{pmatrix} Y_1 \\ Y_2 \\ \vdots \\ Y_n \end{pmatrix}}_{\underline{Y}} = \underbrace{\begin{pmatrix} \omega_1^0 & \omega_1^2 & \cdots & \omega_1^{2n} \\ \omega_2^0 & \omega_2^2 & \cdots & \omega_2^{2n} \\ \vdots & \vdots & \ddots & \vdots \\ \omega_n^0 & \omega_n^2 & \cdots & \omega_n^{2n} \end{pmatrix}}_{\underline{C}} \underbrace{\begin{pmatrix} d_1 \\ d_2 \\ \vdots \\ d_n \end{pmatrix}}_{\underline{d}} \qquad (E.8)$$

which can be simply written as

$$\underline{Y} = \underline{C}\,\underline{d} \qquad (E.9)$$

Then, the collocation constants are

$$\underline{d} = \underline{C}^{-1}\underline{Y} \qquad (E.10)$$

On differentiating the trial function $\underline{Y}$, we get

$$
\underbrace{\begin{pmatrix} \left(\dfrac{dY}{d\omega}\right)_{\omega_1} \\[1.5em] \left(\dfrac{dY}{d\omega}\right)_{\omega_2} \\[1.5em] \vdots \\[1em] \left(\dfrac{dY}{d\omega}\right)_{\omega_n} \end{pmatrix}}_{\underline{DY}} = \underbrace{\begin{pmatrix} \left(\dfrac{d\omega^0}{d\omega}\right)_{\omega_1} & \left(\dfrac{d\omega^2}{d\omega}\right)_{\omega_1} & \cdots & \left(\dfrac{d\omega^{2n}}{d\omega}\right)_{\omega_1} \\[1.5em] \left(\dfrac{d\omega^0}{d\omega}\right)_{\omega_2} & \left(\dfrac{d\omega^2}{d\omega}\right)_{\omega_2} & \cdots & \left(\dfrac{d\omega^{2n}}{d\omega}\right)_{\omega_2} \\[1.5em] \vdots & \vdots & \ddots & \vdots \\[1em] \left(\dfrac{d\omega^0}{d\omega}\right)_{\omega_n} & \left(\dfrac{d\omega^2}{d\omega}\right)_{\omega_n} & \cdots & \left(\dfrac{d\omega^{2n}}{d\omega}\right)_{\omega_n} \end{pmatrix}}_{\underline{R}} \underbrace{\begin{pmatrix} d_1 \\[0.5em] d_2 \\[0.5em] \vdots \\[0.5em] d_n \end{pmatrix}}_{\underline{d}}
$$

$$(E.11)$$

or simple, we can write

$$\underline{DY} = \underline{R}\,\underline{d} \tag{E.12}$$

From Eqs. (E.10) and (E.12), by substituting for $\underline{d}$, we get

$$\underline{DY} = \underline{R}\,\underline{C}^{-1}\underline{Y} \tag{E.13}$$

Let us express

$$\underline{DY} = \underline{A}\,\underline{Y} \tag{E.14}$$

where,

$$\underline{A} = \underline{R}\,\underline{C}^{-1}$$

Also, let us express

$$\nabla^2 Y = \underline{T}\,\underline{d} = \underline{B}\,\underline{Y} \tag{E.15}$$

where

$$
\underline{T} = \begin{pmatrix} \left(\nabla^2\omega^0\right)_{\omega_1} & \left(\nabla^2\omega^2\right)_{\omega_1} & \cdots & \left(\nabla^2\omega^{2n}\right)_{\omega_1} \\[1em] \left(\nabla^2\omega^0\right)_{\omega_2} & \left(\nabla^2\omega^2\right)_{\omega_2} & \cdots & \left(\nabla^2\omega^{2n}\right)_{\omega_2} \\[1em] \vdots & \vdots & \ddots & \vdots \\[1em] \left(\nabla^2\omega^0\right)_{\omega_n} & \left(\nabla^2\omega^2\right)_{\omega_n} & \cdots & \left(\nabla^2\omega^{2n}\right)_{\omega_n} \end{pmatrix}
$$

From Eqs. (E.10) and (E.15), by substituting for $\underline{d}$, we get

$$\nabla^2\underline{Y} = \underline{T}\,\underline{C}^{-1}\underline{Y} = \underline{B}\,\underline{Y} \tag{E.16}$$

Thus,

$$\underline{B} = \underline{T}\,\underline{C}^{-1} \tag{E.17}$$

Also, let

$$F(Y) = \sum_{i=1}^{n+1} C_i \omega^{2(i-1)} \tag{E.18}$$

Then,

$$\omega^2 F(Y) = \sum_{i=1}^{n+1} C_i \omega^{2(i)} \tag{E.19}$$

Therefore,

$$\int_0^{1.0} \omega^2 F(Y)\,d\omega = \int_0^{1.0} \sum_{j=1}^{n+1} C_i \omega^{2(i)}\,d\omega = \text{I.F.} \tag{E.20}$$

The integrated function I.F. will be

$$\text{I.F.} = \underline{W}\,\underline{F}\,\underline{C} \tag{E.21}$$

where

$$\underline{W}\,\underline{F} = \left(\tfrac{1}{3} \quad \tfrac{1}{5} \cdots \quad \frac{1}{2^n + 1} \right) \tag{E.22}$$

and

$$\underline{C} = \begin{pmatrix} C_1 \\ C_2 \\ \vdots \\ C_n \end{pmatrix} \tag{E.23}$$

Also,

$$\underline{F}(Y) = \underline{Q}\,\underline{C} \tag{E.24}$$

where

$$\underline{C} = \begin{pmatrix} \omega_1^0 & \omega_1^2 & \cdots & \omega_1^{2n} \\ \omega_2^0 & \omega_2^2 & \cdots & \omega_2^{2n} \\ \vdots & \vdots & \ddots & \vdots \\ \omega_n^0 & \omega_n^2 & \cdots & \omega_n^{2n} \end{pmatrix} \tag{E.25}$$

From Eq. (E.24), we get

$$\underline{C} = \underline{Q}^{-1}\underline{F}(Y) \tag{E.26}$$

From Eqs. (E.21) and (E.26), the integral can be expressed as

$$\text{I.F.} = \underline{W}\,\underline{F}\,\underline{Q}^{-1}\underline{F}(Y) = \underline{W}\,\underline{F}(Y) \tag{E.27}$$

We can calculate $\underline{W}$ as

$$\underline{W} = \underline{W}\,\underline{F}\,\underline{Q}^{-1} \tag{E.28}$$

Applying the simple case of one internal collocation point ($n = 1$, and ω at this collocation point is ω_1), the Laplacian at the interior collocation point can be written as

$$\nabla^2 Y\big|_{\omega=\omega_2} = B_{11}\,Y(\omega_1) + B_{12}\,Y(\omega_2) \tag{E.29}$$

and the derivative at the surface boundary condition ($\omega = \omega_2$) takes the form

$$\frac{dY}{d\omega}\bigg|_{\omega=\omega_2} = A_{21}\,Y(\omega_1) + A_{22}\,Y(\omega_2) \tag{E.30}$$

and, therefore, the boundary condition can be written as

$$A_{21}\,Y(\omega_1) + A_{22}\,Y(\omega_2) = \alpha(Y_B - Y(\omega_2)) \tag{E.31}$$

Thus, we get

$$Y(\omega_2) = \frac{\alpha Y_B - A_{21}\,Y(\omega_1)}{A_{22} + \alpha} \tag{E.32}$$

Substitution of Eqs. (E.29) and (E.32) into Eq. (E.1) gives

$$A_{11}\,Y(\omega_1) + B_{12}\left[\frac{\alpha Y_B - A_{21}\,Y(\omega_1)}{A_{22} + \alpha}\right] = F(Y(\omega_1)) \tag{E.33}$$

which is a single equation in the single variable $Y(\omega_1)$ at the collocation point ω_1. Solution of this equation gives $Y(\omega_1)$, which can be substituted in Eq. (E.32) to give $Y(\omega_2)$, the surface of Y. The values of $Y(\omega_1)$ and $Y(\omega_2)$ can be used to evaluate the integral as follows:

$$\int_0^{1.0} \omega^2 F(Y)\,d\omega = W_1 F(\omega_1) + W_2 F(\omega_2) \tag{E.34}$$

The values of $\underline{B}$, $\underline{A}$, and $\underline{W}$ for a single internal collocation point using the Legendre polynomial, for a spherical catalyst pellet, are

$$\underline{B} = \begin{pmatrix} -10.5 & 10.5 \\ -10.5 & 10.5 \end{pmatrix}$$

$$\underline{A} = \begin{pmatrix} -2.291 & 2.291 \\ -3.5 & 3.5 \end{pmatrix}$$

$$\underline{W} = \begin{pmatrix} 0.2333 \\ 0.1 \end{pmatrix}$$

The values of $\underline{A}$, $\underline{B}$, and $\underline{W}$ for large number of collocation points are given in a number of references (1,2).

E.2 ORTHOGONAL COLLOCATION TECHNIQUE APPLICABLE TO PARABOLIC PARTIAL DIFFERENTIAL EQUATIONS

The concept of orthogonal collocation for ordinary differential equations can be easily extended to solve "parabolic partial differential equations." The difference is that the application of orthogonal collocation method on the two-point boundary-value differential equation discussed earlier results in a set of algebraic equations, whereas application of orthogonal collocation method on parabolic partial differential equations results in a set of ordinary differential equations.

Let us assume the parabolic partial differential equation to be of the following form:

$$\frac{\partial Y}{\partial t} + \nabla^2 Y = F(Y) \tag{E.35}$$

The second term of Eq. (E.35) can be replaced as shown in Eq. (E.5), whereas the partial derivative term of Eq. (E.35) can be converted to an ordinary derivative term at each collocation point to give the following set:

$$\frac{dY(\omega_i)}{dt} + \sum_{j=1}^{n+1} B_{ij} Y(\omega_j) = F(Y)_{\omega_i} \tag{E.36}$$

where $i = 1, 2, 3, \ldots, n$ and n is the number of interior collocation points. Thus, Eq. (E.36) will give a set of n ordinary differential equations to be solved in order to solve a parabolic partial differential equation.

For elliptical and hyperbolic partial differential equations, the reader is advised to consult engineering mathematics texts (3–6).

Partial differential equations can also be solved numerically using the most commonly used method, the "method of finite differences" or the

Crank–Nicholson method. The reader is advised to consult engineering mathematics texts (3–6).

For higher-order systems, finite element techniques can be used, but this is beyond the scope of this undergraduate book. Any interested reader can consult Refs. (3–6) to master those techniques.

REFERENCES

1. Villadsen, J. and Stewart, W. E. Solution of boundary-value problems by orthogonal collocation. *Chem. Eng. Sci.* 22, 1483–1501, 1967.
2. Finlayson, B. A. *The Method of Weighted Residuals and Variational Principles.* Academic Press, New York.
3. Kreysig, E. *Advanced Engineering Mathematics,* 8th ed. Wiley, New York, 1998.
4. Masatake, M. *The Finite Element Method and Its Applications.* Macmillan, New York, 1986.
5. Strikwerda, J. C. *Finite Difference Schemes and Partial Differential Equations.* CRC Press, London, 1989.
6. Strauss, W. A. *Partial Differential Equations: An Introduction.* Wiley, New York, 1992.

Appendix F

Some Software and Programming Environments

We provide brief details of some commonly used software and programming environments used for solving the mathematical models and visualizing them. We emphasize that the choice of any particular software or programming environment depends on the user.

F.1 POLYMATH 5.X

POLYMATH (http://www.polymath-software.com) is a proven computational system, which has been specifically created for educational or professional use. The various POLYMATH programs allow the user to apply effective numerical analysis techniques during interactive problem-solving on personal computers. Results are presented graphically for easy understanding and for incorporation into papers and reports. Students, engineers, mathematicians, scientists, or anyone with a need to solve problems will appreciate the efficiency and speed of problem solution.

The main options available in POLYMATH are the following:

- LIN: Linear Equations Solver: Enter (in matrix form) and solve a system of up to 200 simultaneous linear equations (64 with convenient input interface).

- NLE: Nonlinear Equations Solver: Enter and solve a system of up to 200 nonlinear algebraic equations.
- DEQ: Differential Equations Solver. Enter and solve a system of up to 200 ordinary differential equations.
- REG: Data analysis and Regression. Enter, analyze, regress, and plot a set of up to 600 data points.

F.2 MATLAB

F.2.1 Unified, Interactive Language, and Programming Environment

The MATLAB (http://www.mathworks.com) language is designed for interactive or automated computation. Matrix-optimized functions let you perform interactive analyses, whereas the structured language features let you develop your own algorithms and applications. The versatile language lets you tackle a range of tasks, including data acquisition, analysis, algorithm development, system simulation, and application development. Language features include data structures, object-oriented programming, graphical user interface (GUI) development tools, debugging features, and the ability to link in C, C + +, Fortran, and Java routines.

F.2.2 Numeric Computing for Quick and Accurate Results

With more than 600 mathematical, statistical, and engineering functions, MATLAB gives immediate access to high-performance numeric computing. The numerical routines are fast, accurate, and reliable. These algorithms, developed by experts in mathematics, are the foundation of the MATLAB language. The math in MATLAB is optimized for matrix operations. This means that you can use it in place of low-level languages like C and C + +, with equal performance but less programming.

MATLAB mainly includes the following:

- Linear algebra and matrix computation
- Fourier and statistical analysis functions
- Differential equation solvers
- Sparse matrix support
- Trigonometric and other fundamental math operations
- Multidimensional data support

F.2.3 Graphics to Visualize and Analyze Data

MATLAB includes power, interactive capabilities for creating two-dimensional plots, images, and three-dimensional surfaces and for visualizing volumetric data. Advanced visualization tools include surface and volume rendering, lighting, camera control, and application-specific plot types.

F.3 MATHCAD

Mathcad (http://www.mathcad.com) provides hundreds of operators and built-in functions for solving technical problems. Mathcad can be used to perform numeric calculations or to find symbolic solutions. It automatically tracks and converts units and operates on scalars, vectors, and matrices.

The following is an overview of Mathcad's computational capabilities:

- Numeric operators perform summations, products, derivatives, integrals, and Boolean operations. Numeric functions apply trigonometric, exponential, hyperbolic, and other functions and transforms.
- Symbolics simplify, differentiate, integrate, and transform expressions algebraically, Mathcad's patented live symbolics technology automatically recalculate algebraic solutions.
- Vectors and Matrices manipulate arrays and perform various linear algebra operations, such as finding eigenvalues and eigenvectors and looking up values in arrays.
- Statistics and Data Analysis generate random numbers or histograms, fit data to built-in and general functions, interpolate data, and build probability distribution models.
- Differential Equation Solvers support ordinary differential equations, systems of differential equations, and boundary-value problems.
- Variables and Units handle real, imaginary, and complex numbers with or without associated units. A high-performance calculation engine provides speed and sophisticated memory management to help find solutions faster.

F.4 DYNAMICS SOLVER

Dynamics Solver (http://tp.lc.ehu.es/jma/ds/ds.html) solves numerically both initial-value problems and boundary-value problems for continuous and discrete dynamical systems:

- A single ordinary differential equation of arbitrary order
- Systems of first-order ordinary differential equations
- A rather large class of functional differential equations and systems

- Discrete dynamical systems in the form iterated maps and recurrences in arbitrary dimensions

It is also able to draw complex mathematical figures, including many fractals. Dynamics Solver is a powerful tool for studying differential equations, (continuous and discrete) nonlinear dynamical systems, deterministic chaos, mechanics, and so forth. For instance, you can draw phase space portraits (including an optional direction field). Poincaré maps, Liapunov exponents, histograms, bifurcation diagrams, attraction basis, and so forth. The results can be watched (in perspective or not) from any direction and particular subspaces can be analyzed.

Dynamics Solver is extensible: Users can add new mathematical functions and integration codes. It also has all the advantages of Windows programs, including the ease of use, the ability to open several output windows simultaneously and a larger amount of memory, which allows analyzing more complex problems. Moreover, there are many ways to print and export the results.

F.5 CONTROL STATION

Control Station (http://www.ControlStation.com) is both a controller design and tuning tool, and a process control training simulator used by industry and academic institutions worldwide for

- control loop analysis and tuning
- dynamic process modeling and simulation
- performance and capability studies
- hands-on process control training

The *Case Studies* feature provides hands-on training by challenging the user with industrially relevant process simulations. The software lets users manipulate process and controller parameters so that they can "learn by doing" as users explore the world of process control.

The *Controller Library* feature lets users use different controller settings (e.g., P-only, PI, PD, PID, cascade, ratio control, and so forth). *Design Tools* is a powerful controller design and analysis tool that can also incorporate dead times and the *Custom Process* feature can let users implement a process and control architecture to their own specifications.

Index

T - #0214 - 101024 - C0 - 234/156/36 [38] - CB - 9780824709570 - Gloss Lamination